Materials Forming, Machining and Tribology

Series Editor

J. Paulo Davim, Department of Mechanical Engineering, University of Aveiro, Aveiro, Portugal

This series fosters information exchange and discussion on all aspects of materials forming, machining and tribology. This series focuses on materials forming and machining processes, namely, metal casting, rolling, forging, extrusion, drawing, sheet metal forming, microforming, hydroforming, thermoforming, incremental forming, joining, powder metallurgy and ceramics processing, shaping processes for plastics/composites, traditional machining (turning, drilling, miling, broaching, etc.), non-traditional machining (EDM, ECM, USM, LAM, etc.), grinding and others abrasive processes, hard part machining, high speed machining, high efficiency machining, micro and nanomachining, among others. The formability and machinability of all materials will be considered, including metals, polymers, ceramics, composites, biomaterials, nanomaterials, special materials, etc. The series covers the full range of tribological aspects such as surface integrity, friction and wear, lubrication and multiscale tribology including biomedical systems and manufacturing processes. It also covers modelling and optimization techniques applied in materials forming, machining and tribology. Contributions to this book series are welcome on all subjects of "green" materials forming, machining and tribology. To submit a proposal or request further information, please contact Dr. Mayra Castro, Publishing Editor Applied Sciences, via mayra.castro@springer.com or Professor J. Paulo Davim, Book Series Editor, via pdavim@ua.pt

More information about this series at http://www.springer.com/series/11181

J. Paulo Davim
Editor

Welding Technology

Editor
J. Paulo Davim
Department of Mechanical Engineering
University of Aveiro
Aveiro, Portugal

ISSN 2195-0911 ISSN 2195-092X (electronic)
Materials Forming, Machining and Tribology
ISBN 978-3-030-63988-4 ISBN 978-3-030-63986-0 (eBook)
https://doi.org/10.1007/978-3-030-63986-0

This Springer imprint is published by the registered company Springer Nature Switzerland AG
The registered company address is: Gewerbestrasse 11, 6330 Cham, Switzerland

Preface

Nowadays, it is usual define welding as "*a materials joining process in which two (or more) parts are coalesced at their contacting surfaces by the suitable application of heat and/or pressure.*" Welding is a relatively new process, and it is normally connected with metal parts, but the process is also used for plastics. Welding is typically executed on parts made of same material, but some welding operations can be used to join different materials. Weldability has been defined by American Welding Society (AWS) as "*the capacity of a metal to be welded under the fabrication conditions imposed into a specific, suitably designed structure and to perform satisfactorily in intended service.*" For this reason, an improvement of the technical capability to predict the weldability is very important for the modern manufacturing industry.

The purpose of this book is to present a collection of examples illustrating research in welding technology with special emphasis in friction stir welding. Chapter 1 of the book provides friction stir welding. Chapter 2 is dedicated to fundamentals of friction stir welding, its application and advancements. Chapter 3 described modeling of friction stir welding processes. Chapter 4 contains information on an application from a defect—a friction stir channeling approach. Chapter 5 is dedicated to welding of dissimilar metals—challenges and a way forward with friction stir welding. Chapter 6 provides microstructure and texture in welding—a case study on friction stir welding. Chapter 7 is dedicated to tubular structures—welding difficulty and potential of friction stir welding. Chapter 8 describes Industry 4.0 in welding. Chapter 9 contains information on a comparative study of laser weldability of titanium alloys. Chapter 10 is dedicated to novel high-efficiency Keyhole tungsten inert gas (K-TIG) welding—principles and practices. Chapter 11 provides fatigue analysis of dissimilar metal welded joints of 316L stainless steel/Monel 400 alloy using GTAW. Finally, Chap. 12 is dedicated to industrial pipeline welding.

This book can be used as a research book for final undergraduate engineering course or as a topic on manufacturing engineering at the postgraduate level. Also, this book can serve as a useful reference for academics, researchers, mechanical, manufacturing, industrial and materials engineers, professionals in welding technology and related industries. The interest of scientific in this book is evident for many important centers of the research, laboratories and universities as well as industry.

Therefore, it is hoped that this book will inspire and enthuse others to undertake research in welding technology.

The editor acknowledges Springer for this opportunity and for their enthusiastic and professional support. Finally, I would like to thank all the chapter authors for their availability for this work.

Aveiro, Portugal
February 2021

J. Paulo Davim

Contents

About the Editor

J. Paulo Davim is a Full Professor at the University of Aveiro, Portugal. He is also distinguished as honorary professor in several universities/colleges in China, India and Spain. He received his Ph.D. degree in Mechanical Engineering in 1997, M.Sc. degree in Mechanical Engineering (materials and manufacturing processes) in 1991, Mechanical Engineering degree (5 years) in 1986, from the University of Porto (FEUP), the Aggregate title (Full Habilitation) from the University of Coimbra in 2005 and the D.Sc. (Higher Doctorate) from London Metropolitan University in 2013. He is Senior Chartered Engineer by the Portuguese Institution of Engineers with an MBA and Specialist titles in Engineering and Industrial Management as well as in Metrology. He is also Eur Ing by FEANI-Brussels and Fellow (FIET) of IET London. He has more than 30 years of teaching and research experience in Manufacturing, Materials, Mechanical and Industrial Engineering, with special emphasis in Machining and Tribology. He has also interest in Management, Engineering Education and Higher Education for Sustainability. He has guided large numbers of postdoc, Ph.D. and master's students as well as has coordinated and participated in several financed research projects. He has received several scientific awards and honors. He has worked as evaluator of projects for ERC-European Research Council and other international research agencies as well as examiner of Ph.D. thesis for many universities in different countries. He is the Editor in Chief of several international journals, Guest Editor of journals, Books Editor, Book Series Editor and Scientific Advisory for many international journals and conferences. Presently, he is an Editorial Board Member of 30 international journals and acts as reviewer for more than 100 prestigious Web of Science journals. In addition, he has also published as editor (and co-editor) more than 200 books and as author (and co-author) more than 15 books, 100 book chapters and 500 articles in journals and conferences (more than 280 articles in journals indexed in Web of Science core collection/h-index 57+/10500+ citations, SCOPUS/h-index 62+/13000+ citations, Google Scholar/h-index 80+/22000+ citations).

Chapter 1
Friction Stir Welding

Suman Kalyan Das, Suresh Gain, and Prasanta Sahoo

Abstract Friction stir welding (FSW) is a solid-state joining process which is a relatively newer technique but has found multiple utility in various industries across the world. By employing FSW, several issues that conventional fusion welding poses, viz. distortion, low efficiency, greater heat-affected zone (HAZ), etc., can be avoided. FSW in fact for some metals gives joint efficiencies near about 100%. FSW has been highly suitable for welding metals, viz. aluminum and magnesium, which have been previously found to be highly challenging to be welded by conventional welding techniques. For this reason, FSW has been embraced by aerospace and marine industries which require welding of large aluminum panels. Joining of dissimilar metals is one of the interesting aspects that FSW is found to handle quite easily as joining occurs much below the actual melting temperature of the metals. Research is also underway for welding metals having high melting points (steel and titanium-based alloys) using FSW in an efficient manner. FSW however involves higher setup costs and thus still out of the reach of many small-scale industries which refrain from higher initial investment. Thus, cost-effective solutions to tackle this issue are continuously explored by engineers and researchers. With a concerted effort, it is likely that the true potential of FSW technique would unfold, and it's benefits reaped by industries worldwide.

Keywords Friction stir welding · FSW · FSW tool · Solid-state welding

1.1 Introduction

Solid-state welding based on friction heating had been around for some time now. Friction stir welding (FSW) is a modified version of this technique which is effective in joining materials and can be joined without reaching their melting point and at the same time producing an efficient joint. The Welding Institute (TWI), UK, is credited to have invented this particular welding technology which is considered to be one of

S. K. Das (✉) · S. Gain · P. Sahoo
Department of Mechanical Engineering, Jadavpur University, Kolkata 700 032, India
e-mail: skdas.me@gmail.com

J. P. Davim (ed.), *Welding Technology*, Materials Forming, Machining and Tribology, https://doi.org/10.1007/978-3-030-63986-0_1

the latest developments in the last few decades in the field of metal joining [3]. FSW created revolution especially in the case of difficult-to-weld (by traditional methods) metals, viz. aluminum (2xxx and 7xxx series), which could now be joined rather easily and effectively. Besides, FSW gave rise to the possibility of effectively joining dissimilar metals. Because of its huge potential, FSW technology has been readily accepted by the fabricating industry and is now used worldwide.

FSW employs a rotating tool which has a shoulder and a pin extending below it. The pin is plunged into the intended weld line (joint area) between the two plates to be joined till the shoulder contacts the surface of the workpieces. Obviously, the workpieces should be rigidly clamped to prevent their separation due to the generated forces from the rotating tool. Frictional heat generated between the tool shoulder and the workpiece surface softens the material. The pin that enters into the workpiece creates a material flow by stirring action. This way the material from the two workpieces is intermixed and gets solidified as the tool progresses forward forming the joint. As bulk of the material melting is avoided, common issues in fusion welding, viz. solidification and liquation crack, distortion, porosity, etc., can be avoided in FSW.

FSW is a solid-state joining process that is capable of creating high-quality and high-strength joints with minimal distortion. This process can create the commonly encountered butt and lap joints and can also deal in wide range of material thicknesses and lengths. As modern applications demand for stronger, cost-effective and efficient joints using lesser energy, FSW seems to be a well-suited answer. In addition, FSW offers an environmentally benign solution to these challenges and at the same time is a very quick process. This has led to its quick advancement from initiation to industrial acceptance.

1.2 Fundamentals of FSW Process

In FSW, a non-consumable rotating tool is employed to join two objects. The tool comprises three parts, viz. shank, shoulder and pin (refer Fig. 1.1a. The shank connects the tool to the FSW machine, whereas the shoulder rubs against the workpieces providing necessary frictional heating. The pin plunges into the seam line between the workpieces (refer Fig. 1.1a, b). This plunging occurs as a result of downward force combined with the rotation of the pin. The plunging is one of the vital phases of the FSW process, as the thermo-mechanical conditions are initiated in this phase setting the stage for starting of the weld. Once the pin fully penetrates into the workpiece, the shoulder provides the necessary friction heating making the material soft (below its melting point temperature). Penetration of the tool results in plastic deformation in the workpiece and friction that leads to further heat production [4]. The plasticized material is contained by the tool shoulder, and the same flows around the pin by its stirring action. As per Fig. 1.1c, the right portion of the tool where the tool rotation direction (anticlockwise for the present case) and the direction of tool traverse make the same vectorial sense is denoted as the advancing

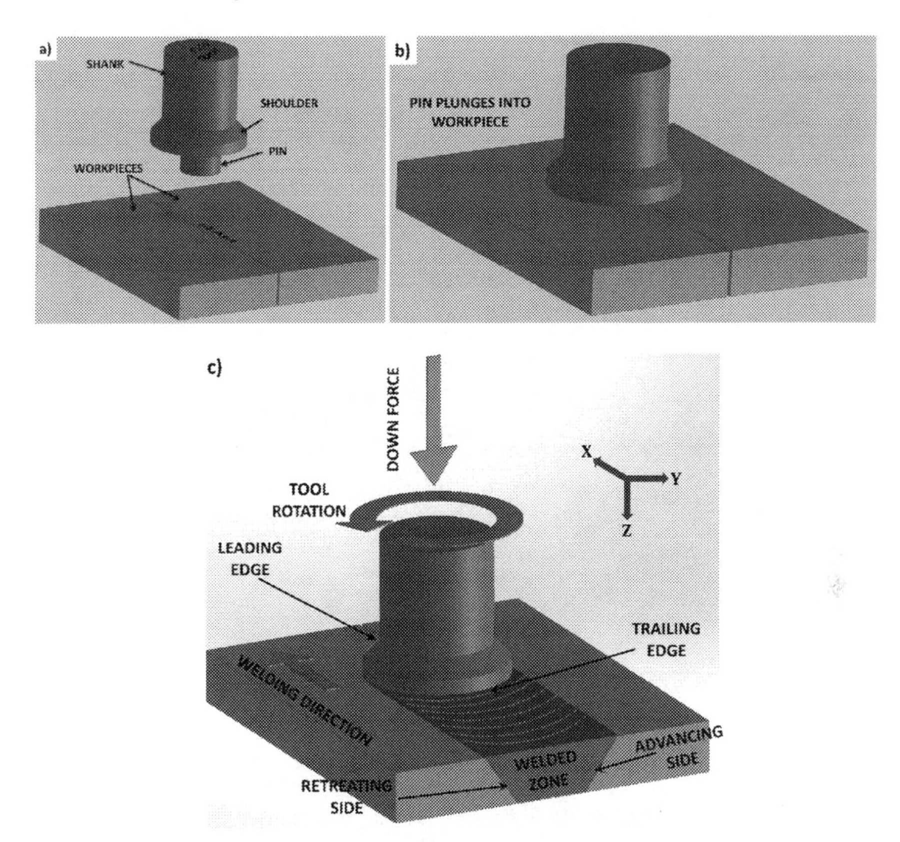

Fig. 1.1 Schematic diagram of FSW: **a** butt joint layout; **b** tool entering into workpiece and **c** welding nomenclature

side. In this portion of the workpiece due to tool advancement, the material tries to flow back but is resisted due to the rotation of the tool pin [39]. Left portion of the tool (direction of tool traverse and tool rotation have opposite vectorial sense) on the other hand is denoted as the retreating side, and the material flow here aided by the rotation of pin is relatively easier. As the tool advances, it is greeted by fresh cold workpiece material on the leading edge (front side), and the material is sweeped to the retreating side which may influence the overall material flow and the appearance of nugget. The back side (trailing edge) of the tool on the other hand continues to supply heat (due to friction) to the vacated weld region influencing the microstructure of the zone. During welding, the portion of the workpiece in contact with the tool can be roughly categorized into various zones as seen in Fig. 1.2a. As per the illustration, the material in front of the tool is at first pre-heated and is then subjected to the initial deformation as the tool nears. Below the shoulder and near the pin, the material suffers severe plastic deformation and is subjected to extrusion by the pin. The shoulder forges down the workpiece material due to the provided tilt at the back

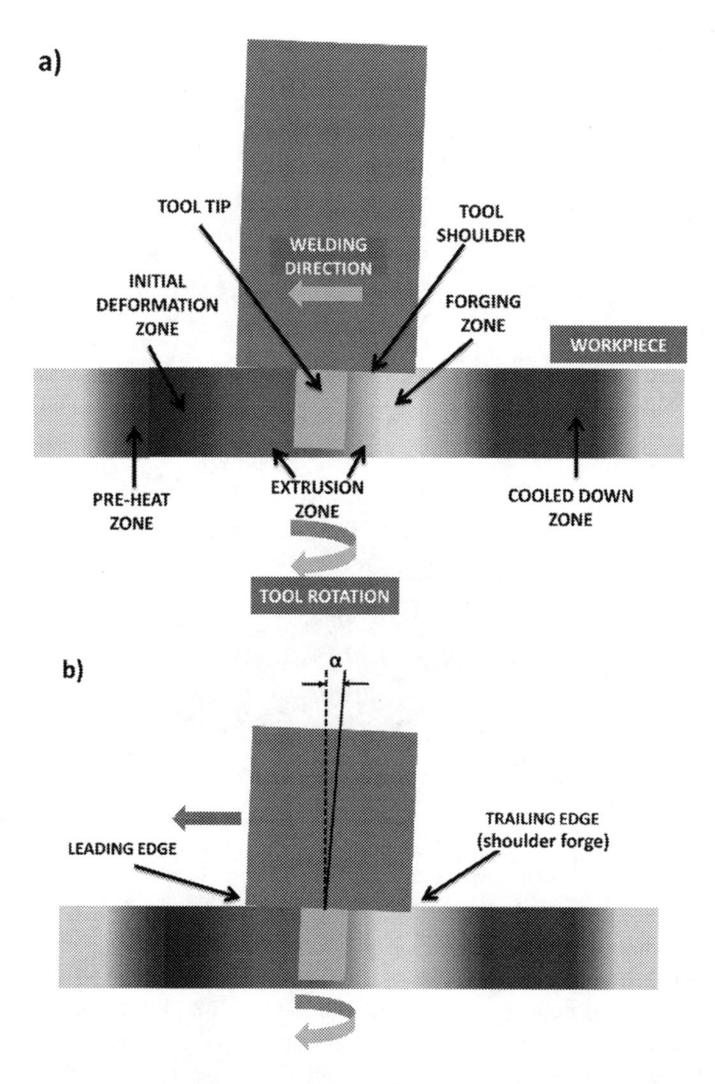

Fig. 1.2 **a** Various heat-affected zones and **b** tilt of FSW tool (exaggerated representation)

side of the pin. And finally, the material cools as it comes out of the influence of the tool. The workpiece material experiences these zones in a cyclic manner as the tool progresses about the intended weld line in the workpieces.

Post-weld, FSW joints exhibit distinct regions which can be observed through microstructural characterizations. The commonly accepted terminology for these zones (refer Fig. 1.3 is as follows [65]:

- Parent material (unaffected zone): This region is never deformed and remains completely unaffected by the welding even though it may have been exposed to

Fig. 1.3 Various regions in a friction stirred weld [65]—reprinted with permission from Taylor & Francis

thermal cycle on a lower level. These zones normally lie remote from the region of weld.

- Heat-affected zone (HAZ): This region is much nearer to the central weld portion, and its microstructure gets affected due to the heat from the welding. Hence, its mechanical properties also undergo a change. However, material is believed not to undergo plastic deformation.
- Thermo-mechanically affected zone (TMAZ)/Weld nugget: The material in this region lies in and around the weld center and is severely affected by heat and plastic deformation. When welding aluminum, there is a possibility that significant plastic strain occurs in TMAZ without the occurrence of recrystallization. When this situation occurs, a thin boundary separating recrystallized and deformed regions within the TMAZ can be observed through a microscope. This recrystallized zone is popularly termed as weld nugget.

1.3 Advantages and Limitations of FSW

The FSW process has a host of advantages which has led to its quick acceptance by the industry. It has an edge above the other welding techniques when considering both the weld microstructure and the joint performance. Besides, FSW does not suffer from common defects in welding, viz. porosity, solidification cracking, etc. Some of the advantages of FSW process are as follows:

- **Good mechanical and metallurgical properties**

FSW operates in solid state of the material, thereby helping in dynamic recrystallization of the weld zone. Due to stir action of the pin, grain refinement phenomenon occurs which leads to the development of fine-grained structure. Hence, the mechanical properties (hardness, tensile strength, fracture strength) of the workpiece get improved along the joint portion.

- **Less distortion (low heat input)**

During FSW, the peak temperature remains lower than the melting point temperature of the workpiece material. Compared to arc welding or any conventional welding, heat input for FSW is low. Less heat input leads to less residual stress which lead to reduced distortion. Due to shoulder pressure and low residual stress in FSW, the joint almost remains smooth eliminating post-weld machining operation.

- **Improved energy efficiency**

As no melting is involved during FSW, the process consumes less energy with no harmful emission as there is no use of any filler material.

- **Improved joint efficiency**

The ratio of strength of the welded joint to that of the base metal defines the efficiency of a welded joint. Joint efficiency of FSW mainly depends on the process parameters. It is important to select suitable process parameters for achieving 100% joint efficiency. For some metals, joint efficiency can be achieved beyond 100% by FSW technique which is impossible to be achieved by other welding techniques.

- **Welding in all positions**

Due to the absence of weld pool, this welding technology can be employed for joining along all positions, i.e., vertical, horizontal, etc. Conventional weld joints, viz. butt joint, lap joint and corner joint, can all be formed using FSW.

- **Increased fatigue life**

For welding of structural components and in transportation industries, fatigue life evaluation is most essential. In aerospace sector too, fatigue strength of the weldment is a major concern. The fatigue life of the welded joint can be improved after FSW due to the reduction in the residual stress and peak temperature. The fatigue strength of FSW joint is higher compared to arc welding.

- **Improved safety and environmentally friendly**

FSW is a fully automatic process. No toxic gas or spatter is produced; due to the absence of toxic gas, this technique is safe for the welder and reduces the chance of causing hazard.

- **Welding of dissimilar materials**

Joining of different metals with different melting temperatures by arc welding is difficult. This is because lower melting temperature metal will start to melt before reaching the melting temperature of high-temperature workpiece. For welding dissimilar material or joining of composite materials with higher strength, this welding technique is suitable. As this is a solid-state welding technique, there have no issue of melting problem.

- **Non-consumable tool**

FSW technique uses a non-consumable tool which is normally made of tool steel, PCBN or W-base for welding. Hence, continuous requirement of filler material is eliminated reducing the associated costs. Besides, chances of production of harmful gases are also reduced.

Some of the limitations of FSW include:

- **Presence of ejection hole/exit hole**

In FSW technique, ejection hole remains on the workpiece when the tool is withdrawn from. In many applications, the presence of this ejection hole is not acceptable, viz. marine industries and aerospace industries. A runoff tab can be used to eliminate this hole or a suitable workpiece being attached at the end of the process so that key hole can be placed on that extra workpiece material. This involves the requirement of extra arrangement increasing the cost.

- **High initial investment**

During FSW, high forces are produced by the machine. The fixture design and the equipment cost increase to handle this level of forces. The tool used for this technique is also very costly. All these make the cost of the machine quite high pushing it beyond the reach of smaller companies. Besides, for manufacturing single product or lesser number of products, this technique may not be economical.

- **Less flexible**

This joining technique is less flexible compared to manual arc welding process, and it is difficult to join different plate thicknesses together.

1.4 Friction Stir Welding Tools

The tool plays an important part in the successful welding through FSW process. As already mentioned, the tool consists of three main parts, viz. the shank, shoulder and the pin. Broadly, the tool has a couple of functions: (a) heating in a localized area and making the material soft and (b) flow of the softened material. Friction generated (at shoulder and pin) due to the rotation of the tool provides local heating that leads to material plasticization (softening) around the tool pin. Rotation of the tool together with its translation (tool traverse) helps in the flow of the soft/plasticized material from the leading side to the trailing side. This transfer of material helps in filling the hole which is generated by the pin, thus forming the joint. The shoulder supplies the majority of the frictional heat and also prevents the escaping of softened material out of the weld zone. Thus, the quality of the weld depends a lot on the design of the shoulder and that of the pin. However, the microstructure and mechanical properties of the stir zone and HAZ zone of FSW also depend on the tool material and its properties. For welding aluminum alloys and soft metals, these techniques gained practical success. However, the design and selection of the tool became very important while welding hard alloys, viz. steel, nickel alloys and titanium alloys, as the life of the tool is shortened considerably and the cost increased. Thus, there has been a concerted effort to develop cost-effective tools for the purpose which at the same time has longer life. Different materials employed for making FSW tools as well as effective tool geometry tool geometry are discussed in the following texts:

1.4.1 Materials for FSW Tool

Tool materials for FSW of high-temperature alloys should have excellent mechanical properties so that it can perform successfully at higher operating temperatures (above 900 °C). The properties include ductility, hardness, thermal conductivity and coefficient of thermal expansion. The selection of the tool materials depends on the recrystallization temperatures of the work piece. Tool materials are required to have necessary strength to survive the process temperature as well have fatigue strength and fracture toughness at the same temperature conditions [37]. Besides, the tool also needs to have good wear characteristics and thermal and chemical stability (inert toward the workpiece material). Apart from this, tool wear and its life are also important factors behind the selection of tool material. And the final deciding factor remains the quality of weld produced by the tool. Steel is the popular choice as tool material for welding of aluminum and magnesium alloys. Polycrystalline cubic boron nitride (PCBN) and W-based tools on the other hand have been used for welding of hard material such as steel and titanium.

1.4.1.1 Tool Material for Low-Temperature Alloys

It is easy to find suitable tool material for processing aluminum alloys as aluminum is softer and has a lower melting point than many of the other metals. For FSW of aluminum alloys, the tool material requires to sustain the process conditions. Besides, it should also possess properties, viz. fracture toughness, long fatigue life, wear resistance, chemical stability and thermal stability. Tool steel (H13 tool steel) and cobalt-based super-alloy (MP159) are among the common tool materials used for FSW of aluminum and its alloys [37, 52]. Prado et al. [50] employed tool steel for FSW of aluminum MMC (aluminum alloys 6061+20%Al_2O_3). They found that the tool was quite effective and suffered minimum wear in case of welding Al 6061 compared to when Al_2O_3 particles were added (refer Fig. 1.4. They further found that the tool suffered maximum wear when operated at 1000 rpm post which the wear decreases but in an irregular manner (see Fig. 1.5. Enhancement in the fluidity of the material and the turbulence in the particle flow beyond 1000 rpm may be the reason for this decrement. FSW for dissimilar metals has also been carried out successfully by various researchers. Chen and Lin [7] have carried out a design of experiment for optimization of the FSW process parameters when welding SS400 low-carbon steel and AA6061 aluminum alloy using AISI 4140 tool steel. Lee et al. [27] successfully achieved a lap joint between Al–Mg alloy and low-carbon steel using tool made of tool steel. They carried out the welding joint without excessive tool wear by putting the Al–Mg alloy plate above the steel plate. In this way, the tool made direct contact

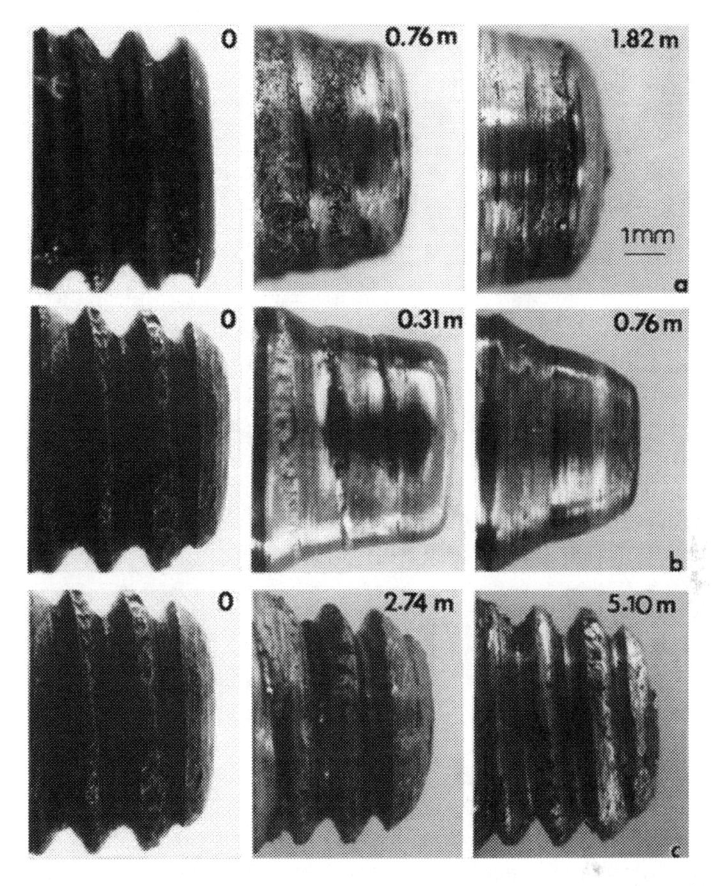

Fig. 1.4 Wear of tool pin (distance welded is indicated in each figure) **a** Al-MMC at 500 rpm, **b** Al-MMC at 1000 rpm and **c** Al 6061 alloy at 1000 rpm [50]—reprinted with permission from Elsevier

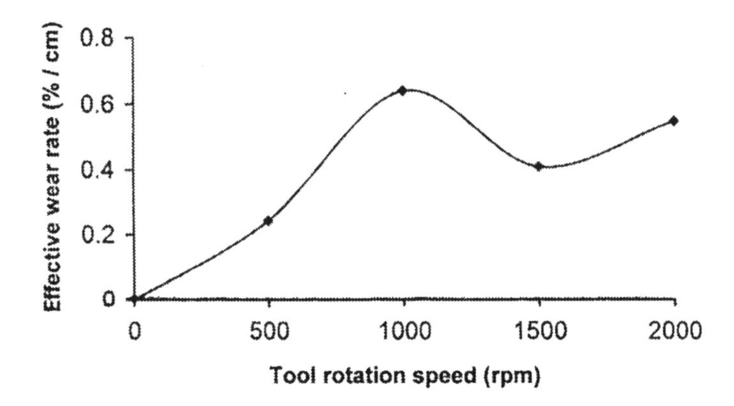

Fig. 1.5 Wear rate of tool versus rotation of tool in rpm [50]—reprinted with permission from Elsevier

with the AL–Mg alloys instead of the steel plate. Meran and Kovan [34] employed X155CrMoV12-1 tool steel (cold worked) for welding copper and brass.

1.4.1.2 Tool Materials for Welding High-Temperature Alloys

PCBN Tool

Harder materials (alloys of steel, nickel and titanium) are always difficult to weld, and selecting proper FSW tool for welding these materials is challenging. PCBN which is only second in hardness to diamond is the preferred tool material for welding these materials [52]. A PCBN tool is known to maintain its strength and hardness at elevated temperature conditions. The wear rate of PCBN is low because of its thermal and chemical stability which is superior to diamond [52]. It is particularly inert to iron. Moreover, PCBN has low coefficient of friction (COF) which yields smooth surface of the weld. However, PCBN tools are very costly due to the requirement of high pressure and temperature during its manufacturing.

Tungsten Based Tool

Tungsten-based composites are one of the popular choices as the tool material for FSW of steel. Pure tungsten has very poor toughness at room temperature, and its wear rate is also high when welding materials, viz. steel and titanium-based alloys. Thus, researchers have employed variety of tools based on tungsten, viz. W-La_2O_3, WC–Co, WC, W-Mo and W-Re for FSW of steel and hard material. These materials possess most of the properties needed for welding of hard materials. The hardness of WC is ~1650HV, and it has excellent toughness [52]. WC and WLa_2O_3 have been utilized as a tool material in most of the investigations for joining sheets of carbon steel and stainless steel by FSW. W–Re (W-25 wt% Re) tool is the most popular among W-based tool material. The addition of rhenium enhances the hot strength of the material, which reduces the deformation of the pin during its insertion into the workpiece. Re also improves the wear resistance of the material. This material has less tendency of fracture because of lower ductile to brittle transition temperature which is approximately 50 °C.

1.4.1.3 Miscellaneous Tool Materials

Superalloys based on cobalt (Co) or nickel (Ni) have been sometimes employed as tool materials for FSW of steel. Si_3N_4 is used as the tool material for welding because it has properties, viz. higher hardness, low coefficient of thermal expansion and higher thermal conductivity. However, the workpiece material may be contaminated with Si and N upon using this tool. Thus, Si_3N_4 tool is provided with a coating of TiC/TiN to prevent this contamination. Lee et al. [29] successfully carried out FSW of titanium with a sintered TiC tool. Also, cooling arrangement was provided so that the excess heat could be extracted from tool to prevent its damage. Molybdenum-based alloy tools are employed to weld titanium and steel workpieces. Tool wear is found to be greater in case of metal matrix composites than soft alloys as the former consisted of hard abrasive particles. The tool rotational speed as well as its traverse speed can affect the life of the tool. Tool life is found to be optimum when lower rotational speed is used with a relatively higher traverse speed.

1.4.2 Geometry of FSW Tool

Heat generation rate, transverse force, torque and the thermo-mechanical aspect of the experiments depend on the tool geometry. Geometry of the tool also affects the material (plasticized) flow of the workpiece. The shoulder diameter, shoulder surface, shape, size, the nature of the tool surface and the pin geometry are the factors which influence the weld quality. Selection of appropriate shoulder with pin geometry is again complicated by the following factors [38]:

- Dimension of workpiece which determines its capacity to acts as heat sink.
- Efficiency of the joint required.
- The forces expected on the tool so that the tool can survive mechanical failures.

1.4.2.1 Shoulder Diameter and Surface Features

The diameter of the FSW tool shoulder is particularly an important parameter as it decides the amount of friction heat produced due to the rubbing of the shoulder surface with the workpiece surface. The thickness of the workpiece decides the shoulder diameter of the tool. This is because for good welding, the material should be softened adequately. Now, as the thickness of the workpiece increases, it requires more heat for softening. This can only be generated by a larger shoulder diameter. Moreover, if the size of the workpiece increases, it acts as larger heat sink, thus requiring more supply of heat for plasticizing of the material. However, it is worth noting here that higher heat generation is also possible by increasing the rotational speed of tool as well as decreasing the feed rate. A suitable shoulder diameter results in splendid joint properties.

Features of the shoulder surface also affect the characteristics of the weld. Traditionally, the shoulder surface can be flat, concave or convex (Fig. 1.6). Among these, the flat face shoulder is the most simple to manufacture. But the flat shoulder fails to contain the plasticized material within the weld zone leading to excessive material flash. Thus, tools with flat shoulder surface are normally operated with a tilt angle. To solve this issue, the concave shoulder was conceptualized which is found to be quite effective and is the most commonly used tool nowadays. The third variety can be tool with convex shoulder which however causes material displacement away from the pin. The benefit of using tool with convex shoulder is that it can handle

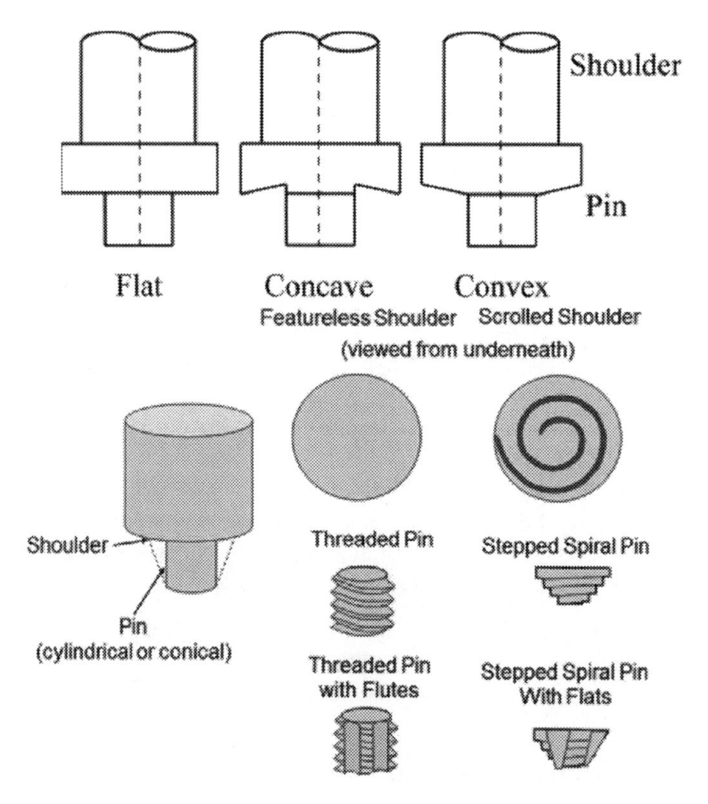

Fig. 1.6 Various types of shoulder and pin geometry [37, 74]—reprinted with permissions from Taylor & Francis and Springer Nature

plates of thickness varying within a certain range as well as plates having issues in flatness. To reduce the material displacement issue, convex shoulders are generally provided with features, viz. scrolls, knurling, ridges, grooves, spirals and circular steps (refer Fig. 1.6). These feature help in increasing the material friction as well as shear and deformation. This leads to better mixing of the softened materials which yields good weld characteristics [74]. Moreover, when a convex shoulder with scrolls are used, in vertical force control mode (machine maintains the specified force) for any increase in depth of penetration, the area of contact between the shoulder and the workpiece increases. This reduces the vertical force, and to maintain it, the depth of penetration decreases returning the tool to its previous position. Similarly, when there is decrease in the penetration depth, the contact area between the shoulder and workpiece decreases. This leads to higher vertical force, for compensating which, the penetration returns to its original value. So, friction stir welding with convex scroll shoulder maintains a constant plunge depth.

Sorensen et al. [60] studied the role of geometry of the tool on the weld characteristics. They selected a tool with convex shoulder having steeped spiral. According to their study, shoulder's radius of curvature and the pitch of the spiral have significant influence of the welding. Changing the shoulder from convex to concave may change the mode of failure of the weld as well as its microstructure. Similar phenomena may occur when flat tool changes to a concave one. Hirasawa et al. [18] studied friction stir welding with cylindrical, triangular, tapered and triangular taper pin geometries and flat, concave, and convex tool shoulders. Their studies show that a concave triangular pin resulted in spot welds of enhanced strength. Li et al. [31] explained that the angle of the shoulder surface influences the vertical force, based on the radius of the tool pin. They reported that convex shoulder with a scroll results in higher stability of the FSW process.

1.4.2.2 Pin Geometry and Features

In FSW, pin penetrates the workpiece and mixes the softened material and hence plays an important role in the process. Thus, shape and size of the pin would influence the quality of the weld. The pin length is selected based on the thickness of the workpiece to be welded. As material softening occurs to some extent beyond the tip of the pin, the pin length is normally taken a bit shorter than the actual workpiece thickness. As a convention, the pin length is taken 0.85–0.95 of the thickness of the workpiece [38]. In case of the diameter of the pin, a smaller diameter is normally preferred because it helps in reducing the HAZ in size. But again a tool with smaller diameter is weak and is vulnerable to failure especially when the tool is plunging down and the vertical force acting on it is at its peak. The pin can also fracture if it faces a higher shear load due to insufficient softening of the material during welding. As a thumb rule, the diameter of the pin is taken as 1/3 of the diameter of the shoulder.

A pin diameter of approximately 1/3 the shoulder diameter is generally used.

The shape of the pin and the features present on it is found to affect the material flow. Cylindrical and truncated conical shapes are the well-known pin shapes [37]. There is no downward push for cylindrical pin when the materials flow from the front to the back of the tool. When the pin is truncated conical, some download force component is generated leading to material flow in downward direction. This is important while welding high-temperature materials. The addition of features to the pin aids the vertical flow of material along with breaking up the flow. Figure 1.6 shows additional features, viz. flutes and flats, which disrupt the material flow further. Various features and geometry of the PCBN tool system are illustrated in Fig. 1.7. Variety of tool designs and their features along with their advantages are summarized in Table 1.1.

1.5 Workpiece Materials Suitable for Friction Stir Welding

Welding aluminum and its alloys are the primary reasons behind the development of FSW technique. This welding technique can be also applied for joining other non-ferrous materials such as copper, magnesium and titanium. Later, FSW is developed

Fig. 1.7 Features and geometry of PCBN tool system [74]—reprinted with permission from Taylor & Francis

further to carry out welding of high-temperature materials like titanium and steel and their alloys. Recently, great focus has been paid to the welding of dissimilar metals and alloys using FSW due to their commercial importance and inability to be joined by conventional welding techniques.

1.5.1 FSW of Aluminum Alloys

Aluminum is one of the most useable alloys in different industries such as aviation, aerospace, marine, automotive and rail. This metal cannot be welded by gas welding because of its poor weldability. Aluminum welding by resistance welding and arc welding are costly and need special techniques. FSW technique has been found to be highly suitable for welding aluminum alloys due to its mechanical properties and low density. Different series of aluminum alloy, viz. 2XXX, 3XXX, 4XXX, 5XXX, 6XXX, 7XXX, can be welded by this technique. Zn, Mg, Cu, Mn and Si are the alloying elements present in pure Al alloys. Work hardening and heat treatment are the two processes through which aluminum alloys get further strengthening. Various dimensions of tool with their features that are used for several aluminum alloys are provided in Table 1.2.

In case of aluminum, FSW is generally processed by tool steel which has high-temperature strength and toughness. The temperature achieved during welding aluminum alloys remains approximately around 400–500 °C. Another parameter is the maximum downward force reached during insertion of tool. The peak force during tool plunging is the combination of vertical force and transverse force. The choice of process parameters is highly important for producing defect-free weld and high joint efficiency in case of aluminum alloys. Another major factor for FSW of Al alloys is the welding time which is very important for the shipbuilding and aviation industry.

Table 1.1 Features of tool and their advantages [37]—reprinted with permission from Springer Nature

Tool feature	Design variation		Advantage
Shoulder	Convex with scroll	Curved geometry	Possible to join workpieces of different thicknesses. The scrolls help in containing material in the weld zone
	Concave (standard design)	Tapered geometry	Possible to join workpieces of different thicknesses. The scrolls help in containing material in the weld zone
		Smooth surface	Due to concavity, material contains in weld zone. Tilt of the tool is needed
		Scrolled surface	The normal force on the tool is reduced. The thickness of weld zone reduces. Tool tilt is not required
Pin	Cylindrical	Flat bottom	Very common. Machining is easy
		Round bottom	Wear of the tool reduces during penetration. Joint strength at the root of the weld improves
		With flats	Plastic strain as well as temperature increases in nugget zone
	Conical	Threaded cone	Lateral force generation lower than cylindrical pins
		Stepped spiral	Employed for high melting point metals. Good strength
		With flats	Plastic strain as well as temperature increases in nugget zone
	Whorl		Lower transverse force on tool
	MX Triflute		Refined version of whorl pin
	Threadless		Reduces tool wear in extreme situations
	Retractable		Exit hole can be eliminated

1.5.2 Magnesium Alloys

Sheets of magnesium alloys are generally manufactured by casting or die casting process, and these alloys have high formability. It is challenging to weld cast magnesium alloys sheet due to a relatively high coefficient of thermal expansion that causes large distortion and deformation weld. The formation of porosity by arc welding

Table 1.2 Dimensions of tool used for welding different aluminum alloys [38]—reprinted with permission from Springer Nature

Shoulder diameter (mm)	Pin diameter (mm)	Shoulder diameter/pin diameter	Material welded	Reference
20	7	2.9	6061 T6	[14]
16	6	2.7	2024 T6	[17]
18	5	3.6	5083 O	[49]
10	4	2.5	5083 H18	[16]
10	4	2.5	6111 T4	[16]
23	8.2	2.8	2024 T3	[62]
19.1	7.87	2.4	7050 T7	[15]
10	3	3.3	6016 T4	[56]

or gas welding makes the welding defective. So, solid-state welding process is the optimum choice for joining cast magnesium alloy sheets. Researchers have conducted FSW on cast AZ91, AZ61, AZ1, AM50 and AM60 sheets. Nakata et al. [45] have studied FSW on AZ91 sheet of 6.4 mm sheet thickness. They have successfully carried out FSW in case of butt joint and have optimized the process parameters. At optimum condition, combination of 50 mm/min traverse speed and 1240–1750 rpm of tool rotational speed is found to yield good quality weld. They suggested that welding speed higher than the optimum level as well as tool rotation speed lower than the optimum level may result in lack of bonding and formation of voids. The FSW of magnesium alloys at varying plate thickness, tool material, spindle speed and welding speed is summarized in the Table 1.3.

1.5.3 Copper Alloys

Conventional fusion welding techniques find it difficult to weld copper alloys. This is because of the high thermal diffusivity of copper which is about 10–100 times higher than that of steel and nickel alloys [38]. Higher thermal diffusivity needs higher heat input leading to low welding speeds (traverse speed). Several researchers have attempted to join pure copper and its alloys through friction stir welding technique. Lee and Jung [28], Park et al. [47], Nakata [44] and Mishra and Jeganathan [41] have successfully conducted FSW of 1.5–50-mm-thick copper plates at variable process parameters.

Table 1.3 FSW of magnesium alloys at different plate thickness, tool material, spindle speed and welding speed

workpiece material	Thickness (mm)	Tool material	Spindle speed (rpm)	Welding speed (mm/min)	Reference
AM60	20	Tool steel	2000	120	[13]
AZ31	4	65Mn steel	375–2250	20–375	[69]
AZ31	6.3	–	800–3500	100	[70]
AZ31B	4	Tool steel	1250	250–750	[33]
AZ31B	3.2	H13	1200–1750	305–762	[10]
AZ31B-H24	4	D2 (SKD11)	1250–2500	87–342	[30]
AZ31B-H24	3.2	–	1500, 2000	78–204	[46]
AZ61A	6	High-carbon steel	1200	90	[53]
ZK60	3, 6	–	600–2000	100–400	[36]
AMX60	4	SKD61 tool steel	1100	200–400	[6]

1.5.4 Steel

Nowadays, the FSW technique is being focused on joining various types of steel as it being the most important engineering material. Steel constitutes the major components in various structures, buildings, machines, etc. FSW studies have been conducted on mild steel (AISI 1010), low-carbon steel, austenitic stainless steel (316L, 304L), super-austenitic stainless steel, high-strength low-alloy (HSLA-65) steel, super duplex stainless steel (SAF 2507), duplex stainless steel (DH-36), creep resistance steel (P91, P92). The process parameters and the tool materials depend on the particular variety of steel as well as the plate thickness. For FSW of steel, shielding by an argon gas environment is provided to protect both the weld zone and tool from oxidation. As already mentioned, FSW produces much lower heat compared to fusion welding. Thus, the chances of metallurgical changes in the HAZ of the material are lesser in case of the former. Moreover, chances of distortion and residual stresses in steel are also lesser in case of FSW which makes it suitable for welding of thicker plates as demanded in shipbuilding and heavy manufacturing industries. The issue of hydrogen cracking in steels is also non-existent in case of FSW as it is a solid-state joining process. FSW of different types of steel at different plate thickness, tool material, spindle speed and welding speed is given in Table 1.4.

1.5.5 Titanium Alloys

It is possible to weld a variety of titanium-based alloys to be welded by conventional fusion-based technique, viz. gas tungsten arc welding (GTAW). However,

Table 1.4 FSW of different types of steel at various plate thickness, tool material, spindle speed and welding speed

workpiece material	Plate thickness (mm)	Tool material	Spindle speed (rpm)	Welding speed (mm/min)	Reference
0.45% carbon steel	2	WC tool	250–1000	100	[20]
Medium-carbon low-alloy S45C	2	WC tool	400	100, 200, 300, 400, 500	[19]
st37 steel	2	Tungsten alloy bar	450–560	50–160	[24]
High Ni steel (24 wt%),	1.6	WC tool,	200, 300, 400	400	[42]
AISI 304L SS	3.2	W-alloy tool	300, 500	102	[55]
AISI 316L austenitic stainless steel	3	W tool doped with 1% La_2O_3	400–800	45	[59]
AISI 409 M ferritic stainless steel	4	W-alloy tool	1000	90	[25]
Super-austenitic stainless steel S32654 sheets	2.4	W–Re tool	300, 400	100	[32]
Duplex stainless steel SAF 2205	2	WC-based material	400	50	[12]
Super duplex stainless steel SAF 2507	4	PCBN tool,	450	60	[58]
API-5LX80 steel plates	12	PCBN tool	300, 350, 500	100	[57]
API-X100 pipeline steel	6.5	WRe tool	600	50	[9]
HSLA-65 steel	6.45	PCBN tool	300- 600	50–200	[68]
Naval grade HSLA steel	5	W-La_2O_3 tool,	600	30	[51]
High-nitrogen stainless steel	2	Si_3N_4-based tool	400	50, 100, 200, 300	[43]
DH36 steel	6	W-Re-PCBN tool,	450	250	[66]
Oxide dispersion-strengthened (ODS) steel MA956 plates	4	PCBN tool	25,50,100	500,400	[5]
18% Ni maraging steel	5.2	PCBN tool inserter in WC shank,	250	25	[35]
Dissimilar steels of grade 304 stainless steel and st37 steel	3	WC-Co tool	400, 800	50	[21]
P91 steel plate	3	PCBN-coated tool	1000	100	[23]

Table 1.5 FSW of various titanium alloys using different tool material, spindle speed and welding speed

Workpiece materialtool	Tool material	Spindle speed (rpm)	Welding speed (mm/min)	Reference
Ti–6Al–4V	Mo-based tool	300–600	60	[72]
Ti–6Al–4V	PCBN	200	50	[73]
Ti–6Al–4V	W-25%Re	150	100	[48]
Ti–6Al–4V	W-La_2O_3	150	100	[11]
Timetal 21S	W-25%Re	200	50–300	[54]

post-welding the joint requires heat treatment which lengthens process step and increases production costs. FSW can eliminate this requirement of post-weld heat treatment in titanium alloys. Little information is available on FSW of titanium alloys. Researchers have suggested that during FSW of Ti alloys, peak temperature in the nugget zone exceeds 1000 °C which is above the β-transus temperature of 995 °C. In general, it has been mentioned that the peak temperature in the heat-affected zone remains below the β-transus temperature. Table 1.5 shows various titanium alloys welded with different tool materials, spindle speed and welding speed.

1.5.6 Dissimilar Alloys and Metals

A variety of applications require weldment between two different metals in order to get an optimized design. Now, joining two different metals having different compositions, melting point temperatures, recrystallization temperatures, etc., is difficult and challenging. Now, through FSW technique, joints between two dissimilar metals, viz. Al and Cu, can be achieved quite efficiently. Based on the placement of the materials, i.e., on the advancing side or the retreating side, the material flow pattern at the joint changes and is a subject of research interest. The target is to achieve a joint strength equivalent to that of material having lower strength. Larsson et al. [26] have joined Al-6061 to Cu by FSW. They found that if the weaker material is placed on the advancing side, FSW yields better quality welds.

1.6 Friction Stir Welding Setup and Joining Configurations

Most of the traditional joint configuration in welding is possible to be achieved through friction stir welding process. These include (a) butt, (b) lap and (c) fillet joints. By combining these configurations, most of the joints encountered in practice can be realized. FSW can be applied for both linear and circular welding.

1.6.1 Linear Welding

1.6.1.1 Dedicated Setups for FSW

Machines for carrying out FSW is available in the market. These setups are equipped with proper bed, suitable jig and fixtures for fixing the workpieces as well as the welding tool. The tool can normally be given motion in three directions (X, Y and Z). There are sensors to measure forces encountered by the tool in various directions and to monitor temperature (at the weld zone) and a host of other parameters, viz. spindle speed, displacement of the tool, etc. There is also provision to provide inclination to the tool. Cooling systems are provided for safeguarding the machine components which have direct metallic contact with the tool which will be subjected to a very high temperature. The entire machine is generally controlled numerically by a computer system. The advantage of using such machines normally reflects in higher quality welds.

Now, the four important process parameters controlled by FSW are as follows:

(a) Position of the tool
(b) Orientation of the tool
(c) Load applied
(d) Rotational speed of tool
(e) Travel speed of tool

The selection of the equipment is based on the range of these parameters that are defined for the materials that are planned to be welded. Figure 1.8 displays a photograph of a linear FSW setup.

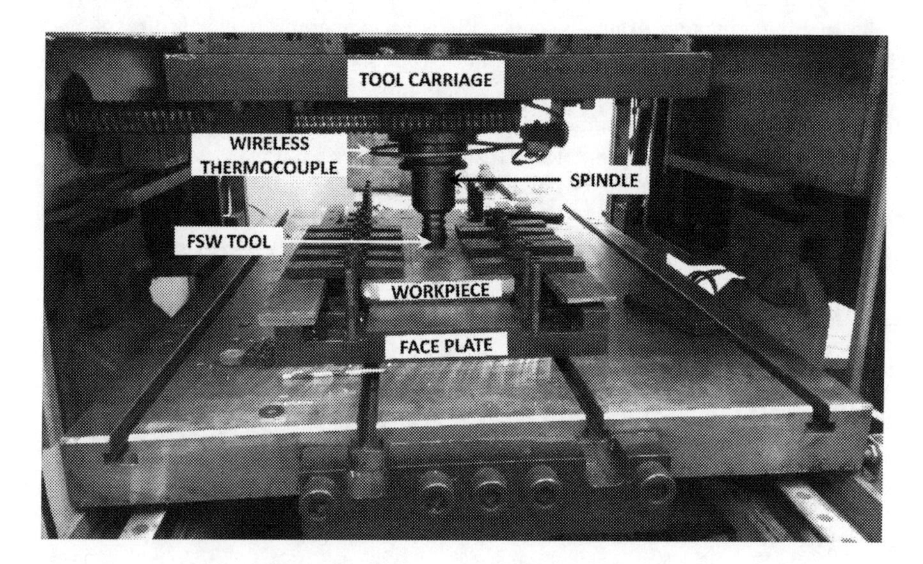

Fig. 1.8 Photograph of a linear FSW setup

1.6.1.2 Using a Milling Machine—Low-Cost Setup

The issue with the dedicated setups is their higher cost, and hence, these are limited to those industries which have sufficient funds to invest in them. Thus, technologists have aimed to create a cost-effective solution so that the benefits of FSW process can be reaped by industries worldwide. The configuration of the milling machine seemed suitable for its adaptability toward the FSW process. As milling machine is a very common machine available almost everywhere, it seemed a viable solution to tackle the huge investment cost required to setup dedicated FSW machines. However, the spindles in milling machines are not designed to endure the radial and thrust loads expected during FSW process. Moreover, traditional milling machines do not offer precise load control or load monitoring. Thus, high-quality weld is difficult to produce through the milling machine setup. Many researchers however use this type of setup to gain knowledge about the FSW process.

Some companies have also developed FSW units which can be retrofitted to any conventional milling machine to convert it into a FSW setup. These type of setups cannot be a 100% replacement for dedicated FSW setups. Nevertheless, these converted machines can produce a decent weld quality and are simple to operate. Figure 1.9 shows a milling machine converted to a FSW setup.

Fig. 1.9 Vertical milling machine converted to FSW setup for linear welding [67]—reprinted under Creative Commons license

1.6.2 *Circular Welding*

Circular welding is mainly used for welding pipes. Although several configurations have been devised, the basic fundamental of welding pipe is to make use of a precision mandrel which can be inserted into the pipe pair, thus producing a seam over it. As of now, there are primarily two setups for circular welding which are elaborated as follows:

1.6.2.1 Customized Setup

Again as in case of linear welding, customized setups for pipe welding can be availed from the major players in the field. These setups provide fast and precision solution for pipe welding which can be beneficial to relevant fabricators.

1.6.2.2 Attaching a Rotary Mandrel Unit

An attachment consisting of a rotary turntable and mandrel can convert an existing setup of linear welding to a circular welder. Sometimes, the attachment consists of an expanding mandrel which can hold pipes with diameters varying within a certain range. Generally, in these types of setups, the FSW tool remains fixed and the rotating mandrel turns the pipes, thus producing the weld. It is obvious that the rotation of the mandrel is precisely controlled. Figure 1.10 shows a typical FSW linear machine fitted with a rotary mandrel for pipe welding.

1.6.2.3 Portable Setups—Larger Scale Welding

For carrying out welding of larger and longer pipes, portable welding machines are also available. These setups have the advantage of being installed in the field. In these setups, normally, the arrangement is such that the tool moves along the weld seam producing the weld as it is difficult to rotate the huge and heavy mandrel supporting the pipes.

1.6.2.4 Issue of the Presence of Ejection Hole

As already discussed (to be discussed previously), that FSW tool leaves an ejection hole on the workpiece at the end of the weld. This may not be problem in case of jobs requiring linear welding as some allowance for this can be incorporated in the job and which can be later cut off. However, in case of circular welding which warrants a sort of continuity, the presence of an ejection hole will make the joint vulnerable and

Fig. 1.10 Circular welding **a** rotary mandrel fitted to the existing linear FSW setup and **b** pipes fitted on the mandrel

hence unacceptable. Thus, some special arrangement has been devised to eliminate this problem.

Designing a Wedge to Transfer the Hole

This has been one of the preferred ways of getting rid of ejection hole in case of pipe welding. In this case, an extra wedge block is attached by temporary means to the pipe preferably at the starting point of the weld track (refer Fig. 1.11a). Now, once the welding is complete and the tool again gets to the starting point, a tangential motion is given to the tool which makes it gradually shift to the attached block, thus leaving the ejection hole on the wedge rather than the pipe. The attachment is then removed by hammering or minor machining. A zoomed view of the wedge connected to the pipe is illustrated in Fig. 1.11b.

Retractable Pin Configuration

In fixed pin configuration, the FSW tool is a single unit consisting of the shoulder and the pin. This is the most common tool used in FSW and is the most easiest to manage from control perspective. But nevertheless, there is no choice but to have an exit hole when using this tool. Thus, researchers have come up with a retractable pin type tool in which the pin and the shoulder are assembled as separate units and arrangement is made such that the pin can move relative to the shoulder (refer Fig. 1.12). The advantage with this modified tool is that it can be used for multi-depth welding of workpieces [1, 71]. This is particularly useful when the thickness of the workpiece is not very precise. In those circumstances, the distance of the back surface of the weld to the tip of the pin can be maintained even though the shoulder remains on the part surface. Additionally, retractable pin type tool can be used to close the exit hole. However, the use of such tool makes the system complex with additional considerations in the control schema.

1.6.3 Some of the Distinct FSW Setups Used Around the World

Apart from the FSW setups discussed previously, there are various setups developed for use in specific applications around the world. Some of the notable equipment currently in use is described below [71]:

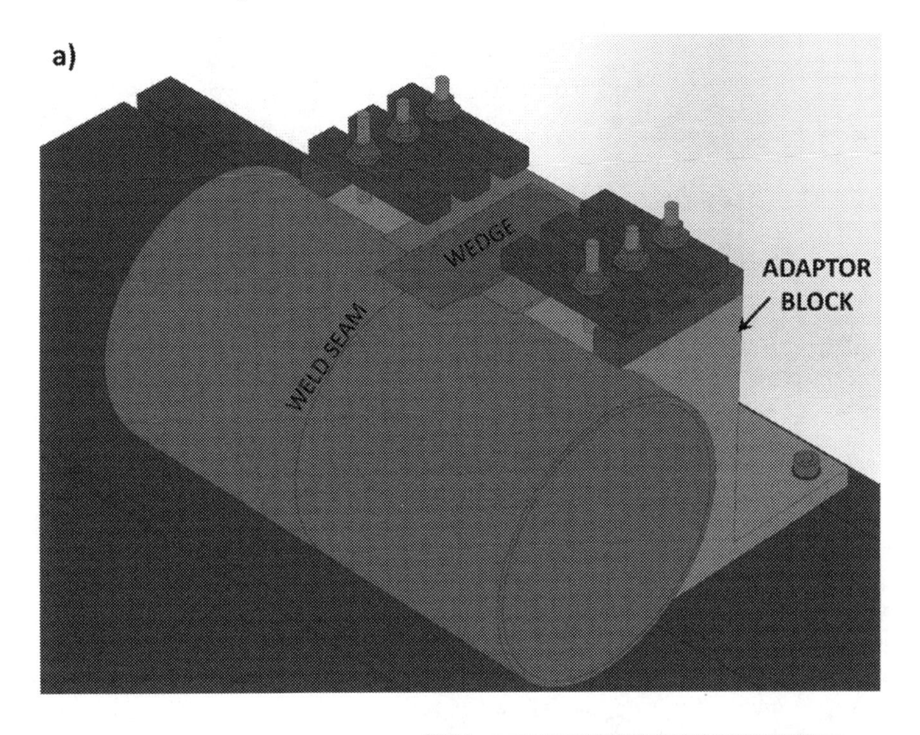

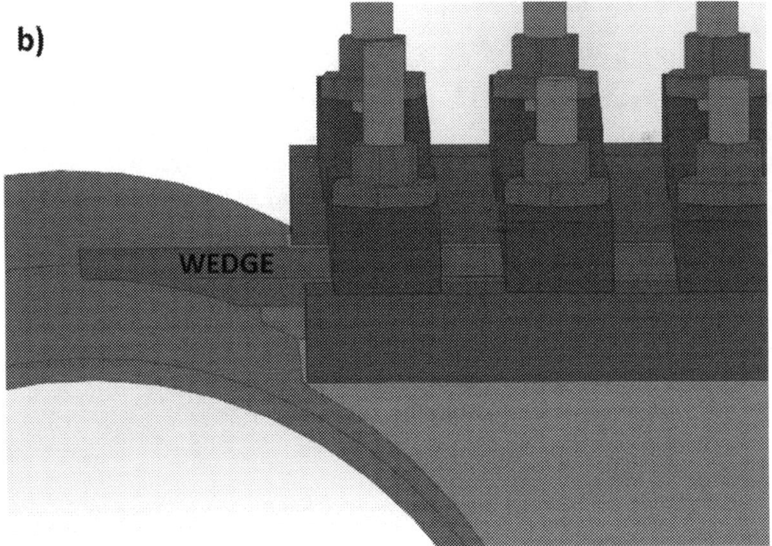

Fig. 1.11 **a** Schematic scheme for transfer of exit hole and **b** zoomed view of the wedge where the exit hole will be transferred

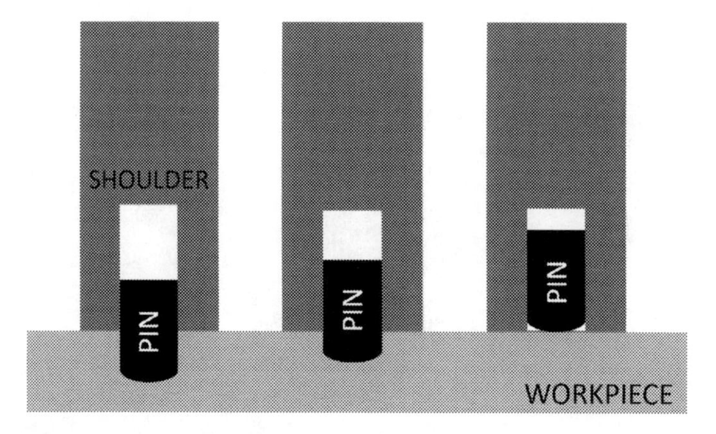

Fig. 1.12 Illustration of retractable pin

1.6.3.1 Robotic Weld Tools

NASA is a pioneer in the field of FSW. It uses FSW for fabricating the cryogenic tank for its Space Launch System (SLS) project. Apart from the conventional FSW, NASA also makes use of self-reacting FSW which eliminates the requirement of back plate support for the welding as well as ensuring welding throughout the plate thickness. Robotic weld tool is being used by NASA for welding of curved gore panels for the upper portion of the liquid hydrogen tank used on the upper stage of NASA Constellation Ares I. The configuration used by NASA is of horizontal boring machine (mill) type (refer Fig. 1.13).

A serial/articulate arm-based configuration is also employed by some companies for carrying out FSW (Fig. 1.14). Their applications encompass productions ranging from low to high volumes and variable job thickness. These types of sophisticated configurations are mainly limited to welding moderate to lower melting point alloys. The FSW robot technology can be used to effectively replace conventional fusion welding and assembly line processes, viz. riveting. These systems are also equipped to handle thin-gauged applications and weld along complex paths and curvatures.

1.6.3.2 Process/Production Development System

Various FSW setups have been used for various process/production development systems. In fact, customized solutions for production development system are available. They are designed to handle wide range of applications varying from 2-D to 3-D thin gage to thick plate (0.75–30 mm) [71].

Fig. 1.13 Ares I upper stage hydrogen dome weld in one of the largest FSW facility in the world—image credit NASA/MSFC

Fig. 1.14 Articulate arm-based robotic configuration for FSW doing panel welding—reprinted under Wikimedia commons license

1.6.3.3 Low-Cost FSW Solutions

Low-cost FSW solutions have been used for large-scale welding processes, viz. fabrication of stiffened panels from aluminum extrusions, to be used in the construction of ships used by the US Navy [71]. The configuration of the machine used by them was of C-Frame type with traveling workpiece. Besides, there has been effort to enhance the monitoring capability of the existing milling machine-based setups. LOSTIR is such a project funded by the European Commission which has led to the development of a low-cost FSW monitoring system that can be retrofitted to milling machines to facilitate their application to FSW [2]. This device consists of a force monitoring system, FSW tool holder, control and logging systems and suitable FSW tools. Currently, the system is suitable for joining low-melting point alloys, viz. aluminum, and is able to provide fairly accurate monitoring of forces, torque and tool temperature during friction stir welding.

1.6.3.4 Panel Production Plant

FSW equipment has been developed for the fabrication of large aluminum panels that have various uses, viz. for shipbuilding structures and decks in marine industry, tractor trailer flatbeds, railcar manufacture, etc. The advantage of using FSW in this case is that it vastly reduced the reworking required to correct the warping and distortion commonly observed in case of fusion welding. Manufacturing Technology, Inc. and ESAB AB are the companies actively involved in the development of FSW systems suitable for panel welding. These systems can join multiple extrusions into wider panels, simultaneously perform dual sided welds on thicker panels, join multiple single-wall extrusions to stiffeners, etc. Besides, they have large weld capacity and can join sheets up to four feet in width and 1.5 inch in thickness. Photograph of one of the panel production plants utilizing FSW is displayed in Fig. 1.15.

1.7 Important Parameters in FSW

FSW provides good quality weld when executed under suitable combination of process parameters and precise control. During welding, retreating side is the portion through which most of the material flow occurs and the welded joint is achieved when transportation of plasticized material occurs behind the tool. Important process parameters that significantly affect the welded joint include traverse speed, spindle speed, tool depth, tool tilt angle, spindle torque, forces acting on the tool (X-force, Y-force, Z-force) during welding, etc. Variation of these parameters can affect the

Fig. 1.15 FSW of aluminum panels for marine application [71]—reprinted with permission from Elsevier

temperature distribution and material flow. This directs the evolution of microstructure which in turn governs the characteristics of the welded joint. Various parameters that affect the weld characteristics are listed in Table 1.6. Some of the process parameters are deliberated in the following sections:

Table 1.6 Important FSW process variables [8]—reprinted with permission from Elsevier

Tool design variables	Machine variables	Other variables
Shoulder and pin materials	Welding speed	Anvil material
Shoulder diameter	Spindle speed	Anvil size
Pin diameter	Plunge force or depth	workpiece size
Pin length	Tool tilt angle	workpiece properties
Thread pitch		
Feature geometry		

1.7.1 Rotational Speed of Tool/Spindle

It is seen that a variety of parameters affect the FSW joint parameters. Thus, their interrelationship makes the situation quite complex and difficult to understand. According to the inventor of FSW Wayne Thomas, the three most important welding parameters are [64] tool rotational speed (along with the direction, i.e., CW or CCW), traverse speed of tool and tilt angle of tool. The rotation of the tool has several implications. Firstly, frictional heating from the rubbing of the shoulder surface to the workpiece is affected. Secondly, the tool rotation also affects the mixing of the softened material. Higher tool rotation results in greater generation of heat and more material softening. The spread of the HAZ is also quite an extent affected by the tool rotation. The spindle speed depends on the material type as well as thickness of the workpiece being used.

1.7.2 Traverse Speed

Tool traverse speed also impacts the weld characteristics. Material is transferred from the leading edge to the trailing edge due to the translation of the tool. Traverse speed along with tool rotational speed decides whether enough heat input is being supplied to the processed material which is favorable for development of desired metallurgical and mechanical properties. Lower welding speed generates excessive heat and makes the welding slow. At high speed, heat generation is low which may not be desirable as it puts higher loads on the front side of the advancing tool (material is not softened enough). Thus, an optimum traverse speed needs to be selected based on a particular case of FSW depending on the type of material and thickness of the workpiece being used. A smooth material flow around the tool pin is always desired. In fact the choice of feed rate along with rotational speed decide the heat generation and ensure the material flow which also leads to reduced forces acting on the tool. Stevenson et al. [61] identify the high cost and low lifespan of FSW tools as well as low welding speed as the major problems resisting commercialization of FSW for steel.

1.7.3 Tilt Angle of Tool

Tool tilt angle is another important FSW process parameter which influences the weld characteristics. The angle of inclination of spindle w.r.t. the workpiece surface in case of linear welding is known as the title of the tool. The tool is normally given a backward tilt (refer Fig. 1.2b) which gives more open space in front of the tool while forging of the workpiece material is done by the back of the shoulder behind the pin. This is particularly true for featureless shoulders, which because of the tilt, can effectively contain the stirred material. Transfer of material from leading to the trailing edge is also aided by the tilt of the tool. In general, a tilt angle between 0 and 3° is selected for optimum weld quality [40]. Takahara et al. [63] reported that the tensile strength of the welded joint is sufficiently maintained by limiting the backward tilt up to 3°.

1.7.4 Other Parameters

1.7.4.1 Forces

Considering the forces in three direction w.r.t. the tool in a FSW system (refer Fig. 1.1c), the down force acting on the tool (Z-force) is the most important one which affects the welding. It is to be noted here that normally, FSW machines can be operated in two modes, viz. position control and force control. In case of the former, the machine tries to maintain the position command given to the tool (plunge depth) while recording the forces as a response. In the force control mode, the machine tries to maintain the force given as the input. As already mentioned, Z-force value is normally specified while the plunge depth can vary. Higher Z-force results in increased plunge depth as well as greater heat generation, thus playing its part in the material softening and the corresponding phenomena. The other two forces, viz. longitudinal force or transverse force (X-directional force) and lateral force (Y-force), are normally recorded as response. However, these forces give an indication regarding the quality of the weld produced. In some cases, defects in the weld can be predicted by observing the trends in the force values recorded during the FSW process.

1.7.4.2 Plunge Depth

The plunge depth is also an essential parameter for FSW. Whether the tool has fully penetrated into the workpiece is ensured by the plunge depth. Moreover, it ensures necessary vertical force needed for the heat generation. Excessive plunge depth may cause the pin to reach up to the reverse side of the workpiece resulting in its welding with the backing plate.

Fig. 1.16 Super liner Ogasawara built by FSW—reprinted under Wikimedia commons license

1.8 Applications of FSW

1.8.1 Marine and Shipbuilding Industries

The marine and shipbuilding industries are one of the sectors which has found immediate application of the FSW process post its development. In fact, the first commercial use of FSW was recorded at Sapa (Sweden) in 1996 when this process was utilized to weld hollow aluminum panels for freezer tanks for fish. Apart from this, FSW has been found suitable for fabricating sections of boats and ships as well as marine structures. In fact, pre-fabricated aluminum panels made with FSW are commercially available for water vessels, viz. ferry boats, cruise ships as well as offshore platforms employed for oil mining. Thus, constructing these structures is now much quicker and easier. FSW has been used to build the Japanese Super Liner Ogasawara Fig. 1.16 which can carry 740 persons and has a maximum speed of 42.8 knots [22].

1.8.2 Aerospace and Aviation Industry

FSW has been widely accepted by the aerospace the aviation industry. Welding of various components in aircrafts including skin to various stiffeners, viz. spars, ribs, etc., is done with FSW. FSW is found to reduce manufacturing costs and weights compared to riveting. They also eliminate the need of machining (from solids) in some cases. TWI has created the first commercial aircraft Eclipse 500 (Fig. 1.17) by replacing the traditional riveting and bonding processes by FSW.

Fig. 1.17 Eclipse 500-first commercial aircraft built by TWI—reprinted under Wikimedia commons license

FSW has been tremendously used by NASA for manufacturing large tanks for their satellite launch vehicles. Many of the fuels tanks for satellites are now fabricated with the help of FSW much quickly and efficiently. Hence, FSW technique is normally considered for fabricating the following and their parts:

- Cryogenic fuel tanks for space vehicles
- Aviation fuel tanks
- Wings, fuselages
- Military and scientific rockets

1.8.3 Railway Industry

FSW in railway industry is primarily used for production of aluminum panels for rolling stock. These panels are used for the side walls, under carriages as well as the roof panels of trains. By using FSW, many of the limitations of traditional fusion welding, viz. low life, distortion and warping, in the panels could be avoided. By realizing these benefits, many of the railway manufacturing industries have employed FSW for producing modern passenger railcars from longitudinal aluminum extrusions with integrated stiffeners.

1.8.4 Automobile Industry

FSW has also been embraced by the automotive industry for production of door panels. Many big shots in the automotive companies, viz. Ford, General Motors, Volvo, etc., have experimentally undertaken the use of FSW for manufacturing of drive shafts and space frames [22]. A new variant of FSW known as friction stir spot welding (FSSW) is increasingly used for production of aluminum components for automotive applications.

1.8.5 Power Plants

One of the applications of FSW here includes the welding of pipes intended for use in supercritical and ultra-supercritical thermal power plants. In these power plants, fluid at high pressure and temperature needs to be carried through pipes which need to be joined with near about 100% efficiency. Thus, FSW offers a potential solution in this case. Moreover, FSW can weld effectively creep resistant steel used for making these pipes. Now, these materials utilized in the power plants should exhibit high strength up to approximately 600 °C temperature. By employing FSW technique, the welded zone contributes to higher thermal fatigue life. This allows them to increase the operating temperature to a higher level and thus increase the overall efficiency of the power plant.

1.9 Closure

FSW has really come up as a boon for the fabricating industries. Much sought after lightweight metals, viz. aluminum, magnesium, etc., which were deemed un-weldable previously can now be welded quite easily and efficiently by employing FSW. Thus, it has been readily accepted by aerospace and marine industries which majorly deal with these metals. The cost-effective application of FSW for high-temperature materials, viz. steel, nickel and titanium-based alloys, is also assessed by various researchers and technocrats worldwide. The joining of dissimilar metals is another field where FSW seems to come up handy. Thus, the future seems to be quite bright for this welding technique. Hopefully, the small limitations of FSW are ironed out, and we see more and more applications of this technique by smaller engineering firms as well as by developing countries.

References

1. Aissani M, Gachi S, Boubenider F, Benkedda Y (2010) Design and optimization of friction stir welding tool. Mater Manuf Process 25(11):1199–1205. https://doi.org/10.1080/10426910903536733
2. Beamish K, Ezeilo A, Smith S, Lewis P, Cheetham P, Blignault C, Welding I (2006) National Research Council of C development of a low cost FSW monitoring system. In: International symposium; 6th, friction stir welding. Cambridge, 2006. TWI Ltd; pp 17–17
3. Besharati Givi MK, Asadi P (2014) 1—General introduction. In: Givi MKB, Asadi P (eds) Advances in friction-stir welding and processing. Woodhead Publishing, pp 1–19. https://doi.org/10.1533/9780857094551.1
4. Bisadi H, Rasaee S, Farahmand M (2014) Effects of pin shape on the tool plunge stage in friction stir welding. Trans Indian Inst Met 67(6):989–995. https://doi.org/10.1007/s12666-014-0421-8
5. Brewer LN, Bennett MS, Baker BW, Payzant EA, Sochalski-Kolbus LM (2015) Characterization of residual stress as a function of friction stir welding parameters in oxide dispersion strengthened (ODS) steel MA956. Mater Sci Eng A 647:313–321. https://doi.org/10.1016/j.msea.2015.09.020
6. Chen J, Fujii H, Sun Y, Morisada Y, Kondoh K, Hashimoto K (2012) Effect of grain size on the microstructure and mechanical properties of friction stir welded non-combustive magnesium alloys. Mater Sci Eng A 549:176–184. https://doi.org/10.1016/j.msea.2012.04.030
7. Chen TP, Lin WB (2010) Optimal FSW process parameters for interface and welded zone toughness of dissimilar aluminium–steel joint. Sci Technol Weld Joining 15(4):279–285. https://doi.org/10.1179/136217109X12518083193711
8. Colligan KJ (2010) The friction stir welding process: an overview. In: Lohwasser D, Chen Z (eds) Friction stir welding. Woodhead Publishing, pp 15–41. https://doi.org/10.1533/9781845697716.1.15
9. Cui HB, Xie GM, Luo ZA, Ma J, Wang GD, Misra RDK (2016) The microstructural evolution and impact toughness of nugget zone in friction stir welded X100 pipeline steel. J Alloy Compd 681:426–433. https://doi.org/10.1016/j.jallcom.2016.03.299
10. Darras BM (2012) A model to predict the resulting grain size of friction-stir-processed AZ31 magnesium alloy. J Mater Eng Perform 21(7):1243–1248. https://doi.org/10.1007/s11665-011-0039-5
11. Edwards P, Ramulu M (2010) Identification of process parameters for friction stir welding Ti–6Al–4V. J Eng Mater Technol 132(3). https://doi.org/10.1115/1.4001302
12. Emami S, Saeid T, Khosroshahi RA (2018) Microstructural evolution of friction stir welded SAF 2205 duplex stainless steel. J Alloy Compd 739:678–689. https://doi.org/10.1016/j.jallcom.2017.12.310
13. Esparza JA, Davis WC, Murr LE (2003) Microstructure-property studies in friction-stir welded, Thixomolded magnesium alloy AM60. J Mater Sci 38(5):941–952. https://doi.org/10.1023/A:1022321107957
14. Fahimpour V, Sadrnezhaad SK, Karimzadeh F (2012) Corrosion behavior of aluminum 6061 alloy joined by friction stir welding and gas tungsten arc welding methods. Mater Des 39:329–333. https://doi.org/10.1016/j.matdes.2012.02.043
15. Fuller CB, Mahoney MW, Calabrese M, Micona L (2010) Evolution of microstructure and mechanical properties in naturally aged 7050 and 7075 Al friction stir welds. Mater Sci Eng A 527(9):2233–2240. https://doi.org/10.1016/j.msea.2009.11.057
16. Gan W, Okamoto K, Hirano S, Chung K, Kim C, Wagoner RH (2008) Properties of friction-stir welded aluminum alloys 6111 and 5083. J Eng Mater Technol 130(3). https://doi.org/10.1115/1.2931143
17. Genevois C, Deschamps A, Denquin A, Doisneau-cottignies B (2005) Quantitative investigation of precipitation and mechanical behaviour for AA2024 friction stir welds. Acta Mater 53(8):2447–2458. https://doi.org/10.1016/j.actamat.2005.02.007

18. Hirasawa S, Badarinarayan H, Okamoto K, Tomimura T, Kawanami T (2010) Analysis of effect of tool geometry on plastic flow during friction stir spot welding using particle method. J Mater Process Technol 210(11):1455–1463. https://doi.org/10.1016/j.jmatprotec.2010.04.003
19. Imam M, Ueji R, Fujii H (2015) Microstructural control and mechanical properties in friction stir welding of medium carbon low alloy S45C steel. Mater Sci Eng A 636:24–34. https://doi.org/10.1016/j.msea.2015.03.089
20. Imam M, Ueji R, Fujii H (2016) Effect of online rapid cooling on microstructure and mechanical properties of friction stir welded medium carbon steel. J Mater Process Technol 230:62–71. https://doi.org/10.1016/j.jmatprotec.2015.11.015
21. Jafarzadegan M, Feng AH, Abdollah-zadeh A, Saeid T, Shen J, Assadi H (2012) Microstructural characterization in dissimilar friction stir welding between 304 stainless steel and st37 steel. Mater Charact 74:28–41. https://doi.org/10.1016/j.matchar.2012.09.004
22. Kallee SW (2010) Industrial applications of friction stir welding. In: Lohwasser D, Chen Z (eds) Friction stir welding. Woodhead Publishing, pp 118–163. https://doi.org/10.1533/9781845697716.1.118
23. Kalvala PR, Akram J, Misra M, Ramachandran D, Gabbita JR (2016) Low temperature friction stir welding of P91 steel. Defence Technol 12(4):285–289. https://doi.org/10.1016/j.dt.2015.11.003
24. Karami S, Jafarian H, Eivani AR, Kheirandish S (2016) Engineering tensile properties by controlling welding parameters and microstructure in a mild steel processed by friction stir welding. Mater Sci Eng A 670:68–74. https://doi.org/10.1016/j.msea.2016.06.008
25. Lakshminarayanan AK, Balasubramanian V (2012) Assessment of fatigue life and crack growth resistance of friction stir welded AISI 409M ferritic stainless steel joints. Mater Sci Eng A 539:143–153. https://doi.org/10.1016/j.msea.2012.01.071
26. Larsson H, Karlsson L, Stoltz S, Bergqvist EL (2000) Joining of dissimilar Al alloys by friction stir welding. In: 2nd international symposium on friction stir welding, Gothenburg, Sweden, TWI Ltd.
27. Lee CY, Choi DH, Yeon YM, Jung SB (2009) Dissimilar friction stir spot welding of low carbon steel and Al–Mg alloy by formation of IMCs. Sci Technol Weld Joining 14(3):216–220. https://doi.org/10.1179/136217109X400439
28. Lee W-B, Jung S-B (2004) The joint properties of copper by friction stir welding. Mater Lett 58(6):1041–1046. https://doi.org/10.1016/j.matlet.2003.08.014
29. Lee W-B, Lee C-Y, Chang W-S, Yeon Y-M, Jung S-B (2005) Microstructural investigation of friction stir welded pure titanium. Mater Lett 59(26):3315–3318. https://doi.org/10.1016/j.matlet.2005.05.064
30. Lee WB, Yeon YM, Jung SB (2003) Joint properties of friction stir welded AZ31B–H24 magnesium alloy. Mater Sci Technol 19(6):785–790. https://doi.org/10.1179/026708303225001867
31. Li H, Mackenzie D, Hamilton R (2010) Parametric finite-element studies on the effect of tool shape in friction stir welding. Proc Inst Mech Eng Part B J Eng Manuf 224(8):1161–1173. https://doi.org/10.1243/09544054jem1810
32. Li H, Yang S, Zhang S, Zhang B, Jiang Z, Feng H, Han P, Li J (2017) Microstructure evolution and mechanical properties of friction stir welding super-austenitic stainless steel S32654. Mater Des 118:207–217. https://doi.org/10.1016/j.matdes.2017.01.034
33. Liao J, Yamamoto N, Nakata K (2009) Effect of dispersed intermetallic particles on microstructural evolution in the friction stir weld of a fine-grained magnesium alloy. Metallur Mater Trans A 40(9):2212–2219. https://doi.org/10.1007/s11661-009-9921-2
34. Meran C, Kovan V (2008) Microstructures and mechanical properties of friction stir welded dissimilar copper/brass joints. Materialwiss Werkstofftech 39(8):521–530. https://doi.org/10.1002/mawe.200800278
35. Meshram SD, Paradkar AG, Reddy GM, Pandey S (2017) Friction stir welding: An alternative to fusion welding for better stress corrosion cracking resistance of maraging steel. J Manuf Process 25:94–103. https://doi.org/10.1016/j.jmapro.2016.11.005

36. Mironov S, Motohashi Y, Kaibyshev R, Somekawa H, Mukai T, Tsuzaki K (2009) Development of fine-grained structure caused by friction stir welding process of a ZK60A magnesium alloy. Mater Trans 50(3):610–617. https://doi.org/10.2320/matertrans.MRA2008192
37. Mishra RS, De PS, Kumar N (2014a) Friction stir welding configurations and tool selection. In: Friction stir welding and processing: science and engineering. Springer International Publishing, Cham, pp 95–108. https://doi.org/10.1007/978-3-319-07043-8_4
38. Mishra RS, De PS, Kumar N (2014b) FSW of aluminum alloys. In: Friction stir welding and processing: science and engineering. Springer International Publishing, Cham, pp 109–148. https://doi.org/10.1007/978-3-319-07043-8_5
39. Mishra RS, De PS, Kumar N (2014c) Fundamentals of the friction stir process. In: Friction stir welding and processing: science and engineering. Springer International Publishing, Cham, pp 13–58. https://doi.org/10.1007/978-3-319-07043-8_2
40. Mishra RS, De PS, Kumar N (2014d) Introduction. In: Friction stir welding and processing: science and engineering. Springer International Publishing, Cham, pp 1–11. https://doi.org/10.1007/978-3-319-07043-8_1
41. Mishra RS, Jeganathan V (2017) Experimental investigation on friction stir welding of copper alloys
42. Miura T, Ueji R, Fujii H (2015) Enhanced tensile properties of Fe–Ni–C steel resulting from stabilization of austenite by friction stir welding. J Mater Process Technol 216:216–222. https://doi.org/10.1016/j.jmatprotec.2014.09.014
43. Miyano Y, Fujii H, Sun Y, Katada Y, Kuroda S, Kamiya O (2011) Mechanical properties of friction stir butt welds of high nitrogen-containing austenitic stainless steel. Mater Sci Eng A 528(6):2917–2921. https://doi.org/10.1016/j.msea.2010.12.071
44. Nakata K (2005) Friction stir welding of copper and copper alloys. Weld Int 19(12):929–933. https://doi.org/10.1533/wint.2005.3519
45. Nakata K, Inoki S, Nagano Y, Hashimoto T, Johgan S, Ushio M (2001) Proceedings of 3rd international friction stir welding symposium : 27–28 Sept 2001, Kobe, Japan. In: International friction stir welding symposium, [Cambridge], 2001. TWI Ltd.
46. Pareek M, Polar A, Rumiche F, Indacochea JE (2007) Metallurgical evaluation of AZ31B-H24 magnesium alloy friction stir welds. J Mater Eng Perform 16(5):655–662. https://doi.org/10.1007/s11665-007-9084-5
47. Park HS, Kimura T, Murakami T, Nagano Y, Nakata K, Ushio M (2004) Microstructures and mechanical properties of friction stir welds of 60% Cu–40% Zn copper alloy. Mater Sci Eng A 371(1):160–169. https://doi.org/10.1016/j.msea.2003.11.030
48. Pasta S, Reynolds AP (2008) Residual stress effects on fatigue crack growth in a Ti-6Al-4V friction stir weld. Fatigue Fract Eng Mater Struct 31(7):569–580. https://doi.org/10.1111/j.1460-2695.2008.01258.x
49. Peel M, Steuwer A, Preuss M, Withers PJ (2003) Microstructure, mechanical properties and residual stresses as a function of welding speed in aluminium AA5083 friction stir welds. Acta Mater 51(16):4791–4801. https://doi.org/10.1016/S1359-6454(03)00319-7
50. Prado RA, Murr LE, Shindo DJ, Soto KF (2001) Tool wear in the friction-stir welding of aluminum alloy 6061+20% Al_2O_3: a preliminary study. Scripta Mater 45(1):75–80. https://doi.org/10.1016/S1359-6462(01)00994-0
51. Ragu Nathan S, Balasubramanian V, Malarvizhi S, Rao AG (2016) An investigation on metallurgical characteristics of tungsten based tool materials used in friction stir welding of naval grade high strength low alloy steels. Int J Refract Metal Hard Mater 56:18–26. https://doi.org/10.1016/j.ijrmhm.2015.12.005
52. Rai R, De A, Bhadeshia HKDH, DebRoy T (2011) Review: friction stir welding tools. Sci Technol Weld Joining 16(4):325–342. https://doi.org/10.1179/1362171811Y.0000000023
53. Razal Rose A, Manisekar K, Balasubramanian V (2012) Influences of welding speed on tensile properties of friction stir welded AZ61A magnesium alloy. J Mater Eng Perform 21(2):257–265. https://doi.org/10.1007/s11665-011-9889-0
54. Reynolds AP, Hood E, Tang W (2005) Texture in friction stir welds of Timetal 21S. Scripta Mater 52(6):491–494. https://doi.org/10.1016/j.scriptamat.2004.11.009

55. Reynolds AP, Tang W, Gnaupel-Herold T, Prask H (2003) Structure, properties, and residual stress of 304L stainless steel friction stir welds. Scripta Mater 48(9):1289–1294. https://doi.org/10.1016/S1359-6462(03)00024-1
56. Rodrigues DM, Loureiro A, Leitao C, Leal RM, Chaparro BM, Vilaça P (2009) Influence of friction stir welding parameters on the microstructural and mechanical properties of AA 6016–T4 thin welds. Mater Des 30(6):1913–1921. https://doi.org/10.1016/j.matdes.2008.09.016
57. Santos TFA, Hermenegildo TFC, Afonso CRM, Marinho RR, Paes MTP, Ramirez AJ (2010) Fracture toughness of ISO 3183 X80M (API 5L X80) steel friction stir welds. Eng Fract Mech 77(15):2937–2945. https://doi.org/10.1016/j.engfracmech.2010.07.022
58. Sato YS, Nelson TW, Sterling CJ, Steel RJ, Pettersson CO (2005) Microstructure and mechanical properties of friction stir welded SAF 2507 super duplex stainless steel. Mater Sci Eng A 397(1):376–384. https://doi.org/10.1016/j.msea.2005.02.054
59. Shashi Kumar S, Murugan N, Ramachandran KK (2016) Influence of tool material on mechanical and microstructural properties of friction stir welded 316L austenitic stainless steel butt joints. Int J Refract Metal Hard Mater 58:196–205. https://doi.org/10.1016/j.ijrmhm.2016.04.015
60. Sorensen C, Nielsen B, Minerals M, Materials S (2009) Exploring geometry effects for convex scrolled shoulder, step spiral probe FSW tools. In: Symposium; 5th, friction stir welding, Warrendale, 2009. Materials Society, pp 85–92
61. Stevenson R, Toumpis A, Galloway A (2015) Defect tolerance of friction stir welds in DH36 steel. Mater Des 87:701–711. https://doi.org/10.1016/j.matdes.2015.08.064
62. Sutton MA, Yang B, Reynolds AP, Yan J (2004) Banded microstructure in 2024-T351 and 2524-T351 aluminum friction stir welds: Part II. Mechanical characterization. Mater Sci Eng A 364(1):66–74. https://doi.org/10.1016/S0921-5093(03)00533-1
63. Takahara H, Tsujikawa M, Chung SW, Okawa Y, Higashi K, Oki S (2008) Optimization of welding condition for nonlinear friction stir welding. Mater Trans 49(6):1359–1364. https://doi.org/10.2320/matertrans.L-MRA2008807
64. Thomas WM, Dolby RE (2002) Friction stir welding developments. In: 6th International conference on trends in welding research, Georgoa, USA. ASM International
65. Threadgill PL, Leonard AJ, Shercliff HR, Withers PJ (2009) Friction stir welding of aluminium alloys. Int Mater Rev 54(2):49–93. https://doi.org/10.1179/174328009X411136
66. Tingey C, Galloway A, Toumpis A, Cater S (2015) Effect of tool centreline deviation on the mechanical properties of friction stir welded DH36 steel. Mater Des 1980–2015(65):896–906. https://doi.org/10.1016/j.matdes.2014.10.017
67. Tolephih MH, Mashloosh KM, Waheed Z (2011) Comparative study of the mechanical properties of (FS) and MIG welded joint in (AA7020-T6) aluminum alloy. Al-Khwarizmi Eng J 7(2):22–35
68. Wei L, Nelson TW (2012) Influence of heat input XE "heat input" on post weld microstructure and mechanical properties of friction stir welded HSLA-65 steel. Mater Sci Eng A 556:51–59. https://doi.org/10.1016/j.msea.2012.06.057
69. Xunhong W, Kuaishe W (2006) Microstructure and properties of friction stir butt-welded AZ31 magnesium alloy. Mater Sci Eng A 431(1):114–117. https://doi.org/10.1016/j.msea.2006.05.128
70. Yang J, Wang D, Xiao BL, Ni DR, Ma ZY (2013) Effects of rotation rates on microstructure, mechanical properties, and fracture behavior of friction stir-Welded (FSW) AZ31 magnesium alloy. Metallur Mater Trans A 44(1):517–530. https://doi.org/10.1007/s11661-012-1373-4
71. Zappia T, Smith C, Colligan K, Ostersehlte H, Kallee SW (2010) Friction stir welding equipment. In: Lohwasser D, Chen Z (eds) Friction stir welding. Woodhead Publishing, pp 73–117. https://doi.org/10.1533/9781845697716.1.74
72. Zhang Y, Sato YS, Kokawa H, Park SHC, Hirano S (2008a) Microstructural characteristics and mechanical properties of Ti–6Al–4V friction stir welds. Mater Sci Eng A 485(1):448–455. https://doi.org/10.1016/j.msea.2007.08.051
73. Zhang Y, Sato YS, Kokawa H, Park SHC, Hirano S (2008b) Stir zone microstructure of commercial purity titanium friction stir welded using pcBN tool. Mater Sci Eng A 488(1):25–30. https://doi.org/10.1016/j.msea.2007.10.062

74. Zhang YN, Cao X, Larose S, Wanjara P (2012) Review of tools for friction stir welding and processing. Can Metall Q 51(3):250–261. https://doi.org/10.1179/1879139512Y.0000000015

Chapter 2
Fundamentals of Friction Stir Welding, Its Application, and Advancements

Atul Kumar Choudhary and Rahul Jain

Abstract In friction stir welding (FSW) the material is welded without melting the base material. Aerospace, automobile, shipbuilding and electronics are some of the industrial sectors which use FSW for manufacturing of components. It is primarily used because of higher weld efficiency as compared with fusion joining processes. In this book chapter, the fundamental aspects of the FSW have been briefly discussed. The chapter also discusses the effect of input responses on the output responses, application of FSW in various industries, research advances in welding of similar and dissimilar material using FSW. The FSW tool has two major features namely, the shoulder and pin, as its shape and dimensions govern the material flow behavior. In general, a cylindrical tool is used for FSW. Tool rotational speed, tilt angle, welding speed, and plunge depth are the input conditions which affect the dynamics of the process. Changes in these parameters have an effect on axial force, spindle torque, temperature distribution/heat generation and strain. In addition, it ultimately changes the microstructural, hardness and weld strength. The choice of the workpiece and tool material for FSW is another big challenge as it has its own limitations. Application of FSW started with welding of aluminum and magnesium and its alloy but eventually, it gained traction in welding high-density metals like copper, titanium and steel. Welding of these materials requires a hard tool material with the better thermomechanical property as compared to the workpiece to avoid premature tool wear/failure. FSW is not only limited to similar metal welding but research is going for welding of material in the dissimilar configuration like aluminum with copper, magnesium with aluminum, and steel with aluminum etc. The successful evolution of FSW has led to the emergence of a few new variants of this technology such as friction stir additive manufacturing, friction stir processing, friction stir spot welding, and micro FSW.

Keywords Friction stir welding · Solid-state joining · Industrial application · Tool design

A. K. Choudhary · R. Jain (✉)
Department of Mechanical Engineering, Indian Institute of Technology Bhilai, Raipur, India
e-mail: rahul@iitbhilai.ac.in

J. P. Davim (ed.), *Welding Technology*, Materials Forming, Machining and Tribology, https://doi.org/10.1007/978-3-030-63986-0_2

2.1 FSW Introduction

2.1.1 History and Background

Human needs and aspirations are increasing and this has led to an increase in the demand for fuel. Further, the growing population and human ambition of luxurious life have made the use of the automobile, and air travel a necessity. These have triggered research to explore the possibility of using material with a lower density such as aluminum, magnesium, etc. for the manufacturing of components instead of a high-density material such as steel to improve fuel efficiency. This will reduce the mass of the component and improve fuel efficiency without compromising on the quality of the product. Working with aluminum/magnesium and its alloy is challenging, as its weldability is poor with conventional fusion welding methods such as MIG, TIG, EBW and Laser welding, etc. [1]. Problems associated with fusion welding of Al/Mg are high distortion and residual stress because of the expansion and contraction of the material due to the thermal cycle. This phenomenon is more severe in aluminum as the coefficient of thermal expansion is twice the steel and thrice of titanium. Also, hydrogen entrapment in the molten pool is a serious issue which reduces strength due to the formation of brittle hydrides. Besides Mg and its alloy are highly flammable which restricts it from fusion welding.

Difficulties associated with fusion welding of aluminum are:

Fusion welding (TIG/MIG):

- Higher heat input because of high heat dissipation because of higher thermal conductivity
- Formation of metal oxide which is brittle and causes embrittlement. It also has a high melting point
- Various solidification defects like porosity, blowholes, and solidification cracks

Electron Beam Welding (EBW):

- Vaporization, loss of material, weld crack in most of the heat treatable aluminum alloys
- The requirement of vacuum at the welding space leads to a high cost [2]
- High voltage leads to high-energy input

Laser welding:

- Porosity and loss of material after solidification of some heat-treatable aluminum alloys apart from solidification cracking
- High reflectivity of Laser for aluminum, limits the usage of this technique [3]
- Low efficiency and high initial cost

FSW is a non-fusion welding process invented in TWI [4] and its schematic is shown in Fig. 2.1. This technology has shown tremendous potential to weld similar and dissimilar materials. It produces superior mechanical properties as compared with fusion welding and therefore it has gained substantial interest in the joining of aluminum and other light metal alloys [5, 6]. Thus, it has immense potential to replace the existing fusion welding to produce a better joint.

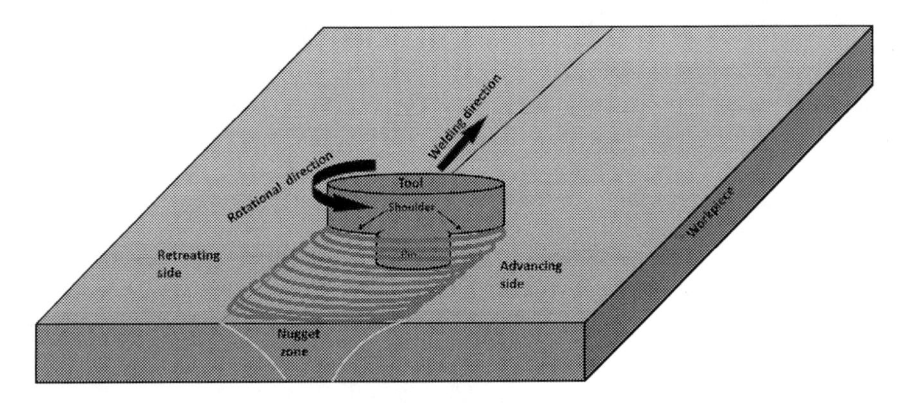

Fig. 2.1 Schematic of FSW and some important nomenclature

2.1.2 Nomenclature of FSW Process

Numerous terminologies used for FSW are shown in Fig. 2.2. Meaning and its implication of each term are mentioned below:

- **Tool shoulder**: It is the feature of the tool that makes contact with the upper surface of the workpiece. It contributes approximately 80 to 92% of the total heat produced. Heat generation takes place due to friction at the interface of the tool-workpiece and workpiece deformation.
- **Tool pin**: Protruded part of the tool that impinges into the material and is responsible for the material stirring in the thickness direction. Its contribution to heat generation is about 10 to 20%. The heat generation is majorly due to stirring that leads to plastic deformation between the pin surface and workpiece and partially due to frictional heat as well.

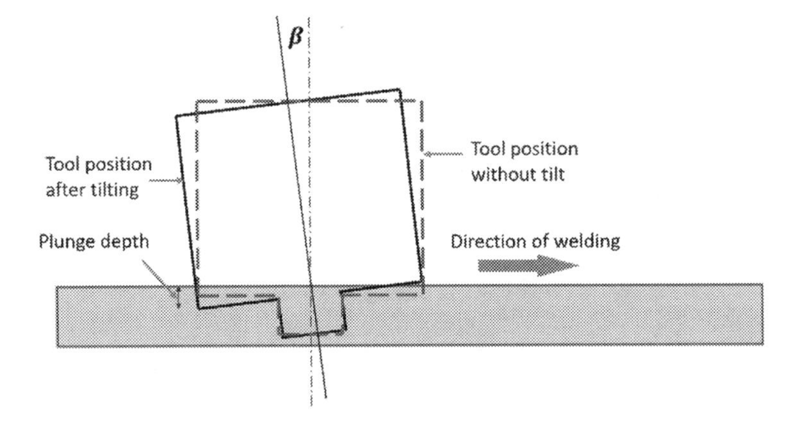

Fig. 2.2 Nomenclature used in the FSW process

- **Advancing side (AS)**: Side of the workpiece, where the direction of tool welding and rotation speed is identical.
- **Retreating side (RS)**: The other side of the workpiece, where the direction of tool rotation and welding is the opposite. The differences in the relative velocity on either side of the workpiece make the FSW process an asymmetric process. This is the reason for higher plastic deformation and temperatures towards AS as compared with RS.
- **Leading-edge (LE) and trailing edge (TE)**: The front and back portion of the tool is called LE and TE, respectively.
- **Plunge depth**: After plunging, it is the distance between the top surface of the material and tool shoulder. It increases the contact area in more heat generation on the workpiece. Plunge depth normally ranges between 0.05–0.5 mm.
- **Tilt angle (*β*)**: Angle by which the tool is tilted towards the backside as shown in Fig. 2.2. It is provided to increase the compressive force on the material and reduces defect and flash formation. Its range is generally 0°–3°.

2.1.3 Stages of FSW

FSW is completed in three different stages. A detailed explanation of each stage is explained below.

Plunging A rotating tool plunges along the adjoining surface of the work-piece (similar to drilling operation). The tool plunges until the tool shoulder touches/impinges inside the workpiece top surface as indicated in Fig. 2.3a. This results in the generation of heat because of friction at the interface of the tool and workpiece, and plastic deformation of the material. The former is the major contributor of the heat generation and contributes about 70–80% of total heat generation [7]. As a result of the development of heat, the plasticized metal starts flowing in an upward direction and eventually gets compressed by the tool shoulder.

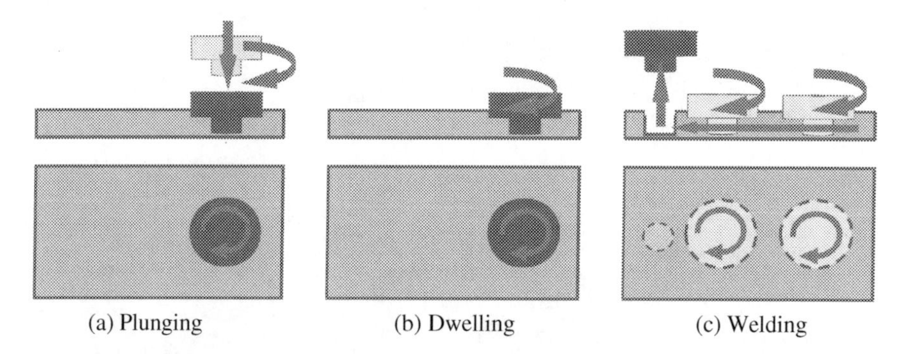

Fig. 2.3 Schematic of plunging, dwelling, and welding phases in FSW. Here dark and light blue colors indicate the final and initial/intermittent location, respectively

Dwelling In this stage, the already plunged tool continues to rotate about its axis to further enhance the temperature for efficient stirring and material mixing. The schematic of the dwelling stage is shown in Fig. 2.3b. The material underneath the tool shoulder is further plastically deformed and softened by the generated heat due to tool rotation. This is an optional stage and its duration depends on the requirement of heat generation.

Welding In this stage, the rotating tool translates along the abutting edge to transport the material from leading to the trailing edge of the tool to produce a joint. A schematic of the same is shown in Fig. 2.3c. In this stage, actual welding is performed and it is reported that axial force, spindle torque data attains a steady-state value and hence an average value of this stage can be calculated for the analysis [8].

2.2 Fundamentals of FSW

2.2.1 Material Flow and Mechanism of Bond Formation

Researchers have conducted various experiments to study the material flow behavior using various techniques like marker material [9, 10], tracer material [11], steel balls [12], and Cu foils [10, 13] in the faying surface. Generally, Cu foil, steel balls, or tungsten is employed as a tracer material to reveal the material flow history by metallography inspection of the flow patterns. Copper is preferred as a marker material because its particles entering in Al matrix can be very easily seen as a difference of color contrast [14]. Also, tungsten material is easy to inspect during material flow visualization. Material flow is a complex phenomenon and most of the research has been carried out on cylindrical or threaded pins. Apart from these square [15], triangular [16], trivex/triflat [17], and conical pins [17, 18] have also been studied for material flow behavior.

The material flow along with all the phases of FSW are studied. When the tool impinges into the material, the tool pin has the initial contact at the faying surface. As the pin impinges into the material, it starts flowing in an upward direction till it eventually gets compressed by the shoulder and finally, the material partially spills out around the shoulder as it plunges into the workpiece. The spilled material is known as flash. It is formed to compensate for the volume of the shoulder occupied in the workpiece [10]. During welding, the material flow takes place due to the relative motion between the rotating-translating tool and the workpiece and the flow path is circular around the pin. The material encountering the pin rotates along with it and follow a circular trajectory before getting deposited on the trailing edge of the tool. The material flow patterns are different in AS and RS [19]. The difference in the relative velocity on both the side has an impact on the flow patterns. Colligan [13] and Jain et al. [20] studied the mechanism of material flow in the different regions using threaded cylindrical and conical pins, respectively. Colligan used steel

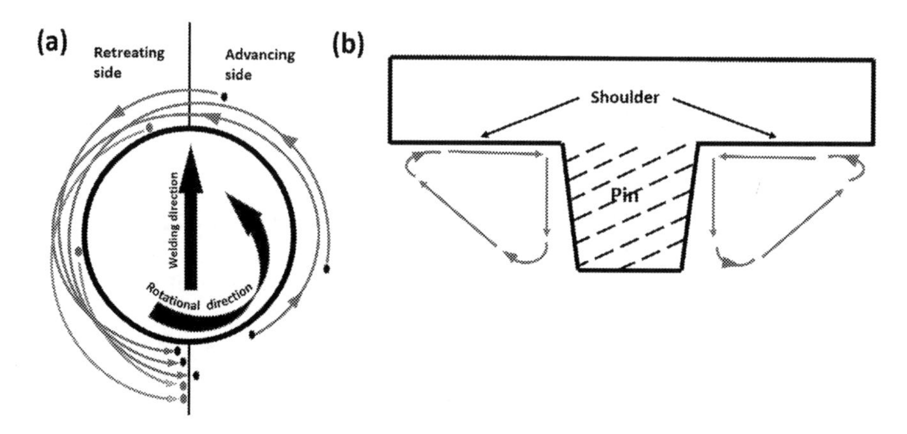

Fig. 2.4 **a** Top view beneath shoulder showing the material flow along the Advancing Side (Red lines) and the Retreating Side (Green lines), **b** material flow pattern along the thickness direction

balls (0.38 mm diameter) embedded in the grooves that are manufactured in the workpiece. Steel balls in the grooves were located in different positions and heights along the direction of the weld centreline. After the series of experiments, it was reported that the most of material from the front end of AS stirs around the pin and gets deposited towards the RS behind the pin while entering into the rotational zone by following an arc-shaped trajectory [21]. The material from RS meeting the pin is swept by the pin and is deposited towards the same RS as shown in Fig. 2.4a. As the viscous and deformed material rotates around the tool it ends up in an incomplete trajectory as compared to AS. The tracer steel balls along the thickness direction show a vertical motion at different depths. The vertical motion reduces when the rotating shoulder meets the faying surface. The extend of material vertical motion depends on the degree of plastic deformation, rotation speed, and welding speed [22]. It shows material beneath the shoulder (as shown in Fig. 2.4b) are moving in a vertical direction along the thickness direction which is absent in a smooth pin profile [18, 23]. Also, the threaded profile leads to higher material velocity as compared to a smooth profile [17]. Since the material is extruded from the leading to the trailing edge of the pin, FSW is also compared with the extrusion process [12, 24].

Dickerson and Shercliff [11] examined the material flow by placing annealed copper foil of 0.1 mm thickness in different orientations in the workpiece during welding. It acts as a marker material as it gets embedded in the weld matrix for better visualization of flow patterns during welding. The material flow patterns around the weld region were inspected using metallurgical polishing and X-ray radiography which revealed 2D material flow visualization of copper along the weld line. Also, a 3D material flow path was developed from tomography images which gave a proper material visualization around the tool. It is very challenging to investigate the flow behavior in NZ by the experimental method [14]. Guerra et al. [13] investigated the effect of threaded pin profiles on the material flow. A copper foil of 0.1 mm thickness as a marker material was employed for the investigation of material flow patterns around the threaded pin profile. The workpiece material within the rotational zone causes the bulk of the vertical movement. Continuous vorticity developed due to the forward and backward material movement along the threads of the tool. Material entering this zone thus follows an incomplete helical path. The vorticity because of the continuous rotational and the translational of the tool pin forms a helical path [25, 26]. The material movement is influenced by the tool shoulder instead of pin threads at the upper surface of the joint [27]. The contact between the material moving in this upper zone and the lower thread cause difference in the degree of material movement. This gradient is a dominating factor in weld formation [28]. In a threaded pin, the plasticized material rotates around the thread of the pin and continues to move in a downward direction towards the bottom of the weld region. The vertical downward motion results in a pile-up of the material per unit time leading to an increase in transfer velocity. This results in more material transfer near the bottom portion making material particles to attain a sufficient transfer height towards the TMAZ. The plastically deformed material touching the pin undergoes continuous pressure due to the extrusion process [24]. The material displacement acts perpendicular to the thread surface of the pin. The above material flow patterns ultimately end up in a bond formation [27].

2.2.2 *Microstructural Zones in FSW*

Various microstructural zones are formed during the FSW process and the same is shown in Fig. 2.5. They are nugget zone (NZ)/stirred zone, thermo-mechanical affected zone (TMAZ), heat-affected zone (HAZ), and base material (BM).

NZ is also called stir zone (SZ), is subjected to high-temperature and strain because of the stirring action of the pin. A localize high-temperature heating and cooling cycle coupled with severe plastic deformation results in dynamic recrystallization to form fine equiaxed grains. The extent of deformation and thermal cycle majorly affects the mechanical properties and grain size of SZ. Higher plastic deformation and lower peak temperatures produce finer grain size [29]. Adjacent to NZ is TMAZ, which experiences plastic deformation due to continuously rotating tool and thermal cycle

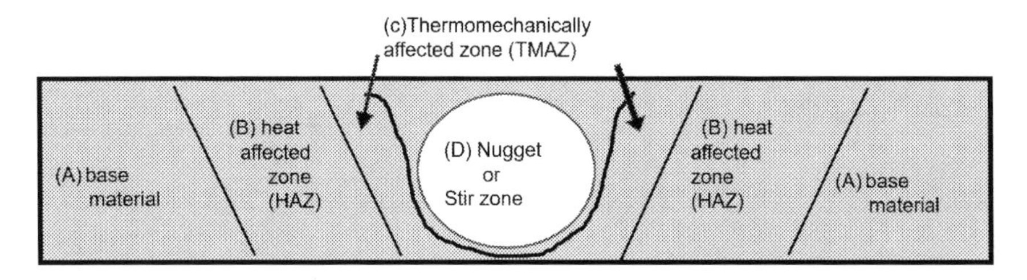

Fig. 2.5 Schematic of the section of FSW specimen in transverse direction **a** base material, **b** heat-affected zone (HAZ), **c** thermomechanical affected zone (TMAZ) **d** Nugget or stir zone (SZ)

resulting in partially recrystallized grains. Here the grains undergo a considerable change of shape and its elongation is followed by bending during the FSW process [30]. Adjacent to TMAZ is HAZ formed due to prolonged heating and cooling cycle leading to the generation of coarser grains as the time required for cooling is high [8].

2.2.3 Advantages of FSW

1. Solidification defects are eliminated in FSW as it does not involve melting and re-solidification [31].
2. For low-density material, the tool wear is almost negligible resulting in lower tooling cost.
3. FSW is an energy-efficient process as it consumes about 2.5% of the energy of laser/Electron beam welding [14].
4. Dynamic recrystallization leads to fine equiaxed grain resulting in higher mechanical and metallurgical properties of the weld. With the proper choice of input process parameters, FSW can result in 70–90% weld efficiency.
5. It is green technology because of the absence of the generation of any toxic gases/fumes.
6. Shielding gases are not required.
7. No filler materials are required as welding takes place in solid-state.
8. It can successfully weld dissimilar metals i.e. steel and aluminium, magnesium, and aluminum.
9. FSW results in a good weld appearance thus it eliminates the need for post-weld machining.

2.2.4 Disadvantages of FSW

1. An exit hole is generated at the end of the process.
2. FSW tool undergoes severe stress and high-temperature exposure. This leads to tool wear and loss of strength resulting in low performance of the tool for high strength material such as steel, titanium alloys, etc. [32].
3. A very high degree of reactive force is generated during the plunging operation and hence requires a machine with high stiffness.

2.3 Effect of Input Process Parameters

2.3.1 Tool Design

Tool design is crucial as it is responsible for maintaining both joint quality and the number of welds that can be done with the same tool. The material of the tool should possess high toughness, strength, and hot hardness, good oxidation resistance and thermal conductivity to dissipate heat while welding [33]. Tool design influences the material flow and bond formation between the workpiece surfaces. Its geometry and design are responsible for the required heat generation. The optimum ratio between the shoulder and pin diameter ranges between 3 and 4 [13]. The tool shoulder is responsible for retaining material beneath it to reduce the flash formation. There exists a temperature gradient in the radial and thickness direction with maximum temperature being generated beneath the shoulder [34]. The tool pin should neither be too long nor too short as the former result in buckling failure and the latter may lead to improper penetration resulting in insufficient material mixing. The tool pin contributes about 10–20% of total heat generation [35]. The main functionality of the tool pin is to provide stirring/material mixing along with the thickness of the material. The pin design is one of the crucial aspects of FSW and research has been carried out for different profiles such as circular, square, triangle, triflute, triflat, etc. as shown in Fig. 2.6 to understand their performance. Among all triflute pin showed the best weld properties at high welding speed followed by the square pin. It is because of the three flutes helical crest which increases the plastic deformation and increases in material transportation rate leading to better material stirring. The square pin profile has good weld efficiency and produces sound mechanical properties because of the pulsating and dynamic orbiting effect. The flat surfaces of the square pin result in better bulk movement of plasticized material instead of layer by layer as in the case of circular pin [27]. Manufacturing of the cylindrical pin is easiest as compared to other pin profiles. The threaded profile is also preferred for better vertical material flow resulting in the defect-free weld.

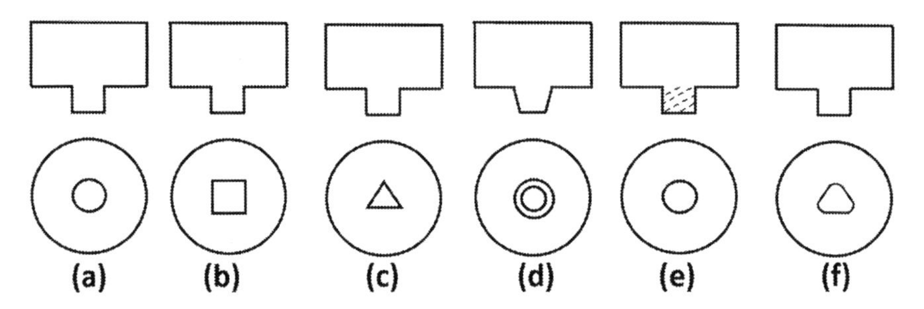

Fig. 2.6 Cross-sectioned view and bottom view of different pin profiles. **a** Straight cylindrical, **b** square, **c** triangle, **d** tapered cylindrical, **e** threaded cylindrical, **f** triflat

Elangovan et al. [16] have compared different tool pin profiles in terms of heat generation and material flow. Pin profiles including cylindrical, tapered cylindrical (conical), square, triangle, and threaded cylindrical shapes (as shown in Fig. 2.6) with shoulder diameter of three different dimensions i.e. 15, 18, and 21 mm have been studied. It was found that the defect-free stir region after welding is dependent on the shoulder diameter and tool pin profile.

Pin profile controls the mechanical properties of weld formed after the FSW process. The joint made from a straight cylindrical pin and conical pin profile with an 18 mm shoulder diameter resulted in defect-free weld. It has shown better properties as compared to the other two shoulder diameters. Also, welding performed by the square and triangular pin profile with an 18 and 21 mm shoulder diameter resulted in the defect-free weld. Square pin profile produced a weld efficiency of about 95%. Also, the tool with an 18 mm shoulder diameter produced 190 MPa of weld strength which was higher as compared to the other two. The reason could be because of the optimum heat generation for 18 mm shoulder. As higher heat generation may lead to slipping of tool shoulder on the material surface leading to insufficient material flow. The square and triangular pins have flat faces and that is responsible for generating dynamic pulsating effect. The better material flow was produced by triflute pin followed by the square and cylindrical pin. Trilflute and cylindrical pin offer the maximum and minimum dynamic area, respectively among the above three-pin profiles. So, the highest and lowest plastic deformation was produced by triflute and cylindrical pin, respectively. Also, the triflute pin has the highest bulk material transportation efficiency along with shearing action. Triflute pin also recorded high peak hardness [36].

The correlation is associated with the dynamic volume and static volume governs of the material flow path of the plasticized weld material for different pin profiles. Material displacement takes place due to the rotating tool from the leading edge to the trailing edge. The triangular and square pin having flat faces produces a dynamic pulsating action during the stirring of the material. The pulsating action is absent in other pin profiles like cylindrical and tapered ones [16]. Pulsating action results in more homogenous material flow and localized mixing of soft plasticized material in the stir zone, better material transportation takes place due to which pore, voids, and tunnel defects are also eliminated [28].

The selection of the shoulder diameter and its profile is another important parameter as it is responsible for the generation of heat and material holding. Three commonly used shoulder geometries are concave, convex, and flat surface. The shoulder profile should be able to contain the plasticized material to avoid/reduce flash formation. The viscous flow capacity of the material depends on the extent of surface area and shoulder geometry, which is in contact with the workpiece. Flat shoulders are simple in design and easy to manufacture. The tool shoulder with a flat surface cannot hold the material efficiently beneath it and leads to the formation of flash [32]. For efficient welding, the flat shoulder should be accompanied by tilt angle. Concave shoulder performs better to contain the material as compared with the flat shoulder. The concavity of the shoulder acts as a reservoir for the material beneath it and does necessary forging action on the material. While moving forward the tool forces material into the concave cavity and pushes the existing material behind the pin. It also increases the tool life by reducing locally generated stress during welding. During welding, the material contained within the shoulder participates in the joint formation as no material is lost in the form of flash [37]. The problem associated with the convex surface is that it further displaces away from the plasticized material from the weld center.

2.3.2 *Tool Material*

The tool material considered for FSW needs to be rigid and strong as compared to the workpiece. Also, the mechanical property of the tool should be on the higher side including its strength and thermal properties. These are the significant and important factor for proper and uniform dissipation of the heat [38]. Tool materials used for joining soft materials like aluminum and magnesium alloy are easy to fabricate with required properties. High-speed steel, low carbon steel, and H13 tool steel are some common commercially available tools for them. These are also effective material to weld in different input process parameters [14]. The above tools perform well without undergoing much wear and material loss. Tool steel H13 is a widely accepted material for the welding of aluminum and magnesium alloys ranging from 0.5 to 50 mm thickness [39, 40]. A tool having a longer life and satisfactory weld properties has a great demand for structural applications especially, in the defense sector, aerospace industry, oil and gas companies, and heavy manufacturing industries. Tool materials for harder alloys like steel and titanium alloys still need further research and development. A few advanced tool materials like tungsten carbide and polycrystalline cubic Boron Nitride (PcBN) is widely used for welding of material having high strength like steel, copper, and titanium. The tool needs to be cost-effective, durable, and structurally sound for commercial FSW applications [32]. Material selection and performance analysis of tools along with the quality of weld and cost is a major concern.

2.3.3 *Commonly Used Tool Materials*

2.3.3.1 Tool Steel

Material like aluminum, magnesium, and other soft materials can be welded using steel tools. H13 grade is mostly used in both lap and butt configuration [41]. Tool steel is preferred for the welding of similar and dissimilar material. It maintains high strength and a significant low wear rate during the service condition. The wear rate increases with an increase in rotational speed and lowering travel speed. It is found that steel tools wear slowly with threads in it and also produce a sound weld [33].

2.3.3.2 PcBN Tools

PcBN is a ceramic-based tool used for the welding of several hard materials and alloys ranging from structural steels, titanium and nickel-based alloys. PcBN retains a high hardness and strength at a higher temperature. Also, PcBN results in a good quality weld with high weld efficiency with reduced distortion. The manufacturing of this tool is difficult because of its low machinability. Diamond-coated tool insert is generally preferred for machining of PcBN. In general, the cost of the tool manufactured by PcBN is very high. It is also prone to fail during the initial phase of plunging because of low fracture strength.

2.3.3.3 Tungsten-Based Tools

Tungsten-based tools are strong at elevated temperatures and retain their high hardness. Due to which it leads to higher wear resistance during service condition. These tools are used for steel and titanium alloy. The titanium-based combat vehicle structures for the defense industry and circumferential segments of steel pipes were successfully welded in a single pass of FSW without disruption with negligible tool degradation [42]. Tungsten alloys, i.e., WC, WRe, and WC-Co are the most promising tool materials because of high-temperature strength, fracture toughness. The crack resistance and wear resistance of this alloy can be improved by adding a definite amount of Rhenium. The commercially available Re based FSW tools are W25%Re, W20%Re and W25%Re-4%HfC [43].

2.3.4 Tool Rotational Speed and Welding Speed

Tool rotation and welding speed are other crucial input parameters in FSW. Productivity and weld quality depends on how fast the tool rotates and drives off the viscous material to create a joint. Establishing a relationship among them is slightly complicated as both are interrelated with each other. An increase in rotational speed and reduction in welding speed leads to higher frictional heat between the tool and workpiece resulting in faster softening of the workpiece. The former increases the relative velocity at the interface and in the case of the latter, higher contact time at the interface leads to increase peak temperature.

Elangovan and Balasubramanian [44] studied the effect of rotational speed for different output parameters. Rotational speed was varied from 1500, 1600, and 1700 rpm at a welding speed of 46 mm/min. They obtained a gradual decrease in axial force and spindle torque due to an increase in developed heat. In addition, higher rotational speed leads to more stirring of deformed material resulting in the formation of surface defects. Elangovan and Balasubramanian [45] studied the effect of the welding speed by varying it from 22 to 75 mm/min at a rotational speed of 1600 rpm. An increasing trend in axial force and spindle torque has been observed.

The reason is higher welding speed results in lower heat input causing the lack of material transportation and mixing and low welding speed cause higher temperature and slow cooling rate resulting in grain growth and formation of precipitates [46].

Jha et al. [47] investigated the influence of welding speed on the microstructure. Increasing welding speed from 40 to 100 mm/min led to a decrease in the grain size at the SZ. The nugget zone corresponding to the highest welding speed of 100 mm/min showed a very fine microstructure of size 5 μm. At low welding speed, more time is available for the recrystallization process to form larger grains [48]. Microstructure observed at a fixed tool welding speed of 40 mm/min with varying rotational speeds of 800, 1000, and 1200 rpm. It shows the extent of grain growth is directly proportional to the tool rotation. It resulted in a coarser microstructure (20 μm) at 1200 rpm and smaller grains (3-4 μm) for 800 rpm. They concluded that larger grain growth at higher rotational speed is because of a higher heat generation. Dynamic recrystallization happens as a result of excessive plastic deformation and thermal cycle. This results in the development of equiaxed grains through nucleation followed by grain growth. So, recrystallized grain size depends on the peak temperature, thermal cycle, and the degree of deformation. When the rotational speed increases, both the factors increase resulting in larger grain growth [49].

2.3.5 *Tool Tilt Angle and Plunge Depth*

Plunge depth is also an important parameter for the successful weld. This is provided to have proper forging load and contact area which helps to further increase the heat generation. An excessive plunge depth could lead to higher flash formation and deteriorates the weld quality due to defects [50]. Experiments were conducted for plunge depth varying from 0.15 to 0.3 mm. Increasing plunge depth from 0.15 to 0.2 mm, the forming limit of FSWed specimen has increased. It is due to the evolution of the thickness gradient in the FSW sample [51]. Zheng et al. [52] studied the effect of plunge depth for 0, 0.1, 0.3, and 0.5 mm. Its effect on grain size and mechanical properties were studied for 1200 rpm and 40 mm/min. At a zero mm plunge depth, the joint failed during the sample preparation. For 0.1 mm, the layer beneath the shoulder skinned off forming two hooks towards the AS and RS. This is very important as it responsible for mechanical interlocking between the two joining surfaces and improves weld quality. As the plunge depth was increased further to 0.3 mm the layer was worn out into pieces due to plastic deformation. This resulted in a non-uniform distribution of these broken particles in the SZ. When the depth was increased, the number of shattered particles increased. When plunge depth increased to extreme 0.5 mm, the interface of the weld could not be filled leading to the formation of the void. It resulted in flash formation throughout the weld length without an appreciable weld formation. Results show that an increase in plunge depth increased in weld strength of FSW specimen. The weld strength of 7.9 kN was obtained for a plunge depth of 0.3 mm.

Tool tilt angle is another important parameter that governs the forging action during FSW, minimizes flash formation and improves mechanical properties of the weld [53]. A non-zero angle facilitates the better flow of the plasticizing material around the tool. Increasing the tilt angle increases the downward force and frictional heat during welding, resulting in higher plasticization and material softening [54]. An optimized tool tilt angle governs the weld quality [55]. A higher tilt angle may damage the pin from the weld root resulting in defective or damaged welds. So, an optimized tool tilt angle leads to efficient transport of the material from the front side to the rear side of the pin resulting in the defect-free weld. Therefore, the selection of appropriate tool tilt angle value along with a suitable combination of other input parameters ie. tool rotation and welding speed is necessary. It is also observed that the tool tilt angle increases stress towards the leading edge of the tool within the workpiece because increasing tilt angle results in the increase in temperature and decrease in stress value is seen as material flow in the rear advancing side is observed. It also improves material stirring action at trailing edge towards the advancing side [56]. Chauhan et al. [57] studied the effect of the tool tilt angle on defect formation. They varied the tilt angle from 0° to 2°. A 0° tilt angle resulted in a defective weld. A tilt angle of 1° resulted in a tunnel defect along the weld line. The tilt angle of 2° resulted in a defect-free weld because of an increase in the axial force value. Weld efficiency for 0° and 1° was 35 and 50%, respectively due to defects. A weld efficiency of 90% was recorded for a 2° tilt angle. Mehta and Badheka [58] studied the influence of tool tilt angle during the FSW of aluminum alloy with copper. Tilt angles varied from 0° to 4° at an interval of 1° at a constant rotational speed and welding speed of 1300 rpm and 40 mm/min, respectively. The tilt angle of 0° and 1° resulted in defective weld while 2°, 3°, and 4° showed a defect-free weld. A maximum weld strength of 117 MPa was obtained at a 4° tilt angle. Maximum macro-hardness 186 HV was reported for 4° tilt angle while minimum macro-hardness 58 HV for 0° tilt angle.

2.4 Output Mechanical Responses of FSW

2.4.1 Temperature Distribution/heat Generation

The study of evolution of temperature in and nearby the NZ has a direct impact on the weld microstructure [59, 60]. It influences grain size, nature of new grains, coarsening and mixing of precipitates with the base metal and resultant mechanical properties of the welds [61, 62]. Temperature measurement in the stirred zone with a thermocouple is difficult as plastic deformation of the material will break and wash away the thermocouple [63]. Therefore, temperature distribution has been either estimated from the history of the weld microstructure or measuring it near to the shoulder, i.e., outside the stirred zone [64]. Microstructural observation related to grain size and its distribution reveals the thermal history. Fine-grain size is an outcome of higher welding speed resulting in a lower temperature. Similarly, the coarse grain is formed at a higher temperature because of high rotational speed and low welding speed. Also, the temperature range is predicted by using thermocouple at different zone or by simulation [65, 66]. During the initial phase of the plunging temperature increase because of the frictional and plastic deformation heat at the tool-workpiece interface. It continues to increase until the tool approaches the thermocouple location where it attains a peak value and then gradually reduces while moving away from the thermocouple as shown in Fig. 2.7a. The differences in relative velocity on either side of the workpiece result in a higher temperature of 2–5 °C towards the AS as compared with the RS of the as shown in Fig. 2.7b.

Literature suggests that temperature in FSW ranges from 0.7 to 0.9 * T_m [14, 39]. The temperature never goes beyond the above range. This is because of the loss in traction between the tool and workpiece with an increase in temperature and also a reduction in the heat generation rate due to material softening, leading to thermal stabilization [67].

Heat is generated due to two reasons, firstly friction between the rotating tool and workpiece and secondly due to plastic deformation of the material by the tool [68]. The heat produced results in the softening of the region between the faying surface resulting in the intermixing of two metals to form the joint. A continuous heating and cooling cycle leads to the development of thermal stresses in the workpiece which can be tensile or compressive [69]. The nature of stress depends upon the exposure of the workpiece to the environment or the backing plate. The contribution of thermal stress in FSW is a major concern for effective joint [46]. The thermal conductivity of the baking plate governs effective heat dissipation [70]. If it is higher then it dissipates more heat from the workpiece resulting in the lower temperature of the workpiece. This can lead to reduced plasticization of material and an increase in axial force and spindle torque value. Therefore, an optimum value of thermal conductivity is advisable for the backing plate.

During, dwelling and welding stage, an increase in temperature is observed so that the material can be plasticized for mixing. It is also observed that the maximum temperature was in the SZ and it keeps on decreasing away from stir zone [71, 72]. There is an increase in temperature at the edge of the stir zone from the bottom to the

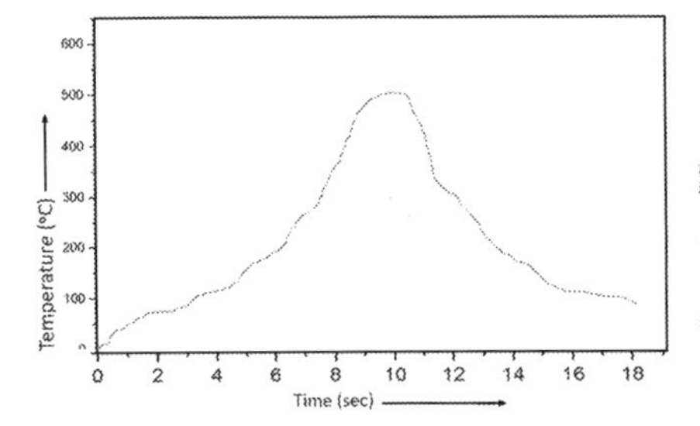

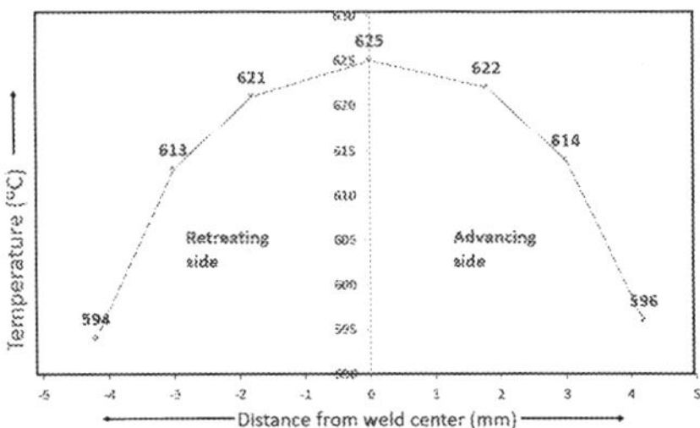

Fig. 2.7 Temperature distribution during FSW. **a** Temperature evolution with time. **b** Temperature profile towards the advancing side and retreating side

top surface of the workpiece. The maximum temperature is always noticed on the top surface and at the corner of the stir zone [73]. The mechanical properties of welds are completely dependent on temperature, material flow, and rate of heat generation. Due to which FSW is intensely a coupled thermo-mechanical based phenomenon [74].

2.4.2 Axial Force and Spindle Torque

Axial Force (Fz) is the component of the force that is acting normal to the workpiece surface. It is the reactive force generated during the plunging and welding phase. Welding Force (F_x) is the component of force acting during the welding stage. Welding force is the resistance offered by the material on the rotating tool. Spindle torque is the moment developed by the rotating tool while deforming the workpiece. The direction of different forces is shown in Fig. 2.8.

These forces are always acting from different directions during the welding process. The force acting in the direction of the tool axis is higher as compared to the force acting in the welding direction. During FSW, forces and spindle torque values can be monitored in real-time using a different method as shown in Fig. 2.9. Dynamometers are used for the measurement of torque and forces. Its usage is restricted due to higher costs. A low-cost method to measure forces and torque is by associating electrical signals of driving motors and then calibrating this information. The output torque corresponding to *X* and *Z*-axis servo motors of the FSW equipment were investigated. The three-phase AC induction motor of the main spindle was observed and logged in real-time. After obtaining these parameters a proper correlation is established to get axial force, traverse force, and tool torque [75, 76].

Axial force influences the stiffness and capacity of the FSW machine structure. Proper prediction and monitoring of force are necessary for optimized tool design and

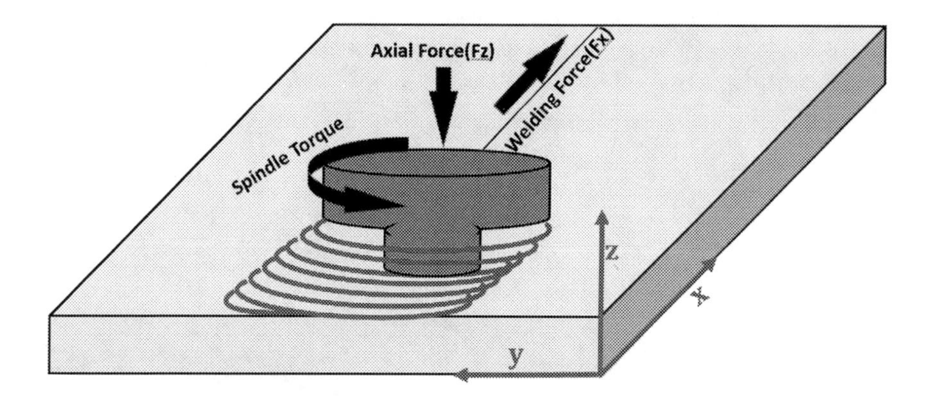

Fig. 2.8 Directions of axial force, welding force, and spindle torque

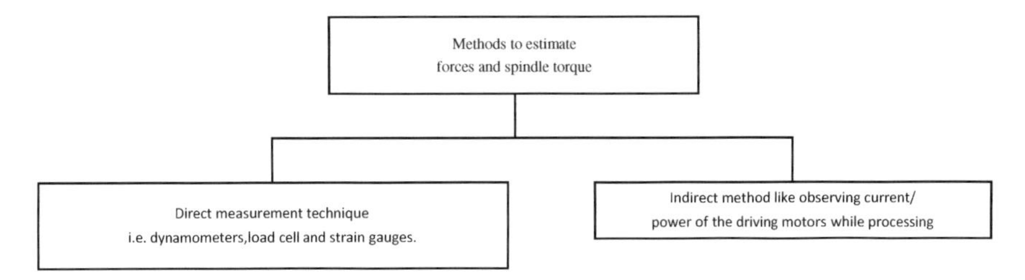

Fig. 2.9 Methods to measure the axial force and spindle torque

longer tool life. Forces are influenced by different input parameters such as the shape and size of the tool, rotational speed, and welding speed [23]. Increasing the welding speed corresponds to an increase in welding force because more material is coming in contact with the tool per unit time which acts as a resistance for tool travel [37]. When the pin diameter is increased, it leads to a higher contact surface of the tool resulting in higher welding force. An axial force is the most important output parameter as governs the weld quality. Axial force greatly influences the FSW process. Spindle torque is another important governing factor. It indicates the power consumption during the welding process and is the main driving factor for the selection of motor capacity for performing FSW. Spindle torque is directly proportional to the shoulder diameter and is inversely proportional to the tool rotational speed [75].

Variation of axial force during the FSW process is shown in Fig. 2.10a. It is observed that at the initial phase during plunging the force rapidly increases (0–1 s) because the pin impinges into the workpiece top surface which is at room temperature. Afterward, it decreases because of the softening effects due to an increase in temperature. Thereafter, the force attains a steady value during plunging (1.5–4.5 s.) until the shoulder meets the workpiece. As soon as the shoulder encounters the workpiece a sudden spike is observed because the shoulder has a larger area as compared to the pin. Then the force value decreases during the dwelling operation because of a further increase in the temperature. After this, it attains a steady-state condition throughout the welding process (10 s onwards). A drop of around 40% in force value is observed from peak value to the welding phase. For spindle torque,

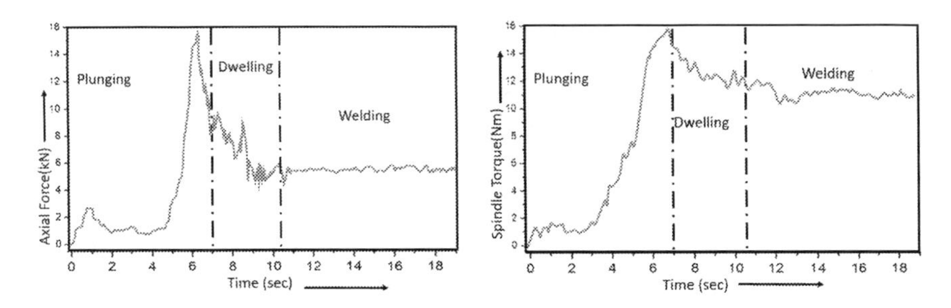

Fig. 2.10 Variation of **a** axial force and **b** spindle torque with respect to time for different phases of FSW

it shows a continuous increase from plunging phase until the shoulder comes in contact. When the shoulder impinges the workpiece, it attains a peak value. During dwelling, the temperature continuously increases, and the torque reduces gradually. After this, torque attains a constant and steady value in the welding phase as shown in Fig. 2.10b.

The tool is susceptible to damage/failure during initial plunging, as the axial force is very high. The threaded cylindrical pin is susceptible to less damage compared to smooth cylindrical because of the former induces more material deformation due to the presence of threads resulting in the higher temperature of the workpiece and hence lowering of axial force [76].

2.4.3 Types of defects

Defects are formed during the FSW process due to improper selection of input parameters. These include tool design and its dimensions, tool rotation, welding speed, tool plunge depth, and tool tilt angle. Defects arise when the temperature is low, and the workpiece material has not softened enough. Very high-temperature is also undesirable during FSW as it changes the weld microstructure or other mechanical properties of the material. Also, there is a loss of traction towards the leading and trailing side of the tool which will lead to improper material transportation [56, 77]. Some of the major defects are discussed below.

Ribbon Flash It is observed as a ribbon-like structure formed due to the expulsion of the material on the top surface. Ribbon flash as shown in Fig. 2.11a is formed due to excessive plunge depth resulting in the generation of excessive heat. Also,

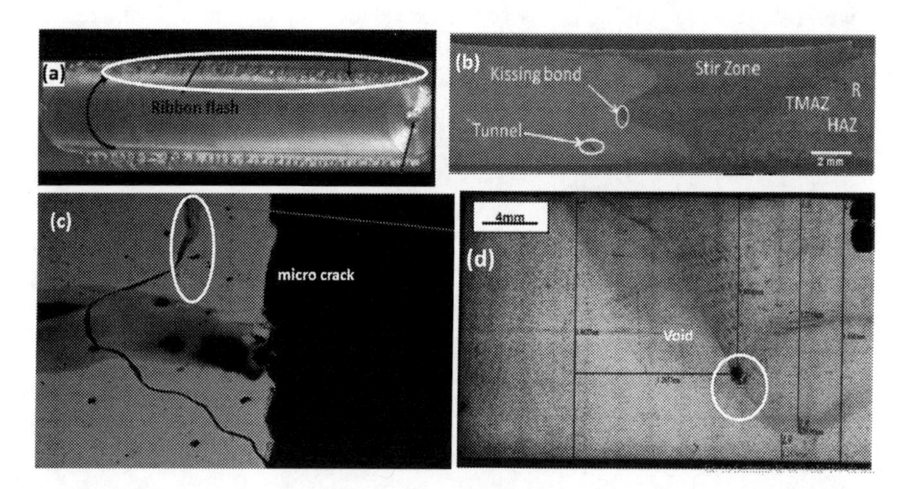

Fig. 2.11 Major defects in FSW **a** ribbon flash [78], **b** kissing bond and tunnel defect [81], c microcrack [80], **d** void [80] (All figures are reproduced with the permission from the publisher)

high rotational speed along with low travel speed is another possible cause of flash formation. This defect leads to material thinning. Selecting appropriate plunge depth coupled with a tilt angle or concave shaped shoulder should help in eliminating it [78].

Kissing Bond They are formed at the interface of the pin offset and the workpiece or towards root side where the material is not bonded [79]. This is due to insufficient stirring and low heat input within the workpiece reducing the flowability of plasticized material resulting in poor mechanical properties as shown in Fig. 2.11b. The identification of this type of defect is challenging. It is investigated by employing non-destructive tests like ultrasound. Also, it needs careful observations by using optical or SEM scanning through a series of careful polishing followed by etching of the sample [80, 81]. The kissing bond can be eliminated by selecting the proper combination of tool rotation and welding speed to generate sufficient heat for bond formation.

Tunnel Defect This defect is generated due to the improper combination of input parameters. Generally, lower rotation speed and high welding speed yields tunnel defects as shown in Fig. 2.11b. In such cases, the workpiece nugget zone is not properly stirred, resulting in a tunnel formation along the weld. To avoid these a proper combination of rotational and welding speed should be given to the tool [29, 81].

Microcracks These are as shown in Fig. 2.11c and is observed along the welding direction between the plunge and steady-state region. The reason could be attributed to a lack of heat dissipation coming from the lower tool rotation speed. Also, an unexpected increase in welding speed, which results in a lack of material flow forming these microcracks. The proper combination of tool rotation speed and welding speed can result in the elimination of these defects [80].

Void It is formed due to the lack of material stirring because of high tool welding speed. Also, it was found that higher welding speed results in the faster cooling of the material leading to an insufficient stirring of the material resulting in void formation [82]. Figure 2.11d shows a void formed in the weld region. Welding speed should be optimized according to the thickness of the base metal to be welded to reduce void formation [80].

Root Flaws It starts from the bottom of the workpiece near the backing plate. These are the irregular planes toward the welded zone that forms when there is an incorrect plunge depth of the tool or pin height. It generally occurs towards the advancing side of the workpiece. Due to the root flaw, ductility decreases [83]. Bobbin tool eliminates this defect or selecting a proper plunge depth or pin height can reduce this defect in the workpiece [84, 85].

Surface Defects These appear as groves, irregularity, or surface galling at the faying/top surface of the weld after processing. These can arise due to an incorrect set of input process parameters. This defect is formed as a result of excessive/deficient heat input or irregular stirring of the workpiece. This defect can be avoided by careful selection of a set of input parameters.

Oxide Entrapment At a higher temperature during the FSW process the eutectic temperature (550 °C) of aluminum alloy can exceed, this may cause localized melting within the workpiece resulting in this defect. In addition to that oxidation of the newly sheared planes and high-temperature exposure of flowing metal to air would lead to the formation of oxides at inner volume called oxide entrapment [86].

The thermal and mechanical properties of the welded specimen should be observed. The welding temperature and welding speed should be governed properly to avoid any changes in parameter and defect generation. The combination of tool rotation and tool traverse speed is responsible for the peak temperature generated during FSW. This can reduce the defects in the weld.

2.5 Mechanical Property

2.5.1 Tensile Property and Hardness Variation

After the FSW the welded material generally shows a decrease in the tensile properties due to the reduction in dislocation density during the recrystallization process resulting in lower stress to distort the material [87].

Figure 2.12 shows the hardness variation in the transverse direction of the weld. In stir zone hardness reduces as compared to base metal due to the process of dissolution

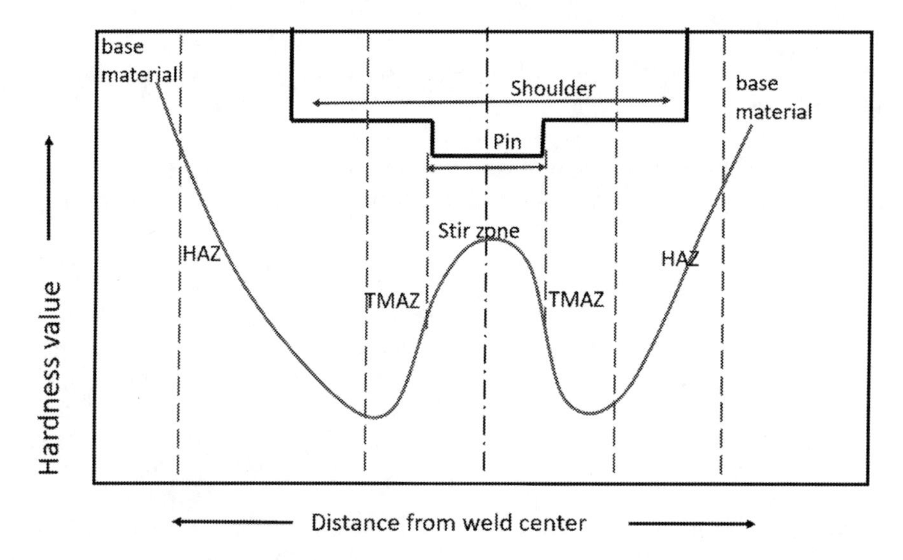

Fig. 2.12 General trend of hardness vs different zones of FSW welded sample

of the precipitates at that rotational speed with the significant amount of grain refinement. This continues to soften till TMAZ/HAZ due to the softening of the material. Thereafter, the reoccurrence of fine precipitates results in an increase in hardness till the base metal. The tensile properties of the weld seam decrease in comparison to the base material due to the softening effect at HAZ leading to failure of material from that region. The region next to TMAZ is HAZ where the effect of plastic deformation is very less. It only experiences a significant amount of thermal cycle. Due to thermal effects, it results in the occurrence of fine precipitates in an alloy. Also, a reduction in hardness profile investigated while increasing the tool shoulder diameter. It is due to higher heat dissipation resulted in grain growth and coarser microstructure [88]. Hardness distribution reveals that with an increase in rotational speed hardness of the weld decrease.

2.6 Metallurgical Aspect (Microstructure and Grain Size)

Fine grains are produced due to continuous dynamic recrystallization during the welding stage. Circular onion ring continuously forms due to the rotational-translation tool motion in the entire welding phase [8]. Finer grains are possible as a result of at low rotational speed and high welding speed or because of faster cooling [29].

A base metal microstructure is always a mixture of coarse and fine grains. After FSW, the weld nugget zone contains fine equiaxed grains. Grains have undergone a dissolution of around half of precipitates present in alloy into the base metal matrix during the welding process. In the stirred zone a continuous grain boundary misorientation occurs along the weld line because of the continuous dynamic recrystallization process which is the primary mechanism for the formation of fine equiaxed nugget grains [89]. In HAZ, grains do not undergo plastic deformation but it only undergoes thermal cycle this results in the growth of grains without a change in shape during FSW. In the intervening time, the grains in TMAZ show a significant change in shape indicating the possibility of plastic deformation caused by the tool. Also, the thermomechanical affected zone (TMAZ) has experienced a similar but lesser extent in the dissolution of precipitates whereas it was observed that significant coarsening of precipitates in the HAZ takes place [83].

2.7 Applications

2.7.1 Industrial

The first unit developed by FSW machine was cooling apparatus honeycomb panel for the fast chilling of fish in fishing containers [90]. FSW machines are commercially available for various industrial applications such as welding of large aluminum hollow panels, shipbuilding, sea platform, high-speed ship deck superstructures etc. [5]. Marine aluminum (Norway) make a pressed aluminum section for shipbuilding and railway rolling.

2.7.2 Aerospace

United launch alliance employed FSW for Atlas V, Delta II and Delta IV for joining its inter-stage superstructure module. Then came Boeing, which employed FSW for welding of shipment wall beams for the Boeing 747 cargo freighter, toenails of Boeing C-17 [91]. Longitudinal and circumferential diameter in propellant tanks of the delta group of rockets had been produced in the USA by using FSW as shown in Fig. 2.13b [92]. Shifting from arc welding to FSW in the manufacturing of propellant tanks of space rocket has resulted in an enormous cost reduction. Ariane 5 is a space take-off vehicle for massive weight lifting developed for the European Space Agency (ESA) is welded using FSW [93]. Boeing successfully welded the aircraft superstructures, they did lap welding of Al7075 sections [94]. FSW is also used for joining deflectors of fighter planes. The Boeing company has done several projects for the US military for the components of the base plate of the cargo section of the C17 Globemaster III aircraft. NASA build the structure of the outer propellent-fuel tank of the Ares

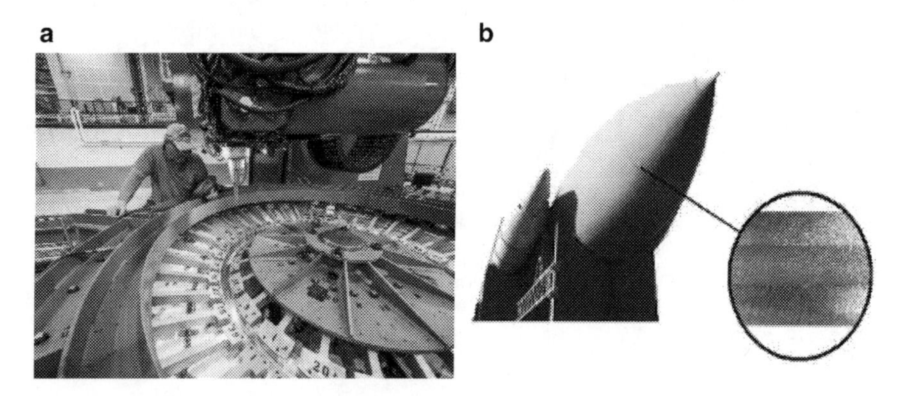

Fig. 2.13 **a** Weld of Orion Exploration Mission-1 Crew Module [95], **b** fuel tank of rocket [96]

I, Orion Crew Vehicle (Fig. 2.13a), and shuttle64 by 2195 aluminum–lithium alloy [95].

Germany based Airbus GmbH company and the Daimler/Chrysler aerospace agency have shown interest in FSW. It is because of advantages i.e. weight reduction of the structure, better joint strength, enhanced fatigue strength, and reduction in cost. They used FSW for making the components of the airframe-structure ie. the fuselage, passenger cabins, wings, and vertical stabilizers. The finishing assembly of the narrow-fuselage aircraft mostly for the A3 series of Airbus A380 [97]. It is to produce lengthwise joints in the fuselage, outer-link, and central container of the wing with AA 2xxx and 7xxx aluminum alloy series. FSW is also used for making panels for Airbus A400M, longitudinal and circumferential welds of Falcon 1 and Falcon 9 rocket fuel-promoter tank at the SpaceX factory [98]. Fokker space company developed the main cryogenic stage engine body of the Arian-5 rocket, its nose comprises of 12 integral flat panels of AA7075 with stiffeners. The aviation company Eclipse Aviation was involved in the manufacturing of 5 seater airliner of the private class Eclipse 500. It's rigid stringer panels are made of AA2xxx (plating) and 7xxx (stringers) groups of aluminum alloys [99].

ZE41A Mg alloy along with AA2024-T3 results in higher strength and is mainly used for aircraft production and towards defense areas joining by dissimilar FSW [100]. Also, ZE41A Mg is an anticorrosive alloy used for airplane gearbox and generator casings in military choppers. As these are exposed to corrosive environments during service conditions [101]. The dissimilar joining of magnesium and aluminum alloys are problematic because of significant differences in physical properties and development of intermetallic compounds i.e. $Al_{12}Mg_{17}$ and Al_3Mg_2 in SZ [102]. Indian space research organization (ISRO) has made propellant tanks constructed using FSW techniques [103].

2.7.3 Automobile

Companies are using FSW for the manufacturing of spatial frames of car structure, components of the chassis, bodies of bus structure, fuel tanks, and other transportation media [104]. Ford GT has efficiently welded aluminum extruded bend sheets and fuel tank housing. In the automotive industries, it is used in the manufacturing of hood panels, suspension links, and driveshafts [105]. FSW is extensively used in the automotive industries and its outreach has reached beyond materials like aluminum and magnesium. Few more application from automobile sector include rear seat frame of 2001 Volvo V70), suspension components of 2003 Lincoln Town Car L, hood and rear doors of 2004 Mazda RX-8, space frame parts of 2005 Ford GT and 2007 Audi R8, hood and trunk of 2006 Mazda MX-5 Miata, rear hatch parts of 2010 Toyota Plus and front subframe parts of 2013 Honda Accord (Fig. 2.14a) [102].

The most widely employed aluminum alloy in the automotive industry is AA7075 comprising of aluminum, zinc, magnesium, and copper. It is strongest of all aluminum alloy group and is comparable to steel. It weighs about only one-third of the steel and

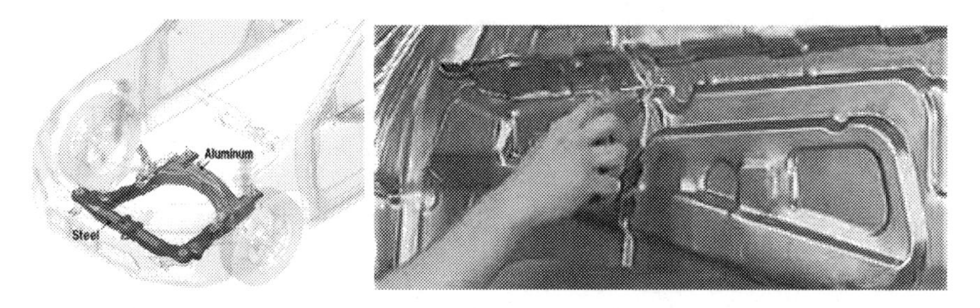

Fig. 2.14 **a** Dissimilar cast aluminum and steel FSW(2013 Honda Accord engine cradles) [102] (Reproduced with permission from the publisher), **b** tailor welded blank (Land Rover and BMW body panel) from 6xxx series aluminum [107] (Reproduced with permission from the publisher)

is used for various automobile components. AA5454 alloy having good corrosion resistance is used in wheel rims and suspension components. AA5182 having good formability and stress corrosion cracking performance used for body panels and interior. AA5154 is used for wheels, drivetrain components, and underbody components. AA5022 is used for bonnets, oil pans, floors, rear fenders, roofs, and doors [106]. In 1998, TWI fabricated FSW tailor welded blank for Land Rover and BMW body panel [107].

Magnesium alloys have been extensively used in the automobile industry for engine bracket of Chevrolet Corvette and front-end of Jaguar XJ351 but due to the high cost of magnesium alloys it restricts for mass production i.e. for a typical GM vehicle, the consumption weight ratio of magnesium/aluminum is 4:123 [102].

2.7.4 Others Application I.E. (Nuclear, Shipbuilding, Electronics)

Application of FSW in shipbuilding includes the joining of ship-panels and outer body parts [108]. Honeycomb panels with the added advantage of high noise-absorbing factors are used for ship cabin walls and railway wagons. Other industrial applications include joints for the sailboat, refrigerators, helicopters landing platforms, oil-takeout platforms, ship's mast, supports in yachts, and rolling stock for railway transport, etc. [5]. The Shinkansen express (Japan) used finned panels floor and wagon walls assembled using FSW [109].

Prefabricated panels through FSW is being used in the Freedom-class Littoral Combat ships and the sea fighter of Nichols Bros. Special dedicated FSW machine developed under joint treaties of the University in Adelaide (Australia) and TWI for joining of curved joints at the nose part of a cruise ship (Patent WO98/38084, Austria). The assembly comprises 5 mm thick AA5083 sheets. Hydro Marine fabricated friction stir welded roof panel as shown in Fig. 2.15 for delivery to Alstom LHB (Germany) [107]. Another application of FSW is Siemen's X-ray vacuum vessel. FSW can weld 3D printed materials with a weld joint efficiency of 83.3%

Fig. 2.15 Hydro Marine aluminum produced roof panel that was delivered to Alstom LHB (Germany) [107] (Reproduced with permission from the publisher)

[110]. Nuclear applications include joints in the encapsulation of nuclear waste (SKB company) as shown in Fig. 2.16a in 50 mm copper containers [111].

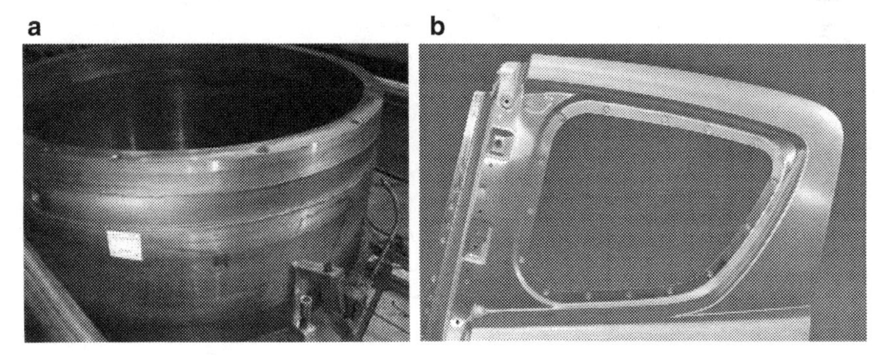

Fig. 2.16 **a** Nuclear waste encapsulation SKB in 50 mm thick copper container [111], **b** outer rear door panel (Mazda RX-8) [107] (Reproduced with permission from the publisher)

2.8 Research Area

Investigation on the friction stir welding for similar and dissimilar metal has been done extensively by the researchers. It has been successfully adopted in the field of automobile spare part joining and aerospace field. It has gathered immense potential in research and development. FSW has been used for a variety of materials ie. aluminum, copper, magnesium, titanium, and steel. It is also used for a dissimilar material like aluminum-magnesium, magnesium-copper, copper–aluminum, etc. Research has been carried out in different domains of FSW i.e. material flow, temperature distribution, microstructural characterization, and improvement in the mechanical property of the weld. A brief overview of the literature review for the application of FSW for different materials is explained in the following sections. Also, tool wear during FSW has been briefly elaborated as it is important in terms of tool design and tool life predictions in context for the optimized input process parameter. In addition, a few variants of FSW have been explained with important literature. For a detailed review, the reader should look for more papers in the relevant areas.

2.8.1 Overview of Friction Stir Welding for Similar Material

2.8.1.1 Aluminum Welding

Friction stir welding has progressed as a front runner for the welding of aluminum and its alloy for high weld strength. Since its inception, extensive research efforts have been carried out for the FSW. An investigation has been done to understand temperature distribution, material flow, the effect of the input parameter, and pin profile on the welding of aluminum alloy [112].

Hou et al. [113] investigated FSW of 6061-T6 aluminum alloy at a fixed welding speed of 150 mm/min. The microstructure and mechanical properties of the joints were observed for different rotational speeds. In SZ fine recrystallized grains of 1 μm were formed at 400 rpm. Maximum joint strength of 188 MPa with a weld efficiency of 66% was obtained. They reported that defect-free joints were observed between 400 and 600 rpm and defects such as void were reported at higher rotation speeds of 700 and 800 rpm. Liu et al. [114] FSWed 7.5 mm thick 2219-T6 aluminum alloy. A concave shoulder with a threaded conical pin tool with a tilt angle of 2.5° and a plunge depth of 0.3 mm was used for joining. The highest tensile strength of 340 MPa with 79% weld efficiency was obtained at 800 rpm. Defect-free weld was observed at 600 rpm and 200 mm/min.

Marzbanrad et al. [115] investigated the performance of the tool pin profile on weld properties of 3 mm thick AA5083. Cylindrical and square pin profiles with a 2° tilt angle tool were employed. Weld parameters of 1120 rpm and 30 mm/min were selected for the analysis. A strain of 50 and 40 mm/mm was observed for the square and cylindrical profile, respectively. The peak temperature of 480 and 455 °C was obtained with a weld efficiency of 93% and 89% for square and cylindrical pin, respectively. Fine and uniform recrystallized grains were observed in the case of a square pin due to an increased degree of plastic deformation.

Mohammadi-pour and Khodabandeh [116] FSWed 4 mm thick 7075-T6 aluminum alloy. A comparative study of various tool pin profiles of threaded conical, conical, and triangular pins made of H13 tool steel was investigated for different input parameters. Fine equiaxed grains of 2.5 μm, 12.5 μm, and 10 μm were observed at 1600 rpm and 63 mm/min, for respective tool pin profiles. A maximum weld strength of 370 MPa was obtained for a conical threaded tool profile. Hardness value at SZ for threaded conical, triangular and conical pins were 180 HV, 175 HV, and 165 HV respectively.

Jamalian and Farahani [117] joined 6 mm thick 5086-H34 aluminum via FSW. Tunneling and wormhole defects were obtained at lower rotational speeds due to insufficient heat generation and lower material flow rates. These defects were also observed at high rotational speeds because of turbulence in the weld. Weld parameter of 1250 rpm and 80 mm/min produced sound weld with a weld efficiency of 85%.

Saravanan et al. [118] studied the performance of shoulder to pin diameter (D/d) ratio (2, 3 and 4) on weld properties for welding of 2xxx series aluminum alloy. The maximum weld strength of 356 MPa and a hardness value of 151 HV at SZ was observed at 1200 rpm and 12 mm/min with the D/d ratio of 3.

Tufaro et al. [88] investigated the performance of tool shoulder diameter on heat distribution during FSW of 3 mm thick AA5052-H32 alloy. Tools of 10, 12, 14, and 16 mm shoulder diameter were used for joining in butt configuration. Sound weld was produced at 514 rpm and 98 mm/min. It was reported that increasing shoulder diameter resulted in a defect-free weld in the SZ with better material flow characteristics.

Grujicic et al. [119] investigated the FSW of AA5059. AISI H13 tool with conical threaded was selected as a tool. Material flow behavior was observed towards AS and RS. The maximum extent of material mixing was investigated by varying input parameters ie. tool tilt angle and tool pin size. Tool tilt ranges of 3.5°, 2.5°, and 2.0° were investigated and 2.5° was reported as an optimized parameter to produce a sound weld.

Zhang et al. [120] investigated the heat generation phenomenon during FSW. Results show that slip rates are lower on the retreating and the front sides as compared to the advancing and trailing side. This is because of the better material flow in those regions and this also results in higher heat flux on trailing and advancing sides. A significant decrease in equivalent plastic strain is observed due to the lowering of stirring action in the thick plate.

Buffa et al. [121] investigated varying included cone angle with a cylindrical pin (0°) to conical pin up to 40°. AA7075 was selected as workpiece material. Investigation of the material flow behavior and grain size distribution in the joints were carried for different cone angles. Results show that the extent of plastic deformation, strain rate, and size of the weld zone in SZ, and HAZ increases with an increase in pin angle. A combination of 100 mm/min and pin angle of 40° was reported as an optimized parameter to get sound weld.

2.8.1.2 Cu Welding

Pure copper or copper-based alloy is the third most widely used material after iron/steel and aluminum alloys. It has excellent electrical conductivity and good corrosion resistance. In addition to this, it has good ductility, toughness, non-magnetic behavior, and catalytic properties that have attracted considerable interest in various industries. It has wide applications in the field of space, defense, electronics, and military. Also, copper alloy welded with FSW has shown excellent formability and good resistance to oxidation [122].

Jha et al. [47] investigated the microstructural investigation for Cu0.8Cr0.1Zr alloy. Welding speed of 100 mm/min showed a fine grain size in the weld nugget zone. At lower travel velocity, larger grains were observed, the reason could be a long time offered for the recrystallization process. Defect-free weld was produced between 800 to 1200 rpm and welding speed from 40 to 100 mm/min. The thermocouple is fixed at a distance of 2 mm and 12 mm from the shoulder in the transverse direction. In both cases, a temperature difference of around 250 °C is observed. Also, the temperature difference of 2–3 °C is reported between AS and RS it is due to the difference in relative velocity components between AS and RS. The weld efficiency was reported to be 60%.

Guan et al. [123] investigated the FSW of Cu alloy with a thickness of 20 and 50-micron thickness. The peak temperature attained is 256 °C which enables the recrystallization temperature of copper. Moving from 800 to 900 rotational speed resulted in enhanced interface bonding. The EDS analysis indicated no impurities were present in the weld. For 50 μm thickness foil, process parameters of 800 rpm and 25 mm/min resulted in a 56 N fracture capacity.

Lee and Jung [124] studied FSW of copper at 1250 rpm and 61 mm/min and it resulted in the defect-free weld. They used tool steel material with a 3° tilt angle. The average grain size of base metal was 210 μm, while HAZ and SZ have 230 μm and 100 μm, respectively. Weld zone hardness was lower as compared to the base metal. The result shows the hardness distribution near the weld zone was independent of grain size. The base metal hardness was around 110 HV, but hardness near the weld region has reduced significantly and varied from 60 to 90 HV. The tensile test result showed that fracture happened around the HAZ due to lower hardness. A joint efficiency of 87% is reported for the weld.

Hwang et al. [125] investigated FSW in butt configuration for joining of C11000 with 3.1 mm thickness. The K-type thermocouple was employed for temperature measurements at different locations. Also, a 1° tilt was provided for better forging. Successful FSW weld was formed between the temperature of 460 and 530 °C. Also, the temperature towards AS is a little higher than the RS. Tensile strength and the hardness variation (near TMAZ) were reduced to about 60% as compared to the original sample.

Khodaverdizadeh et al. [15] investigated the influence of tool pin shape on grain size and mechanical properties for FSW of copper. Threaded cylindrical and square pin were used at 600 rpm and 75 mm/min. Temperature history was measured using a K-type thermocouple. Square pin profile resulted in a fine recrystallized grain and better mechanical outcome as compared to the threaded cylindrical pin. The reason could be credited to the high eccentricity present in square pin and dynamic pulsation effect which results in a higher amount of plastic deformation and greater heat dissipation in the weld. At the nugget zone, square and threaded cylindrical pin resulted in fine recrystallized grains of 10 and 15 microns, respectively.

Heidarzadeh et al. [126] developed the optimized process parameter windows for the welding of copper alloy. Rotational speed (700–1100 rpm), welding speed (50–100 mm/min), and axial force (1.5–2.5 kN) produced defect-free weld and better strength. Hardness value showed a direct relationship with welding speed and inversely varies with rotational speed and axial force.

Sun et al. [49] welded CuNiCrSi and CuCrZr alloy in butt configuration. A fixed welding velocity of 150 mm/min with a varying rotating speed of 800, 1100, 1400, 1700, and 2100 rpm was used. They reported that 1400 and 1700 rpm resulted in a sound quality joint. Cr precipitates in the weld matrix show proper dissolution to increase the mechanical and microstructural properties of the stir zone. Microhardness at the SZ and weld strength decreased with the increase in rotational speeds due to the formation of larger grain size.

Sahlot et al. [127] studied FSW of pure copper. A tungsten carbide tool with 12% cobalt with a flat shoulder and tapered cylindrical profile was used. Heat dissipation was studied during the process. Results show that tool rotation speed directly influences the peak temperature and weld strength.

2.8.1.3 Mg Welding

Welding of magnesium alloy is difficult because of high flammability. In recent years, FSW has been effectively applied for the welding of magnesium alloy [128]. Pan et al. [129] FSWed Mg-5Al-1Sn alloy in butt configuration. The influence of tool rotational speed on weld properties has been investigated. The H13 tool steel with a concave shoulder and a threaded cylindrical pin was employed for welding with a tilt angle of 2.5°. Microstructural observation reveals equiaxed and fine grains at the nugget because of adequate heating and intense plastic deformation. Hardness distribution showed the lowest value between NZ and TMAZ, due to the extended deformed grains. Weld strength of 258 MPa was obtained at 800 rpm resulting in 91% of weld efficiency.

Pareek et al. [130] FSWed 3.175 mm thick AZ31-H24 magnesium alloy. The influence of rotational and welding speed on metallurgical evolution was studied. It reveals that dynamic recrystallized grains and the beginning of growing grains in some regions of the weld nugget. Partially recrystallized grains were also identified towards (TMAZ). During the tensile test, all weld specimens failed from the advancing side of the workpiece. Weld efficiency of 75% was obtained at a combination of 2000 rpm and 204 mm/min.

Wang et al. [131] investigated the welding of AZ31 using FSW. The experiment was done by varying rotational speeds from 800 to 1600 rpm at a constant welding speed of 120 mm/min. Hardened H13 tool steel with a threaded pin was used. Microstructural investigation reveals that grain size in the SZ increases with an increase in tool rotation. A joint efficiency of 90.2% was obtained at 1200 rpm. It showed an increase in hardness value within the SZ because small grains attributes to higher grain boundary and act as an obstacle for slip dislocation. This dislocation acts as resistance to deformation.

Also, the peak temperature at SZ was reported as 495 °C. It was calculated by the finite element model as it is difficult to obtain experimentally because of the wash away of thermocouple in the stir zone.

Forcellese et al. [132] investigated the effect of input parameters on vertical force and temperatures during FSW of magnesium alloy AZ31 sheets. Experiments were carried from 1200 to 2500 rpm and 30 to 100 mm/min. Forces and temperature were measured by a dynamometer and K-type thermocouple, respectively.

Mironov et al. [133] studied material flow behavior during FSW of AZ31 magnesium alloy. Also, the influence of temperature on material flow was examined. It employed tool steel with a concave shoulder having a diameter of 15 mm with a threaded cylindrical pin. Increasing the tool plunge depth increases welding temperature resulting in more stirring and material flow beneath the shoulder. Observation of transitional material flow due to the presence of a small temperature gradient within the stir zone showed that it resulted in a weld defect near the weld root.

Asadi et al. [134] investigated FSW of magnesium alloy plates using a FE model. The developed model can predict the nucleation process and grain growth through dynamic recrystallization. A combination of rotational and traverse speeds of (710 rpm with 50 mm/min) and (1400 rpm with 25 mm/min) was employed as input weld parameters to produce joints. The result shows that new grains nucleate and grow from the region of high dislocation density because of temperature evolution and plastic deformation.

Richmire et al. [135] studied FSW of magnesium alloy with a triangular tapered pin. The magnesium AM60 cast alloy applicable to frames for the seat, steering wheels, instrument panels, and brackets for the automobile company. It is ductile and has significant toughness as compared to AZ91 Mg alloy. Microstructural evolution revealed the presence of equiaxed fine grains in SZ and elongated tilted grains at TMAZ. The results could be attributed to dynamic recrystallization, followed by the grained refinement. Various defects like porosity and void are observed at a low rotational speed. Also, insufficient heat resulted in inadequate material flow. High rotation speed and low travel resulted in too much heat dissipation resulting in local melting and grain growth due to softening. The presence of intermetallic i.e. $Al_{12}Mg_{17}$ is one of the possible reasons for void defects.

2.8.1.4 Steel

FSW is also used for welding hard alloys like steel and titanium. Steel is widely used as a structural component and it is a universal material. FSW application on steel is limited due to its high hardness, strength, high loading force, and severe wear of the tool. Also, the manufacturing of tool materials for steel is complex and expensive.

Generally, polycrystalline cubic Boron nitride or tungsten carbide could be employed as a good tool material. Tool material should retain a high-temperature strength and toughness, excellent wear resistance and good microstructural stability at elevated temperature [32]

Al-Moussawi and Smith [80] studied defects generated during FSW of steel. Steel grades of DH36 and EH46 were used for joining. Experiments were conducted at different combinations of welding parameters on as-received material. It yields a microcrack due to lower linear speed between the plunge and the steady-state region. Welding speed for steel was optimized, as higher speed resulted in inadequate material flow to form joint. The formation of microcrack was attributed to the elemental precipitation of alloying elements in steel due to higher rotational speed. At high rotational speed, the peak temperature rises and formed TiN. Therefore, optimum speed was found between 200 and 500 rpm. These elemental precipitation/segregation resulted in the formation of microcracks which later act as the initiation of the stress concentration region leading to the reduction of mechanical properties of the weld.

Dinda and Ramakrishnan [136] have studied FSW of 12.7 mm high-strength martensitic steel plates. FSW tools were made of W-25Re with 2% HfC. FSW process was carried out in water-assisted cooling. The tilt angle for the tool was maintained at 2° for all experiments. An optimized tool rotation speed of 740 rpm and a welding speed of 12.7 mm/min resulted in the defect-free weld. Also, increasing the welding speed further resulted in a wormhole in the stir zone. An exhaustive microstructural characterization of all the regions was investigated by the author. SZ had 45% bainite, 48% martensite, and rest retained austenite by volume %age resulting in the higher hardness of the SZ as compared to other regions of the weld. The TMAZ revealed a combination of 35% ferrite, 34% bainite, and rest as martensite phases. TMAZ consists of alternate layers of bright equiaxed ferrite phase due to continuous dynamic recrystallization of the pro eutectoid ferrite phase into the ferrite and austenite region. The tempered martensite in HAZ was formed from martensite steel due to the FSW process.

Li and Yang [137] investigated the FSW of modified 9Cr–1Mo steel. The tool was made of W–Re alloy. It resulted in defect-free weld at optimized welding parameters of 300 rpm and 400 rpm for a fixed welding speed of 50 mm/min, respectively. Microstructural observation revealed the development of quenched martensite within SZ and HAZ along with M23C6 carbide precipitate. It showed a significant drop in the grain size of SZ as compared to the base metal. Hardness value has increased to twice that of base metal of the SZ. A weld efficiency of 98% was obtained in the experiment. Post weld heat treatment showed a substantial effect on the hardness profile and microstructure of the FSW joints.

Avinish Tiwari et al. [138] investigated FSW of DH36 a shipbuilding grade steel. Tungsten carbide tools have been used for welding in butt configuration. Two different grades of WC tools were used with 6 wt% (Tool A) and 10 wt% Co (Tool B). Their wear properties were analyzed. The wear mechanism depends on tool configuration and the input welding parameters. Tool A has undergone intergranular failure of tungsten carbide grains, which resulted in cracks initiation. Progressive wear was investigated in tool B. Adhesion at the edge was due to compression load between the tool and faying surface at high temperature. The shoulder surface was highly influenced by abrasion and adhesion wear. The porous and brittle oxide was formed on the tool due to oxidation.

2.8.1.5 Titanium Alloys

An investigation has been done on the welding of titanium alloy with the help of FSW. Different literature has been discussed based on the FSW approach. Application of titanium alloys in the automobile and aerospace sector is due to its superior strength-to-weight ratios, its capability to withstand high melting point, and outstanding corrosion resistance. Apart from these titanium alloys are also used because of their smaller thermal expansion coefficient, fire resistance, non-magnetic nature, and short radioactive half-life. Important applications include the aerospace industry for the manufacturing of turbine jet engines as it can withstand high temperatures and non-corrosive nature of the titanium alloy [139]. The difficulties associated with FSW of titanium are its high strength and sufficiently high hardness. This makes titanium difficult to soften and undergo appreciable deformation to form joint. The tool needed for titanium alloy is of a higher grade as compared to the soft alloys like aluminum and magnesium [139].

Gianluca Buffa et al. [140] investigated different tool materials for titanium alloy based on refractory metals with cobalt added as a binder. As it significantly reduced the cost and has a wide industrial application. It also limits due to the high wear and deformation rate during the processing of titanium alloy. The best results were obtained with the W25Re and WC tool in terms of both tool life and joint quality without any evidence of degradation during the entire process. A special machine is to be developed for the welding of titanium alloy because of very high reactive loads in a range of 85–140 kN.

Sanders et al. [141] conducted experiments with a combination of higher rotational and lower welding speed. It leads to a coarse grain microstructure because of high temperature and higher exposure time. FSW of Ti–6Al–4V sheets has been investigated to reveal microstructural change, hardness variation, and mechanical properties. Microstructure at the SZ results in high grain refinement and it increases with the depth of the weld. During microstructural evolution of FSW at the SZ results in the formation of fine equiaxed alpha grains surrounded by zigzag grain boundaries. It is a result of beta to alpha ($\beta \rightarrow \alpha$) allotropic transformation during the cooling cycle. Fine alpha grains showed an increase in hardness as related to the base metal. The hardness values at the stir zone vary with depth along the transverse direction. It changed by about 5 HV for each successive traverse depth.

Seighalani et al. [53] studied FSW of titanium alloy assisted with compressed air jet to provide a cooling medium to lower the temperature of the tool. It is due to the higher heat generation between the tungsten carbide tool and the titanium. This will lead to lower degradation and wear of the tool. In addition, a 1-degree tilt angle was reported as an optimized parameter and increasing the tilt angle up to 3° resulted in surface defect because of the increase in forging action. FSW of titanium was carried out in a shielded environment of argon gas because titanium and oxygen react together to form the titanium oxide resulting in increased wear rate.

Lauro [142] investigated that the tool material selection is an important aspect of titanium alloy. The tool should have sufficiently high-temperature resistance, better tensile strength, excellent high-temperature hardness, high compressive strength with exceptionally good toughness, and should be oxidation resistance at the higher temperature. Mostly polycrystalline cubic boron nitride has wide application for industrial application. The application limits due to extremely high cost, its brittleness nature, and an unexpected failure during continuous service.

2.8.2 Dissimilar Welding Al–Cu, Al–Mg, Al–Steel

As a growing advancement, FSW has also evolved in the joining of dissimilar metal. Dissimilar joints have applications in industrial, aerospace, and automobile to amalgamate two different metal properties in the components. This review has been focused on the dissimilar joining of aluminum with copper, steel, and magnesium.

2.8.2.1 FSW of Aluminum and Copper

Dissimilar FSW of aluminum to copper has gained the utmost attention in different fields. The dissimilar weld has great advantages in terms of cost and weight reduction. Due to which joining these materials is attractive for the aerospace, chemical, electronic industries, and automobile sector.

Sinha et al. [143] have successfully FSWed Cu and Al. FSW joining of aluminum alloy with copper was studied with varying rotational speed from 150 to 900 rpm. Experiments were carried out in steps of 150 rpm at a fixed welding speed of 60 mm/min. Its microstructure and mechanical properties were studied which clearly showed the formation of all the zones within the weld similar to FSW of the same material. Also, it showed the formation of intermetallic phases Al_2Cu_{39}, Al_2Cu, AlCu, and Al_4Cu at the interface of the stir zone.

Li et al. [144] investigated the FSW of pure copper with aluminum alloy. The pin was shifted towards the aluminum side through a pin-offset method. Defect-free weld was observed at 1000 rpm and 80 mm/min without the formation of intermetallic phases. It resulted in a complex microstructural zone in the nugget zone with the formation of a vortex-like pattern and lamella structure was found but the absence of any intermetallic compounds in the weld nugget. Also, the hardness towards the copper side of the nugget was higher as compared to aluminum alloy. The hardness observed on the bottom side of the nugget is found to be higher than in other areas. During the tensile test, the fracture surface at the joint of the dissimilar region failed in a ductile–brittle mixed fracture mode. The weld strength and elongation were found to be 152 MPa and 6.3%, respectively.

Carlone et al. [145] investigated dissimilar FSW of 2 mm thick AA2024-T3 rolled sheet with pure Cu10100 at 2° tilt angle. The tool was made up of Cr–Mo steel. Experiments were carried out at 1000 rpm and 80 mm/min. Tool offset towards aluminum in the advancing side was kept at 1.3 mm for efficient material mixing. Microstructural examination showed that the weld nugget zone comprised of a combination of the recrystallized aluminum matrix along with twinned and deformed copper scrap. Small copper dispersed within Al matrix was also detected. It resulted in the formation of intermetallic compounds like AlCu, Al_2Cu, and Al_3Cu_4. These intermetallics significantly change hardness distribution within the weld zones.

Bakhtiari et al. [146] studied dissimilar FSW of 4 mm thick 5754-H114 aluminum alloy with commercially pure copper in butt configuration with H13 threaded tapered tool. The aluminum plate was positioned on the advancing side with 1 mm offset for better material flow and material mixing. FSW was carried out at a different input parameter to get optimized results. Micrograph revealed the formation of intermetallic compounds ie. Al_4Cu_9 and Al_2Cu within FSW joints. Good results were found at an optimized rotational and welding speed of 1000 rpm with 100 mm/min. A maximum weld efficiency of 84% was achieved with a weld strength of 219 MPa. The peak value of microhardness was about 120 HV at the SZ.

Akinlabi [147] investigated the effect of the tool shoulder diameter on dissimilar friction stir weld of AA5754 with C11000 copper. H13 tool steel with 2° tilt was employed for welding. The shoulder diameter varied from 15, 18, and 25 mm for rotational speed between 600 and 1200 rpm and the welding speed between 50 and 300 mm/min. The optimized weld parameter was obtained with 18 mm shoulder diameter and at 950 rpm and 50 mm/min. A joint efficiency of 86% was obtained for this parameter. Average Vickers microhardness was obtained at 95HV towards the copper side. Al_4Cu_9 and Al_2Cu were the intermetallic reported that were detected at the weld interface.

2.8.2.2 FSW of Aluminum and Magnesium

Dissimilar FSW of Al with Mg is investigated as it has gained a huge interest due to lightweight of magnesium which later resulted in a reduction of weight of the entire welded components. These dissimilar FSW joined components are applicable to mostly automobile and aerospace parts to have a weight reduction factor [148].

Vahid Firouzdor [149] FSWed AA6061 with AZ31 to study the effect of process parameters on joint properties of the weld. During dissimilar FSW, precipitates of Al_3Mg_2 and $Al_{12}Mg_{17}$ were found at the stir zone resulting in a reduction of joint strength. Magnesium was kept on the retreating side with tool offset to produce the defect-free weld. As magnesium is a softer material, its alloy resulted in superior flow behavior and continue to stir behind the tool followed by mixing with Al alloy. At slower welding speed and with aluminum on AS, higher heat is generated. While putting Mg, on the AS at slow welding speed can enhance the weld strength by decreasing the heat input.

Zhao et al. [150] investigated dissimilar FSW of aluminum alloy 6013 and magnesium alloy AZ31. The joints were made in underwater and conventional conditions. H13 tool steel with a concave shoulder and 2.5° tilt angle was used for welding. The optimized parameter was found at 1200 rpm and 80 mm/min. Water-assisted FSW resulted in better joint quality and improved the joint strength close to 152 MPa which was higher as compared to air-assisted which was 131 MPa. Water leads to faster cooling of the FSW process resulting in lower thermal gradients. It showed no cracks as faster dissolution of intermetallic inclusions in the weld. Also, in water appearance of the joint was smoother and brighter than in air. In case of air cooling, microhardness was higher as compared to the water cooling due to a strong stirring effect and better thermal cycles.

Somasekharan et al. [151] FSWed 1.75 mm thick plates of AA6061 and Mg AZ91D was carried at 800 rpm and 90 mm/min. Intermetallic phases of Mg with Al were observed as a result of swirls and vortexes within the weld zone. It is due to dynamically recrystallized regions formed due to extensive plastic deformation and high temperature. SZ consists of a fine homogeneous microstructure. Due to the presence of dislocation substrates and precipitates, it resulted in high weld strength.

Sato et al. [152] conducted dissimilar joint of Mg alloy AZ31with AA1050 using FSW. It resulted in an excessive quantity of intermetallic compound $Al_{12}Mg_{17}$. Due to which an appreciably higher hardness was observed in the weld nugget. The experiment was carried out at 2450 rpm with 90 mm/min and 3° tool tilt angle. The author has reported that intermetallic was formed because of constitutional liquation during dissimilar welding.

2.8.2.3 FSW of Aluminum and Steel

Tanaka et al. [153] FSWed 3 mm thick mild steel sheet with AA1050 aluminum alloy at 1800 rpm and 100 mm/min. Aluminum was placed at RS. The steel plate was kept on AS with a probe contact width of 0.2 mm for better material flow. K-type thermocouple was employed to reveal thermal history. Investigation of the intermetallic compound was extensively studied as they later resulted in defects and cause a significant reduction in weld strength. The intermetallic compound is formed due to excessive heat generation.

Chen [154] carried out FSW on 6 mm thick AA6061-T651 with low-carbon SS400 steel using a tool tilt angle of 3°. Weld at 550 rpm and 54 mm/min, yielded excellent impact strength and a satisfactory tensile strength. This weld specimen has been bent to 150° without breaking.

Derazkola and Khodabakhshi [155] carried out underwater FSW of 5 mm thick AA5083 with A441 AISI steel. Tungsten carbide tool with a frustum probe with a conical diameter was employed for joining the base metals. The tool tilt angle of 2° with a 0.4 mm plunge depth and tool offset of 1.5 mm towards the aluminum alloy side was provided. The welding took place by immersing the workpiece into the water at three different temperatures of 0, 25, and 50 °C. Optimized input processing parameters of 1500 rpm and 25 mm/min was set for all conditions. Increasing cooling rates brought the peak temperature down to 400 °C and reduced the possibility of intermetallic formation in the weld nugget. Weld efficiency of 97% was reported.

Habibnia et al. [156] investigated mechanical properties and microstructure of FSW of 3 mm thick AA5050 with 304 stainless steel. Positioning Al alloy towards the retreating side with variable tool offsets ranging from 0, 0.8, and 1.5 mm. Tungsten carbide tool with a conical probe was employed with a 2° tilt angle. Defect-free and improved joints were formed while decreasing the rotational speed from 710 to 500 rpm and increasing welding speed from 40 to 80 mm/min at a tool offset of 1.5 mm. Also, intermetallic compounds such as $Al_{13}Fe_4$ and Al_5Fe_2 were observed in the stir zone.

Karakizis et al. [157] investigated FSW of 2 mm thick AA5754 and H114 strain hardened mild steel. The tool pin geometry of a star-shaped polygon was used for joining. Steel was kept at the retreating side with a tool offset of 0.2 mm. Experiments were carried out at distinct input parameters. The rotational speed of 750 rpm and a welding speed of 30 mm/min showed the best result. An increase in microhardness at the SZ was reported because of the grain refinement.

Lan et al. [158] successfully welded high-strength TRIP 780 steel with aluminum alloy AA 6061-T6 using FSW. Steel was kept toward the advancing side in butt configuration. Tool material was tungsten carbide with 10% cobalt and pin geometry was an non-threaded conical pin. The formation of the intermetallic Fe_3Al or FeAl layers thickness close to 1 μm was observed.

Yazdipour and Heidarzadeh [159] FSWed AA 5083-H321 with 316L stainless steel plates in lap configuration. H13 tool steel with a tilt angle of 2.5° was employed for joining two metals. The maximum weld strength of 238 MPa was obtained at 280 rpm and 160 mm/min. At lower tool rotational of 180 rpm resulted in tunnel defect at the joint interface.

2.8.3 Tool Wear

Tool wear rate is a critical factor of tool design. It is majorly dependent on tool rotation speed, welding speed, workpiece and tool material, and axial force. It is also reported that the influence of these parameters is noteworthy only during the plunging and the initial phase of the welding and tool wear drops considerably during the later stage of welding [160]. The axial force is significantly high during the plunging and this is one of the primary reasons for the tool failure/wear during that stage [8]. Generally, tool wear is reported to be directly proportional to the tool rotational speed and welding speed. The tool undergoes self-optimization because, during the initial stage, high-stress concentration gradients regions undergo maximum wear followed by tool attaining a shape where the stress gradients reduce across the tool [41].

Investigation of the deformation behavior and load-bearing capacity of the tool pin during the FSW process is important, as the tool pin is susceptible to more deformation/failure as compared to the shoulder. The pin is the weakest module of the tool as it undergoes severe stress conditions at elevated temperature and that resulted in both bending moment and torsional load resulting in shear stress. So, the selection of tool material and deciding the optimum tool pin geometry is crucial for efficient welding. Arora et al. [161] reported that tool pin having a smaller length and larger diameter can withstand higher stress, torsional load, and undergoes less wear as compared to a longer pin with a smaller diameter. The wear rate of the tool is correlated with the stress generated due to the forces acting on the tool [162]. Tool welding force (F_x) reduces with the combined effect of an increase in rotational speed along with a lower plunge depth at a fixed welding speed [163]. As tool rotation increases, it leads to a higher heat generation rate that increases softening effect while increasing plunge force resulting in higher friction force on the shoulder that leads to an increase in wear rate. Increasing welding speed lowers the heat generation per unit length leading to insufficient softening of the material and hence an increase in traverse force [72]. Due to this temperature gradient the pin experiences low temperature at the bottom portion leading to higher flow stress while it is deforming the workpiece material towards its free end. This results in greater resistance to the material motion and leads to a higher welding force. Tool shoulder is more influential in heat production and softening than pin diameter. Therefore, a pin with a larger length having a fixed pin diameter would need superior tool material for better functioning without much tool pin wear.

Cui et al. [43] have observed that the tool materials for FSW of steels are an important concern as it is hard to weld. It needs a specially designed tool with superior mechanical properties. PcBN and W10–25%Re alloy are the two promising materials that have good wear resistance, superior strength, and fracture toughness. Tool material for steel with tapered threaded probe and concave shoulder is mostly preferable for sound joint and to avoid high wear while processing. Also, the wear and deformation characteristics of the tools are different for PcBN and WRe alloy. They are used for welding distinct grades of steel. A tool associated with a chamfer pin shows improved wear resistance than a helical pin. Also, increasing plunge depth or providing a pilot hole can increase tool life. Park et al. [164] reported that the PcBN tool mainly fails due to chemical reactions between the nitrogen, chromium, and boron elements with iron during the FSW process and tungsten-based tools mainly fail due to the inclusion of intermetallic in SZ. The wear of tools can be supervised by proper optimization of input process parameters and the tool shape designs.

Selecting tool material for steel is a critical aspect of the FSW process. The tool should withstand severe working conditions during the welding process. Also, the tool material should not undergo any chemical reactions with the base material and it should retain its high-temperature strength and toughness. The extent of tool wear was investigated for the FSW process of HSLA-65 steel using a W-25%Re tool material [43, 165].

Wang et al. [166] analyzed tool wear during FSW of Ti-6Al-4V. Three different tool materials namely, W-1.1%La_2O_3 and WC-8%Co and WC-11%Co were studied. Tool wear classification was done by investigating the pin profile using photographic method, weight loss measurement, and microscopic observations. Adhesive wear was the main mechanism of tool wear because of the chemical affinity between Ti-6Al-4V and WC-Co alloy. W-La_2O_3 tool outperforms the rest two profiles in terms of tool wear because the adhesion mechanism influences the heat transfer rate resulted in a narrower HAZ. While comparing the performance between the WC-Co grade tool, 11% Co was better as compared with the tool having 8% Co. Yusuf et al. [167] investigated wear mechanism by evaluating the tribological and mechanical properties of the tool. A physical cathodic arc vapor-deposited AlCrN coating on spark plasma sintered W25%Re-HfC tool was studied for wear analysis. Both AlCrN coated and uncoated tool samples were investigated by pin-on-disk tests to estimate the wear properties. Results show that the wear rate of the coated specimen is ten times lesser than the uncoated sample. The coated specimen has undergone a reduced wear characteristic due to the formation of Cr_2O_3 and Al_2O_3 oxides layers. Improved wear property attributed to better mechanical properties i.e. improved cohesive strength and reduced coefficient of friction. Adesina et al. [168] studied the wear properties of the AlCrN (through physical vapor deposition(PVD)) coated FSW tool for joining 6061-T6 aluminum alloy and 4140 alloy grade steel in a butt configuration. It showed full penetration without defects and a reduction in the weight loss of around 87% as compared with the uncoated specimen. Improved cohesive and adhesive load of the coating to around 19.12 N and 10.70 N, respectively. Siddiquee and Pandey [169] have investigated the deformation and wear behavior of the WC tool during FSW of austenitic stainless steel AISI 304. A conical tool with a 1.5° tilt angle was employed for welding. Elevated temperature and severe stress conditions resulted in deformation and tool wear during welding. The wear at pin bottom show groove formation because of the diffusion during FSW. A significant reduction of microhardness was observed for tool due to high-temperature exposure and wear of tools under stress conditions. Wear of tool by protruding of the shoulder resulted from an increased pin cone angle. The tool is also susceptible to buckle at pin when it is plunged with higer axial force, when tool rotational speed is sufficiently lower as shown in Fig. 2.17.

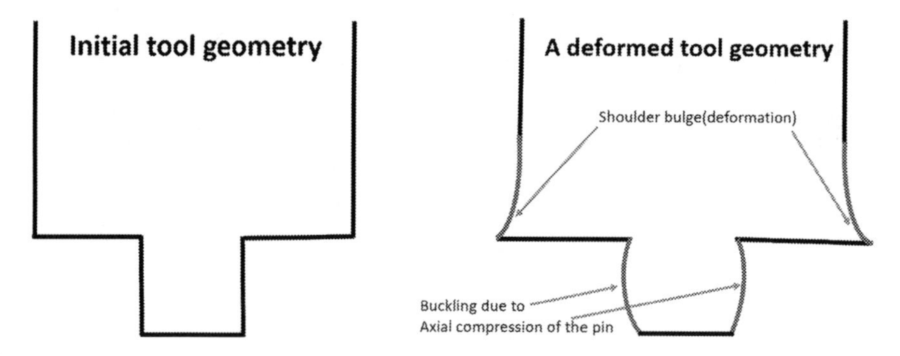

Fig. 2.17 The tool configuration before welding and after tool wear

2.8.4 Variants of Friction Stir Welding

With the successful development friction stir welding led to the emergence of different variants. This variant can cater to joining, surface modification, surface composite manufacturing, microstructural modification, and additive manufacturing. Some of the popular variants are discussed below.

2.8.4.1 Friction Stir Processing (FSP)

FSP is a surface amendment process carried out to change the surface properties of the material due to localized plastic deformation and heat generation. Friction stir processing is shown in Fig. 2.18 and is similar to FSW, with a minor difference that the tool impinges into the single material block instead of two [170]. The tool supports heat generation and material mixing simultaneously. The stirring motion intermixes the two materials together to enhance its surface properties. It also produces microstructural refinement, homogeneity, and densification of the processed zone. FSP helps in matrix grain modification, porosity elimination, dissolution of precipitates to get a fine, uniform, and equiaxed grains. In addition, FSP can also be used to increase certain material properties such as wear resistance, ductile strength, hardness, formability, corrosion behavior, and fatigue property [171].

Reddy and Rao [172] did a comparative analysis of the FSP method for making composites by surface modification in ZM21 magnesium alloy. Particles such as boron carbide and silicon carbide were placed in the groves manufactured on the top surface of the workpiece. Composite processed with boron carbide gave better mechanical properties as compared to silicon carbide. Boron carbide produced a higher hardness of 140 HV as compared to 100 HV in case of silicon carbide. The base metal hardness was 60 HV. Ma [173] has worked on the application of the FSP method that has been adopted widely in different sectors. It has applications in fabrication, processing, and synthesis of metallic components. FSP offers very excellent potentials for commercial applications due to its environment-friendly nature, efficient technology, and versatility.

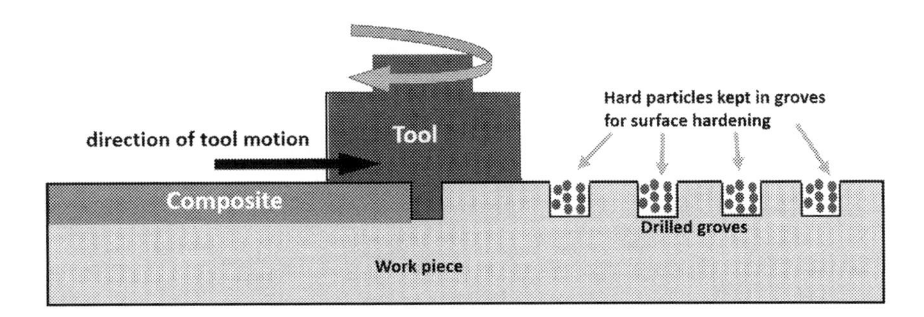

Fig. 2.18 Schematic diagram of Friction stir processing

2.8.4.2 Friction Stir Spot Welding (FSSW)

The FSSW process was developed by Mazda Motor Corporation in 1993 [174] and the schematic is shown in Fig. 2.19. It has the potential to replace welding processes like resistance spot welding and rivet joining. It comprises of three phases: plunging, stirring, and withdraw. In the first stage, a rotating tool is plunged into the top surfaces of the workpiece in lap configuration. Plunging is followed by the stirring stage where the tool continues to revolve in the same location to further raise the temperature. During plunging and stirring stages enough heat is generated to weld the overlapped material.

The joint formed has an unavoidable keyhole resulting in lower mechanical properties. FSSW results in microstructural zones that are similar to the FSW process [174]. It is an efficient and environmentally friendly process adopted by many industries for different applications like aerospace, automobile, and military sector.

Mubiayi and Akinlabi [175] investigated the FSSW of 1060 aluminum alloy with C11000 copper by employing H13 tool steel. Two different tool geometries were used to produce the spot welds, a cylindrical pin with a flat shoulder, and a conical pin with a concave shoulder. Experiments were carried out at two different rotational speeds (800 and 1200 rpm) with a 0.5 mm plunge depth. It resulted in the formation of intermetallic compounds like Al_4Cu_9, $AlCu_3$, Al_2Cu_3, and Al_2Cu. The failure load of 5.2 kN and 4.6kN was obtained for a cylindrical flat pin and conical concave tool, respectively. Wang et al. [176] explained the difficulty that occurs as keyhole while joining light metals. However, to achieve high weld strength in case of refilled FSSW is challenging because of the hooking defect. In this investigation, an innovative technique was suggested to employ graphene nanosheets to reinforce the edge of the hook defect. The spot weld made of aluminum alloy AA2014 with 0.6%wt graphene nanosheets was synthesized using replenished FSSW. The results show an increase in joint strength by 31%. It also resulted in improved fatigue life.

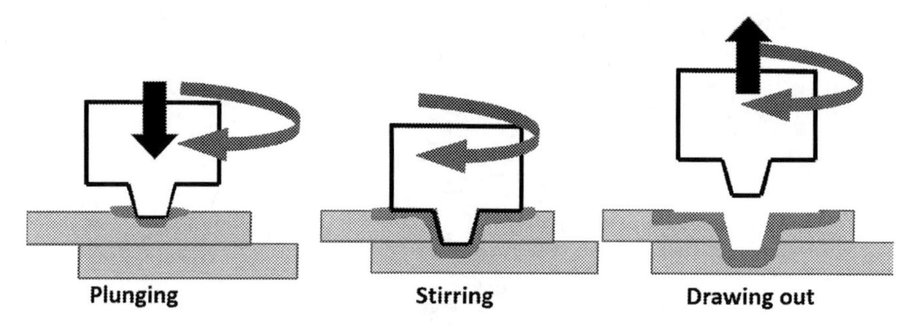

Fig. 2.19 Schematic diagram of a Friction stir spot welding

2.8.4.3 Micro Friction Stir Welding

When welding is performed for the material with a thickness less than a millimeter than it is called micro-FSW. This process can be performed for different materials having a sufficiently lower thickness for similar and dissimilar configuration [177]. Micro-FSW is used for joining micro-mechanical components, electronic components, and micro-electrical components. Optimization of input welding factors i.e. tool rotation speed, welding speed, axial force, and tool tilt angle is also significant to deliver a good and efficient joint [178]. Huang et al. [179, 180] investigated μFSW of 0.5 mm thick Al-6061 at a plunge depth of 0.05 mm. Higher or lower plunging depth leads to various welding defects i.e. unfilled groove, flashes, and reduction in thickness. An increase in rotational speed from 1500 to 2000 resulted in enhanced surface properties and eliminated kissing bond defects within the workpiece. Maximum weld efficiency of 90% was obtained. They also investigated a tool with a concave shoulder and triflat pin for similar configuration. A lower plunge depth of 0.02 mm resulted in a lack of root penetration. This defect starts appearing at the bottom of the joint. The possible reason could be unsatisfactory frictional heat resulting in lowering the thermo-mechanical behavior of the joint. Further increase in plunging depth to 0.08 mm led to flash formation as a result of higher frictional heat.

Papaefthymiou et al. [181] welded ZnTiCu sheets of 700μm by micro-Friction Stir Welding (μFSW). The process temperature reached up to 320°C which equivalent to the $0.75T_m$. Process parameter of 1000 rpm and 318 mm/min provided the refined grain structure at nugget consisting of equiaxed small grains of (0.17 μm) due to dynamic recrystallization as compared to 2 μm of base metal. It differs from columnar large grains formed during the fusion welding. The $TiZn_{15}$ β-phase particles are disintegrated during the μFSW procedure and it shows a reduction in size in the stir zone. Regardless of the effective μFSW joints, three main defects associated with input parameters were reported and they are root opening at the bottom, thinning of the workpiece, and kissing bond.

2.8.4.4 Friction Stir Additive Manufacturing (FSAM)

FSAM process is a non-fusion additive manufacturing process and is a remarkably simple technique. It involves a rotating non-consumable tool designed with shoulder and pin. The tool is plunged into stacked sheet/plates to be joined and consequently navigated lengthwise the weld line. The multi-layered shape is obtained by successively stacking layers one onto the other. Therefore, FSAM is a method of manufacturing parts by the sequential joining of sheet layers to build up an object. Also, the mechanical property associated with this process is better as the FSAM occurs in solid-state. Due to this, it has numerous advantages over the other additive manufacturing process [182].

Palanivel et al. [183] manufactured a multi-layered stack of Mg-based WE43 alloy using FSAM. This multi-layered stack consists of four sheets each having 1.7 mm thickness. Tool steel with a spiral pin along with a concave featureless shoulder was employed for joining with a 1.5° tilt angle. Two different tool rotation speed (800 and 1400 rpm) with a constant welding speed of 102 mm/min was defined. At 1400 rpm, Mg_2Y intermetallic was observed at the interface of the tool and workpiece. However, 800 rpm resulted in the dissolution of precipitates due to increased strain at the non-contacting region. Dynamic recrystallization resulted in a finer grain size of 2–3 μm within the SZ of stacked layers. Such fine grain size along with precipitates resulted in good mechanical properties. The strength of the weld significantly increased to 400 MPa and 17% ductility as compared to base material after subjected to aging.

2.9 Conclusion

FSW is a solid-state welding process and has promising aspects in welding different materials in various configurations. It is also efficient in joining similar as well as dissimilar material. Being a solid-state joining process it offers various advantages such as defects related to solidification is eliminated, higher strength due to finer microstructure, lower energy etc. The selection of the input process parameter is an important aspect of efficient welding. The choice of input parameter should lead to optimum heat generation leading to efficient material flow. FSW finds application in all sectors of industries starting from aerospace, automobile, electronics etc. Though FSW can efficiently weld high strength material, tool wear remains one of the major concerns in that direction. Apart from welding the mechanism of bond formation during FSW has led to evolution in different variants of the process such as friction stir processing, spot welding, solid-state additive manufacturing. All of them provide great versatility in the manufacturing of the component.

References

1. Kah P, Rajan R, Martikainen J, Suoranta R (2015) Investigation of weld defects in friction-stir welding and fusion welding of aluminium alloys. Int J Mech Mater Eng 10(1)
2. Rao SRK, Reddy GM, Rao KS, Kamaraj M, Rao KP (2005) Reasons for superior mechanical and corrosion properties of 2219 aluminum alloy electron beam welds. Mater Charact 55:345–354
3. Pastor M, Zhao H, Martukanitz RP, Debroy T. Porosity, Underfill and magnesium loss during continuous wave Nd :YAG laser welding of thin plates of aluminum alloys 5182 and 5754. Weld Res Suppl 207–216
4. Thomas CJ, Nicholas MW, Needham JC Murch MG, Templesmith P, Dawes CJ (1995) Friction stir butt welding. GB Patent Application 9125978.8, no. US Patent 5460317
5. Nicholas ED (2000) Friction stir welding Da decade on. In: IIW Asian Pacific international congress. Sydney
6. Dawes CJ, Thomas WM (1996) Friction stir welding for aluminum alloys. Weld 3(75):41–45
7. Threadgill PL, Leonard AJ, Shercliff HR, Withers PJ (2009) Friction stir welding of aluminium alloys. Int Mater Rev 54(2):49–93

8. Astarita A, Squillace A, Carrino L (2014) Experimental study of the forces acting on the tool in the friction-stir welding of AA 2024 T3 sheets. J Mater Eng Perform
9. Reynolds AP (2004) Visualisation of material flow in autogenous friction stir welds. Sci Technol Weld Join 5(2):120–124
10. Huang Y, Wang Y, Wan L, Liu H, Shen J (2016) Material-flow behavior during friction-stir welding of 6082-T6 aluminum alloy. Int J Adv Manuf Technol 1115–1123
11. Dickerson T, Shercliff HR (2015) A weld marker technique for flow visualization in FSW. In: 4th International symposium on friction stir welding
12. Colligan K (1999) Material flow behavior during friction stir welding of aluminum. Weld J (Miami, Fla) 78(7):229-s
13. Guerra M, Schmidt C, Mcclure JC, Murr LE, Nunes AC (2003) Flow patterns during friction stir welding. Mater Charact 49:95–101
14. Mishra RS, Mahoney MW, Sato Y, Hovanski Y (2016) Friction stir welding and processing VIII. Frict Stir Weld Process VIII 50:1–300
15. Khodaverdizadeh H, Heidarzadeh A, Saeid T (2013) Effect of tool pin profile on microstructure and mechanical properties of friction stir welded pure copper joints. Mater Des 45:265–270
16. Elangovan K, Balasubramanian V, Valliappan M (2008) Influences of tool pin profile and axial force on the formation of friction stir processing zone in AA6061 aluminium alloy. Int J Adv Manuf Technol 285–295
17. Kumar A, Biswas P (2017) Effect of tool pin profile on the material flow characteristics of AA6061. J Manuf Process 26:382–392
18. Mohanty H, Mahapatra MM, Kumar P (2012) Study on the effect of tool profiles on temperature distribution and material flow characteristics in friction stir welding. Inst Mech Eng 226(9):1527–1535
19. Kumar K, Pancholi V, Bharti RP (2018) Material flow visualization and determination of strain rate during friction stir welding. J Mater Process Tech 255(July 2017):470–476
20. Jain R, Pal SK, Singh SB (2018) Finite element simulation of pin shape influence on material flow, forces in friction stir welding. Int J Adv Manuf Technol 1781–1797
21. Pourahmad P, Abbasi M (2013) Materials flow and phase transformation in friction stir welding of Al 6013/Mg. Trans Nonferrous Met Soc China 23(5):1253–1261
22. Sadeghian B, Taherizadeh A, Atapour M (2018) Simulation of weld morphology during friction stir welding of aluminum-stainless steel joint. Elsevier B.V.
23. Jain R, Pal SK, Singh SB (2016) A study on the variation of forces and temperature in a friction stir welding process: a finite element approach. J Manuf Process 23:278–286
24. Buffa G, Donati L, Fratini L, Tomesani L (2006) Solid state bonding in extrusion and FSW: process mechanics and analogies. J Mater Process Technol 177:344–347
25. CL, Yajie Li ZW, Qin F (2016) Effects of friction stir welding on microstructure and mechanical properties of magnesium alloy Mg–5Al–3Sn. Metals (Basel) 110:266–274
26. Ouyang J, Yarrapareddy E, Kovacevic R (2006) Microstructural evolution in the friction stir welded 6061 aluminum alloy (T6-temper condition) to copper. J Mater Process Technol 172(1):110–122
27. Zhao Y, Lin S, Qu F, Wu L (2006) Influence of pin geometry on material flow in friction stir welding process. Mater Sci Technol
28. Yuqing M, Liming K, Fencheng L, Yuhua C, Li X (2016) Effect of tool pin-tip profiles on material flow and mechanical properties of friction stir welding thick AA7075-T6 alloy joints. Int J Adv Manuf Technol
29. Heidarzadeh A, Jabbari M, Esmaily M (2015) Prediction of grain size and mechanical properties in friction stir welded pure copper joints using a thermal model. Int J Adv Manuf Technol 1819–1829
30. Prangnell PB, Heason CP (2005) Grain structure formation during friction stir welding observed by the 'stop action technique.' Acta Mater 53:3179–3192
31. Nicholas ED (1998) Developments in the friction-stir welding of metals. In: ICAA-6 6th international conference on aluminum alloy. Japan

32. Rai R, Bhadeshia HKDH, Debroy T (2011) Review: friction stir welding tools. Sci Technol Weld Join 16(4):325–342
33. Zhang YN, Cao X, Larose S, Wanjara P (2012) Review of tools for friction stir welding and processing. Can Metall Q 51(3):250–261
34. Mohanty HK, Mahapatra MM, Kumar P, Biswas P, Mandal NR (2012) Effect of tool shoulder and pin probe profiles on friction stirred aluminum welds—A comparative study. J Mar Sci Appl 11:200–207
35. He X, Gu F, Ball A (2014) A review of numerical analysis of friction stir welding. Prog Mater Sci 65:1–66
36. Emamian S, Awang M, Yusof F (2017) A review of friction stir welding pin pro file. Awang M (ed) In: 2nd International conference on mechanical, manufacturing and process plant engineering. Springer, Singapore.
37. Trimble D, Donnell GEO, Monaghan J (2014) Characterisation of tool shape and rotational speed for increased speed during friction stir welding of AA2024-T3. J Manuf Process
38. Liu FC, Ma ZY (2008) Influence of tool dimension and welding parameters on microstructure and mechanical properties of friction-stir-welded 6061-T651 aluminum alloy. Miner Met Mater Soc ASM Int
39. Nandan R, Debroy T, Bhadeshia HKDH (2008) Recent advances in friction-stir welding—Process, weldment structure and properties. Prog Mater Sci 53:980–1023
40. Prado HF, Murr RA, Shindo LE, Soto DJ (2001) Tool wear in the friction stir welding of aluminium alloy 6061+20% Al_2O_3: a preliminary study. Scr Mater 75–80
41. Sahlot P, Jha K, Dey GK, Arora A (2017) Quantitative wear analysis of H13 steel tool during friction stir welding of Cu-0.8%Cr-0.1%Zr alloy. Wear 378–379:82–89
42. Thompson B (2011) Tungsten-based tool material development for the friction stir welding of hard metals. Miner Met Mater Soc 1–8
43. Cui L, Zhang C, Liu X, Li DWH (2018) Recent progress in friction stir welding tools used for steels. J Iron Steel Res Int 25(5):477–486
44. Elangovan K, Balasubramanian V (2007) Influences of pin profile and rotational speed of the tool on the formation of friction stir processing zone in AA2219 aluminium alloy. Mater Sci Eng A 459:7–18
45. Elangovan K, Balasubramanian V (2007) Influences of tool pin profile and welding speed on the formation of friction stir processing zone in AA2219 aluminium alloy. J Mater Process Technol 163–175
46. Soundararajan V, Zekovic S, Kovacevic R (2005) Thermo-mechanical model with adaptive boundary conditions for friction stir welding of Al 6061. Int J Mach Tools Manuf 45:1577–1587
47. Jha K, Kumar S, Nachiket K, Bhanumurthy K, Dey GK (2018) Friction stir welding (FSW) of aged CuCrZr alloy plates. Metall Mater Trans A Phys Metall Mater Sci 49(1):223–234
48. Shen JJ, Liu HJ, Cui F (2010) Effect of welding speed on microstructure and mechanical properties of friction stir welded copper. Mater Des 31(8):3937–3942
49. FX, Youqing Sun RL, He D (2018) Effect of tool rotational speeds on the microstructure and mechanical properties of a dissimilar. Metals (Basel)
50. Devanathan C, Babu AS (2014) Effect of plunge depth on friction stir welding of Al 6063. In: 2nd International conference on advanced materials and modern manufacturing automation. March 2013
51. Ramulu PJ, Narayanan RG (2013) Forming limit investigation of friction stir welded sheets: influence of shoulder diameter and plunge depth. Int J Adv Manuf Technol 2757–2772
52. Zheng Q, Feng X, Shen Y, Huang G, Zhao P (2016) Effect of plunge depth on microstructure and mechanical properties of FSW lap joint between aluminum alloy and nickel-base alloy. J Alloys Compd
53. Seighalani KR, Givi MKB, Nasiri AM, Bahemmat P (2010) Investigations on the effects of the tool material, geometry, and tilt angle on friction stir welding of pure titanium. J Mater Eng Perform 19(October):955–962

54. Kumar K, Kailas SV (2008) The role of friction stir welding tool on material flow and weld formation. Mater Sci Eng A 485:367–374
55. Zhang S, Shi Q, Liu Q, Xie R, Zhang G, Chen G (2018) Effects of tool tilt angle on the in-process heat transfer and mass transfer during friction stir welding. Int J Heat Mass Transf 125:32–42
56. Dialami N, Cervera M, Chiumenti M (2019) Effect of the tool tilt angle on the heat generation and the material flow in friction stir welding. Metals (Basel)
57. Chauhan P, Jain R, Pal SK, Singh SB (2018) Modeling of defects in friction stir welding using coupled Eulerian and Lagrangian method. J Manuf Process 34(November 2017):158–166
58. Mehta KP, Badheka VJ (2014) Materials and manufacturing processes effects of tilt angle on properties of dissimilar friction stir welding copper to aluminum. Mater Manuf Process January 2015:37–41
59. Sorger G, Sarikka T, Vilaça P, Santos TG (2018) Effect of processing temperatures on the properties of a high-strength steel welded by FSW. Weld World 62:1173–1185
60. Yau YH, Hussain A, Lalwani RK, Chan HK, Hakimi N (2013) Temperature distribution study during the friction stir welding process of Al2024-T3 aluminum alloy. Int J Miner Metall Mater 20(8):779–787
61. Tang W, Guo X, McClure JC, Murr LE, Nunes A (1998) Heat input and temperature distribution in friction stir welding. J Mater Process Manuf Sci 7(2):163–172
62. Ramanjaneyulu K, Madhusudhan Reddy G, Venugopal Rao A (2014) Role of tool shoulder diameter in friction stir welding: an analysis of the temperature and plastic deformation of AA 2014 aluminium alloy. Trans Indian Inst Met 67(5):769–780
63. Fehrenbacher A, Schmale JR, Zinn MR, Pfefferl FE (2014) Measurement of tool-workpiece interface temperature distribution in friction stir weiding. J Manuf Sci Eng 136
64. Fehrenbacher A, Duffie NA, Ferrier NJ, Pfefferkorn FE, Zinn MR (2014) Effects of tool—workpiece interface temperature on weld quality and quality improvements through temperature control in friction stir welding. Int J Adv Manuf Technol 165–179
65. Ozturk F, Jarrar F, Evis Z (2016) Thermal history and microstructure during friction stir welding of Al–Mg alloy. Int J Adv Manuf Technol 1071–1081
66. Chao YJ, Qi X, Tang W (2003) Heat transfer in friction stir welding—Experimental and numerical studies. Trans ASME 125(February 2003):138–145
67. Zuo L, Zuo D, Zhu Y, Wang H (2018) Effect of process parameters on surface topography of friction stir welding. Int J Adv Manuf Technol 98(5–8):1807–1816
68. Su H, Wu CS, Pittner A Rethmeier M (2014) Thermal energy generation and distribution in friction stir welding of aluminum alloys. Energy 1–12
69. Buglioni L, Tufaro LN, Svoboda HG (2015) Thermal cycles and residual stresses in FSW of aluminum alloys: experimental measurements and numerical models. Procedia Mater Sci 9:87–96
70. Zybin I, Trukhanov K, Tsarkov A, Kheylo S (2018) Backing plate effect on temperature controlled FSW process. In: MATEC web conference, vol 01084
71. Hwang Y, Kang Z, Chiou Y, Hsu H (2008) Experimental study on temperature distributions within the workpiece during friction stir welding of aluminum alloys. Int. J. Mach. Tools Manuf. 48:778–787
72. Zhu R, Gong W, Cui H (2020) Temperature evolution, microstructure, and properties of friction stir welded ultra-thick 6082 aluminum alloy joints. Int J Adv Manuf Technol 331–343
73. Khandkar MZH, Khan JA, Reynolds AP. Prediction of temperature distribution and thermal history during friction stir welding : input torque based model. Sci Technol Weld Join 165–174.
74. Silva ACF, De Backer J, Bolmsjö G (2017) Temperature measurements during friction stir welding. Int J Adv Manuf Technol 2899–2908
75. Shahi P, Barmouz M, Asadi P (2014) Force and torque in friction stir welding. Adv Frict Stir Weld Process 459–498
76. Trimble D, Monaghan J, O'Donnell GE (2012) Force generation during friction stir welding of AA2024-T3. CIRP Ann-Manuf Technol 61(1):9–12

77. Grujicic M, Ramaswami S, Snipes JS, Avuthu V, Galgalikar R, Zhang Z (2015) Prediction of the grain-microstructure evolution within a friction stir welding (FSW) joint via the use of the Monte Carlo simulation method. J Mater Eng Perform
78. Baghdadi A, Sajuri Z, Kokabi AH (2018) Weldability and mechanical properties of dissimilar Al–MgSi to pure aluminium and Al–Mg using friction stir welding process. J Teknol Weld
79. Ruzek R, Kadlec M (2016) Friction stir welded structures : kissing bond defects Friction stir welded structures : kissing bond defects. Int J Terrasp Sci Eng
80. Al-Moussawi M, Smith AJ (2018) Defects in friction stir welding of steel. Metallogr Microstruct Anal 7(2):194–202
81. Zaman N, Noor A, Khan ZA, Shihab SK (2015) Investigations on tunneling and kissing bond defects in FSW joints for dissimilar aluminum alloys. J Alloys Compd 648:360–367
82. Zhao Y, Han J, Domblesky JP, Yang Z, Li Z (2019) Investigation of void formation in friction stir welding of 7N01 aluminum alloy. J Manuf Process 37(September 2018):139–149
83. Arora KS, Pandey S, Schaper M, Kumar R (2010) Microstructure evolution during friction stir welding of aluminum alloy AA2219. J Mater Sci Technol 26(8):747–753
84. Zhou L et al (2017) Effect of rotation speed on microstructure and mechanical properties of self-reacting friction stir welded Al–Mg–Si alloy. Int J Adv Manuf Technol 89(9–12):3509–3516
85. Zhao S, Bi Q, Wang Y, Shi J (2017) Empirical modeling for the effects of welding factors on tensile properties of bobbin tool friction stir-welded 2219-T87 aluminum alloy. Int J Adv Manuf Technol 90(1–4):1105–1118
86. Schneider J, Chen P, Nunes AC Jr (2019) Entrapped oxide formation in the friction stir weld (FSW) process. Metall Mater Trans A 50(1):257–270
87. Khodaverdizadeh H, Mahmoudi A, Heidarzadeh A, Nazari E (2012) Effect of friction stir welding (FSW) parameters on strain hardening behavior of pure copper joints. Mater Des 35:330–334
88. Tufaro LN, Manzoni I, Svoboda HG (2015) Effect of heat input on AA5052 friction stir welds characteristics. Procedia Mater Sci 8:914–923
89. Buffa G, Fratini L, Shivpuri R (2007) CDRX modelling in friction stir welding of AA7075-T6 aluminum alloy: analytical approaches. J Mater Process Technol 191:356–359
90. Midling O (1998) Friction stir welding aluminium—Process and applications. In: 7th INA conference, Cambridge
91. Polt W. A little friction at Boeing. In: Boeing Front. Online, vol 3, issue 5
92. Shtrikman MM (2008) Current state and development of friction stir welding Part 3. Industrial application of friction stir welding. Weld Int 22(11):806–815
93. Landmann P. Ariane 5 cryogenic tank production. Sci. Photo Libr
94. Talwar JBR, Bolser D, Lederich R (2000) Friction stir welding of airframe structures. In: 2nd international symposium. FSW. Gothenbg
95. First weld of orion exploration Mission-1 crew module. NASA/Radislav Sinyak
96. Gatwick Technologies special joining and forming processes. FSW Mach Distribution Co.
97. Kallee SW, Nicholas ED, Thomas WM. Indusrtialization of friction stir welding for aerospace structures. TWI Ltd., Cambridge, UK
98. Sherherd G (2003) The evaluation of friction stir welded joints on airbus aircraft wing structure. In: 4th International symposium on friction stir welding. Utah (USA)
99. Higgins S, Christner B, McCoury J (2003) Development and testing of friction stir welding as a joining method for primary aircraft structure. In: 4th International symposium on friction stir Welding. Utah, 2003.
100. Paradiso V, Rubino F, Carlone P, Palazzo GS (2017) Magnesium and aluminium alloys dissimilar joining by friction stir welding. Procedia Eng 183:239–244
101. Neil WC, Forsyth M, Howlett PC, Hutchinson CR, Hinton BRW (2009) Corrosion of magnesium alloy ZE41—The role of microstructural features. Corros Sci 51(2):387–394
102. Haghshenas M, Gerlich AP (2018) Joining of automotive sheet materials by friction-based welding methods: a review. Eng Sci Technol Int J 21(1):130–148

103. Research areas in space by ISRO. AI, respond office capacity building program. ISRO HQ, Bengaluru
104. Thomas WM, Kallee SW, Staines DG, Oakley PJ (2006) Friction stir welding—Process variants and developments in the automotive industry (TWI Ltd.) In:SAE world congress. Cobo Center, Detroit, Michigan, USA
105. Toros S, Ozturk F, Kacar I (2008) Review of warm forming of aluminum—magnesium alloys. J. Mater Process Technol 7:1–12
106. Miller WS et al (2000) Recent development in aluminium alloys for the automotive industry. Mater Sci Eng A280
107. Kallee SW. Industrial applications of friction stir welding. Woodhead Publishing Limited
108. Oma S, Midling OT, Kvale JS (2000) Application of prefabricated friction stir welded panels in catamaran building. In: 4th International forum aluminum ships. New Orleans
109. Kallee SW, Davenport J, Nicholas ED (2002) Railway manufacturers implement friction stir welding. Weld 81:47–50
110. Bakar SSSA, Sharif S, Faridh M (2019) Assessment of friction stir welding on aluminium 3D printing materials. Int J Recent Technol Eng 4:10975–10980
111. Ribton CN, Andrews RE (2001) Canister sealing for high level encapsulation TWI Ltd. In: International high-level radioactive waste management conference, Las Vegas, Nevada, USA, 29 Apr–3 May 2001
112. Deqing W, Shuhua LIU (2004) Study of friction stir welding of aluminum. J Mater Sci 9:1689–1693
113. Hou JC, Liu HJ, Zhao YQ (2014) Influences of rotation speed on microstructures and mechanical properties of 6061-T6 aluminum alloy joints fabricated by self-reacting friction stir welding tool. Int J Adv Manuf Technol 1073–1079
114. Liu H, Zhang H, Pan Q, Yu L (2012) Effect of friction stir welding parameters on microstructural characteristics and mechanical properties of 2219-T6 aluminum alloy joints. Int J Mater Form 235–241
115. Marzbanrad J, Akbari M, Asadi P, Safaee S (2014) Characterization of the influence of tool pin profile on microstructural and mechanical properties of friction stir welding
116. Mohammadi-pour M, Khodabandeh A (2016) Microstructure and mechanical properties of joints welded by friction-stir welding in aluminum alloy 7075-T6 plates for aerospace application. Rare Met
117. Jamalian HM, Farahani M (2016) Study on the effects of friction stir welding process parameters on the microstructure and mechanical properties of 5086-H34 aluminum welded joints. Int J Adv Manuf Technol 611–621
118. Saravanan V, Rajakumar S, Banerjee N, Amuthakkannan R (2016) Effect of shoulder diameter to pin diameter ratio on microstructure and mechanical properties of dissimilar friction stir welded AA2024-T6 and AA7075-T6 aluminum alloy joints. Int J Adv Manuf Technol 3637–3645
119. Grujicic M et al (2012) Computational analysis of material flow during friction stir welding of AA5059 aluminum alloys. J Mater Eng Perform 21(September):1824–1840
120. Zhang HWZ, Chen JT, Zhang ZW (2011) Coupled thermo-mechanical model based comparison of friction stir welding processes of AA2024-T3 in different thicknesses. J Mater Sci 5815–5821
121. Buffa G, Hua J, Shivpuri R, Fratini L (2006) Design of the friction stir welding tool using the continuum based FEM model. Mater Sci Eng A 419:381–388
122. Ebrahimi M, Par MA (2019) Twenty-year uninterrupted endeavor of friction stir processing by focusing on copper and its alloys. J Alloys Compd 781:1074–1090
123. Guan W, Shen Y, Yan Y, Guo R, Zhang W (2018) Fabrication of ultra-thin copper foil pressure welding using FSW equipment. J Mater Process Tech 251(February 2017):343–349
124. Lee WB, Jung SB (2004) The joint properties of copper by friction stir welding. Mater Lett 58(6):1041–1046
125. Hwang YM, Fan PL, Lin CH (2010) Experimental study on friction stir welding of copper metals. J Mater Process Technol 210(12):1667–1672

126. Heidarzadeh A, Saeid T (2013) Prediction of mechanical properties in friction stir welds of pure copper. Mater Des 52:1077–1087
127. Sahlot P, Kumar A, Vishvesh S, Amit JB (2019) Friction stir welding of copper: numerical modeling and validation. Trans Indian Inst Met 72(5):1339–1347
128. Singh K, Singh G, Singh H (2018) Review on friction stir welding of magnesium alloys. J Magnes Alloy 000:1–18
129. Pan F, Xu A, Ye J, Tang A, Jiang X, Ran Y (2017) Effects of rotation rate on microstructure and mechanical properties of friction stir-welded Mg–5Al–1Sn magnesium alloy. Int J Adv Manuf Technol 389–397
130. Pareek M et al (2007) Metallurgical evaluation of AZ31B-H24 magnesium alloy friction stir welds. J Mater Eng Perform 16(October):655–662
131. Wang W, Deng D, Mao Z, Tong Y, Ran Y (2017) Influence of tool rotation rates on temperature profiles and mechanical properties of friction stir welded AZ31 magnesium alloy. Int J Adv Manuf Technol 174:2191–2200
132. Forcellese A, Martarelli M, Simoncini M (2016) Effect of process parameters on vertical forces and temperatures developed during friction stir welding of magnesium alloys. Int J Adv Manuf Technol 595–604
133. Mironov S, Sato YS Kokawa H. Influence of welding temperature on material flow during friction stir welding of AZ31 magnesium alloy. Metall Mater Trans A 50(6):2798–2806
134. Asadi P, Kazem M, Givi B Akbari M (2015) Simulation of dynamic recrystallization process during friction stir welding of AZ91 magnesium alloy
135. Richmire S, Sharifi P Haghshenas M (2018) On microstructure, hardness, and fatigue properties of friction stir-welded AM60 cast magnesium alloy. Int J Adv Manuf Technol 2157–2172,
136. Dinda GP, Ramakrishnan A (2019) Friction stir welding of high-strength steel. Int J. Adv Manuf Technol 4763–4769
137. Li S, Yang X (2019) Microstructural characteristics and mechanical properties of friction-stir-welded modified 9Cr–1Mo steel. J Mater Sci 54(8):6632–6650
138. Avinish Tiwari AG, Pankaj P, Biswas P, Kore SD (2019) Tool performance evaluation of friction stir welded shipbuilding grade DH36 steel butt joints. Int J Adv Manuf Technol 1989–2005
139. Kapil Gangwar MR (2017) Friction stir welding of titanium alloys: A review. Mater Des
140. Gianluca Buffa LS, Fratini L, Micari F (2020) On the choice of tool material in friction stir welding of titanium alloys. In: Proceedings of NAMRI/SME, vol 40
141. Sanders DG, Edwards P, Cantrell AM Gangwar K (2015) Friction stir-welded titanium alloy Ti–6Al–4V : microstructure, mechanical and fracture properties. Miner Met Mater Soc 67(5)
142. Lauro A (2012) Friction stir welding of titanium alloys. Weld Int December 2014:37–41
143. Sinha VC, Kundu S, Chatterjee S (2016) Microstructure and mechanical properties of similar and dissimilar joints of aluminium alloy and pure copper by friction stir welding. Perspect Sci 8:543–546
144. Li XW, Zhang DT, Qiu C, Zhang W (2012) Microstructure and mechanical properties of dissimilar pure copper/1350 aluminum alloy butt joints by friction stir welding. Trans Nonferrous Met Soc China (English Ed.) 22(6):1298–1306
145. Carlone P, Astarita A, Palazzo GS, Paradiso V, Squillace A (2015) Microstructural aspects in Al–Cu dissimilar joining by FSW. Int J Adv Manuf Technol 1109–1116
146. Bakhtiari F, Ali A, Seyyed S, Mirsalehi E (2018) Dissimilar joining of pure copper to aluminum alloy via friction stir welding. Acta Metall Sin (English Lett) 31(11):1183–1196
147. Akinlabi ET (2012) Effect of shoulder size on weld properties of dissimilar metal friction stir welds. J Mater Eng Perform 21(July):1514–1519
148. Shah LH, Othman NH, Gerlich A (2017) Review of research progress on aluminium–magnesium dissimilar friction stir welding. Sci Technol Weld. Join 1718(August)
149. Vahid Firouzdor SK (2010) Al-to-Mg friction stir welding : effect of material position, travel speed, and rotation speed. Miner Met Mater Soc ASM Int

150. Zhao Y, Jiang S, Yang S, Lu Z, Yan K (2016) Influence of cooling conditions on joint properties and microstructures of aluminum and magnesium dissimilar alloys by friction stir welding. Int J Adv Manuf Technol 673–679
151. Somasekharan LEMAC Characterization of complex, solid-state flow and mixing in the friction-stir welding (FSW) of aluminum alloy 6061-T6 to magnesium alloy AZ91D using color metallography. J Mater Sci 5365–5370
152. Sato YS, Park SHC, Michiuchi M, Kokawa H (2004) Constitutional liquation during dissimilar friction stir welding of Al and Mg alloys. Scr Mater 50:1233–1236
153. Tanaka T, Masayuki Nezu TH, Uchida S (2020) Corrosion mechanism of intermetallic compound formation during the dissimilar friction stir welding of aluminum and steel. J Mater Sci 55(7) 3064–3072
154. Chen T (2009) Process parameters study on FSW joint of dissimilar metals for aluminum–steel. J Mater Sci 2573–2580
155. Derazkola HA, Khodabakhshi F (2019) Underwater submerged dissimilar friction-stir welding of AA5083 aluminum alloy and A441 AISI steel. Int J Adv Manuf Technol 4383–4395
156. Habibnia M, Shakeri M, Nourouzi S (2015) Microstructural and mechanical properties of friction stir welded 5050 Al alloy and 304 stainless steel plates. Int J Adv Manuf Technol 819–829
157. Karakizis PN, Pantelis DI, Dragatogiannis DA, Bougiouri VD, Charitidis CA (2019) Study of friction stir butt welding between thin plates of AA5754 and mild steel for automotive applications. Int J Adv Manuf Technol 3065–3076
158. Lan S, Liu X, Ni J (2016) Microstructural evolution during friction stir welding of dissimilar aluminum alloy to advanced high-strength steel. Int J Adv Manuf Technol 2183–2193
159. Yazdipour A, Heidarzadeh A (2016) Dissimilar butt friction stir welding of Al 5083-H321 and 316L stainless steel alloys. Int J Adv Manuf Technol 3105–3112
160. Ma Z, Li Q, Ma L, Hu W, Xu B (2019) Process parameters optimization of friction stir welding of 6005A–T6 aluminum alloy using taguchi technique. Trans Ind Inst Met 72:1721–1731
161. Arora A, Mehta M, Debroy T (2012) Load bearing capacity of tool pin during friction stir welding. 911–920
162. Sorensen CD, Stahl AL (2007) Experimental measurements of load distributions on friction stir weld pin tools. Miner Met Mater Soc ASM Int 38(June):451–459
163. Cemal Meran OEC (2011) The effects of tool rotation speed and traverse speed on friction stir welding of AISI 304 austenitic stainless steel. Int J Mat Res 102
164. Park SHC, Sato YS, Kokawa H (2009) Boride formation induced by pcBN tool wear in friction-stir-welded stainless steels. Miner Met Mater Soc ASM Int 40(March)
165. Barnes SJ, Bhatti AR, Steuwer A, Johnson R, Altenkirch J (2012) Friction stir welding in HSLA-65 steel : Part I. Influence of weld speed and tool material on microstructural development. Miner Met Mater Soc ASM Int
166. Wang J, Su J, Mishra RS, Xu R, Baumann JA (2014) Tool wear mechanisms in friction stir welding of Ti–6Al–4V alloy. Wear
167. Yusuf A, Iqbal Z, Al-badour FA, Gasem ZM (2018) Mechanical and tribological characterization of AlCrN coated spark plasma sintered W—25% Re–Hfc composite material for FSW tool. Integr Med Res 1–11
168. Adesina AY, Gasem ZM, Al-badour FA (2017) Characterization and evaluation of AlCrN coated FSW tool: a preliminary study. J Manuf Process 25:432–442
169. Siddiquee AN, Pandey S (2014) Experimental investigation on deformation and wear of WC tool during friction stir welding (FSW) of stainless steel. Int J Adv Manuf Technol 479–486
170. Kumar RA, Kumar RGA, Ahamed KA, Alstyn BD, Vignesh V (2019) ScienceDirect review of friction stir processing of aluminium alloys. Mater Today Proc 16:1048–1054
171. Boopathiraja KP, Ramamoorthi R, Vickram VVK, Yuvaraj KP (2020) Characterization and surface modification on composites by friction stir processing—A review. Mater Today Proc 1–
172. Reddy GM, Rao AS (2013) Friction stir processing for enhancement of wear resistance of ZM21 magnesium alloy. Trans Ind Inst Met 66(February):13–24

173. Ma ZY (2008) Friction stir processing technology : a review. Miner Met Mater Soc ASM Int
174. Yang XW, Fu T, Li WY (2014) Friction stir spot welding : a review on joint macro- and microstructure, property, and process modelling. Adv Mater Sci Eng
175. Mubiayi MP, Akinlabi ET (2016) Evolving properties of friction stir spot welds between AA1060 and commercially pure copper C11000. Trans Nonferrous Met Soc China 26(7):1852–1862
176. Wang S et al (2020) Strengthening and toughening mechanisms in refilled friction stir spot welding of AA2014 aluminum alloy reinforced by graphene nanosheets. Mater Des 186:108212
177. Sithole K, Rao VV (2016) Recent developments in micro friction stir welding: a review. Mater Sci Eng
178. Sen M, Shankar S, Chattopadhyaya S (2019) Micro-friction stir welding—A review. Mater Today Proc
179. Huang Y, Meng X, Zhang Y, Cao J, Feng J (2017) Micro friction stir welding of ultra-thin Al-6061 sheets. J Mater Process Tech
180. Huang Y, Meng X, Lv Z, Huang T, Zhang Y (2018) Microstructures and mechanical properties of micro friction stir welding of 6061-T4 aluminum alloy. J Mater Res Technol 1–8
181. Papaefthymiou S, Goulas C, Gavalas E (2014) Micro-friction stir welding of titan zinc sheets. J Mater Process Tech
182. Padhy GK, Wu CS, Gao S (2018) Friction stir based welding and processing technologies - processes, parameters, microstructures and applications: a review. J Mater Sci Technol 34:1–38
183. Palanivel S, Nelaturu P, Glass B, Mishra RS (2015) Friction stir additive manufacturing for high structural performance through microstructural control in an Mg based WE43 alloy. Mater Des 65:934–952

Chapter 3
Modeling of Friction Stir Welding Processes

N. Bhardwaj, R. Ganesh Narayanan, and U. S. Dixit

Abstract A review on the modeling of friction stir welding (FSW) is presented. Modeling of FSW involves two main tasks—(1) modeling of heat transfer and (2) modeling of plastic deformation. A few noteworthy analytical models are discussed; however, these models make several approximations. Numerical modeling of FSW can carry out reasonably accurate simulations of thermal, mechanical and a metallurgical phenomenon. Fundamentals of numerical models are described, and their implementation is illustrated through an example of modeling of friction stir spot welding process using a commercial software, viz. DEFORM-3D. Typical simulation results are presented to demonstrate the capability of modeling in predicting thermal and mechanical response of a FSW process.

Keywords Friction stir welding · Modeling · Finite element method · Analytical modeling · Welding · Simulation

3.1 Introduction

The friction stir welding (FSW), a solid-state welding technique, is a significant development of the 1990s in the field of joining. In this process, the heat generated due to friction and plastic deformation caused by a rigid tool rotating on and beneath the surface of the workpiece is utilized to form the weld. Harmful gases, spatter and arc flash get avoided in this process. It is more energy-efficient and economical compared to conventional fusion welding techniques. The joints also benefit from low thermal distortions and reduced hot cracking. Hot cracking refers to formation of shrinkage cracks during solidification of weld metal. FSW occurs at a temperature lower than the workpiece melting point, thus avoiding problems associated with solidification of metal or excessive thermal expansion at high temperature [57]. It is considered as a sustainable welding technique with increasing applications in automobile, aerospace, shipbuilding and railways industries [12, 45].

N. Bhardwaj · R. Ganesh Narayanan · U. S. Dixit (✉)
Department of Mechanical Engineering, Indian Institute of Technology, Guwahati, India
e-mail: uday@iitg.ac.in

J. P. Davim (ed.), *Welding Technology*, Materials Forming, Machining and Tribology, https://doi.org/10.1007/978-3-030-63986-0_3

Friction stir welding involves combination of several physical phenomena like frictional heat due to tool-work surface interaction, heat due to large plastic deformation, dynamic microstructural evolution and material stirring [40]. Multiple process parameters influence the quality of weld produced. The fine-tuning of these parameters helps in producing defect-free welds by controlling the heat input during welding. Modeling and simulation of the FSW process help to choose the best process parameters without undergoing extensive experimental trials, making the FSW process economical and less time-consuming. Moreover, modeling of the process using various numerical methods and computational tools helps in understanding the process and improving it. Analysis and visualization of material flow, stress and strain evolution, residual stresses after welding and temperature field in the cross-section during welding can be obtained easily by modeling [2]. Modeling does not have the physical limitations of experimental observation,thus, it can be effective in the understanding of the effect of process parameters during welding. There has been a drastic improvement in the modeling of multiphysics problems, including FSW, in recent years. However, the studies are limited, and there is a requirement of further studies in the modeling of FSW process, especially for reducing simulation time. The present article presents a survey on different approaches taken for modeling of FSW and its future scope. A case study of a finite element (FE) simulation of two AA6061 sheets welded using FSW is also presented as an example.

3.2 Friction Stir Welding Process

In 1891, Bevington patented friction welding where frictional heat produced by relative spinning of two axisymmetric parts is utilized to form the joint between the two parts [11]. The modifications over a period of time, led to development of a variant of friction welding, are known as the friction stir welding, which removed the limitation of welding only axisymmetric parts. Thomas [82] patented the FSW process first in 1991. It was mainly developed for butt welding of aluminum alloys that are usually hard to join using conventional welding. The process expanded to other joint configurations (lap joints and T-joints) and also to different materials like magnesium alloys, copper as well as steel and titanium [45]. The commercial applications of FSW in industries have increased in the last two decades owing to its economic and environmental advantages over conventional fusion welding. FSW welds showed higher strength in both bending and tensile tests as compared to MIG and TIG welds of AA5086-H32 [81]. A higher fatigue strength was also observed in FSW joints [28]. A brief overview of the FSW process is presented in the next section to pave the way for discussion on its modeling in detail.

3.2.1 Working Principle

In FSW, the heat required for the weld is derived from interaction of a rigid rotating tool and the workpiece. The tool consists of a shoulder and a pin (also called probe) as shown in Fig. 3.1a. In a typical friction stir welding, the non-consumable rotating tool is brought in contact with two sheets/plates in contact with each other in butt configuration. The tool applies large axial force as it gets plunged in the joint line. In order to restrict the movement of the two base materials during plunging of the rotating tool, they are clamped using strong fixtures. A baseplate placed below the workpiece provides the reaction to large downward axial forces applied by the tool. It also helps in dissipating the heat out of the workpiece and may have significant effect on the weld quality. The FSW process takes place in four stages—(i) plunging, (ii) dwelling, (iii) traverse/welding and (iv) retracting. Plunging involves inserting a rotating tool into the joint line, initiating the frictional heat and plastic deformation. The tool is rotated, without translation, for a few seconds to increase the temperature and plasticize the material in the dwelling stage. Next is the traverse or welding stage in which the tool moves ahead on the joint line while the shoulder encloses the heat softened plasticized workpiece material to form the weld. The final step is retracting the tool out of the workpiece, which leaves a pinhole on the workpiece. Figure 3.1b shows the schematic of an FSW process.

The mechanism of joint formation can be understood by studying the material flow in the weld. The understanding of the material flow can help in minimizing the defects and in having a better control over the process. The workpiece material is assumed to extrude around the pin and get deposited toward the trailing edge of the tool during its forward motion [22, 35, 51, 65]. The stirring of the material takes place in the joint region aided by the rotation of the pin and the shoulder of the tool. By studying the material flow, Reynolds [65] suggested that FSW is a process undergoing in situ extrusion. The base material outside the weld zone, where temperature is much lower, acts as an 'extrusion chamber' and material within gets extruded. FSW joint is not symmetric about its centerline. On the advancing side (AS), the traverse velocity of the rotating tool and tangential velocity are in the same

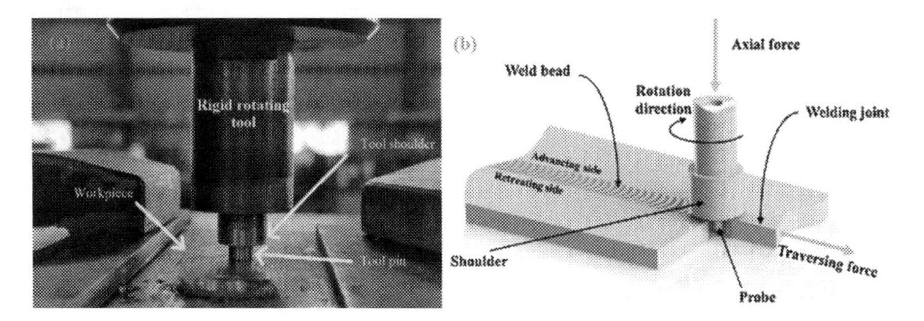

Fig. 3.1 FSW process experiment and schematic: **a** tool resting on the workpiece (with permission from [14], Copyright Elsevier), **b** a schematic (with permission from [29], Copyright Elsevier)

direction. On the retreating side (RS), the traverse velocity and tangential velocity are in opposite directions. As the relative velocity between the tool and workpiece is higher on the advancing side, it shows higher temperature [20, 49, 71]. This results in microstructural differences and difference in physical properties [54]. Thus, the thermal cycle and stresses in various parts of the weld are important aspects, and modeling can help in predicting the quality of weld produced.

3.2.2 Parameters Affecting FSW Process

The important parameters that affect the weld quality in FSW are traverse speed, rotational speed, tool tilt angle, plunge depth and tool offset. Modeling of the FSW process can be helpful in determining the optimum combination of these process parameters for producing quality welds with minimum defects. Rotational speed determines the amount of heat input into the weld and also affects the plasticization of the workpiece material. Traverse speed of the tool determines the peak temperature produced in the weld. Greater the traverse speed, less is the time for heat to dissipate into the workpiece. Plunge depth is an important parameter controlling the depth of the stir zone as well as the forging force on the joint. The forging force is also affected by the tool tilt angle. On the other hand, tool offset determines the heat distribution in the joint, especially in case of dissimilar material welding.

3.2.3 Defects in Welding

The formation of defect-free welds depends on the amount of heat input during FSW. Improper process parameters cause upward movement of material from the lower sheet into the upper sheet, especially in welding of dissimilar materials. This is known as hook defect formed mainly at the thermo-mechanically affected zone (TMAZ). Deficient material consolidation due to low heat input results in tunneling defect left behind the pin as the tool moves forward. Low heat input resulting in ineffective plastic deformation also causes kissing bond defect, where there is no metallurgical bond between interfaces. Kissing bonds mainly appear in the stirring zone (SZ). On the other hand, incomplete root penetration is caused due to insufficient tool plunge depth, inappropriate tool design or inadequate length of pin. The understanding of the thermal cycle, stresses and material flow using modeling can greatly help in minimizing these defects.

3.3 Basic Governing Equations of the Process

As discussed earlier, modeling is an important aspect of analysis of a process in order to gain understanding as well as for optimization of the process. Process modeling and optimization help reduce expensive and time-consuming experiments. Two aspects should be modeled in FSW—plastic deformation and heat transfer. This section discusses both the aspects.

3.3.1 *Modeling of Plastic Deformation*

In order to formulate plastic deformation problems, the governing equations used are discussed in this section. When a body undergoes plastic deformation, it retains the deformed configuration after removal of load and cannot go back to its original form. Unlike the elastic region, the stress–strain relationship does not follow a linear relation in plastic region. The material behavior is also different in case of loading and unloading. It is generally observed that for small deformation, a material behaves elastically at the initial stage, and with further deformation, the material behaves plastically. A yield criterion is set to determine the end of elastic behavior and onset of plastic behavior. For continued further plastic deformation after (initial) yield point, additional stress is required suggesting that the yield condition keeps on changing with plastic deformation. This is known as strain hardening. A criterion for subsequent yielding is required for modeling the hardening behavior. An accurate definition of material behavior is required for mathematical modeling of plastic analysis. Mathematically, yield criterion, in the form of a scalar function, is represented as

$$f(\sigma_{ij}) = 0, \tag{3.1}$$

where f is a yield function, and σ_{ij} are Cauchy stress tensor components. Two commonly used yield criteria for representing material behavior in the context of FSW are discussed, viz. von Mises and Tresca yield criteria [26]. Along with yield criteria, hardening associated with plastic deformation is also discussed.

3.3.1.1 The Maximum Distortion Energy Yield Criterion

The maximum distortion energy (or von Mises) yield criterion is represented as

$$f(\sigma_{ij}) \equiv 3J_2 - \sigma_Y^2 = 0, \tag{3.2}$$

where σ_Y is the yield stress, and J_2 is an invariant of the Cauchy stress tensor's deviatoric component [27]. J_2 is defined as

$$J_2 = \frac{1}{2}\sigma'_{ij}\sigma'_{ij}, \tag{3.3}$$

where σ_{ij} is a typical component of a Cauchy stress tensor, and σ'_{ij} is its deviatoric component defined as

$$\sigma'_{ij} = \sigma_{ij} - \frac{1}{3}\sigma_{kk}\delta_{ij}. \tag{3.4}$$

Here, δ_{ij} is Kronecker's delta, whose value is 1, when indices i and j are equal; otherwise, it is zero. In Eqs. (3.3) and (3.4), Einstein's summation convention is used. In terms of principle stresses, σ_1,σ_2 and σ_3, the von Mises yield criterion is represented by

$$(\sigma_1 - \sigma_2)^2 + (\sigma_2 - \sigma_3)^2 + (\sigma_3 - \sigma_1)^2 - 2\sigma_Y^2 = 0. \tag{3.5}$$

3.3.1.2 Tresca Yield Criterion

On the basis of experiments of metal extrusion through different dies, Tresca proposed a yield criterion [26]. According to the criterion, 'yielding starts when the maximum shear stress at a point exceeds a critical value'. At a given point, the maximum shear stress σ_s is expressed as

$$|\sigma_s|_{\max} = \frac{1}{2}\max\{|\sigma_1 - \sigma_2|, |\sigma_2 - \sigma_3|, |\sigma_3 - \sigma_1|\}. \tag{3.6}$$

When principle stresses are used to represent the Tresca yield criterion, the expression becomes

$$[(\sigma_1 - \sigma_2)^2 - \sigma_Y^2][(\sigma_2 - \sigma_3)^2 - \sigma_Y^2][(\sigma_3 - \sigma_1)^2 - \sigma_Y^2] = 0. \tag{3.7}$$

When the yield function is plotted in the 3D space of σ_1,σ_2 and σ_3, it represents the yield surface. The yield surface of the von Mises criterion is a right circular cylinder with axis along the line $\sigma_1 = \sigma_2 = \sigma_3$ and radius $2\sigma_Y/\sqrt{3}$. The yield surface of the Tresca yield criterion can be represented as a hexagonal prism inside the von Mises yield cylinder. Plane $\sigma_1 + \sigma_2 + \sigma_3 = 0$ is a deviatoric plane called π plane; on this plane, hydrostatic part of the stress, i.e., the mean of direct stresses, is zero. The intersection of yield surface with the π plane is a curve called yield locus.

3.3.1.3 Strain Hardening During Plastic Deformation

In solving plastic deformation problems, two strain hardening laws are widely used—isotropic hardening and kinematic hardening. As strain increases after the plastic

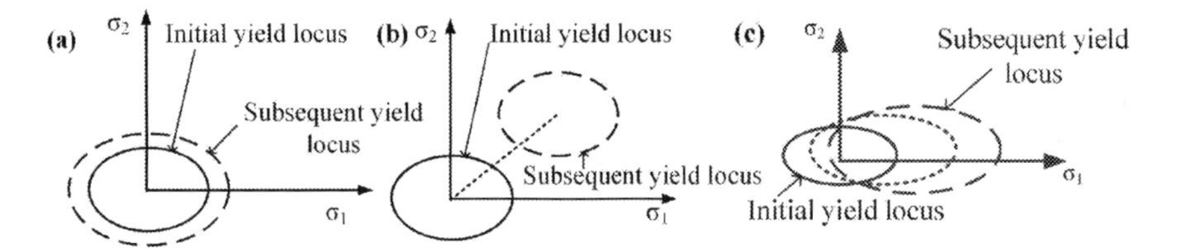

Fig. 3.2 Hardening laws: **a** isotropic hardening, **b** kinematic hardening, **c** combined hardening

state is reached, the size of yield locus increases while its shape remains intact during isotropic hardening as shown in Fig. 3.2a. Isotropic strain hardening in case of uniaxial tension is expressed as

$$\sigma = H(\varepsilon^p), \tag{3.8}$$

where ε^p represents the plastic strain, σ is the true stress, and H represents a scalar hardening function. By using equivalent stress and strain in Eq. (3.8) in place of uniaxial stress and longitudinal strain, respectively, the equation can be used for three-dimensional case as well. Some hardening functions commonly used for modeling plastic deformation are as follows [26]:

1. Holloman's law:

$$\sigma = K(\varepsilon^p)^n, \tag{3.9}$$

2. Swift's law:

$$\sigma = \sigma_Y(1 + K\varepsilon^p)^n, \tag{3.10}$$

3. Ludwik's law:

$$\sigma = \sigma_Y + K(\varepsilon^p)^n, \tag{3.11}$$

4. Ramberg–Osgood equation:

$$\varepsilon^p = \frac{\sigma}{E}\left\{1 + \alpha\left(\frac{\sigma}{\sigma_Y}\right)^{n-1}\right\}, \tag{3.12}$$

5. Prager's law:

$$\sigma = \tanh\left(\frac{E\varepsilon^p}{\sigma_Y}\right), \tag{3.13}$$

where n is a hardening exponent, K is a strength coefficient, E represents Young's modulus of elasticity, and α is another material parameter; the true stress–strain curve of the material is used to determine these constants by fitting the function to the curve.

In case of kinematic hardening, with increase in plastic strain, the yield locus translates in the direction of incremental plastic strain. The kinematic hardening model also includes the Bauschinger effect. For subsequent yielding, yield criterion is represented as

$$f(\sigma_{ij} - \alpha_{ij}) = 0, \tag{3.14}$$

where α_{ij} is called the back stress. It denotes the yield locus's incremental translation while retaining its shape Fig. 3.2b. After implementing kinematic hardening, the yield criterion for further yielding according to the maximum distortion energy yield criterion is expressed as

$$\left(\sigma'_{ij} - \mathrm{d}\alpha_{ij}\right)\left(\sigma'_{ij} - \mathrm{d}\alpha_{ij}\right) - \frac{2}{3}\sigma_Y^2 = 0, \tag{3.15}$$

where $\mathrm{d}\alpha_{ij}$ represents the incremental back stress. According to Prager's hardening law, translation of yield locus occurs in the direction of plastic strain. However, Ziegler refined the hardening model by stating that the yield locus translates along the line joining the stress tensor and the center of the yield locus. Ziegler's hardening law in linear form is given as

$$\mathrm{d}\alpha_{ij} = \frac{H'}{\sigma_Y}\left(\sigma_{ij} - \alpha_{ij}\right)\mathrm{d}\varepsilon^p_{\mathrm{eq}}, \tag{3.16}$$

where H' is the derivative of stress with respect to equivalent plastic strain, $\left(\sigma_{ij} - \alpha_{ij}\right)$ is a vector along the incremental translation, and incremental equivalent plastic strain $\mathrm{d}\varepsilon^{\mathrm{p}}_{\mathrm{eq}}$ is expressed as

$$\mathrm{d}\varepsilon^{\mathrm{p}}_{\mathrm{eq}} = \sqrt{\frac{2}{3}\mathrm{d}\varepsilon^{\mathrm{p}}_{\mathrm{ij}}\mathrm{d}\varepsilon^{\mathrm{p}}_{\mathrm{ij}}}. \tag{3.17}$$

Figure 3.2c shows a schematic of combination of isotropic and kinematic hardening, often called as combined hardening.

Two different approaches are taken when formulating the governing equations of a mathematical model for evaluating stresses, strains, displacement or other parameters. One is the Lagrangian approach, in which the region of analysis keeps changing continuously; the other is the Eulerian approach where analysis is performed in a fixed region of space. Eulerian formulation assumes a fixed space known as the control volume, where the entire analysis takes place. It is assumed that the deforming material flows through the control volume behaving like a non-Newtonian fluid. This approach is also referred to as flow formulation to imply the flowing nature of the deformed material. It is essential to compute the material boundary at each time increment, since material and element boundaries in Eulerian approach do not correspond to each other. Processes like rolling, wire drawing and extrusion that involve flow of material continuously through a control volume can be modeled using the Eulerian approach [26]. The mesh in Eulerian approach never changes,therefore, there is no problem of excessive mesh distortion in processes with high plastic strains [89]. However, this approach is not much suitable for modeling free boundary surfaces. The elemental and boundary nodes may not coincide and can be used only if the surface boundaries that are deformed are known [61]. An impression of the tool shape needs to be created on the workpiece in order to implement this formulation in FSW because plunging stage cannot be modeled using Eulerian formulation. Eulerian formulation is not suitable for unsteady problems as well. Due to these disadvantages and difficulty in tracking surfaces and boundary conditions, a purely Eulerian approach for modeling FSW is rarely used.

In Lagrangian formulation, the analysis is performed on a set of particles, which keep moving with the deformed material. The path of the particle is traced from the original configuration to the material's new deformed state. The primary variable in Lagrangian configuration can be incremental displacement as the deformation takes place incrementally with respect to time. This approach is generally used for processes like forging, deep drawing or processes involving interaction of solid boundaries [26, 41]. It is also easier to use boundary conditions in this approach. In this approach, the material and element boundary coincide with each other [52]. However, with large deformation, there is distortion and tangling of mesh leading to difficulty in convergence. Although Lagrangian approach requires complex remeshing, its ability to model heat from friction at the contacting surface of tool and workpiece, material deformation in the workpiece and material flow makes it a suitable and widely used approach for modeling FSW process [2, 44]. The reference for the current deformed state can be set as the reference frame at $t = 0$ or the deformed configuration from previous state,the former is known as total Lagrangian formulation, and the latter is known as the updated Lagrangian formulation [27]. In the updated Lagrangian formulation, it is assumed that the deformations are finite. It is also assumed that solution at time t (previous state) is known for evaluation of unknown variables for the state at $t + \Delta t$. The following are the governing equations [26]:

1. Incremental strain–displacement relations:

$$ {}_t\Delta\varepsilon_{ij}^L = \begin{cases} \ln({}_t\Delta\lambda_i) & \text{if } i = j \\ 0 & \text{if } i \neq j \end{cases}, \tag{3.18} $$

$$ {}_t\Delta U_{ij}^2 = ({}_t\Delta F)_{ik}^T ({}_t\Delta F)_{kj}, \tag{3.19} $$

$$ {}_t\Delta F_{ij} = \delta_{ij} + {}_t\Delta u_{i,j}, \tag{3.20} $$

where ${}_t\Delta\varepsilon_{ij}^L$ denotes incremental logarithmic strain tensor caused by incremental displacement ${}_t\Delta u$ and ${}_t\Delta u_{i,j}$ represent the derivative of the incremental displacement with respect to position ${}^t x$. The incremental right stretch tensor ${}_t\Delta U$ is derived from the incremental deformation gradient tensor ${}_t\Delta F$ by its polar decomposition. The eigenvalues for the tensor ${}_t\Delta U$ are given by ${}_t\Delta\lambda_i$. The measure of finite deformation as well as rotation F is expressed as

$$F_{ij} = \frac{\partial x_i}{\partial (x_0)_j} = x_{i,j}, \tag{3.21}$$

where x_0 is the position vector at initial configuration. All the time-dependent variables are denoted with the subscript t.

2. The different stress–strain relations:

All materials follow different constitutive relations before and after yielding. The governing equations for most of the metals are given as.

a. *Post yielding*:

$$_t\Delta\sigma_{ij} = \int\limits_{t}^{t+\Delta t} {}^tC^{EP}_{ijkl}\mathrm{d}\left({}_t\Delta\varepsilon^L_{kl}\right) - \int\limits_{t}^{t+\Delta t} {}^tC^{EP}_{ijkl}\alpha\delta_{kl}\mathrm{d}({}_t\Delta T), \tag{3.22}$$

where ${}^tC^{EP}_{ijkl}$ is a fourth-order tensor given by [42]

$${}^tC^{EP}_{ijkl} = \frac{\left(\frac{2v}{1-2v}\right)G\delta_{kl}\delta_{ij} + 2G\delta_{ik}\delta_{jl} - \frac{9G^2}{2}{}^ts'_{ij}{}^ts'_{kl}}{({}^tH' + 3G)^t\sigma^2_{eq}}, \tag{3.23}$$

and ${}^ts'_{ij}$ given by

$${}^ts'_{ij} = \begin{cases} {}^t\sigma'_{ij} & \text{for isotropic hardening} \\ {}^t\sigma'_{ij} - {}^t\alpha'_{ij} & \text{for kinematic hardening} \end{cases}, \tag{3.24}$$

and

$${}^t\sigma_{eq} = H\left({}^t\varepsilon^p_{eq}\right). \tag{3.25}$$

Further, v is the Poisson's ratio, G is the modulus of rigidity, α is the coefficient of thermal expansion, and ${}_t\Delta T$ is temperature rise at time t with respect to ambience.

b. Before yielding and after unloading:

$$_t\Delta\sigma_{ij} = \int\limits_{t}^{t+\Delta t} {}^tC^{E}_{ijkl}\mathrm{d}\left({}_t\Delta\varepsilon^L_{kl}\right) - \int\limits_{t}^{t+\Delta t} {}^tC^{E}_{ijkl}\alpha\delta_{kl}\mathrm{d}({}_t\Delta T), \tag{3.26}$$

where

$$
{}^{t}C_{ijkl}^{E} = \left(\frac{2v}{1-2v}\right) G\delta_{kl}\delta_{ij} + 2G\delta_{ik}\delta_{jl} \tag{3.27}
$$

3. Updating scheme:

It is important to use the stress tensors in objective form to nullify the effect of rigid body rotation. The tensors are made objective by making them invariant with change in reference frame. In case of finite deformation in an updated Lagrangian formulation, the procedure to convert the stress tensor into objective is given as

$$
{}^{t+\Delta t}\sigma = ({}_{t}\Delta R)\left({}^{t}\sigma\right)({}_{t}\Delta R)^{\mathrm{T}} + {}_{t}\Delta\sigma, \tag{3.28}
$$

where ${}_{t}\Delta R$ represents a finite incremental form of rotation tensor.

4. Equilibrium equations:

$$
\frac{\partial^{t+\Delta t}\sigma_{ij}}{\partial^{t+\Delta t}x_{j}} = 0, \tag{3.29}
$$

where ${}^{t+\Delta t}\sigma_{ij}$ and ${}^{t+\Delta t}x_{j}$ are stress tensor components and position vectors, respectively, at updated time.

Both Eulerian and Lagrangian approaches have their own advantages and disadvantages. There have been efforts to combine the advantages of the two formulations in a single approach. Arbitrary Lagrangian–Eulerian (ALE) is one such formulation that takes advantage of the two formulations in the same part mesh. As seen in Fig. 3.3, in case of Lagrangian formulation, the mesh gets distorted after deformation. In case of Eulerian approach, material moves through the mesh as the workpiece gets deformed while the mesh stays fixed in its position. In case of ALE, the mesh adjusts according to the deformed body such that uniformity of mesh is maintained

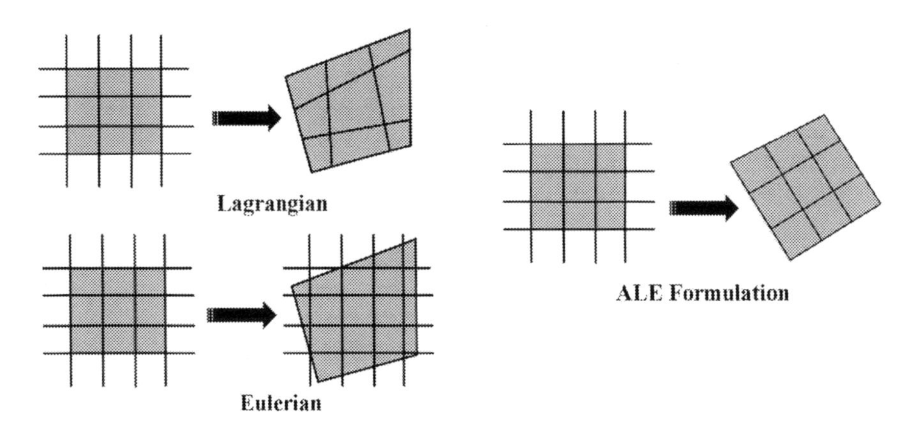

Fig. 3.3 Graphical representation of Lagrangian, Eulerian and ALE formulations (shaded portion is material)

while deformation of the body takes place. In ALE, instead of material configuration, motion is described by the reference configuration in Lagrangian formulation and spatial configuration in Eulerian formulation. It consists of two phases—Lagrangian phase and Eulerian phase, connected by some convective terms. The mesh and material movements are identical in Lagrangian phase, while in the Eulerian phase, the mesh undergoes an arbitrary motion independent of material motion while keeping the mesh undistorted. The approach involves analyzing each time step by Lagrangian phase until convergence, followed by application of Eulerian phase to keep mesh configuration undistorted. All the dependent variables (stress, strain, etc.) are convected through the Eulerian phase since there is relative displacement between material and mesh.

There are three domains in ALE formulation: material Ω_0, spatial Ω and reference or ALE domain $\hat{\Omega}$. The transformation equations mapping the domain are [48]

$$\boldsymbol{x} = \boldsymbol{x}(\boldsymbol{X}, t), \boldsymbol{x} = \boldsymbol{x}(\boldsymbol{\chi}, t), \tag{3.30}$$

which give spatial position of material point $\boldsymbol{X}$ and grid point $\boldsymbol{\chi}$, respectively. The relative motion between the mesh and material defined as convective velocity is

$$\boldsymbol{c} = \boldsymbol{v} - \hat{\boldsymbol{v}}, \tag{3.31}$$

where $\hat{\boldsymbol{v}}$ is the mesh velocity, and $\boldsymbol{v}$ is the material velocity. Similarly, acceleration $\boldsymbol{a}$ is

$$\boldsymbol{a} = \left.\frac{\partial \boldsymbol{v}}{\partial t}\right|_{x} = \left.\frac{\partial \boldsymbol{v}}{\partial t}\right|_{\chi} + (\nabla \boldsymbol{v})\boldsymbol{c} \equiv \left.\frac{\partial \boldsymbol{v}}{\partial t}\right|_{\chi} + (\boldsymbol{c} \bullet \nabla)\boldsymbol{v}, \tag{3.32}$$

where $\left.\frac{\partial v}{\partial t}\right|_{\chi}$ is called the local acceleration, and $(\nabla \boldsymbol{v})\boldsymbol{c}$ is called the convective acceleration; $\nabla \boldsymbol{v}$ is the spatial gradient of velocity. Using $\boldsymbol{a}$ from Eq. (3.32) and substituting in momentum equation in Lagrangian formulation gives

$$\sigma_{ji,j} + \rho b_i = \rho a_i \equiv \rho \left.\frac{\partial v_i}{\partial t}\right|_{\chi} + \rho c_j v_{i,j}, \tag{3.33}$$

where b is the body force per unit mass, σ is the Cauchy stress, and ρ is the density. Assuming that loads are applied slowly and inertia forces are much smaller, acceleration can be omitted which results in the equilibrium equation expressed as

$$\sigma_{ji,j} + \rho b_i = 0. \tag{3.34}$$

Equation (3.34) is common for Lagrangian, Eulerian and ALE formulations as there are no convective terms in the equation.

The material rate of stress in nonlinear solid mechanics is dependent on current state of stress and deformation history. A convective term is added for the relation between material and reference rate of stresses as

$$\left.\frac{\partial \boldsymbol{\sigma}}{\partial t}\right|_{x} = \left.\frac{\partial \boldsymbol{\sigma}}{\partial t}\right|_{\chi} + (\boldsymbol{c} \bullet \nabla)\boldsymbol{\sigma}. \tag{3.35}$$

This approach helps to reduce element distortion in parts undergoing large deformations [52]. Some amount of mesh distortion still occurs in the tool which can be solved by using adaptive remeshing tools [36]. ALE has been used to model temperature distribution, material flow and residual stresses in friction stir welded joints [16, 84].

Another approach that takes advantage of the combination of strengths of Eulerian and Lagrangian approaches is the coupled Eulerian–Lagrangian (CEL) formulation. In CEL approach, the body undergoing large deformation is meshed exclusively with Eulerian element, and the stiffer body is meshed exclusively using Lagrangian elements. In case of FSW, the tool is meshed using Lagrangian elements while the workpiece is meshed using Eulerian elements. The kinematic constraint for Eulerian formulation is implemented by using the velocity of the Lagrangian boundary. On the other hand, surface forces on the Lagrangian domain are calculated using stresses within the Eulerian cell [8, 9, 15]. In addition to the advantages of ALE, CEL approach can simulate void formation and defects in FSW [2]. However, this approach is not widely used since it requires very powerful computational facility [56]. For more details on the CEL approach, the article by Skrzat [77] may be referred. A comparison of the formulations is summarized briefly in Table 3.1.

Table 3.1 Salient points of formulations for modeling FSW

Formulation	Salient point
Eulerian approach	• Assumes fixed control volume for analysis • Capable of simulating material flow • Material and element boundaries do not correspond to each other
Lagrangian approach	• Position of particle under analysis keeps changing with deformation • Can simulate material flow, heat developed due to friction and material deformation • In case of severe deformation, frequent complex remeshing is required
ALE approach	• Combines advantages of Eulerian and Lagrangian approaches • Mesh adjusts according to the deformed body such that uniformity of mesh is maintained • Element distortion is reduced for parts undergoing severe plastic deformation
CEL approach	• Tool is formulated using Lagrangian approach, while the workpiece is formulated using Eulerian approach • Combines advantages of Lagrangian and Eulerian formulations. No problem of mesh distortion • High computation cost

3.3.2 Thermal Modeling

The thermal history of a welding process determines the thermal stresses and microstructure, which leads to determination of the strength, hardness, fatigue behavior and elongation of the joined product. As such, modeling of heat source is an important aspect, and the basics of modeling conduction in moving heat source is briefly described in this section. A detailed section on thermal modeling of FSW process is provided later.

The convective diffusion equation is

$$\frac{\partial \rho u}{\partial t} + \frac{\partial \rho h V}{\partial x} = \nabla \cdot (k \nabla T) + \dot{q}, \tag{3.36}$$

where ρ represents density, k is the thermal conductivity, u is the internal energy, h is the enthalpy, $\dot{q}$ is volumetric heat source, V is the speed of heat source, and T is the temperature. The tool is considered stationary. The velocity of the moving workpiece is in x-direction, y lies on the plane of workpiece top surface perpendicular to x, and z is perpendicular to the workpiece top surface as shown in Fig. 3.4.

If $V = 0$, and $\mathrm{d}u = C_v \mathrm{d}T$, Eq. (3.36) becomes

$$C_v \frac{\partial \rho T}{\partial t} = \nabla \cdot (k \nabla T) + \dot{q}, \tag{3.37}$$

where C_v is the heat capacity. Using the continuity equation, the steady-state equation for constant velocity V can be expressed as [18]

$$\frac{\partial \rho}{\partial t} + \frac{\partial \rho V}{\partial x} = 0. \tag{3.38}$$

Further considering $\mathrm{d}u = \mathrm{d}h = C_v \mathrm{d}T$, as there is no significant volumetric expansion, Eq. (3.37) becomes [86]

$$\rho C_v(T) V \frac{\partial T}{\partial x} = \nabla \cdot \{k(T) \nabla T\} + \dot{q}. \tag{3.39}$$

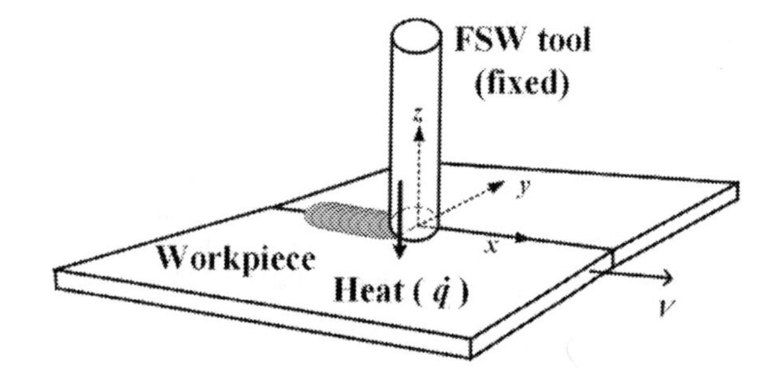

Fig. 3.4 Reference frame for heat transfer during welding

The temperature distribution due to a moving point heat source was given by Carslaw and Jaeger [17] as

$$T = T_0 + \frac{P_L}{4\pi k R}\exp\{-V(R + x)/2\kappa\}, \tag{3.40}$$

where T_0 represents ambient temperature, P_L *is* applied power, R is distance from the source of heat where $\left(R^2 = x^2 + y^2 + z^2\right)$, and κ is the thermal diffusivity. The axis x is considered to be along the heat source velocity, y is on the plane of workpiece top surface and perpendicular to x, and z is also perpendicular to x along the thickness. However, in case of FSW, the equation has to be modified to account for the heat source being non-uniform.

3.4 Analytical Modeling of FSW

Analytical modeling of FSW includes mathematical expressions that help in understanding the process better by arriving at relationship between different parameters (tool dimensions, process parameters, etc.) and related outputs (heat generated, torque produced, etc.) involved during welding. Since FSW is a complex multi-physics problem with transient nature, there have been very few analytical studies on the process. The analytical modeling of FSW has been limited to its thermal aspect only, dealing mainly with heat generation during welding and corresponding temperature distribution. Modeling of material flow, mechanical properties, defects and residual stresses have been undertaken using numerical approach which will be discussed in the next section. Some significant works on analytical modeling of FSW are discussed in this section.

Frigaard et al. [32] were among the first to describe an analytical approach for calculating the heat generated during FSW. The heat generated was expressed as a function of axial load and torque required. The torque applicable on the tool during rotation is given by

$$M = \int_0^R \mu P 2\pi r^2 \mathrm{d}r = \frac{2}{3}\mu\pi P R^3, \tag{3.41}$$

where M is the torque required, P is the axial pressure assumed to be constant, μ is the Coulomb's coefficient of friction between the FSW tool and workpiece, and R denotes tool radius. A differential ring-shaped area of width dr on the circular tool is considered at a distance r from the center of the tool. It is assumed that all the shear work is converted into heat. The average rate of heat generation per unit area is given by

$$q_0 = \int_0^R \omega \mu P 2\pi r^2 \mathrm{d}r, \tag{3.42}$$

where ω is the angular velocity in rad/s. If the tool rotates at N revolutions per second (rps), then $\omega = 2\pi N$ and

$$q_0 = \int_0^R 4\pi^2 \mu P N r^2 \mathrm{d}r = \frac{4}{3}\pi^2 \mu P N R^3, \tag{3.43}$$

Frigaard et al. [32] considered the tool to be a flat shoulder pinless tool, thus ignored the contribution from the pin in heat generation. Schmidt et al. [72] considered the contributions in heat generation from different parts of tool separately. The schematic diagram of the FSW tool with its different parts is shown in Fig. 3.5a. The tool has a shoulder radius of R_{sh}, pin radius of R_{pin}, pin height of H_{pin} and cone angle of α. The cone angle is the angle made by the concave shoulder surface with the horizontal.

The total heat produced by the tool is the summation of contributions of the shoulder portion Q_1, pin side Q_2 and pin bottom Q_3 as

$$Q_{\text{total}} = Q_1 + Q_2 + Q_3. \tag{3.44}$$

The mechanical power input is converted into heat generated at the tool as per the following relation:

$$\mathrm{d}Q = \omega r \tau_{\text{contact}} \mathrm{d}A, \tag{3.45}$$

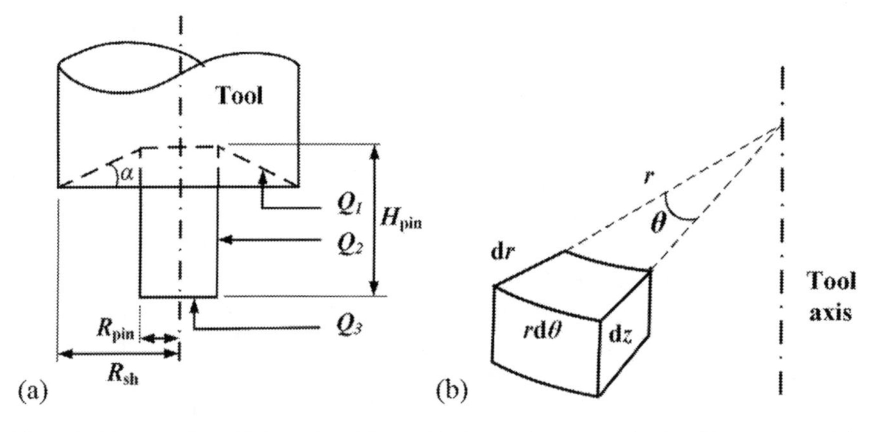

Fig. 3.5 Diagram of an FSW tool (**a**) with contributions to heat generation by different parts of the tool (**b**) infinitesimal element on the tool (drawn based on explanations of [72])

where r is the radius at which a differential area dA exists with contact shear stress of τ_{contact}. The differential area is calculated for each portion by considering an element of height dz, width dr in the radial direction subtending an angle θ at the center and length rdθ in the circumferential direction Fig. 3.5b. Using Eq. (3.45), dQ is integrated over the entire surfaces to arrive at the different heat contributions expressed as

$$Q_1 = \int_0^{2\pi}\int_{R_{\text{pin}}}^{R_{\text{sh}}} \omega\tau_{\text{contact}} r^2(1+\tan\alpha)\mathrm{d}r\mathrm{d}\theta = \frac{2}{3}\pi\tau_{\text{contact}}\omega\left(R_{\text{sh}}^3 - R_{\text{pin}}^3\right)(1+\tan\alpha), \tag{3.46}$$

$$Q_2 = \int_0^{2\pi}\int_0^{H_{\text{pin}}} \omega\tau_{\text{contact}} R_{\text{pin}}^2 \mathrm{d}z\mathrm{d}\theta = 2\pi\tau_{\text{contact}}\omega R_{\text{pin}}^2 H_{\text{pin}}, \tag{3.47}$$

$$Q_3 = \int_0^{2\pi}\int_0^{R_{\text{pin}}} \omega\tau_{\text{contact}} r^2 \mathrm{d}r\mathrm{d}\theta = \frac{2}{3}\pi\tau_{\text{contact}}\omega R_{\text{pin}}^3. \tag{3.48}$$

From Eq. (3.44)

$$Q_{\text{total}} = \frac{2}{3}\pi\tau_{\text{contact}}\omega\left\{\left(R_{\text{sh}}^3 - R_{\text{pin}}^3\right)(1+\tan\alpha) + R_{\text{pin}}^3 + 3R_{\text{pin}}^2 H_{\text{pin}}\right\}. \tag{3.49}$$

For a flat shoulder tool ($\alpha = 0$), which simplifies Eq. (3.49) to arrive at

$$Q_{\text{total}} = \frac{2}{3}\pi\tau_{\text{contact}}\omega\left(R_{\text{sh}}^3 + 3R_{\text{pin}}^2 H_{\text{pin}}\right). \tag{3.50}$$

However, while considering the force and torque on the inclined shoulder surface, Schmidt et al. [72] split up the contributions from projected vertical and horizontal areas of the inclined area. Perhaps a more accurate approach would have been to consider the area of the inclined plane for force and torque calculations. This aspect needs more investigation. Depending on the surface interaction condition, τ_{contact} varies as follows:

(a) *Sticking condition*

By using von Mises criterion,

$$\tau_{\text{contact}} = \tau_{\text{yield}} = \frac{\sigma_{\text{yield}}}{\sqrt{3}}. \tag{3.51}$$

From Eq. (3.49)

$$Q_{\text{total}} = \frac{2}{3}\pi\frac{\sigma_{\text{yield}}}{\sqrt{3}}\omega\left\{\left(R_{\text{sh}}^3 - R_{\text{pin}}^3\right)(1 + \tan\alpha) + R_{\text{pin}}^3 + 3R_{\text{pin}}^2 H_{\text{pin}}\right\}. \tag{3.52}$$

(b) *Sliding condition*

$$\tau_{\text{contact}} = \tau_{\text{friction}} = \mu p, \tag{3.53}$$

where p denotes the axial contact pressure. From Eq. (3.49) using Eq. (3.53)

$$Q_{\text{total}} = \frac{2}{3}\pi\mu p\omega\left\{\left(R_{\text{sh}}^3 - R_{\text{pin}}^3\right)(1 + \tan\alpha) + R_{\text{pin}}^3 + 3R_{\text{pin}}^2 H_{\text{pin}}\right\}. \tag{3.54}$$

Considering a flat tool ($\alpha = 0$) and $p = \frac{F}{A_{\text{projected}}}$, where $A_{\text{projected}} = \pi R_{\text{sh}}^2$ is the projected area where pressure is applicable; Eq. (3.54) can be further simplified as

$$Q_{\text{total}} = \frac{2}{3}\mu F\omega\left(R_{\text{sh}} + 3\frac{R_{\text{pin}}^2 H_{\text{pin}}}{R_{\text{sh}}^2}\right). \tag{3.55}$$

Equation (3.55) is similar to Eq. (3.43) obtained by Frigaard et al. [32]. However, it shows a directly proportional relationship between Q and F, which is experimentally found to be not true in case of FSW. Thus, sliding friction model alone does not give the true nature of the problem.

(c) *Partial Sticking/Sliding condition*

In this condition, the interaction may not be purely sticking or sliding. A variable δ is defined as

$$\delta = \frac{\text{v}_{\text{mat}}}{\text{v}_{\text{tool}}} = 1 - \frac{\dot{\gamma}}{\text{v}_{\text{tool}}}, \tag{3.56}$$

$$\dot{\gamma} = \text{v}_{\text{tool}} - \text{v}_{\text{mat}}, \tag{3.57}$$

where $\dot{\gamma}$ is the slip rate, v_{mat} is the velocity of the material under the tool, and v_{tool} is the position-dependent tangential velocity of the tool. $\delta = 1$ implies sticking condition, and $\delta = 0$ implies sliding condition. $0 < \delta < 1$ implies partial sticking/sliding condition. Combining Eqs. (3.52) and (3.54), we get

$$\begin{aligned} Q_{\text{total}} &= \delta Q_{\text{total,sticking}} + (1 - \delta) Q_{\text{total,sliding}} \\ &= \frac{2}{3}\pi\left(\delta\tau_{\text{yield}} + (1 - \delta)\mu p\right)\omega\left\{\left(R_{\text{sh}}^3 - R_{\text{pin}}^3\right)(1 + \tan\alpha) + R_{\text{pin}}^3 + 3R_{\text{pin}}^2 H_{\text{pin}}\right\}. \end{aligned} \tag{3.58}$$

Mijajlovic and Milčic [55] proposed heat transfer efficiency η_Q as opposed to 100% conversion efficiency considered by Schmidt et al. [72] and gave heat generation as

$$Q_{\text{generated}} = \eta_Q P, \tag{3.59}$$

where P is the mechanical power, and $Q_{\text{generated}}$ is the heat generated by the tool, and the typical value of η_Q varies from 0.6 to 1. A median value of 0.865 for the complete weld cycle and 0.9 for the welding stage was obtained during the study.

Based on Schmidt et al.'s [72] work, Salimi et al. [68] developed a mathematical relation to calculate the temperature distribution in the weld zone. Heat flux generated beneath the shoulder q_1, pin side q_2 and pin bottom surface q_3 were given as

$$q_1 = \omega r \tau_c, \tag{3.60}$$

$$q_2 = \omega R_{\text{pin}} \tau_c, \tag{3.61}$$

$$q_3 = \omega r \tau_c, \tag{3.62}$$

where τ_c represents equivalent shear stress, which depends on the contact condition. The temperature field is described by heat conduction equation. Although Salimi et al. [68] nicely described the heat conduction phenomenon, their expressions for rate of heat generation per unit volume denoted by the function $g(x, y, z, t)$ were mathematically not correct. Here, the correct expressions are provided using Dirac delta function. Dirac delta function converts heat flux into heat per unit volume. Assuming that there are four heat flux sources, the function g is split into four parts, i.e.,

$$g(x, y, z, t) = g_1(x, y, z, t) + g_2(x, y, z, t) + g_3(x, y, z, t) + g_4(x, y, z, t). \tag{3.63}$$

Thus, FSW tool is considered to be a three-dimensional non-uniform moving heat source that supplies has three heat sources; the rate of heat generation by ith sources is $g_i(x, y, z, t)$. Expressions for rate of heat supplied due to rubbing of the shoulder on the top surface of the plates are given as (see Fig.3.6 for hint)

$$g_1(x, y, z, t) = \begin{cases} \omega r \tau_c \delta(z - h) & \begin{array}{l} \left(\bar{x}(t) - \sqrt{R_{\text{sh}}^2 - (y - \bar{y}(t))^2}\right) < x \\ < \left(\bar{x}(t) + \sqrt{R_{\text{sh}}^2 - (y - \bar{y}(t))^2}\right), \\ (\bar{y}(t) - R_{\text{sh}}) < y < (\bar{y}(t) + R_{\text{sh}}); \end{array} \\ 0 & \text{otherwise,} \end{cases} \tag{3.64}$$

where $\delta(\cdot)$ is the delta function of Dirac, and centers of tool are denoted by $\overline{x}(t)$ and $\overline{y}(t)$. However, expression given by Eq. (3.64) gives non-zero heat in the shoulder area where there is pin and consequently no heat generation. To nullify it, negative heat generation is taken at that portion. Thus,

$$g_2(x, y, z, t) = \begin{cases} \omega r \tau_c \delta(z-h) & \left(\bar{x}(t) - \sqrt{R_{\text{pin}}^2 - (y-\bar{y}(t))^2}\right) < x \\ & < \left(\bar{x}(t) + \sqrt{R_{\text{pin}}^2 - (y-\bar{y}(t))^2}\right), \\ & \left(\bar{y}(t) - R_{\text{pin}}\right) < y < \left(\bar{y}(t) + R_{\text{pin}}\right); \\ 0 & \text{otherwise.} \end{cases} \quad (3.65)$$

To take into account the heat due to rubbing of pin bottom surface, the expression $g_3(x, y, z, t)$ is given as

$$g_3(x, y, z, t) = \begin{cases} \omega r \tau_c \delta\left(z - h + H_{\text{pin}}\right) & \left(\bar{x}(t) - \sqrt{R_{\text{pin}}^2 - (y-\bar{y}(t))^2}\right) < x \\ & < \left(\bar{x}(t) + \sqrt{R_{\text{pin}}^2 - (y-\bar{y}(t))^2}\right), \\ & \left(\bar{y}(t) - R_{\text{pin}}\right) < y < \left(\bar{y}(t) + R_{\text{pin}}\right); \\ 0 & \text{otherwise,} \end{cases} \quad (3.66)$$

where H_{pin} is the pin length. Finally, heat generation from cylindrical surface of the pin is accounted for by the following expression:

$$g_4(x, y, z, t) = \begin{cases} \omega r \tau_c \, \delta\left(\sqrt{(x-\overline{x}(t))^2 + (y-\overline{y}(t))^2} - R_{\text{pin}}\right) & \text{for } \left(h - H_{\text{pin}}\right) < z < h, \\ 0 & \text{otherwise.} \end{cases} \quad (3.67)$$

These expressions are for zero plunge depth, i.e., the shoulder just touches the top surface of the plates. If there is some plunge depth, the aforesaid expressions will get modified slightly. With x_0 and y_0 as the initiation points and v as the welding speed

$$\overline{x}(t) = x_0 + \text{v}t, \quad (3.68)$$

$$\overline{y}(t) = y_0. \quad (3.69)$$

The boundary conditions used to solve the heat conduction equation are:

$$T_i = T(x, y, z, 0) = 25\ ^\circ\text{C}, \quad (3.70)$$

$$\frac{\partial T(a, y, z, t)}{\partial x} = \frac{\partial T(0, y, z, t)}{\partial x} = 0, \quad (3.71)$$

$$\frac{\partial T(x, b, z, t)}{\partial y} = \frac{\partial T(x, 0, z, t)}{\partial y} = 0, \quad (3.72)$$

$$k\frac{\partial T(x, y, h, t)}{\partial z} = h_1[T_1 - T(x, y, h, t)], \quad (3.73)$$

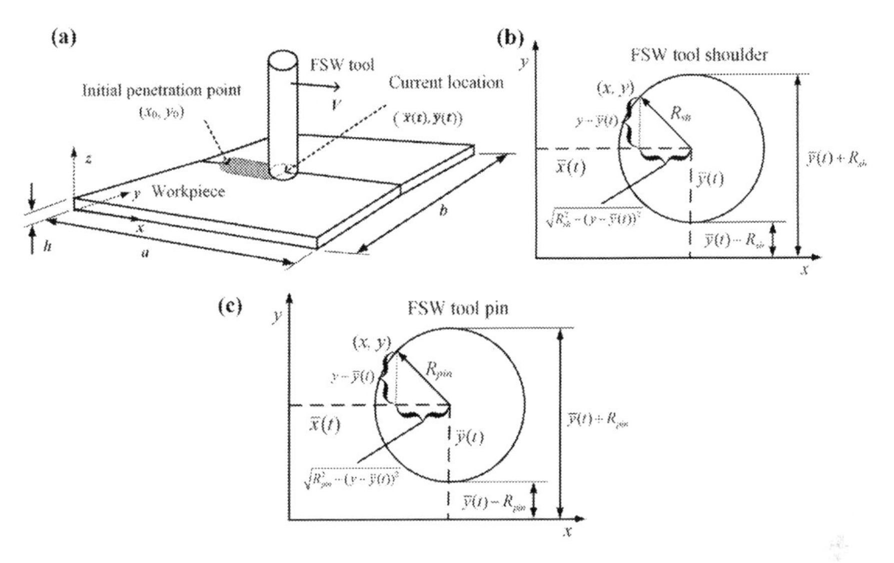

Fig. 3.6 Coordinate system used by Salimi et al. [68] for describing 3D heat flux during FSW **a** initial position and current position of tool, workpiece dimensions, **b** FSW tool shoulder, **c** FSW tool pin

$$k\frac{\partial T(x, y, 0, t)}{\partial z} = -h_0[T_0 - T(x, y, 0, t)], \tag{3.74}$$

where $h_0$is the coefficient of convective heat transfer for heat interaction of the workpiece and the backing plate considering equivalent amount of heat transfer through conduction, T_1 is the ambient temperature, T_0 is the workpiece temperature on the surface, and h_1 is the heat transfer coefficient. On the other hand, T_i is the initial temperature. The length of the workpiece is considered as a, the width is considered as b, and the thickness is considered as h Fig. 3.6a. For the complete solution, the work of Salimi et al. [68] may be referred.

3.5 Numerical Modeling

Numerical modeling of the FSW process is carried out to improve the understanding of the technology. Numerical simulations help in capturing the complex nature of the process by including different aspects like thermal aspects, thermo-mechanical aspects, contact conditions and material behavior in a single problem. The computational models are especially helpful in visualizing material flow, temperature distribution, stress and strain evolution during the FSW process. Most of the numerical approaches use finite element method (FEM) for modeling. Different commercially available codes are used nowadays for numerical simulation which is based on FEM.

A brief introduction to FEM is given in the next section followed by its implementation into FSW modeling for thermal, material flow and mechanical properties analysis. An introduction to cellular automata finite element (CAFE) is also given, which is gaining popularity for predicting microstructure evolution during FSW.

3.5.1 Basics of Finite Element Modeling

A method used to numerically solve differential as well as integral equations by discretizing a body into small finite elements is referred to as finite element method (FEM). The method is useful for numerically solving both ordinary and partial differential equations. The solution is arrived at by assuming a piecewise continuous function and obtaining parameters of the function such that error in the solution reduces. It can be applied to solve physical problems when the governing equations of the physical phenomenon are available. The steps involved in FEM are as follows: (i) pre-processing that involves mesh generation, (ii) obtaining the assembled system of equations, for which the elemental matrices and vectors are to be evaluated, (iii) determining the boundary conditions and applying them, (iv) solving the system of equations and (v) post-processing. Finite difference approximation is taken for solving time-dependent problems by treating the derivative with time [25].

In FEM, the continuum body is discretized into subdivisions called finite elements that are connected to each other via joints called nodes/nodal points. The field variables (e.g., displacement, temperature, stress, etc.) are approximated using a simple function (approximating function or interpolation models). The approximating functions are defined using the nodal field variables. A polynomial form is considered for the solution or interpolation model. The new unknowns in the field equations for the whole continuum will be values of the field variables at the nodes. After solving the set of finite element equations, the nodal field variables will be known which will lead to approximation of the field variables in the whole continuum. Since a complicated physical problem is converted to a simpler problem, only an approximate solution is obtained. However, the approximate solution can be refined and improved by reducing the error by using some computational efforts. The method is explained with the following simple example.

A linear differential equation can be represented as

$$L\boldsymbol{u} + \boldsymbol{q} = 0, \tag{3.75}$$

where $\boldsymbol{u}$ is the vector of primary variables of the problem as a function of the coordinates, $\boldsymbol{q}$ is the vector of known functions, and L represents a differential operator. The differential equation is subjected to natural and essential boundary conditions. The conditions that are sufficient for completely solving a differential equation are called essential boundary conditions. However, the natural boundary conditions contain higher-order derivatives and cannot solve the differential equation completely. For example, considering the differential equation

$$\frac{\mathrm{d}^2}{\mathrm{d}x^2}\left(EI\frac{\mathrm{d}^2\Omega}{\mathrm{d}x^2}\right) - q = 0. \tag{3.76}$$

The differential equation can be completely solved by specifying Ω and $\mathrm{d}\Omega/\mathrm{d}x$ at both ends or $\mathrm{d}^2\Omega/\mathrm{d}x^2$ and/or $\mathrm{d}^3\Omega/\mathrm{d}x^3$ as boundary conditions. However, out of the four boundary conditions, two must be as follows:

i. Ω is specified at both ends.
ii. Ω is specified at one end and $\mathrm{d}\Omega/\mathrm{d}x$ at any end.

Here, Ω and $\mathrm{d}\Omega/\mathrm{d}x$ constitute essential boundary conditions and $\mathrm{d}^2\Omega/\mathrm{d}x^2$ and $\mathrm{d}^3\Omega/\mathrm{d}x^3$ constitute natural boundary conditions.

The variables in FEM are generally approximated using Galerkin formulation and Ritz formulation. In case of Galerkin formulation, the approximation of the primary variable is by using a continuous function inside the element. A residue is obtained depending on the approximating function when primary variable u^e is substituted in Eq. (3.75), i.e.,

$$Lu^e + q = R \tag{3.77}$$

The residue should be zero everywhere; however, it is difficult to approximate the true value. Therefore, the weighted residual is made equal to zero, i.e.,

$$\int_D wR\mathrm{d}A = 0 \tag{3.78}$$

where w is the weight function. In order to weaken the requirement on the differentiability of the approximating function, Eq. (3.78) is integrated by parts to redistribute the order of derivative in w and R. The weight function is of the same form as the approximating function in Galerkin method, which is usually some algebraic function. The unknown coefficients of the function are replaced by unknown nodal degrees of freedom as

$$u^e = [N]\{u^{ne}\} \tag{3.79}$$

where $\{u^{ne}\}$ represents nodal degrees of freedom, and $[N]$ represents the shape functions matrix.

In Ritz formulation, using calculus of variation, the differential equation Eq. (3.75) is converted in its integral form. Equation (3.79) is substituted in the integral equation, and it is partially differentiated with respect to $\{u^{ne}\}$ in order to extremize the form.

Assembly of the elemental equations is performed in the next step. The equations of each element are written in global form and added to each similar equation of all the elements, i.e., all first equations of each element are added to obtain the assembled first equation and so on. After the boundary conditions are applied, the equations are solved by suitable solver. It is followed by post-processing [25].

Physical boundary value problems can be categorized as (i) steady state or equilibrium, (ii) eigenvalue and (iii) transient problems. The steady-state problems are time-independent and involve evaluation of displacement or stress distribution, temperature and heat flux distribution, or velocity and pressure distribution in spatial directions. Eigenvalue problems involve determination of critical values of certain parameters in addition to equilibrium configurations, for example, determination of buckling loads or natural frequencies in structural problems and stability of laminar flows in fluid mechanics. Transient problems include analyzing a body under time-dependent force or heating/cooling of a particular point with respect to time [64].

Different commercial software applications are available for solving FEM problems in engineering like ABAQUS, ANSYS, DEFORM-3D, etc. The software applications perform the FE analysis in three steps: (1) In pre-processing, the geometry of the parts, loads applicable, material properties and boundary conditions are specified. Finite element mesh is generated by an in-built automatic mesh generation module where the type, size and remeshing criteria are specified. (2) In numerical analysis, the stiffness matrices (element characteristics) and load vectors are generated and assembled to generate the system of equations. The specified boundary conditions are implemented, and the nodal field variables (e.g., displacement) are obtained by solving the equations to compute the element resultants (e.g., stress and strains). (3) In post-processing, the solutions are displayed either in tabular form or graphically. It is extremely important to verify the results obtained from FE analysis from the software applications by comparing them with known solutions.

3.5.2 *Thermal Modeling*

Thermal model governs the mechanical and microstructure models. Understanding of the heat generation is vital to understanding the process and its related phenomena. Material flow, which depends on flow stress, is dependent on the thermal cycle during welding. One of the first thermal models is given by Chao and Qi [19]. It was a three-dimensional heat transfer model that assumed sliding friction only to generate a constant heat flux with Coulomb's law for friction. Pressure was assumed to be constant. However, the study only used heat generated by the tool shoulder. Frigaard et al. [32] improved the model by incorporating heat generated at the interfaces of FSW tool shoulder-workpiece and the pin-workpiece. Smith et al. [78] and Bendzsak et al. [10] used finite difference method for heat transfer modeling as well as material flow in FSW assuming workpiece material to be a non-Newtonian fluid. Gould and Feng [34] developed a model using the Rosenthal equation for heat transfer during FSW. Russell and Shercliff [67] also used the Rosenthal equation to incorporate heat input as a point source or line source. Khandkar and Khan [46] and Khandkar et al. [47] proposed a torque-based heat input model. The heat generated is from contact friction and plastic deformation [13]. Bhardwaj et al. [13] used different lubricants during FSW to reduce the friction at the tool-workpiece interface and analyzed the heat contribution from friction and plastic deformation using results from both experiments and simulations. The process was modeled using DEFORM-3D. The modified steady-state energy conservation equation used by Nandan et al. [58] is expressed as

$$\rho C_p \frac{\partial(u_i T)}{\partial x_i} = -\rho C_p U_1 \frac{\partial T}{\partial x_1} + \frac{\partial}{\partial x_i}\left(k \frac{\partial T}{\partial x_i}\right) + Q_i + Q_b, \quad (3.80)$$

where u_i is the velocity of plastic flow, U_1 is the tool velocity, Q_i is the heat generation rate per unit volume due to frictional heat, and Q_b is the heat generation rate per unit volume contributed by plastic deformation. To compute the temperature distribution, the heat conduction equation in matrix form is

$$[C(t)]\{\dot{T}\} + [K(t)]\{T\} = \{Q(t)\}, \quad (3.81)$$

where $[K(t)]$ is conductivity matrix, $[C(t)]$ is capacitance matrix, and $\{Q(t)\}$ is heat vector, which are time-dependent. $\dot{T}$ is the time derivative of nodal temperature vector T. Solving for $\dot{T}$ yields.

$$\{\dot{T}\}_i = [C]^{-1}(\{Q\} - [K]\{T\}_i). \quad (3.82)$$

For nodal temperature rate, using forward difference integration gives.

$$\{\dot{T}\}_i = \frac{\{T\}_{i+1} - \{T\}_i}{\Delta t_{i+1}}, \quad (3.83)$$

Rearranging Eq. (3.83)

$$\{T\}_{i+1} = (\Delta t_{i+1})\{T\}_i + \{T\}_i. \quad (3.84)$$

Substituting Eq. (3.82) in Eq. (3.84) gives the nodal temperature as

$$\{T\}_{i+1} = (\Delta t_{i+1})[C]^{-1}([Q] - [K]\{T\}_i) + \{T\}_i. \quad (3.85)$$

The contact conditions play a crucial role in thermal modeling of FSW process. However, there are only a few dedicated researches on friction in FSW. A sliding friction model using Coulomb's law was used by Chao and Qi [19] to model the contact condition by a trial and error technique to obtain close match between experimental and computed temperatures. A constant Coulomb's coefficient of friction ($\mu = 0.4$) was used by Frigaard et al. [32]. Coulomb's friction law is widely used, which gives frictional shear stress τ as

$$\tau = \mu p, \quad (3.86)$$

where p represents axial contact pressure, and μ represents coefficient of friction. Schmidt et al. [72] defined a term called contact state variable, which is the ratio of velocity of workpiece material to velocity of its contact point on the tool. Based on the contact state variable, the contact condition at the interface may be (i) sliding, (ii) sticking or (iii) combination of both sticking and sliding. Another widely used friction model is the constant shear (Tresca) friction model given as

$$\tau = mk, \tag{3.87}$$

where $k = \frac{\sigma_y}{\sqrt{3}}$ with σ_y is the yield stress of a material following von Mises yield criteria, and m is the friction factor. Apart from these two laws, Norton law of friction is also used by a few researchers [31, 37, 38]. It is expressed as

$$\tau = -\mu K|V|^{q-1}V, \tag{3.88}$$

where K is the consistency, and V is the relative velocity [52]. Bhardwaj et al. [14] gave a comprehensive method to inversely evaluate the contact condition during friction stir spot welding (FSSW). The contact condition (friction factor and coefficient of friction) at the interface of tool and workpiece was evaluated by adjusting these parameters to achieve close match between computed and experimentally obtain torque. A temperature-dependent hybrid friction model was proposed which incorporated both Coulomb's friction and Tresca friction models Fig. 3.7. The details of modeling by Bhardwaj et al. [14] are discussed separately as an example with typical results derived from the model in Sect. 3.6.

Heat dissipation can occur in three ways: (a) heat dissipated into the tool, (b) heat dissipated to the backing plate and (c) heat loss to the atmosphere. Loss of heat to the tool is very insignificant, which can be estimated by considering temperatures at two points on the tool axis [73]. A convective heat transfer coefficient is assigned to account for equivalent heat transfer through conduction during experiment. The heat dissipation Q_{wt} to tool is given as.

$$Q_{\mathrm{wt}} = K_{\mathrm{wt}}\frac{\partial T}{\partial z} = h_{\mathrm{t}}(T - T_{\mathrm{t}}), \tag{3.89}$$

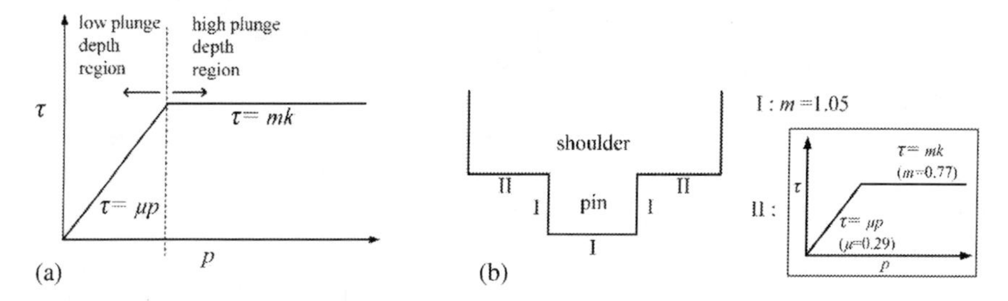

Fig. 3.7 Temperature-dependent hybrid friction model **a** incorporation of Coulomb's friction at low plunge depth and constant shear friction at high plunge depth, **b** different friction conditions at tool pin and shoulder [13], under a Creative Commons license)

where $K_{wt}$and h_t are the conductance and convective heat transfer coefficient, T_t is the tool temperature, and T is the workpiece temperature. The heat dissipation from the top surface to the environment consists of both convective and radiation losses expressed as.

$$Q_{wa} = \sigma_b \varepsilon_b \left(T^4 - T_a^4\right) + h_b(T - T_a), \quad (3.90)$$

where Q_{wa} is heat loss through top surface to surrounding, σ_b is the Stefan–Boltzmann constant, h_b is the heat transfer coefficient at the top surface, $T_a$is the ambient temperature, and ε_b is the emissivity. The heat loss at the workpiece bottom surface and the backing plate Q_{wb} is modeled by considering a suitable convective heat transfer coefficient with equivalent conductive heat transfer during experiment. The heat dissipation to backing plate is expressed as

$$Q_{wb} = K_{wb} \frac{\partial T}{\partial z} = h_b(T - T_b), \quad (3.91)$$

where $K_{wb}$and h_b are the conductance and convective heat transfer coefficient for the heat interaction of workpiece and the backing plate, and T_b is the temperature of the backing plate.

3.5.3 *Material Flow Modeling*

In order to obtain high structural efficiency of the welds and for optimal tool design, it is necessary to understand the material flow. Computational fluid dynamics (CFD) using commercial software Fluent was employed by Colegrove and Shercliff [21] to model the material flow in FSW. They developed a 'slip' model where the local shear stresses govern the interface conditions. It was assumed that the shear stress experienced by the workpiece is different for different tool materials. The surface shear stress was assigned maximum limiting value. Two different boundary conditions were used: (a) stick condition, where the shear stress was lower than the limiting shear stress, and (b) slip condition, where for stick condition to exist, the shear stress necessary was more than the limiting shear stress,however, the applied shear stress was limited to the maximum shear stress value, thus resulting in slipping at the interface. A three-dimensional thermo-mechanical FE model was developed by Jain et al. [44] to predict the material flow and forces in FSW. The model used Lagrangian formulation with a built-in feature of DEFORM-3D called point tracking tool to analyze the velocity and material flow as shown in Fig. 3.8.

Sellars-Tegart law is also used to model steady-state flow stress. Sellars-Tegart law [74] considers the material to be incompressible viscous non-Newtonian fluid. It puts forward a relation between temperature T and rate of deformation $\dot{\varepsilon}$ by using the Zener-Hollomon parameter as

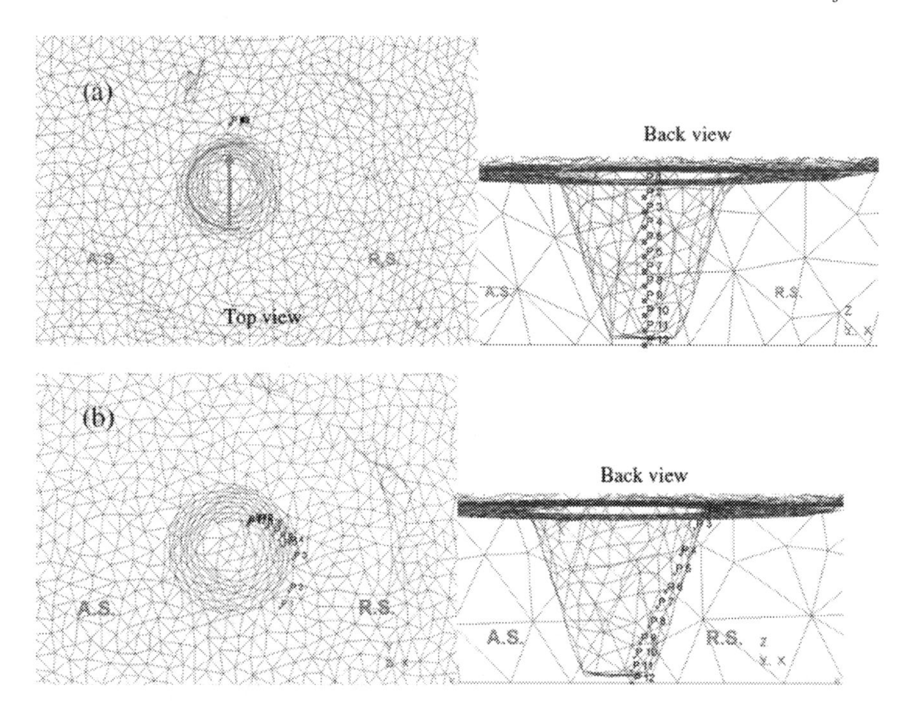

Fig. 3.8 Use of point tracker tool to visualize material flow in DEFORM-3D (with permission from [44]. Copyright Springer Nature)

$$Z = \dot{\varepsilon}\exp\left(\frac{Q}{RT}\right) = A(\sinh\alpha\sigma)^n, \tag{3.92}$$

where R is the gas constant, Q is the activation energy, and α, A and n are material parameters [2]. Some researchers used coupled Eulerian and Lagrangian (CEL) to developed three-dimensional model for predicting voids [2]. CEL was used by Tongne et al. [83] to study the formation of alternate bands of light and dark rings during FSW. The study also analyzed defects in FSW with particular interest in correlation of kissing bonds with banded structures.

3.5.4 Evolution of Mechanical Properties

The thermal cycle and plastic deformation during welding result in alteration of mechanical properties in welded material. Numerical modeling of mechanical properties essentially consists of a thermal model for the thermal cycle and an elastic–plastic mechanical model with temperature-dependent yield stress. The thermal model is discussed already in Sect. 5.2. The differential equation of motion in the analysis of mechanical responses is expressed as

$$\rho \ddot{u} + c_b \dot{u} + k_s u = F_b, \tag{3.93}$$

where c_b represents damping coefficient, k_s represents stiffness coefficient, ρ is the density, and F_b denotes body force. Further, u represents nodal displacement, $\dot{u}$ denotes nodal velocity, and $\ddot{u}$ denotes nodal acceleration [6]. It can be expressed as

$$[M]\{\ddot{u}\} + [C_d]\{\dot{u}\} + [K_s]\{u\} = \{F\}, \tag{3.94}$$

where $[M]$ represents discrete mass matrix, and $[K_s]$ and $[C_b]$ are stiffness and viscous damping matrices, respectively. $\{F\}$ is external force vector. The initial acceleration can be expressed as

$$\{\ddot{u}\}_i = [M]^{-1}(\{F\} - [C_d]\{\dot{u}\}_i - [K_s]\{u\}_i), \tag{3.95}$$

where i is the time step. Using the central difference scheme to discretize the control equation, the acceleration can be expressed as

$$\{\ddot{u}\}_i = \frac{\{\dot{u}\}_{i} + 1/2 - \{\dot{u}\}_{i} - 1/2}{(\Delta t_{i+1} + \Delta t_i)/2}. \tag{3.96}$$

Rearranging Eq. (3.96) gives

$$\{\dot{u}\}_{i+1/2} = \left(\frac{\Delta t_{i+1} + \Delta t_i}{2}\right)\{\ddot{u}\}_i + \{\dot{u}\}_{i-1/2} \tag{3.97}$$

Finally, replacing Eq. (3.95) in Eq. (3.97) gives the nodal velocity as

$$\{\dot{u}\}_{i+1/2} = \left(\frac{\Delta t_{i+1} + \Delta t_i}{2}\right)[M]^{-1}(\{F\} - [C_d]\{\dot{u}\}_i - [K_s]\{u\}_i) + \{\dot{u}\}_{i-1/2} \tag{3.98}$$

The evolution of hardness during FSW is modeled by the Myhr and Grong model [30, 66, 67]. It is applicable to materials with hardening precipitates. The hardness is related to volume fraction of precipitates using kinetics of precipitate dissolution. The dissolved hardening precipitate fraction X_d is related to equivalent heat treatment time $t_{eq} = t/t*$, where the sample is at temperature T for time t, while t^* is the time required for complete dissolution of precipitates at that temperature as

$$\begin{aligned} X_d &= t_{eq}^n, \\ t_{eq} &= \sum_{i=1}^{N_{total}} \frac{\Delta t_i}{t_i^*} = \sum_{i=1}^{N_{total}} \frac{\Delta t_i}{t_{ref} \exp\left[\frac{Q_{eff}}{R}\left(\frac{1}{T_i} - \frac{1}{T_{ref}}\right)\right]}, \end{aligned} \tag{3.99}$$

where n is a material constant obtained experimentally, t_{ref} is the time for complete dissolution of precipitates at reference temperature T_{ref}, Q_{eff} is the activation energy for precipitate dissolution, and the gas constant is denoted by R. The thermal history of the process is discretized into N_{total} small steps where equivalent time for each step is calculated followed by summation of all the time steps to arrive at the total equivalent time. The fraction of hardening precipitates f/f_0 is expressed as.

$$\frac{f}{f_0} = 1 - X_d = 1 - t_{eq}^n, \tag{3.100}$$

where the value of n is usually considered as 0.5 [39]. Linear interpolation between original and fully dissolved state is used to predict the final hardness distribution as

$$HV = (HV_{max} - HV_{min})\frac{f}{f_0} + HV_{min}, \tag{3.101}$$

where HV_{max} and HV_{min} are hardness of the material at fully hardened and softened (original) conditions, respectively. The modeling of mechanical response and hardness accompanied by the thermal modeling was used for prediction of residual stresses by Feng et al. [30]. The prediction of residual stresses improved significantly with inclusion of metallurgical model under hot welding conditions [2].

Apart from the thermal and elasto-plastic mechanical models, a model to analyzing the microstructure of the welded material is also important. Coupling of metallurgical model improves the efficiency of the model at predicting the evolution of mechanical properties. The next section discusses modeling of microstructure during FSW.

3.5.5 *Cellular Automata Modeling of FSW*

A cellular automata (CA) model uses an algorithm to represent discrete spatial and/or temporal evolution of a complex system by application of deterministic/probabilistic transformation rules to the lattice locations [62]. Cellular automata (CA) models are generally used to relate the initial microstructure to evolving microstructure during materials processing techniques such as rolling, extrusion, welding and casting solidification. It is a collection of 'colored' cells on a grid of specified shape that evolves through a number of discrete time steps according to a set of rules based on the states of neighboring cells. The rules are applied iteratively for as many time steps as required. Figure 3.9 shows a schematic of the working of CA with a set of simple rules as an example. Often, CA is combined with finite element method (FEM) to provide robust simulation methodology called the cellular automata finite element (CAFE) method. In this method, CA cells containing properties of micro-features (like dislocation density, initial grain size) are connected to the integration point of elements that are once again related to macro-, micro-outputs (like stress, strain, etc.).

For several years, the applications of CAFE models include modeling mixed microstructures [23], evolution of microstructure during rolling [90], effect of

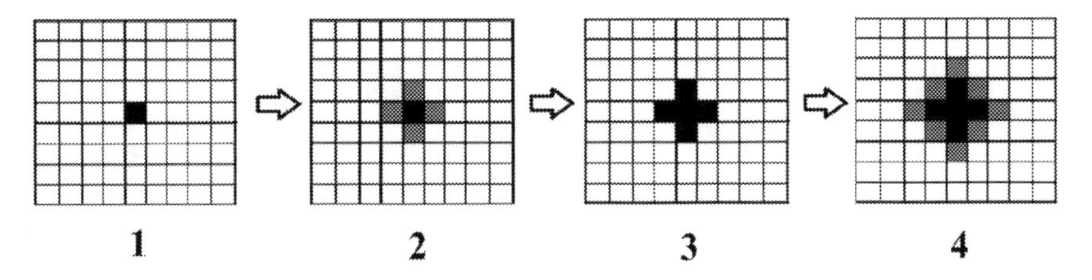

Fig. 3.9 Schematic of working principle of cellular automation with four iterations, assuming the simple rules: (i) every white cell adjacent to a black cell turns gray (ii) all gray cells adjacent to black cells turn black

microstructure on Nb-microalloyed rolled sheet properties [87] and similar studies. Specifically, in materials processing, the notable works are from Gandin and Rappaz [33], Davies [24], Lan et al. [50], Shterenlikht and Howard [76] and Madej et al. [53].

In recent times, CA/CAFE models are applied to FSW for understanding the significance of process parameters on the microstructures in the weld zone. It also helps in designing and optimizing the FSW process and selection of tools to minimize FSW defects and for reducing experimental time and associated costs. For the first time, Saluja et al. [69] used ABAQUS to develop a CAFE model to predict the size of grains during FSW of AA6061-T6 thin sheets. In this attempt, the stirring of materials that takes place physically in real world during FSW is not simulated. Instead, the strain, heat flux and strain-rate developed during the process are calculated using existing analytical equations, which are subsequently provided as inputs to FE and CA cells via user sub-routine. To predict the final size of the grains, a group of CA cells with initial grain size that are distributed following Gaussian distribution are created in three-dimensional fashion. The final grain size (D_{CA}) during FSW has been obtained using the transition rule,

$$D_{\mathrm{CA}} = C\varepsilon^{x}\dot{\varepsilon}^{y}d^{z}\exp\left(\frac{-Q}{RT}\right), \tag{3.102}$$

where ε,$\dot{\varepsilon}$, T are strain, rate of deformation and temperature during FSW, Q is continuous recrystallization activation energy, d is the initial grain size, and R is gas constant. x, y, z are material constants evaluated by error minimization. Later, the tensile behavior of FSW sheet has been predicted by considering the flow stress evolution at element level and at CA cell level.

The CAFE model has been utilized to develop an artificial neural network (ANN) model for predicting the yield strength and grain size during FSW [60]. CAFE model generated yield strength and grain size as a function of FSW parameters, and the outputs are trained by neural network. Valvi et al. [85] extended the CA model to predict the dislocation density (ρ) in the FS zone from grain size (D_{CA}) by using (i) $\rho = \left(\frac{k}{\alpha Eb}\right)^{2} D_{\mathrm{CA}}^{-1}$, where k is a constant, b is the burgers vector, E denotes the elastic modulus, and (ii) fourth-order polynomial equation relating it with D_{CA}. To predict the flow stress evolution during tensile deformation of FSW sheets, four different methods were also designed using the CAFE model. Tool selection with various pin profiles for single- and double-side FSW of sheets was done by Rajpoot et al. [63] using the CAFE method. In this method, grain size is used as output and optimized for different pin profiles to select the appropriate one,however, actually the pin profile used was a different one. The influence of single-side FSW and double-side FSW on grain size distribution was also analyzed with predictions from CAFE model.

There have been several other researches to predict the evolution of microstructure during FSW and other FS variants of various sheet grades. For instance, Shojaeefard et al. [75] optimized the mechanical properties and grain size of FS welds made of AA1100 using Taguchi method, and the microstructure model was built using CA model in DEFORM (a commercial FEM package) environment. They predicted the dislocation density using modified Laasraoui–Jonas (LJ) model in combination with CA model to optimize rotational speed, transverse speed and tilt angle during FSW. LJ model is a dislocation density model used to predict flow stresses and evolution of dislocation densities during a hot working process. The effect of grain boundary migration on dislocation density is also considered in the modified LJ model. The dislocation density given by the modified LJ method is expressed as [2].

$$\mathrm{d}\rho_i = (h - r\rho_i)\mathrm{d}\varepsilon - \rho_i \mathrm{d}V, \tag{3.103}$$

where ρ_i represents the dislocation density for the ith grain; $\mathrm{d}V$ denotes volume swept by the boundaries; ε is the strain; average strain hardening denoted by a parameter h and recovery coefficient r. Akbari et al. [1] established a FE model in combination with CA, LJ and Kocks–Mecking (KM) models to predict the microstructure evolution (such as nucleation and grain growth) during dynamic recrystallization of FSW of AZ91 in DEFORM environment. Besides evaluating the microstructures, the model predicted macro-outputs such as strain, temperature as a function of sheet thickness and rotation and traverse speeds. Asadi et al. [4, 5] made similar attempts. The microstructure evolution by DRX (dynamic recrystallization) in the friction stir blind riveting process [70], prediction of microstructure during friction stir extrusion process [7] and CA model development for FSW of titanium alloy [79] are other notable contributions.

3.6 An Example

Modeling of FSSW carried out by Bhardwaj et al. [14] is described in this section as an example. The FSSW of AA6061 sheet using H13 tool was modeled using a fully coupled temperature-displacement finite element model in DEFORM-3D owing to its efficient remeshing capabilities. Temperature and displacement were calculated at the same time for each node. The formulation used was Lagrangian implicit, and to account for severe plastic deformation and mesh distortion, adaptive remeshing was

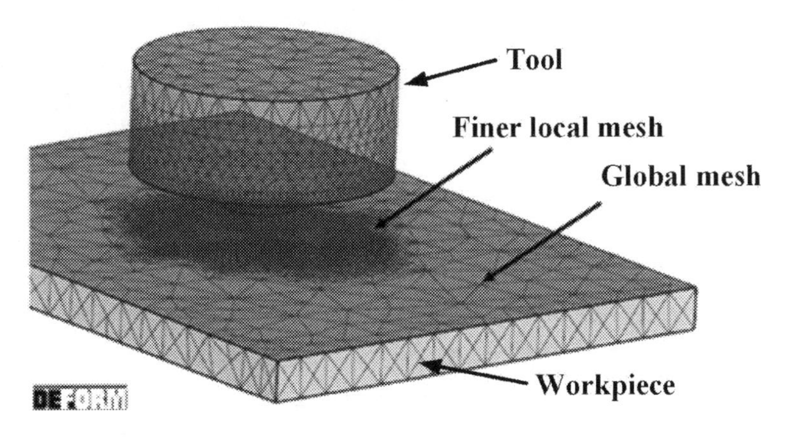

Fig. 3.10 Tool and workpiece in DEFORM-3D with local and global mesh

used. The solver used was sparse, and the iteration method used was direct iteration for convergence.

Geometry and boundary conditions: The tool and the workpiece geometry were designed according to corresponding experimentally used components. A tetrahedral mesh was considered. The mesh size assigned was 2 mm for the global elements and 0.3 mm for local elements at the interacting surfaces Fig. 3.10. This was done in order to obtain good accuracy with reasonable simulation time.

The boundary conditions implemented in the model were: (i) workpiece bottom surface fixed in Z-direction, (ii) tool given translation velocity in the Z-direction along with rotation about Z-axis, (iii) heat transfer coefficient at workpiece bottom and baseplate interface is 200 W/(m^2 K) [3] and (iv) convective heat transfer coefficient for workpiece-environment interaction is 20 W/(m^2 K) [43].

Material model: The workpiece was considered as visco-plastic and followed von Mises yield criterion (Eqs. 3.2–3.5) with isotropic hardening (Eq. 3.8). The flow stress was considered dependent on the strain, temperature and rate of deformation. All the properties were considered rate dependent. The tool was considered rigid.

Thermal model: The heat generated in this model was considered to be from frictional heat. The temperature distribution was obtained by solving finite difference equations (Eqs. 3.81−3.85).

Contact conditions: The interface of pin and workpiece was assigned constant shear friction (Eq. 3.87). On the other hand, the interface of shoulder and workpiece was assigned Coulomb's friction (Eq. 3.86) and Tresca friction model (Eq. 3.87) in the low and high pressure regions of plunge depth, respectively. The friction model was temperature-dependent since yield stress involved in the constant shear friction model was dependent on temperature.

Results: The methodology used in the study was able to inversely obtain the friction factor at the interface of pin and workpiece as $m = 1.05$. For the hybrid model considered at the interface of the shoulder and the workpiece, coefficient of friction was found as $\mu = 0.29$ (at low shoulder plunge depth), and friction factor was

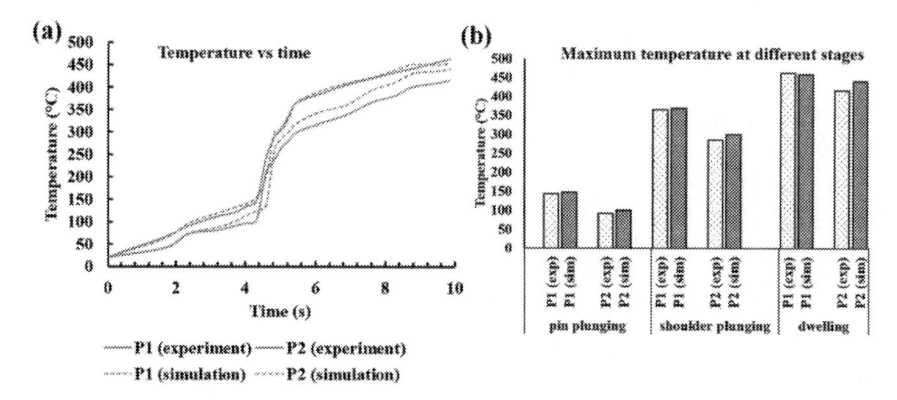

Fig. 3.11 Experimental and computed temperatures: **a** at two points P1 and P2 on the workpiece, **b** maximum temperature during the different stages of welding (Bhardwaj et al. [14], under a Creative Commons license)

found as $m = 0.77$ (at high shoulder plunge depth). The value of m was found to be more than 1 for pin-workpiece interface, which may be due to the asperities getting welded at the tool-workpiece interface resulting in local hardening in that region. These results were implemented as friction model, and validation was carried out by evaluation of temperature and torque by simulation and comparing with experimental results. Temperature at two points (P1 and P2) on the top surface of the workpiece (5 mm and 15 mm from the shoulder edge, respectively) was predicted using the model and compared with experimental results. Temperature versus time plots for the points showed that the model could predict the evolution of temperature at both points with error less than 10% Fig. 3.11a. The maximum temperature developed during each stage of welding was also compared for the two points and shown in Fig. 3.11b. The temperature contour during welding was also generated by simulation using the model Fig. 3.12.

The validation of the model was also checked by using a pair of tool and workpiece with different dimensions; however, material was kept the same. Computed torque and plunge force using the model were found to match with an error <5% and <10%, respectively, with respect to the experimental results Fig. 3.13a, b. The example demonstrates the prediction capabilities of the friction stir spot welding model generated.

3.7 Future Challenges

One of the major challenges in numerical simulation of FSW is the computational time. FSW simulation is a complex multiphysics problem and is computationally very expensive. As the computation capabilities of modern computers have increased, the simulation time has reduced drastically. However, there is a need to further reduce

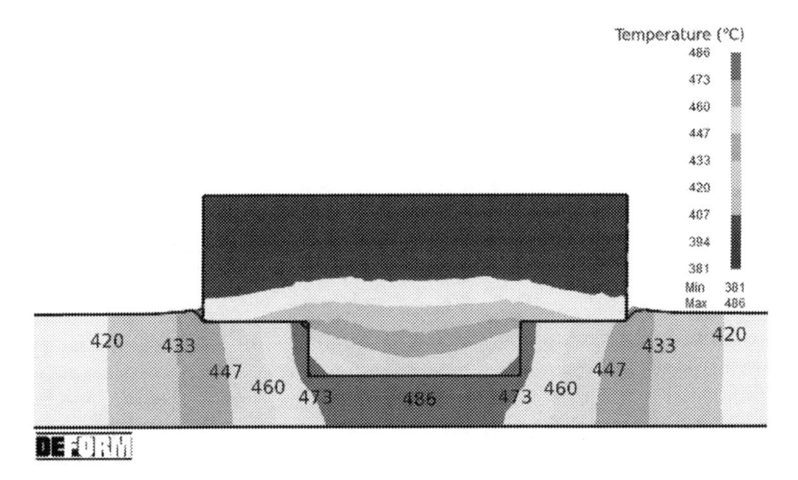

Fig. 3.12 Temperature contour in the weld region obtained from simulation (Bhardwaj et al. [14], under a Creative Commons license)

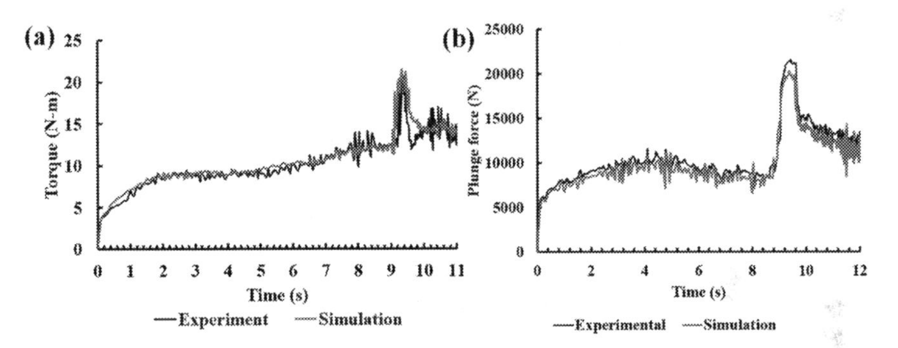

Fig. 3.13 Computed and experimental graphs of **a** torque with respect to time, **b** plunge force with respect to time (Bhardwaj et al. [14], under a Creative Commons license)

the simulation time, which will make it more suitable for repetitive simulations with minor variation in process parameters. Although there have been many numerical studies on FSW, there are very few researches on analytical studies. Most analytical works have extended the work of Schmidt et al. [72] along with the assumptions taken by them, thus inheriting their limitations at the same time. The tool-workpiece interface contact condition is a vital aspect of modeling process and needs specific studies on the friction conditions. The different friction models available are not capable of predicting mechanical responses (torque or plunge force, etc.) accurately when used alone. Hybrid friction models incorporating more than one friction model may be used for better agreement with experimental results.

3.8 Conclusion

A comprehensive study of the modeling of FSW using both analytical and numerical methods has been presented in this work. Lagrangian and Eulerian approaches are primarily used for modeling. In Lagrangian approach, the study is performed on a set of points which keep moving during deformation. On the other hand, Eulerian approach assumes a fixed control volume through which material passes. Combining the two approaches, ALE and CEL are slowly gaining popularity; however, they are limited by their requirement of higher computation capabilities. Different yield functions are used for encompassing the material behavior in modeling, of which von Mises and Tresca yield criteria are most commonly used. It is observed that a very small number of dedicated studies have been conducted on analytical modeling of FSW. Most of the analytical models concentrate on the thermal aspect of FSW. Some available models have related torque and plunge force during welding to heat generated during welding. The work of Schmidt et al. [72] is referred by many researchers in developing heat flux models during FSW. The temperature distribution in the weld is arrived at by using the heat flux generated by the FSW tool. A numerical model of FSW may include thermal, mechanical as well as a metallurgical model. A coupled thermo-mechanical model is more efficient at predicting results closer to experiments. A thermal model, an elastic–plastic model with temperature-dependent flow stress and a metallurgical model together can predict mechanical properties like hardness as well as residual stresses. CAFE model is capable of predicting microstructure evolution during welding. Although advances in modeling of FSW have been taking place, the extensive simulation time remains one of its major challenges. Further research is needed to mitigate this problem, either by developing more accurate analytical models or by enhancing the computational efficiency of the numerical models.

References

1. Akbari M, Asadi P, Givi MB, Zolghadr P (2016) A cellular automaton model for microstructural simulation of friction stir welded AZ91 magnesium alloy. Modell Simul Mater Sci Eng 24(3):035012
2. Al-Badour F, Merah N, Shuaib A, Bazoune A (2013) Coupled EulerianLagrangian finite element modeling of friction stir welding processes. J Mater Process Technol 213(8):1433–1439
3. Asadi P, Mahdavinejad RA, Tutunchilar S (2011) Simulation and experimental investigation of FSP of AZ91 magnesium alloy. Mater Sci Eng: A 528(21):6469–6477
4. Asadi P, Givi MKB, Akbari M (2015) Microstructural simulation of friction stir welding using a cellular automaton method: a microstructure prediction of AZ91 magnesium alloy. Int J Mech Mater Eng 10(1):20
5. Asadi P, Givi MKB, Akbari M (2016) Simulation of dynamic recrystallization process during friction stir welding of AZ91 magnesium alloy. Int J Adv Manuf Technol 83(1–4):301–311
6. Awang M (2007) Simulation of friction stir spot welding (FSSW) process: study of friction phenomena. West Virginia University
7. Behnagh RA, Samanta A, Pour MAM, Esmailzadeh P, Ding H (2019) Predicting microstructure evolution for friction stir extrusion using a cellular automaton method. Modell Simul Mater Sci Eng 27(3):035006

8. Benson DJ (1992) Computational methods in Lagrangian and Eulerianhydrocodes. Comput Methods Appl Mech Eng 99:235–394
9. Benson DJ (1997) A mixture theory for contact in multi-material Eulerian formulations. Comput Methods Appl Mech Eng 140(1–2):59–86
10. Bendzsak GB, North TB, Smith CB (2000) An experimentally validated 3D model for friction stir welding. In: Proceedings of the second international symposium on friction stir welding, Sweden.
11. Bevington J (1891) Spinning tubes mode of welding the ends of wire, rods, etc, and mode of making tubes. US patent 463134
12. Bhardwaj N, Narayanan RG, Dixit US, Hashmi MS (2019) Recent developments in friction stir welding and resulting industrial practices. Adv Mater Process Technol 5(3):461–496
13. Bhardwaj N, Narayanan RG, Dixit US (2020) Effect of lubrication on energy requirement and joint properties during FSW of AA5052-H32Aluminium alloy. In: Vishal S, Sharma et al (eds) Manufacturing engineering: lecture notes on multidisciplinary industrial engineering, Springer, Singapore
14. Bhardwaj N, Narayanan RG, Dixit US, Petrov M, Petrov P (2020) An inverse approach towards determination of friction in friction stir spot welding. Procedia Manuf 47:839–846
15. Brown KH, Burns SP, Christon MA (2002) Coupled Eulerian-Lagrangian methods for earth penetrating weapon applications, Sand Report, Sand2002–1014
16. Buffa G, Campanile G, Fratini L, Prisco A (2009) Friction stir welding of lap joints: influence of process parameters on the metallurgical and mechanical properties. Mater Sci Eng: A 519(1–2):19–26
17. Carslaw HS, Jaeger JC (1959) Conduction of heat in solids. Oxford University Press, Amen House, London
18. Cengel YA, Boles MA (2007) Thermodynamics: an engineering approach 6th editon (SI Units). The McGraw-Hill Companies Inc., New York
19. Chao Y, Qi X (1998) Thermal and thermo-mechanical modeling of friction stir welding of aluminum alloy 6061–T6. J Mater Proc Mfg Sci 7:215–233
20. Chao YJ, Qi X, Tang W (2003) Heat transfer in friction stir welding—experimental and numerical studies. J Manuf Sci Eng 125(1):138–145
21. Colegrove PA, Shercliff HR (2004) Two-dimensional CFD modelling of flow round profiled FSW tooling. Sci Technol Weld Joining 9(6):483–492
22. Colligan K (1999) Material flow behavior during friction welding of aluminum. Weld J 75(7):229s–237s
23. Das S (2010) Modeling mixed microstructures using a multi-level cellular automata finite element framework. Comput Mater Sci 47(3):705–711
24. Davies CHJ (1995) The effect of neighbourhood on the kinetics of a cellular automaton recrystallisation model. Scriptametallurgicaetmaterialia 33(7):1139–1143
25. Dixit US (2009) Finite element methods for engineers. Cengage Learning, Singapore
26. Dixit PM, Dixit US (2008) Modeling of metal forming and machining processes: by finite element and soft computing methods. Springer Science and Business Media
27. Dixit US, Shufen R (2020) Finite element method modeling of hydraulic and thermal autofrettage processes. In: Mechanics of materials in modern manufacturing methods and processing techniques, pp 31–69. Elsevier
28. Ericsson M, Sandström R (2003) Influence of welding speed on the fatigue of friction stir welds, and comparison with MIG and TIG. Int J Fatigue 25(12):1379–1387
29. Eslami S, Ramos T, Tavares PJ et al (2015) Shoulder design developments for FSW lap joints of dissimilar polymers. J Manuf Process 20:15–23
30. Feng Z, Wang X, David SA, Sklad P (2007) Modeling of residual stresses and property distributions in friction stir welds of aluminum alloy 6061–T6. Sci Technol Weld Joining 12(4):348–356
31. Feulvarch É (2005) Modélisation numérique du soudage par friction-malaxage (Friction StirWelding). Université Jean Monnet de Saint-Etienne, These de doctorat

32. Frigaard O, Grong O, Midling OT (1998, April) Modeling of the heat flow phenomena in friction stir welding of aluminum alloys. In: Proceedings of the seventh international conference joints in aluminum—INALCO, vol 98, pp 15–17
33. Gandin CA, Rappaz M (1994) A coupled finite element-cellular automaton model for the prediction of dendritic grain structures in solidification processes. Actametallurgicaetmaterialia 42(7):2233–2246
34. Gould JE, Feng Z (1998) Heat flow model for friction stir welding of aluminum alloys. JMPMS 7(2):185–194
35. Gratecap F, Girard M, Marya S, Racineux G (2012) Exploring material flow in friction stir welding: tool eccentricity and formation of banded structures. IntJ Mater Form 5(2):99–107
36. Grujicic M, He T, Arakere G, Yalavarthy HV, Yen C-F, Cheeseman BA (2010) Fully coupled thermomechanical finite element analysis of material evolution during friction-stir welding of AA5083. Proc IMechE Part B: J Eng Manuf 224:609–625
37. Guerdoux S, Fourment L, Miles M, Sorensen C (2004, June) Numerical simulation of the friction stir welding process using both Lagrangian and arbitrary Lagrangian Eulerian formulations. In: AIP conference proceedings, vol 712(1), pp 1259–1264. American Institute of Physics
38. Guerdoux S, Fourment L (2005) ALE formulation for the numerical simulation of friction stir welding. In: Oñate et al E (ed) Computational plasticity: fundamentals and applications: proceedings of the eighth international conference on computational plasticity, COMPLAS VIII, CIMNE, Barcelona
39. Hattel JH, Sonne MR, Tutum CC (2015) Modelling residual stresses in friction stir welding of Al alloys—a review of possibilities and future trends. Int J Adv Manuf Technol 76(9–12):1793–1805
40. He X, Gu F, Ball A (2014) A review of numerical analysis of friction stir welding. Prog Mater Sci 65:1–66
41. Hossfeld M, Roos E (2013) A new approach to modelling friction stir welding using the CEL method. In: Int. Conf. Adv. Manuf. Eng. Technol. NEWTECH. pp 179–190. https://doi.org/10.18419/opus-8825
42. Hsu TT (2012) The finite element method in thermomechanics. Allen and Unwin, London
43. Jain R, Pal SK, Singh SB (2017) Finite element simulation of temperature and strain distribution during friction stir welding of AA2024 aluminum alloy. J Inst Eng (India): Series C 98(1):37–43
44. Jain R, Pal SK, Singh SB (2018) Finite element simulation of pin shape influence on material flow, forces in friction stir welding. Int J Adv Manuf Technol 94(5–8):1781–1797
45. Khan NZ, Siddiquee AN, Khan ZA (2017) Friction stir welding: dissimilar aluminium alloys. CRC Press
46. Khandkar MZH, Khan JA (2001) Thermal modeling of overlap friction stir welding for Al-alloys. JMPMS 10(2):91–105
47. Khandkar MZH, Khan JA, Reynolds AP (2003) Prediction of temperature distribution and thermal history during friction stir welding: input torque based model. Sci Technol Weld Joining 8(3):165–174
48. Khoei AR, Anahid M, Mofid M (2003) An application of arbitrary Lagrangian-Eulerian method in numerical simulation of forming processes using cap plasticity model. In: 12th international science conference on achievements in mechanical and materials engineering, AMME
49. Kwon YJ, Shigematsu I, Saito N (2004) Mechanical property improvements in aluminum alloy through grain refinement using friction stir process. Mater Trans 45(7):2304–2311
50. Lan YJ, Li DZ, Li YY (2004) Modeling austenite decomposition into ferrite at different cooling rate in low-carbon steel with cellular automaton method. Acta Mater 52(6):1721–1729
51. Liechty BC, Webb BW (2007) The use of plasticine as an analog to explore material flow in friction stir welding. J Mater Process Technol 184(1–3):240–250
52. Lorrain O, Serri J, Favier V, Zahrouni H, Hadrouz ME (2009) A contribution to a critical review of friction stir welding numerical simulation. J Mech Mater Struct 4(2): 351–370
53. Madej L, Hodgson PD, Pietrzyk M (2009) Development of the multi-scale analysis model to simulate strain localization occurring during material processing. Arch Comput Methods Eng 16(3):287–318

54. Mahoney M, Mishra RS, Nelson T (2001) Friction stir welding and processing. TMS, Warrendale, PA, USA
55. Mijajlović M, Milčic D (2012) Analytical model for estimating the amount of heat generated during friction stir welding: application on plates made of aluminium alloy 2024 T351. In: Welding processes. InTech
56. Malik V, Sanjeev NK, Hebbar HS, Kailas SV (2014) Time efficient simulations of plunge and dwell phase of FSW and its significance in FSSW. Procedia Mater Sci 5:630–639
57. Mishra RS, Ma ZY (2005) Friction stir welding and processing. Mater Sci Eng: R: Rep 50(1–2):1–78
58. Nandan R, Roy GG, Lienert TJ, DebRoy T (2006) Numerical modelling of 3D plastic flow and heat transfer during friction stir welding of stainless steel. Sci Technol Weld Joining 11(5):526–537
59. Neto DM, Neto P (2013) Numerical modeling of friction stir welding process: a literature review. Int J Adv Manuf Technol 65(1–4):115–126
60. Patel C, Das S, Narayanan RG (2013) CAFE modeling, neural network modeling, and experimental investigation of friction stir welding. Proc Inst Mech Eng Part C: J Mech Eng Sci 227(6):1164–1176
61. Priyadarshini A, Pal SK, Samantaray AK (2012) Finite element modeling of chip formation in orthogonal machining. Stat Comput Tech Manuf 101–144
62. Qian M, Guo ZX (2004) Cellular automata simulation of microstructural evolution during dynamic recrystallization of an HY-100 steel. Mater Sci Eng: A 365(1–2):180–185
63. Rajpoot YS, Narayanan RG, Das S (2018) Predicting the effect of tool configuration during friction stir welding by cellular automata finite element analyses. Int J Manuf Res 13(4):359–381
64. Rao SS (2017) The finite element method in engineering. Butterworth-heinemann
65. Reynolds AP (2000) Visualisation of material flow in autogenous friction stir welds. Sci Technol Weld Joining 5(2):120–124
66. Richards DG, PrangnellPB WSW, Withers PJ (2008) Global mechanical tensioning for the management of residual stresses in welds. Mater Sci Eng A 489(1–2):351–362
67. Russel M, Shercliff H (1999) Analytical modeling of microstructure development in friction stir welding. In: 1st International symposium on friction stir welding, USA
68. Salimi S, Bahemmat P, Haghpanahi M (2016) Analytical model for the temperature field around a nonuniform three-dimensional moving heat source: friction stir welding modelling. J Eng Math 98(1):71–91
69. Saluja RS, Narayanan RG, Das S (2012) Cellular automata finite element (CAFE) model to predict the forming of friction stir welded blanks. Comput Mater Sci 58:87–100
70. Samanta A, Shen N, Ji H, Wang W, Li J, Ding H (2018) Cellular automaton simulation of microstructure evolution for friction stir blind riveting. J Manuf Sci Eng 140(3)
71. Sato YS, Urata M, Kokawa H (2002) Parameters controlling microstructure and hardness during friction–stir welding of precipitation–hardenable aluminum alloy 6063. Metall Mater Trans A 33(3):625–635
72. Schmidt H, Hattel J, Wert J (2003) An analytical model for the heat generation in friction stir welding. Modell Simul Mater Sci Eng 12(1):143
73. Schmidt H, Hattel J (2005) Modelling heat flow around tool probe in friction stir welding. Sci Technol Weld Joining 10(2):176–186
74. Sellars CM, Tegart WM (1972) Hot workability. . Int Metall Rev 17(1):1–24
75. Shojaeefard MH, Akbari M, Khalkhali A, Asadi P, Parivar AH (2014) Optimization of microstructural and mechanical properties of friction stir welding using the cellular automaton and Taguchi method. Mater Des 64:660–666
76. Shterenlikht A, Howard IC (2006) The CAFE model of fracture—application to a TMCR steel. Fatigue Fract Eng Mater Struct 29(9–10):770–787
77. Skrzat A (2012) Application of coupled Eulerian-Lagrangian approach in metal forming simulations. ZeszytyNaukowePolitechnikiRzeszowskiej. Mechanika 84 [284], nr 4, 25–35

78. Smith CB, Noruk JS, Bendzsak GB, North TH, Hinrichs JF, Heideman RJ, Smith AO (2000) Heat and material flow modeling of the friction stir welding process. NIST SPECIAL PUBLICATION SP, pp 475–488
79. Song KJ, Dong ZB, Fang K, Zhan XH, Wei YH (2014) Cellular automaton modelling of dynamic recrystallisation microstructure evolution during friction stir welding of titanium alloy. Mater Sci Technol 30(6):700–711
80. SonneMR TCC, HattelJH SA, deMeester B (2013) The effect of hardening laws and thermal softening on modeling residual stresses in FSWofaluminumalloy 2024–T3. J Mater Process Technol 213:477–486
81. Taban E, Kaluc E (2007) Comparison between microstructure characteristics and joint performance of 5086–H32aluminium alloy welded by MIG, TIG and friction stir welding processes. KovoveMaterialy 45(5):241
82. Thomas W (1991) International patent Application No. PCT/GB92. GB patent Application No. 9125978
83. Tongne A, Desrayaud C, Jahazi M, Feulvarch E (2017) On material flow in friction stir welded Al alloys. J Mater Process Technol 239:284–296
84. Trimble D, Monaghan J, O'Donnell GE (2012) Force generation during friction stir welding of AA2024-T3. CIRP Annals—Manuf Technol 61(1):9–12
85. Valvi SR, Krishnan A, Das S, Narayanan RG (2016) Prediction of microstructural features and forming of friction stir welded sheets using cellular automata finite element (CAFE) approach. Int J Mater Form 9(1):115–129
86. Van Elsen M, Baelmans M, Mercelis P, Kruth JP (2007) Solutions for modelling moving heat sources in a semi-infinite medium and applications to laser material processing. Int J Heat Mass Transf 50(23–24):4872–4882
87. Wu SJ, Davis CL, Shterenlikht A, Howard IC (2005) Modeling the ductile-brittle transition behavior in thermomechanically controlled rolled steels. Metall Mater Trans A 36(4):989–997
88. Xiao LIU, Li LX, He FY, Jia ZHOU, Zhu BW, Zhang LQ (2013) Simulation on dynamic recrystallization behavior of AZ31 magnesium alloy using cellular automaton method coupling Laasraoui-Jonas model. Trans Nonferrous Met Soc China 23(9):2692–2699
89. Xu S, Deng X, Reynolds AP, Seidel TU (2001) Finite element simulation of material flow in friction stir welding. Sci Technol Weld Joining 6(3):191–193
90. Zheng C, Xiao N, Li D, Li Y (2008) Microstructure prediction of the austenite recrystallization during multi-pass steel strip hot rolling: A cellular automaton modeling. Comput Mater Sci 44(2):507–514

Chapter 4
An Application from a Defect—A Friction Stir Channeling Approach

Pooja Sarkar, Surjya Kanta Pal, Anandaroop Bhattacharya, and Barbara Shollock

Abstract This chapter gives an overview of a solid-state welding process, i.e., Friction Stir Welding (FSW) process. The basic principle, usefulness, applications, and challenges of the process are all described in the chapter. The wormhole defect that occurs during FSW because of improper process parameter selection leads to the development of a new cooling channel manufacturing process called the Friction Stir Channeling (FSC). This chapter gives an extensive idea about the development of FSC and its capability to replace the conventional heat exchanger manufacturing processes. The feasibility of the FSC process to produce a continuous cooling channel is also presented at the end of the chapter through a case study.

Keywords Friction stir welding · Wormhole defect · Friction stir channeling · Heat exchanger

4.1 Introduction to Friction Stir Welding (FSW)

Friction Stir Welding (FSW) is one of the most successful and quickly adapted solid-state welding techniques available globally. The acceptance of products after joining by FSW is much higher than other joining techniques owing to various benefits

---

P. Sarkar
Advanced Technology Development Centre, Indian Institute of Technology Kharagpur, Kharagpur 721302, West Bengal, India
e-mail: sarkarpooja90@gmail.com

S. K. Pal (✉) · A. Bhattacharya
Department of Mechanical Engineering, Indian Institute of Technology Kharagpur, Kharagpur 721302, West Bengal, India
e-mail: skpal@mech.iitkgp.ac.in

A. Bhattacharya
e-mail: anandaroop@mech.iitkgp.ac.in

B. Shollock
King's College, Strand Campus, Strand, London WC2R 2LS, UK
e-mail: barbara.shollock@kcl.ac.uk

J. P. Davim (ed.), *Welding Technology*, Materials Forming, Machining and Tribology, https://doi.org/10.1007/978-3-030-63986-0_4

offered by this process. The process and its benefits are broadly explained in the subsequent sections.

4.1.1 Development of FSW

The history of welding can be traced back to ancient times. In the Bronze Age, humans realized that joining two metals provides the opportunity for new tools and objects that could not be previously manufactured. This opportunity brought out the invention of welding. In the 1800s, with the help of the discovery of the arc, the arc welding process was first invented for the joining of lead plates [1]. Since then, the science and application of metal joining has grown from strength to strength. Many new metal joining techniques have been developed over all the years, both in fusion and solid-state conditions. One of the recent additions to the list of solid-state welding techniques is FSW.

The developmental history of FSW is quite interesting. Around the 1980s, it was observed that aluminium (Al) and magnesium (Mg) alloys, because of their low weight and higher strength to weight ratio, have enormous potential to be used in various parts of machinery to reduce weight and fuel consumption. In a component comprised of steel, substituting steel sub-components with lightweight alloys like Al and Mg reduced the component's overall weight, notably without making a significant compromise in the strength. Researchers realized that using these alloys in various industries could prove to be significantly economic and have environmental value. For example, in automobile industries, vehicles built with lighter alloys would increase fuel efficiency and reduce carbon emissions. The use of mechanical fasteners offset weight reduction. Again, for traditional welding techniques, the high reactivity of these lightweight alloys to oxygen in their molten state was found to be a severe bottleneck that had to be eliminated to facilitate their use. The traditional welding techniques available such as TIG, MIG, and resistance welding, showed limited success, and the joint efficiency was unsatisfactory. These challenges, as well as the defects generated during traditional welding techniques including porosity, voids, distortion, and the formation of brittle intermetallics, motivated the researchers to seek a better welding technique to overcome these problems related to the traditional fusion welding techniques.

In 1991, The Welding Institute (TWI) invented the Friction Stir Welding Process [2], an advanced solid-state joining process that can weld materials of different alloys without reaching their molten state. FSW can weld similar or dissimilar materials, especially the lightweight and reactive materials, that are otherwise hard-to-weld with the available traditional welding processes like TIG and MIG. The FSW technique was found to be superior to these traditional welding processes in other aspects as well. FSW was found to be immune to the ill effects of re-solidification observed in fusion welding processes. It produces zero pollutants, and hence can be termed as '*green welding*' technique. The stress corrosion cracking related to the traditional fusion welding processes is also absent in FSW.

Though the process was primarily developed to join aluminium alloys, the benefits of FSW established its applicability to produce monolithic joints, such as Mg–Mg, Cu–Cu, etc., and to join dissimilar alloys, including Al-steel and Mg-steel. Al-steel, Mg-steel welded parts have considerable applications in the aerospace sector and automobile industries.

The development of FSW has proven to be transformative in paving the way for lightweight Al and Mg alloys in various industries. Replacing metals such as steel with lightweight materials has significantly reduced the weight of different machinery parts. For example, by substituting a conventional steel subframe of a vehicle's chassis with a lap welded Al-Steel subframe, Honda achieved a weight reduction of 25% [3].

In this century, the FSW process has been extensively researched all over the globe. This fact has resulted in the improvement of the technique and widespread use in several new industries. The contemporary applications of FSW include aircraft components, shipbuilding, rail carriages, bridge components, pressure vessels, etc. NASA and SpaceX have broadly used FSW in various aerospace applications [4].

4.1.2 FSW Process

The FSW process is a solid-state welding process where the materials are welded well below their melting point, eliminating numerous defects associated with a molten state. FSW is performed with a non-consumable rotating tool with hardness and strength higher than the materials to be joined. A shoulder and a pin are the essential parts of the tool used during welding. The tool up to a certain depth of the shoulder plunges into the part to be welded and traverses along the joint line. Frictional heat generated because of the contact of the shoulder and pin with the workpiece surface is primarily used to soften the material. The severe plastic deformation (SPD) induced by the movement of the pin and the lower portion of the shoulder inside the material aids in supplementary heating and mixing of the material, providing successful welding. Throughout the process, the plunging force on the softened material given by the shoulder restrains its motion outwards, thus sealing the void created by the pin, forming a macro defect-free joint. Being a type of friction welding, the workpieces joined by FSW should be in close contact with each other, and specially designed fixtures are used to accomplish this.

A vertical milling machine was conventionally used to conduct FSW by mounting the FSW tool on the spindle. More recently, FSW machines with higher degrees of freedom and simple human-machine interface (HMI) makes it an efficient adaptive process for the industries. The introduction of robotic FSW machines will enhance the productivity of the process to a higher level.

4.1.3 Principles of FSW

The heat generated due to the friction between the tool and workpiece and the dynamic recrystallization due to the stirring of the pin leads to the joining by FSW. To achieve welding, FSW undergoes the following stages, as shown in Fig. 4.1:

i. *Stage 1: Plunging*—This is the stage where the pin and a pre-set depth of the shoulder progress into the adjoined surface of the workpieces to be welded. The material is initially at room temperature and hence is not soft enough to reduce the force. Therefore, the axial resistance force encountered by the tool in this stage is the maximum.
ii. *Stage 2: Dwelling*—Sufficient heat must be generated for softening the material preceding the pin to initiate the welding. Initial heating is achieved by rotating the tool in the starting spot for a specific period fixed by the operator.
iii. *Stage 3: Traversing*—The tool then traverses along the weld line of the parts to be joined, prompting the material to flow from the front edge to the back edge of the tool by stirring action. As a consequence, the mixing of the material takes place, and it forms the joint.
iv. *Stage 4: Retracting*—After traversing through the desired distance, the tool is withdrawn from the workpiece with an exit hole at the end of the joint line. Recent developments in the FSW process enables the exit hole to be sealed [5].

In FSW, both frictional and mechanical (SPD) heating cause the temperature to reach around 0.6-0.8 times the melting temperature of the materials to be joined. Since the temperature rise exceeds the recrystallization temperature, the formation of new grains takes place. It is observed from the literature [6] that the action of the pin to stir continuously and the local heating of the material alter the strain, strain rate, and temperature at different sections of the weld zone. In a cross-sectional view of an FSWed sample, shown in Fig. 4.2, the weld region can be partitioned into four

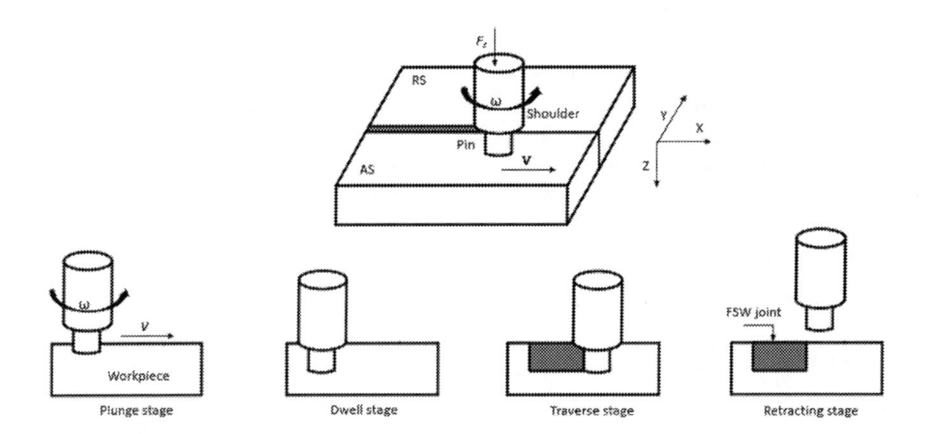

Fig. 4.1 Stages in FSW process

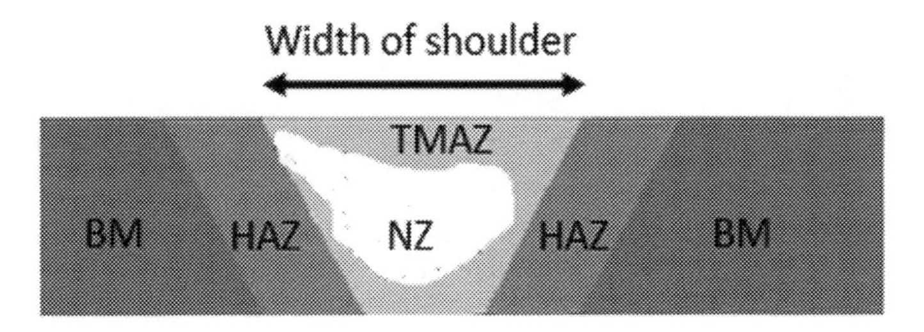

Fig. 4.2 Zones in FSWed workpiece

specific zones, namely the nugget zone (NZ), thermo-mechanically affected zone (TMAZ), heat affected zone (HAZ), and base material (BM). The NZ is formed by the pin where the heat generation is mainly due to friction, as well as the mechanical mixing of the material. This brings about finer, equiaxed, and recrystallized grains at the NZ. TMAZ is next to the NZ, where the effect of heat is noticeable, but minimal mechanical deformation leads to coarser grains compared to that in NZ. The HAZ is entirely affected by the heat generated during the process. Since, in FSW, heat generation is well below the melting point temperature of the workpiece material, the HAZ zone is much smaller compared to that formed in fusion welding techniques.

4.1.4 FSW Terminologies

The following terminologies are the most commonly used in describing aspects of the FSW process. They are shown in Fig. 4.3.

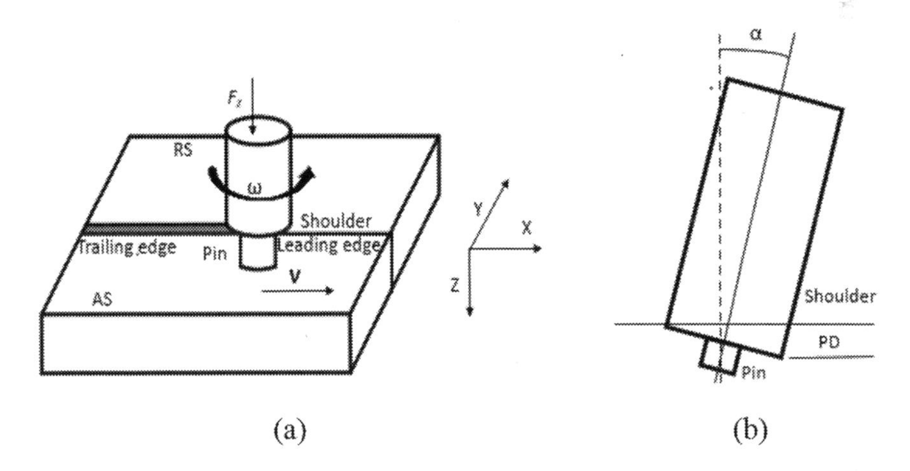

Fig. 4.3 **a**, **b** Showing different terminologies used in FSW

Shoulder: Shoulder is the part of the tool that is in contact with the workpiece surface. This part of the tool generates the maximum amount of frictional heat. For a certain plunge depth (defined later), the shoulder is also responsible for mixing in the top part of the workpiece material.

Pin: It is a small portion that protrudes from the shoulder. During welding, this part of the tool is completely inside the materials to be joined. The pin plastically deforms the material due to its rotation, generating additional heat.

Leading edge: As the name suggests, the leading edge is the edge ahead of the pin along the traverse direction.

Trailing edge: It is the edge of the parts to be joined which is behind the pin along the traverse direction.

During welding, the material moves to the trailing edge from the leading edge of the tool. This material is further compressed by the shoulder and is pushed into the hole created by the pin. Hence, a perfect joint is achieved.

Advancing side (*AS*): It is the side of the parts to be welded, where the direction of the tangential component of the tool rotational speed is similar to that of the traversing speed.

Retreating side (*RS*): The side of the parts to be welded, where the direction of the tangential component of the tool rotational speed is opposite to that of the traversing speed.

Plunge depth (*PD*): The depth of the shoulder up to which it is inserted into the workpiece is called the plunge depth. The plunge depth has different values depending on the material and thickness of plates to be welded.

Tool tilt angle (α): For better mixing of the material and purposefully generate more heat on the high melting point material side during dissimilar material welding, the spindle of the FSW machine is tilted to a particular value. The angle by which the spindle is tilted from the line normal to the weld line is called the tool tilt angle.

Axial force (F_z): F_z is the component of the force generated due to the resistance offered by the workpiece material, which acts perpendicular to the weld traverse direction. F_z is the maximum during the plunging stage as the resistance offered by unsoftened material at room temperature is the highest at that stage. F_z further reduces with the softening of the material.

4.1.5 Important Parameters in FSW

The following process parameters have a vital role in controlling weld quality. For a high-quality weld with good strength, selection of suitable parameters is crucial. The parameters and their effects can be understood from Fig. 4.4.

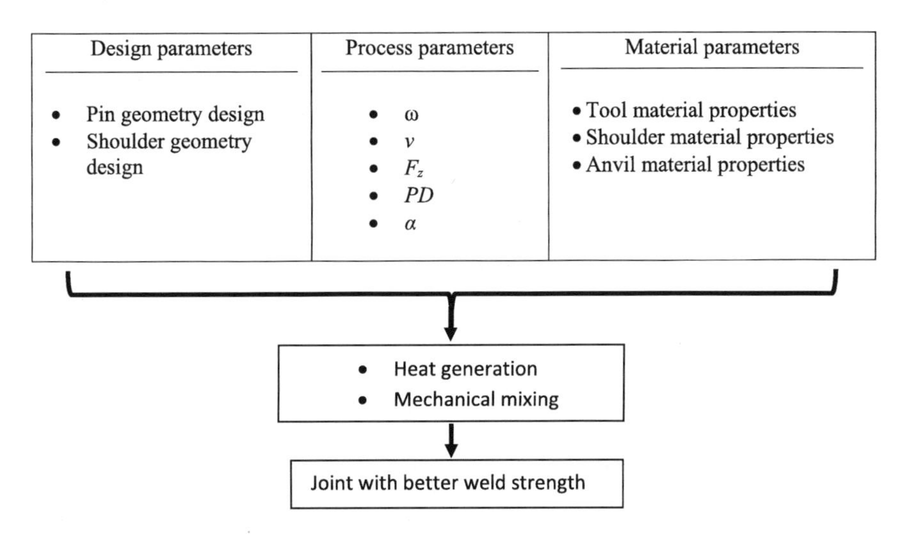

Fig. 4.4 Importance of different parameters in FSW

4.1.5.1 Influence of Tool Design Parameters

The tool design parameters contribute significantly to heat generation and material mixing in the weld zone. An FSW tool has two main parts that control the process: the shoulder and the pin. The geometry of both parts influences the process to no small extent.

The tool shoulder can be flat, concave, or convex. Out of these, the most commonly used shape is a flat shoulder. Concavity and convexity are added for better containment of the material and can control the heat generation as required by the workpieces to be joined. The most commonly used shoulder geometries are shown in Fig. 4.5. During the forward motion of the pin, the concave shoulder forces new material beneath the shoulder, and the already existing material flows along the pin traverse line. Typically, the 6°–10° concavity is given between the corner of the shoulder and

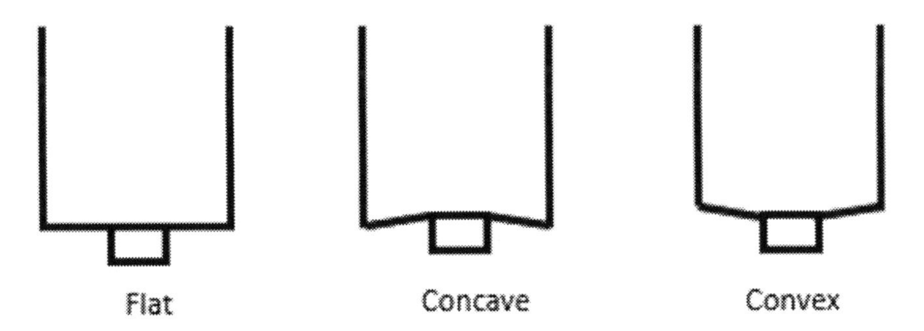

Fig. 4.5 Different FSW shoulders

the pin. A tilt of 2°–4° towards the trailing edge is required to exert the forging force along the weld line properly. Generally, linear welds are produced with a concave shoulder. The demand for maintaining a tool tilt around the concavity makes it difficult for performing non-linear welding with a simple FSW machine [7].

The convex shape propels material off the pin. The convex shoulder is generally used with scrolls that help the flow of material from outer to the inner side of the shoulder and perform sound welding. Due to the convex shape, the user can set the required amount of surface contact of the shoulder by controlling *PD*, thus, a more precise weld line is created [8].

Various features like scrolls, grooves, concentric circles can also be machined into the shoulder (concave, convex, or flat) of an FSW tool [8]. These features enhance the deformation of the materials produced by the shoulder. Consequently, the mixing of material is increased, and hence the weld quality gets improved. The most common feature is the scrolled shoulder, which not only directs the material for better mixing but also helps to avoid the requirement of the tilt angle. The scrolled shoulder can significantly reduce the undercut, flashes, and the amount of tool lift seen when using a concave shoulder. However, precise control of the position of the tool must be maintained in case of scrolled shoulder to avoid excessive flash generation. Scrolled shoulder geometries cannot join two plates with different thicknesses or plates having thickness variations along the weld line.

Various pin geometries have been developed to enhance the process quality and material mixing to increase joint strength and to enable the process to join numerous hard-to-weld materials [7–11]. Initially, FSW was conducted with a round bottom cylindrical threaded pin. The threads present in the pin excavated material from the mating shoulder surface and deposited them at the bottom of the pin. Tool wear, in this case, was less because of the presence of the round dome. The efficiency was the maximum with a dome having a radius equal to 75% of the pin diameter.

However, the low velocity of the round bottom led to the discovery of a flat bottom pin, the velocity of which was increased to approximately 30 times by providing a suitable α [12]. The ease of manufacturing a flat bottom versus a curved pin provides an additional advantage. Both these pins were mainly used for thin plates. The need for welding of thicker plates necessitates further tool developments, and hence truncated cone pin tools were developed [13]. With this pin, aluminium plates with thicknesses greater than 12 mm can be welded with a faster v [14]. Lower F_z and higher moment load at the base of the truncated pin than that of the cylindrical pin made it a better option for joining thicker materials [14].

The need for welding of high-temperature materials further motivated the development of new tools. Threads that were used in the initially developed tools may get distorted by these high-temperature materials. On the other hand, a thread-less pin profile cannot produce enough material flow. The use of a stepped spiral tool was successful for welding high-temperature materials that can sustain temperatures up to 1000 °C [15]. The possibility of grinding the stepped spiral feature into a ceramic tool like polycrystalline cubic boron nitride (PCBN), increases the deformation volume of materials [16].

Further developments focused on increasing the deformation and on providing local turbulent flow of the plasticized material. To achieve this, a machined flat area was incorporated onto a threaded pin, and a truncated coned pin was developed. The addition of flats increased the weld nugget area and reduced the transverse force and tool torque. Further, whorl pins were developed by adding helical ridges on the pin. The introduction of the helical ridge enhanced the downward motion of the material. The use of this pin also decreased the traverse loads and increased the v by reducing the displaced volume. Later, by adding flutes on the helical ridges to enhance the material mixing, MX triflute pins were developed. A helical ridge with three flutes machined into the ridge is the main feature of the pin. The flutes improved the deformation at the weld line. A modification of the triflute pin is the trivex pin. Trivex pins were found to reduce the traverse and the forging forces further compared to the MX triflute pins [14].

Thread-less pins are most commonly used for FSW of thin sheets, such as 0.4 mm thick Mg AZ31, to avoid the weakening of the tool due to the presence of threads. Another version of the thread-less pin is the square pin. Out of five tool profiles studied (straight cylindrical, tapered cylindrical, threaded cylindrical, square, and triangular) for welding AA 6061 aluminium alloy, it was found that the tool with square pin produced defect-free welds for different axial forces used [14].

A void-like defect called the wormhole defect can be seen at the AS of the FSWed workpiece. It forms at low processing temperatures that cause insufficient material flow [9–11]. To avoid this defect, the 'restir' tool was invented by TWI. This tool is capable of reversing its direction of rotation periodically. The latest modification in the FSW tool is the use of the twin tool concept [17]. In the twin tool set up, two tools are mounted on the same spindle, offering better control over mixing. They have different variations, where the rotation can be reversed, and the gap between the tools can be varied as per requirement [18]. These tools are proposed to provide high productivity.

4.1.5.2 Influence of Process Parameters

The process parameters, mainly ω, v, and F_z play essential roles in generating the heat by friction, mechanical deformation, and material mixing. It was stated that the increase in ω increases the heat input causing the material to soften for better mixing. Nevertheless, it was also seen that for a particular material increasing the value of ω beyond a specific limit can lead to higher heat generation that is detrimental for the joint strength. v, on the other hand, mainly dominates the cooling rate. Too low a v can increase the time of contact of the tool and workpiece, generating more frictional heat and also causing a slow cooling rate giving rise to coarser grains. This can cause voids to appear along the joint line. Again, too high v leads to inappropriate material mixing that can lead to void formation. Thus, a perfect range of v must be selected for greater joint strength.

F_z is another critical factor that affects the joint strength significantly. The higher F_z causes the value of friction factor to increase, which increases the frictional heat generation and softens the material for proper mixing. The F_z must be so selected that the stress caused by it becomes higher than the flow stress of the base material for a satisfactory joint.

Along with the above-mentioned process parameters, two other process parameters namely, α and *PD* are also important in producing a good weld. α is used to increase the forging force further and also to decrease the amount of flash formation. For better mixing, α must be inclined towards the trailing edge of the pin. *PD* helps to increase the heat input by providing more contact surface, and also by increasing the downward F_z.

4.1.5.3 Influence of Material Properties

The properties of tool, workpiece, and anvil materials also play essential roles in the process. As the tool touches the workpiece material, it generates frictional and deformational heating that consequently softens the materials and aids in better mixing and joining. The tool further increases the deformational volume by dwelling in one place and creating more heat to soften the material. During welding, the material offers resistance to tool movement. This causes the tool to experience F_z and transverse load. Depending on the type of material to be welded and to ensure a long life of the FSW tool, the tool material must possess properties like strength and stability at high temperatures. Less reactivity to the environment, high fracture toughness, wear resistance are some of the other vital criteria the tool material must fulfil. It should have good machinability to be easily manufactured, and last but not the least, they must be of low cost and readily available. The most frequently used FSW tool to weld Al and other softer materials are AISI H13 hot-working, air-hardened Cr-Mo tool. Table 4.1 describes some of the most used FSW tool materials for welding different workpiece materials.

Again, the properties of the workpiece material like strength, melting point, recrystallization temperature, thermal conductivity, compatibility to join with other materials also influence the process. The thermal conductivity of the anvil material also is a secondary factor for distributing the heat generated to the parts to be welded during the FSW process.

4.1.6 Advantages of FSW Process

Being a solid-state joining process, FSW has many advantages over other traditional welding processes that made it acceptable over the globe.

The absence of melting in FSW, unlike other fusion welding processes, makes the weld-free from defects like porosity, blowholes, cracking, distortion, etc. In addition, it eliminates the use of any protective gas (for Al grades, steel). Further, no fume is generated during the process, thus making it an environment-friendly and green process.

Table 4.1 FSW tool materials for welding different materials [8, 14, 19]

Tool material	Grade of tool	Important features	Application
Tool steel	AISI H13, O1 tool steel, SKD 61 tool steel, Wear resistant steel, AISI 4340	Shows good strength at elevated-temperature They also provide good wear resistance and thermal fatigue resistance	Used for FSW of Al, Mg, Cu alloys and Al–Mg MMC
Nickel-and cobalt-base alloys	IN738LC, IN 939, MAR-M-002, Stellite 12, Nimonic 90, IN 100 and Nimonic 105, PM 3030, Inconel 718, Wasp alloy	High strength and better ductility At high temperature they show good creep resistance. They also have considerable corrosion resistance at high temperature	Mainly used for the FSW of Al, Cu alloys and Al-MMC
Refractory metals (W based tools)	W, W-25%Re, Densimet (90 W–10Fe–Ni), and W-1%LaO_2, W-Mo alloys, Niobium and Tantalum	High-temperature strength, requires grinding but costly Tungsten-based alloy is W-1% LaO_2, which is of low cost and is easily machinable	Mainly used for the FSW of high melting point temperature materials like Ti, Cu, steel etc.
Carbides or cermets	WC, WC-Co based alloys, Ti-C, Ti-Ni-Mo, TiC-Ni-W	Higher resistance to wear At room temperature it shows reasonable fracture toughness	Used for the FSW of Al, Mg and Steel. They are capable of doing FSW of Ti alloys also
CBN	PCBN	Very high hardness, good wear resistance at elevated temperature High cost and low fracture toughness	Used for the FSW of Ti alloys, Ni alloys and steel. Cu and Al-MMC can also be FSWed with PCBN

i. FSW plasticizes the material. Here, the mixing of the workpiece material occurs due to the tool rotation, thus the joint can be made without the use of the filler materials. Many materials e.g. some grades of Al that are very difficult to weld by traditional fusion welding process due to the non-availability of compatible filler materials thus can be welded successfully with the FSW process.
ii. The feasibility of FSW process to join dissimilar materials which have differences in their melting points, to replace high-density materials with low-density material, e.g., Al-Steel, wherever possible, tremendously increased its application in industries.
iii. FSW, being a low-temperature welding process, has shown better fatigue performance compared to mechanical fasteners [20, 21]. This increased usage of FSW in the Eclipse aircraft and also other aerospace applications, reducing its weight further [22, 23]. Issues related to the general corrosion as well as

stress corrosion cracking in Al alloys are reduced due to the low residual stress, and the peak temperature associated with the process.

iv. Since the tool never penetrates through the workpiece completely, FSW produces a very smooth root appearance that hardly requires further machining.
v. This solid-state welding permits the dynamic recrystallization of the grains producing a fine-grained nugget zone that enhances the hardness and strength of the joint up to 97% of the base material [24].
vi. Another benefit of the process is its potential to create a very narrow HAZ. A wide HAZ that is generally observed in other low heat density fusion welding processes can be eliminated in FSW.
vii. The mechanization of the process further eradicates the need for a high skilled worker. Lower automation cost of the machines and operating costs, less employee training, and no need for consumables in this process result in cost savings.
viii. The simplicity of the process and availability of the tool materials (especially for Al material) increase its production rate that makes it easily acceptable for the industries.

4.1.7 Disadvantages of FSW Process

FSW has very few disadvantages, unlike the other fusion welding techniques. They are described below.

i. FSW being a fully mechanized process, increases the capital cost, thus offering a barrier to use this process for small industries. Moreover, the capital cost is justified only if continuous production is done using the process.
ii. Due to the generation of high force and torque during the process, the workpiece must be held very rigidly to avoid any defective welding or accidents. It thus necessitates the highly rigid fixtures that can sustain the load and torque. Moreover, for different jobs, the fixtures must be designed separately.
iii. The absence of the use of filler material in FSW restricts its application to produce fillet shape welding. The autogenous process restricts its ability to join at certain welding positions only.
iv. The existence of keyhole at the end of the joint line is another critical limitation of the process. Some applications like the marine tank joining etc. cannot allow any keyhole at the end of the joint.

The advancements made by various researchers in the FSW process address the above-mentioned issues and are mentioned in the following section.

4.1.8 Variations in FSW Process

Low v and high F_z and torque on the workpiece are some of the major concerns related to the process that limits the use of the process. Use of larger size machinery, wear of tool at high temperature, welding of high-strength application materials, etc. are challenges that restrict its use to its full potential. To address these issues, continuous improvement and development have been made in the field of FSW with every passing day. The issues related to process, tool, and machine are addressed in some of the below-mentioned developed versions of the FSW.

- Stationary shoulder friction stir welding (SSFSW)
- Reverse dual rotation friction stir welding (RDRFSW)
- Bobbin tool friction stir welding (BTFSW)
- Friction stir spot welding (FSSW)
- Refill friction stir spot welding (RFSSW)
- Hybrid FSW processes.

4.1.9 Applications

Shipbuilding and offshore Freezer panels, ship, and oil-rig panels, the hull of an ocean viewer vessel, honeycomb and corrosion resistant panels, deck panels of ships, landing pads and vessels, etc. are commonly welded by FSW in recent times, in different countries like the USA, China, Australia [25–27].

Aerospace Fuel tanks for spacecraft, aircraft fuselages, and wings are also welded by FSW for defect-free joints [25, 28].

Railway Aluminium roof and floor panels and rolling stock, for the railway, are fabricated by FSW [29].

Automotive Closures, tanks, suspensions, and pistons, and trailers used in automotive industries are done by FSW. The solid-state FSW process also joins wheels rims and centre tunnel of fuel tank [25, 30, 31].

Other applications Cathode sheets, motor, and loudspeaker housings, heat sinks, Heating, ventilating, and air conditioning units, backing plates for sputter targets, vacuum vessels, panels are some of the other areas where FSW finds its application. Components for the food industry, copper canisters for nuclear waste, etc. are also welded by FSW recently [25].

The various benefits associated with FSW processes like low distortion, high joint strength, smooth surface finish, flexibility in welding various materials and different thicknesses are the reasons for the popularity of FSW in these sectors.

4.2 Defects in FSW Process

Being a solid-state welding process, the defects generated during FSW process are very few and can be controlled effectively. The defects e.g. flash, voids, wormhole, kissing bond, etc. can be seen in the weld region if the selection of process conditions is not proper. Any material can be deformed mechanically when the resistance offered by the material to deformation is exceeded by supplying sufficient energy. In FSW, this energy is in the form of heat. Again, the resistance offered is dependent on the material properties and these properties are related to the composition of the material. Thus, one should always consider the material and choose the process conditions that are compatible with the material to form a strong joint. Conditions like excessive or insufficient heat generation and also improper mixing of the material due to tool geometry or improper positioning of the tool on the workpiece can lead to the formation of the following defects that severely deteriorate the joint strength. However, defects can be completely eliminated by choosing proper process parameters.

4.2.1 Defects Generated Due to Excessive Heat Generation

Excessive heat generation occurs if ω is high and v is low. The increase in *PD* also adds to the heat generation. Combined effect of all these factors softens the material excessively which lead to the following defects during the process:

Nugget collapse Although the excessive heat generation can lead to the softening of the material around the immediate neighborhood of the pin forming the nugget zone with recrystallized grain, the shoulder continues to apply a torsional effect on the material. However, due to the local drop in temperature for high thermal conductivity materials and also because of the insufficiency of this torsion to recrystallize the material beneath the shoulder, coarser grains are formed than that in the recrystallized NZ. This broadens the TMAZ zone at the cost of collapsing the nugget [32, 33]. This defect reduces the joint strength.

Surface galling Due to severe plasticization of the material, some small blisters start to appear on the surface of the weld line which is called surface galling [34]. This hampers the welding appearance. These blisters can also act as a site for corrosion. Surface galling can be seen in Al alloys.

Onion ring formation The onion ring forms as a band that matches with the forward motion of the tool in a single rotation. Too high ω can increase the spacing between the bands, and at hotter process condition it was observed that fracture occurs along with the band in the weld line [35].

Flash Flashes are generated when the material is over plasticized due to high heat input and the shoulder cannot retain the material under it, expelling them as a form of flash. Too high *PD* also leads to the formation of the excessive flash generation making a concave weld line and thinning of the plate [36]. The excessive flash lowers the product quality and necessitates surface smoothing operations.

Root flaws Root flaws occur because of very hot processing conditions under a constant load. As a result, material gets plasticized severely and the pin penetrates deeper into the material to maintain the constant load. This consequently leads to the adherence of the plasticized material with the backing plate [33]. This defect leads to poor weld quality.

By lowering ω and increasing v, the excessive heat generation can be controlled. The decrease in PD and also by reducing α, heat generation can be further reduced to avoid the above-mentioned defects.

4.2.2 Defects Generated Due to Insufficient Heat Generation

Insufficient heat results from very low ω and very high v. The low value of F_z also leads to a reduction in heat generation. The low amount of heat cannot plasticize the material to its desired value, and hence affects the material mixing and formation of bonds. This can lead to defects like void, kissing bond, tunnel or wormhole defects, which are described below.

Lack of fill or void Due to insufficient heat generation, the material is not plasticized properly, and as a result, the tool is not able to mix the material to diffuse them uniformly. This results in a lack of fill or voids on the surface of the weld line [36]. This reduces the joint strength.

Kissing bond The reactive materials like Al, Mg form oxide in the presence of air, and because of insufficient heat generation, oxide layers break into fragments due to improper mixing and fail to form a bond with the material. The presence of the oxide fragments in the diffusion zone weakens the bond. This defect is called a kissing bond defect [36, 37].

Tunnel or wormhole In the absence of required heat generation, and not having sufficient plunge force from the tool shoulder, the material underneath the shoulder and around the pin does not get mixed properly, and hence cannot fill the void created by the pin at the trailing edge. This leads to the formation of tunnel-like void. The tunnel defect usually occurs inside the weld line and is bigger than the voids that are normally seen on the surface. The presence of a tunnel severely reduces the joint strength [37].

By increasing the value of ω and reducing v, the heat generation can be controlled and the required amount of heat can be obtained for sound welding. Increasing PD and also by increasing α, contributes further to generate more heat. These parameters thus should be chosen accordingly to avoid any defects formed due to insufficient heat generation.

4.2.3 Defects Generated Due to Improper Mixing

Improper mixing is the result of many factors. The low heat generation and improper geometry of the tool and its positioning lead to defects. Some of the usual defects due to improper mixing are oxide entrapment, lack of penetration, excessive flash generation, which are described below.

Oxide entrapment The oxide layer, formed by the material in the presence of air, gets penetrated through the weld material during the welding. This hinders the formation of a bond, which severely decreases the joint strength [33].

Lack of penetration If *PD* is not sufficient enough, the tool is then unable to mix the material at the bottom of the plate thus reducing the joint strength [33].

Excessive flash This occurs if *PD* is very high due to the use of the required longer tool pin. This leads to severe plasticization, as well as expelling of the plasticized material from the stir zone creating thinning of the base plate [33].

By properly cleaning the workpiece material before welding and also selecting the tool pin and shoulder geometry according to the material properties and can solve the above-mentioned problems.

4.3 Recent Market Demands of FSW Process

Initially, with the use of milling machines, the FSW process was estimated to be a laboratory-compatible process only. However, soon with the ample advantages over the conventional joining process and also with constant growth in research in this area, FSW has been proved to be one of the fastest-growing welding processes to be adopted in the industrial sector just after four years, from its invention in 1991. ESAB of Sweden realized the importance of the process and developed the industry-scale FSW machines with many modifications [38]. Many other companies soon followed the path; and now the market is full with a large variety of advanced FSW machines for more complicated joining.

The 2xxx and 7xxx series aluminium alloys, which are considered impossible to be welded with the traditional welding techniques, were successfully welded by FSW. Later, this process proved itself compatible with numerous alloys and materials, including magnesium (Mg), thick copper, high-strength steels, titanium, stainless steel, and pure carbon steel. FSW also successfully joined the dissimilar materials. The invention of FSW changed the fabrication ways for various components used in different industries, namely aerospace, automotive, shipbuilding, etc. [25]. The constant requirement of light-weightiness without compromising the strength due to low carbon emission and less fuel consumption is the prime factor for the increasing demand for the FSW process. This factor is projected to increase the global market of FSW by USD 885.55 Million increasing the CAGR by 6.9% in the forecast period of

2018–26 [39]. With the advent of automation in the welding processes for better weld quality, precision, increased productivity and decreased labor cost also add to the gain in popularity of FSW. This is going to play a significant role in the acceptance of the process in many industries. Moreover, for the various lean initiatives like reduced bottlenecks and enhanced productivity of the process, the competitors are motivated to develop the new technology to gain more benefits.

Fixed product types mainly dominate the FSW market. However, due to its feasibility to join dissimilar material and materials with different thicknesses, the adjustable product type (where different products can be fabricated) FSW market will grow soon to a large extent. Again, with the rising demand for high speed and light-weightiness with higher fuel economy, the automotive sectors are replacing many steel parts with lightweight options like aluminium, magnesium, (Al, Mg), etc. by joining them with steel using FSW. Though in 2017, it was seen that the automotive sectors dominated the FSW market share, but it is predicted that the aerospace sector will grow at a faster rate in the forecast year of 2018–2026 [40]. The reduction in weight by replacing thousands of rivets with FSW joints in wing and fuselage components and the process-capability to weld large tanks are the possible reasons for the growth.

The Asia Pacific Region, being home to large automotive and fabrication industries, generates the highest revenue and has the maximum unit shipment as per the report in 2017. The region itself generated a 35.1% revenue share of the total FSW market worldwide [40]. This trend shall prolong in the forecast year also. China represented the largest individual FSW market in the Asia Pacific, followed by Japan and India. On the other hand, after the economic recession in North America and Europe, the automotive, marine, and aerospace industries in these developed countries showed signs of recovery and are adopting state of the art FSW equipment for more benefits.

4.4 Wormhole Defect—Invention of Friction Stir Channeling

To make a joint successful by FSW process, there must be adequate heat generation and appropriate mixing of the softened material. The heat generation in FSW depends mainly on the process parameters like ω, v and PD. The geometry of the tool also contributes to the generation of heat and primarily on the deformation and mixing of the material. It was seen that to avoid flow-related defects the parameters must be such selected that the temperature creates the stick-slip wiping flow of the material. Moreover, the flow must ensure that the material flowing from the leading edge of the pin is exactly balanced with the material deposited at the trailing edge of the pin [41]. If process parameters are not optimum, heat generation becomes insufficient or there is inappropriate mixing during the process. This leads to the formation of a defect called the *"tunnel defect"* or *"the wormhole defect"*. It has already been mentioned in Sect. 2.2 that the *wormhole defect* is nothing but voids inside the weld line that staggeringly deteriorates the weld quality, as well as strength. However, utility of the said defect is discussed further.

In 2005, researcher showed that by reversing the material flow generally occurring in FSW and by purposefully creating and utilizing the wormhole defect, an integral and continuous tunnel or channel can be created inside a monolithic plate in a single step [42]. This can be possible by selecting the process parameters and tool designs that promote the generation of wormhole defect during FSW. The process was named Friction Stir Channeling (FSC). It was identified that the channels created by FSC have huge potential to be used in the heat exchanger applications.

4.4.1 Importance of Compact Heat Exchanger

The problem of heat generation and removal is present in a myriad of applications and processes across the world necessitating the design and development of novel heat dissipation and exchange devices. From aerospace to power plants, automotive to space, power systems to consumer electronics, the importance of thermal management calls for innovative designs of heat sinks and heat exchangers and processes to manufacture them. Heat Exchanger is a device that promotes heat flow between two or more fluids having a difference in temperature, thus maintaining the temperature in the desired range. Heat exchangers can be classified based on their geometry (tube, plate, extended surfaces), based on transfer phenomenon (direct contact type or indirect contact type), flow arrangement types (parallel, counterflow, cross-flow, etc.), heat transfer mechanisms (single-phase or two-phase), and also surface compactness (micro, mini, macro, etc.) [43].

In this era, a very serious issue like energy depletion must be addressed sincerely. Therefore, researchers are continuously thriving to produce energy-efficient systems. The two most common ways of producing such a system are the integration of lightweight materials and miniaturization or compactness. It was stated that miniaturization in manufacturing and electromechanical systems would be the economic drivers shortly [44]. It was observed that heat exchangers with a high aspect ratio significantly increase the heat transfer efficiency [45]. So, studies on the miniaturization of heat exchangers also gained a considerable thrust. In this regard, two groups of researchers classified the heat exchangers according to the size. According to Mehendale's classification [46] the conventional channels have $D_h > 6$ mm, compact passages have 1 mm $< D_h \leq 6$ mm, meso-channels have 100 μm $< D_h \leq 1$ mm, and the microchannels have 1 μm $< D_h \leq 100$ μm. Again as per the classification by Kandlikar and Grande [47], the cooling channels can be identified as conventional channel with $D_h > 3$ mm, mini channel having 200 μm $< D_h \leq 3$ mm, micro-channel having 10 μm $< D_h \leq 200$ μm, transitional-channel having 0.1 μm $< D_h \leq 10$ μm, and molecular nanochannel $D_h \leq 0.1$ μm. The former classification was based merely on the dimension of the heat exchanger channels, whereas the latter and the most commonly used classification was done by taking the flow into consideration.

It was further reported that very widely used gas to gas, or liquid to gas compact heat exchangers are made with a surface area density greater than about 700 m^2/m^3 [43]. Some examples of their applications are heat exchangers used in automotive vehicles, air conditioning and refrigeration condensers, and evaporators. Oil coolers used in aerospace, automotive radiators, air heaters, intercoolers of compressors, etc. are other important areas where these heat exchangers are used extensively. The compact heat exchangers also frequently find their applications in electronics cooling, energy recovery systems, and cryogenics.

4.4.2 Conventional Techniques and Limitations

Owing to the various advantages of using the compact heat exchangers with mini channels, researchers started to venture into different fabrication techniques for creating continuous cooling channels. Some of the conventional techniques like drilling, milling, electro-discharge machining (EDM), and advanced techniques like chemical etching, diffusion bonding, etc. are the techniques most commonly used for creating cooling channels in heat exchangers. Nevertheless, these techniques have their limitations.

Drilling, for example, is capable of producing straight hole only, that too up to a certain depth because of restriction of the length of the drill bit. Milling can produce non-linear profiled channels, but its capability to produce only open channels adds to its limitations. EDM, on the other hand, can create channels on conductive materials. The high cost of the set up, the addition of consumables, and the use of non-flexible tools again limit its application for creating cooling channels. Other advanced processes like chemical etching have a low material removal rate that reduces productivity. Selective laser melting is used for fabricating channels, but the very high cost associated with this process limits its application to the industry level. Diffusion bonding was also reported to be used for fabricating heat exchangers, but the process requires primary channel creation by some other techniques. These limitations call for a better manufacturing process that can minimize the disadvantages of the processes mentioned above.

4.4.3 Potential of FSC Process

FSC has emerged as a new and non-conventional technique by which, in a single step, continuous and integral channels can be created inside monolithic or dissimilar plates. It is possible to fabricate different shapes of channels by FSC so that the heat transfer rate can be enhanced. The cooling channels fabricated by FSC have potential applications in the automotive cooling system, electronic devices and systems, conformal cooling in the mould industry, and several other domains [48, 49].

The unique features of the process that proves its benefits over other manufacturing processes are as follows:

- Single-step process to create a channel.
- Different positions and dimensions can be attained along the length of the channel by FSC.
- Repeatability and stability in maintaining the dimension of the channel are ensured by the process.
- Environmentally friendly process using no consumables and no coolant.
- Fine-grain size at the NZ of the channel enhances the ceiling strength of the channel.
- High productivity.

4.5 FSC Process

As mentioned previously, the FSC process is derived by creating the *wormhole defect* with low heat generation and the purposeful material extraction from the workpiece during FSW. By controlling the process parameters, this wormhole defect can be made throughout the plate resulting in a continuous channel. Since FSC is a derivative of the FSW process, thus the main steps of the process are almost similar to that of the FSW. The main feature that makes this process unique is the absence of plunging off the shoulder onto the workpiece. The FSC process can be understood from Fig. 4.6.

The initial step of FSC is the insertion of the pin with a given ω and vertical feed. The main difference in this step of FSC with that of FSW is the provision of

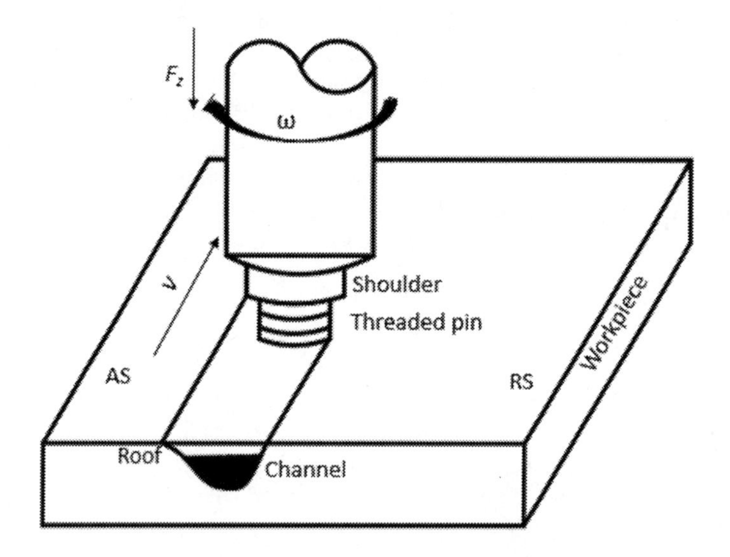

Fig. 4.6 FSC process

clearance, *C* between the shoulder and the workpiece surface. For the conventional FSC process, proper *C* must be maintained for deposition of the material that is extruded out and flows upward during the process. However, in another variant, i.e., the new FSC technique, it is seen that the channel can be formed at zero value of *C* also [50].

The second step, i.e., *dwelling* of the pin is also similar to FSW. But unlike the insertion of shoulder in FSW, since in FSC only the pin is inserted into a certain depth, therefore, in FSC only the pin dwells in the same position to generate heat. Thus, the heat generation, and as a consequence, the amount of material that gets softened in this step, is low as compared to that of FSW.

In the third step, the pin then *traverses* with a given value of v. But unlike FSW, the tool traverses with extracting the material along its length. The extraction is caused because of ω, and the thread orientation of the pin. The extracted material accumulates in the gap or clearance provided and seals the channel roof along the length. The value of v given in FSC is higher than that in FSW so that low heat generation conditions can be fulfilled.

The last stage i.e., the *retracting* of the pin is similar to that of the last step of FSW. Here the pin is pulled out at the end of the channeling line leaving a keyhole at the end.

4.5.1 FSC Principle

The main features of FSC that are different from FSW are as follows:

- The flow of material in FSC is upward and towards the shoulder from the base of the plate. It is obtained by rotating a profiled tool (mainly threaded).
- A suitable gap, i.e., a proper value of *C* is maintained throughout the process so that the material that comes out of the base of the material can be deposited in this gap itself. The geometry, dimension, and continuity of the channel can be adjusted by *C*.

The clockwise rotation of a right-hand threaded (RHT) tool or counterclockwise rotation of a left-hand threaded (LHT) tool produces a force that acts in the upward direction during FSC. The upward force shears the plasticized material from the bottom and around the pin. This material is then deposited in the gap. Unlike FSW, the absence of *PD* of the shoulder in FSC results in lower heat generation. The low heat generation is necessary for the formation of a wormhole defect.

4.5.2 FSC Parameters and Their Influence

The primary parameters that influence the FSC process are also similar to that of FSW process (ω, v, *PD*). With the most common influential parameters in FSW, two other parameters that play a vital role in creating a channel in FSC are C and geometry of the tool. The influence of various parameters that control the FSC process is discussed briefly below:

Clearance between shoulder and workpiece surface, (***C***): Deposition of the extracted material from the NZ at the top of the channel is needed to close the channel roof. Thus, minimum C must be maintained. Higher C leads to open channels, whereas very low value closes the path of the channel.

Tool rotational speed, ($\boldsymbol{\omega}$): Heat generation in FSC is directly proportional to ω. In order to obtain a continuous channel, material should not be too softened, and thus ω is restricted to a reasonably smaller value than that is used in FSW.

Tool traverse speed, ($\boldsymbol{v}$): v also influences heat generation. For faster v, lesser would be the quantity of heat generation, because of insufficient contact time of the pin with the workpiece.

Tool geometry: The pin geometry governs the material removal from the base of the workpiece. The material at the base of the plate is flown upward due to the thread orientation of the pin. Again, use of spiral striate on the shoulder helps in moving the material away from the shoulder. For a plain and smooth shoulder (i.e., not scrolled or with nostriate), the material taken out by the threaded pin is settled down beneath the shoulder, i.e., at the gap.

The FSC parameters cannot be directly assigned from one aluminium alloy to another. For different grades of the same metal, the process parameters vary to give a better-quality channel.

4.6 Evolution of FSC Process

From its invention, the FSC process, because of its huge potential to replace the conventional techniques to create cooling channels and of the various advantages of this process, researchers started working on this field to modify and develop the process further to make it more industry adaptable. A timeline of the evolution of the process is given in Fig. 4.7.

The details of these processes are briefly discussed in the following sub-sections.

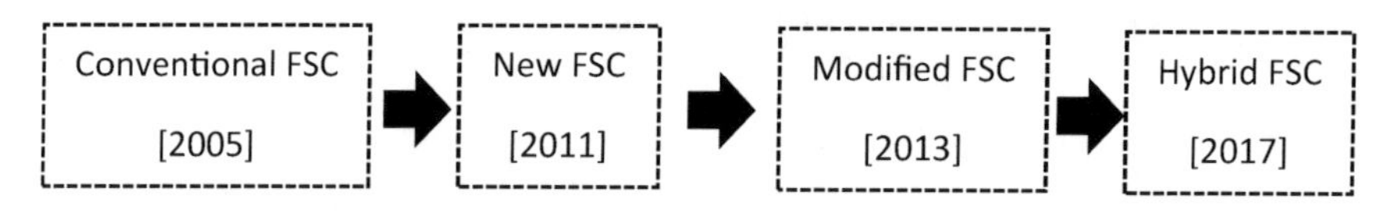

Fig. 4.7 Schematic representation of development of FSC with time

4.6.1 Conventional FSC (CFSC)

CFSC is the first version of the FSC process which is shown in Fig. 4.8. The main features of this technique are:

- No plunging of the shoulder takes place during the process.
- Threaded pin is used.
- No α is used.
- An initial C is maintained.
- Material flowing upward will deposit beneath the shoulder.

In CFSC, it was found that the wormhole defect generated during FSW can be converted to a continuous channel inside a monolithic plate by controlling the process parameters and tool geometry [42]. Preliminary studies on heat transfer and the pressured rop across channels of hydraulic diameters, varied from 0.8 to 1.4 mm, fabricated by FSC was also done to check the performance of the channels [51]. It was found that channels with smaller relative roughness have lower friction factors and it decreased with higher flow rates as the friction factor is inversely proportional to the flow kinetic energy. It was also seen that the pressure drop value raised by lowering the channel size as the area of the channel also decreased. Later, the parametric effect and thread and depth of cut of thread on the channel shape, size, and roughness created by FSC on a commercial Al6061 plate of 5 mm thickness were analyzed [52]. It was reported that channel size was enlarged with increase in v and decrease in ω of the tool. Both the factors contribute to lowering the heat generation, and thus restricts the material flow creating void. In this study, it was also found that the roughness produced in the channel roof had a periodicity which matched with the periodicity of the process pitch. Also, larger thread profile enhances the amount of material extraction and hence increased the channel cross-sectional area. From this study, it was shown that undesired defective open or closed-path channels were formed at cold and too hot process conditions, respectively. The authors also tested the influence of positioning of RS and AS increasing the serpentine profile. It was then concluded that positioning the AS in the inner curve for serpentine profile produced a better-quality channel. The reduced forging force acting on the materials depositing at C due to flow of material from the inner side of the curve to the outer with higher area

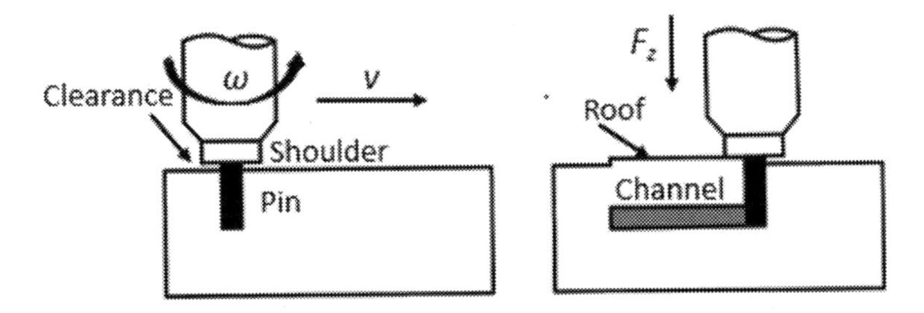

Fig. 4.8 Conventional FSC process

is the possible reason for this [52]. Co-relation of the process forces with the channel processing parameters and channel quality was also studied, and it was concluded that positioning of the total force applied on the tool decides the formation of channel in the NZ [53]. It was seen that for a good quality channel that forms at the nugget, the net resultant force acting upon the pin is positioned between the RS and trailing side of the tool. Again, from a developed mechanistic model for spindle torque and specific energy as a function of ω, v and PD, it was found that the decrease in ω and increase in v both reduce the plastic flow of the material. This is the reason for the increase in the resistance, thus, the specific energy of the material increases [54]. Similarly, the increment in PD increases the contact area of the pin with the material thus increases the resistance which ultimately increases the specific energy.

The material flow study during FSC was done on a 6 mm thick Al6061-T6 plate with the breaking tool technique [55]. The channel cross-section along with the embedded broken pin is studied with the X-ray Computed Tomography. The authors have reported that five distinct regions of material flow that are contributed to the formation of the four walls on the cross-section of the channel. It was observed that the roof of the channel got formed due to the combined effect of the pin influenced and the shoulder-influenced regions; whereas, the channel walls formed at the RS, AS, and the bottom of the channel got formed by the pin influenced-region only. Later, from the thermo-hydraulic studies by passing de-ionized water through the channels, it was found that due to the axial wall conduction along the length of the plate, non-uniform distribution of temperature was recorded [56]. However, in this study, the experimental value of heat transfer characteristics was found to be higher as compared to the theoretical values. The presence of surface roughness inside the channel wall justifies this deviation. The stability of the FSC process by comparing through force control and position-control modes proved that during position-control mode the FSC process underwent severe vibrations, and hence was lesser stable during its traverse [57].

4.6.2 New FSC (NFSC)

A new concept of FSC process was developed in 2011 and is shown in Fig. 4.9. The main features of this technique are:

- No plunging of the shoulder takes place.
- Tool having threaded pin and special featured shoulder containing striates is used.
- The shoulder touches the workpiece surface. Thus, no initial C is given.
- Due to the presence of features in the shoulder, the material flowing upward is flown away from the tool and is removed as a self-detachable flash.

In new FSC, the material taken out by the threaded pin from the base of the workpiece is directly detached as flash. Specially designed shoulder helps to detach the flashes rather than depositing it under the shoulder. This helps to maintain the workpiece surface at the pre-processed level that reduces the use of any surface

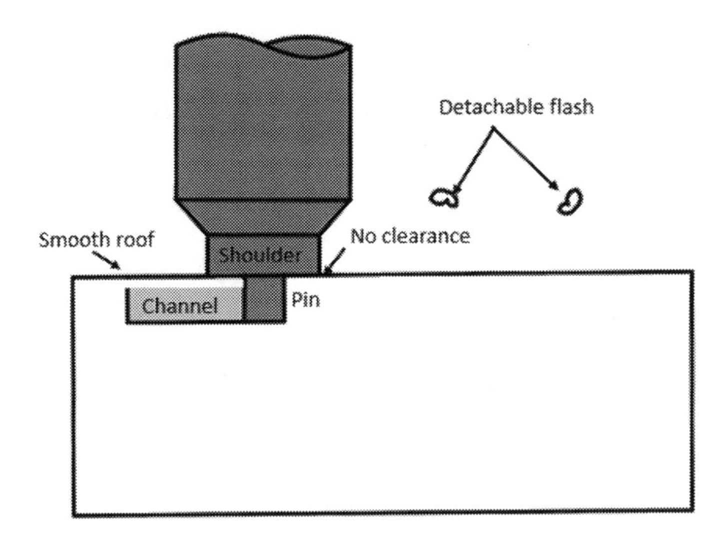

Fig. 4.9 New FSC process

finishing operation after the process [50]. The authors performed the FSC process on a 13 mm thick Al7178-T6 plate which is generally used in structural aircraft application. A conical left-hand threaded pin with a 5 mm bottom diameter is used. The shoulder with two spiral striates with a scrolling angle of 360° having outer and inner diameters of 20 mm and 9 mm, respectively, was used to form the channels. To see the influence of process parameters on the mechanical behavior of the channels in plane bending and internal pressure, tests were conducted. According to the bending tests, the bottom of the channel was more resistant than the roof due to the smoothness and integrity at the bottom. From the internal pressure tests, it was further possible to conclude that with an increase in ω, the minimum pressure, at which leak starts from the ruptured roof surface, increases. This is supported by the thicker channel roof for the processing condition with higher ω that produced better material flow [58]. From the metallographic characterization of the channels made on 13 mm thick Al7178-T6 alloy, it was found that the inner surface of the roof of the channel is comparatively rough and had wavy pattern, but the AS and the channel bottom were smooth. It was also reported that the channel area decreased with higher ω values, and it decreased further with an increase in v. The improper plasticization and inability of the pin to fill the gap at lower ω and higher v can be the possible reason for this [59]. To check the fatigue behavior of the channel, the authors did the fatigue assessment on 15 mm thick Al5083-H111 material. Four-point bending tests at room temperature were carried out to find out the fatigue strength of the channels, and it was concluded the nugget width is very much dependent on the fatigue strength of workpieces. Based on the observations of fracture surfaces, it was concluded that the fatigue-crack always got developed in the nugget/TMAZ interface on the AS because of stress intensity created by the gradients of very fine and coarse grains, further enhanced by channel corner geometry. As compared to results obtained at room temperature, while the tests were performed at 120 and 200 °C, a low value of the fatigue strength of the channels were found. Generally, the yield strength of metal decreases with the increase of temperature. At high temperature, due to the action of the additional factors activated by thermal as well as time-dependent phenomenon with fatigue loading, results in a reduced value of the fatigue strength of the material.

4.6.3 Modified FSC (MFSC)

MFSC was developed in 2013 and is shown in Fig. 4.10 schematically. The main features of MFSC are:

- No plunging of the shoulder takes place.
- Smooth cylindrical pin is used.
- The process can be done with or without α.
- C is given at the leading edge only.
- Material flowing upward will deposit beneath the shoulder to close the channel roof.

FSC channels on 10 mm thick Al6061-T6 plate with unthreaded pin and with α *was* created with a new version of FSC called the MFSC [60]. It was found that the unthreaded pins results in more uniform and regular shaped channel. In this case, α more than 2° is used in place of providing the gap to flow the material from around the pin to the upward direction. It was observed that the channel area produced by MFSC is bigger as compared to that produced by CFSC. Increasing amount of material extraction done by increased value of α in MFSC can be attributed to this. It was also seen that the channel produced during MFSC is more regular in shape and size due to absence of the thread in the pin. Unlike CFSC process, the use of non-threaded tool pin helps in maintaining a constant material flow during MFSC process. A different study with a broken pin technique on a 10 mm plate of Al-5083 plate was also performed by the same authors [61]. A columnar tool with α of 3°

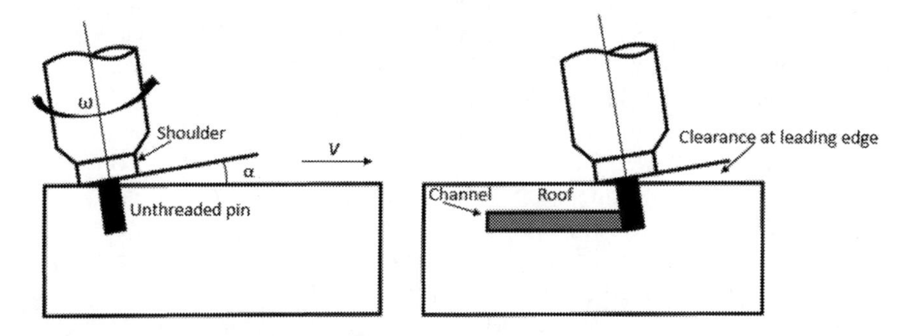

Fig. 4.10 MFSC process

was used in the study. In this study, the authors identified four different regions in the channel, namely the *nugget*, *channel roof*, *channel bottom* and *remaining region*, which was named as '*R*'. *R* was found to be a part of the nugget zone which was filled with materials from RS, the width of which was 25% of the pin diameter. The width of the channel was equal to the remaining 75% of the pin diameter. The influence of ω, v and α on the channel size was studied by observing the material flow with stop action tool method. The study showed that two defects namely *step defect* that is formed as an extended solidified material in the channel line and *groove defect* which forms at the channel roof due to insufficient material flow appeared [62]. The *step defect* was formed at hot process condition due to severe heat generation and can be eliminated by using low values of ω. Again, the *groove defect* was seen at cold process conditions due to tool o wheat generation that was unable to flow the material properly. In that study, a comparison of two-pin profiles to investigate the effect of tool geometry on MFSC was performed. For the investigation, an upper conical pin (UCP) and a straight cylindrical pin (SCP) were used. It was found that at the same process parameters, the UCP tool increased the channel height by 32%, and the hydraulic diameter was increased from 2.90 to 3.50 mm because of reduction in falling material into the channel cavity [63].

4.6.4 Hybrid FSC (HFSC)

HFSC is a new version of the FSC process that can produce a continuous channel, as well as a weld simultaneously by rotating a specially fabricated tool. The tip of the pin is used for joining two metal plates, whereas the rest of the pin is used to extract material from the bottom of the upper plate [50]. The HFSC tool is shown in Fig. 4.11.

The main features of this technique are:

- HFSC allows to get the benefit of FSW and FSC, simultaneously.
- Useful for tailor-made components.
- Specially designed shoulder and pin are used to join the materials as well as to form the channel within it simultaneously.

For the thermal management of chassis component for electronic application, the HFSC was introduced by using a specifically designed tool that is capable of joining, as well as channeling simultaneously in a plate or combination of different material plates in a single pass [64]. To investigate the thermal efficiency of the channel and also to verify the feasibility of the process, HFSC was done on two overlapped AA5754-H111 plates with the thickness of the top plate being 8 mm and a bottom plate of 5 mm thickness. It was found that the HFSC is repeatable and can produce channels with various dimensions and profiles. For better evaluation of the cooling efficiency of the HFSCs, a FEM model of real chassis of electronic components with multiple heat sources was developed. The channel dimensions were made the same as the dimensions being obtained by HFSC. It was found that the

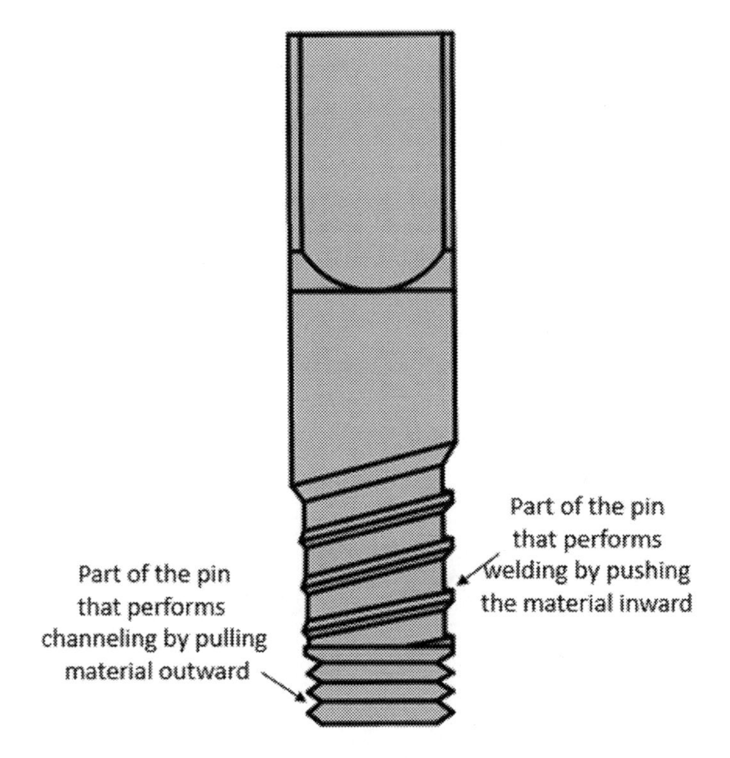

Fig. 4.11 HFSC tool used for simultaneous action of welding and channeling

roughness of the channel allows a significant cooling efficiency even for low flows, such as 0.5 l/min. At lower flow rates, the peak temperature in the chassis was about 29.4 °C. Whereas, for the 6.0 l/min the peak temperature in the chassis was 24.6 °C. Simulated results were also validated by fabricating the curved geometrical chassis component for electronic components and also compared the efficiency with a similar channel fabricated by milling. It was found that the HFSC reduced the steady-state temperature by 30–40%, and at the same time cooling rate during the transient period was enhanced by 33% than those of the milled channels [65].

4.7 Case Study on FSC

FSC being a very newly invented process, the research to understand the process for its better utilization is still in its nascent stage. The process has a very high probability of acceptance by replacing the conventional processes used to manufacture heat exchangers. Thus, the feasibility to produce better-quality channels to obtain higher thermal efficiency and fluid flow must be studied thoroughly. Some studies co-relating the process parameters to the quality of the channels in different versions of the process have been discussed earlier in the chapter.

One of the primary parameters of FSC is the *C*. It controls the channel height which ultimately decides the channel area and its hydraulic diameter. The amount of material excavated by the threaded pin depends on the dimension of the pin. Therefore, one should be very careful in selecting the dimension of the pin (diameter and height) and *C*. If the material excavated is less but the *C* is high, this may lead to open channel. However, if the *C* is too small to accommodate the amount of material pulled out by the pin, the channel produced may be discontinuous or even possibility of no channel creation is also there. The authors of this chapter have done a few experiments and tried to analyze the effect of *C* on the channel size and its fluid flow performance. For creating the cooling channels, a 6 mm thick AA 6063-T6 plate, having a dimension of 100 mm × 75 mm × 6 mm, was chosen because of its usefulness in heat exchanger applications. The chemical composition, mechanical and thermal properties of the material used have been shown in Tables 4.2 and 4.3, respectively.

The tool used to conduct the FSC process is AISI-H13 tool steel. It is the most frequently used tool material in FSW of aluminium due to its good strength at elevated temperature, corrosion resistance, wear resistance, and also easy availability.

FSC to create the linear channels was performed by a NC FSW machine (Make: ETA India, WS004), shown in Fig. 4.12. FSC is done by using the position-control mode. The process parameters, such as ω, v, PD, α, etc. can be controlled with the help of an HMI. The software installed gives the values of force, torque with respect to time, that can be taken by the user in a storage device.

An H13 right-hand threaded cylindrical pin modular tool, shown in Fig. 4.13, is used to create the linear channels. The tool was clamped on the spindle of the FSW machine and given a ω and v to perform the operation. The workpiece was fixed on the table of the machine with a proper fixture so that during the process the workpiece does not experience any displacement.

To investigate the effect of *C* on the channel quality, three experiments with repeatability were conducted. In all of the experiments, the ω and v were kept constant. The process parameters used for the creation of channel are given in Table 4.4.

Table 4.2 Chemical composition of AA 6063-T6

Al	Fe	Mg	Si	Cu
97.5	0.35	0.45–0.9	0.2–0.6	0.1

Table 4.3 Mechanical and thermal properties AA6063-T6

E (GPa)	Yield stress (MPa)	Ultimate tensile stress (MPa)	Elongation (%)	Thermal conductivity (W/m-K)
68.9	214	241	12	200

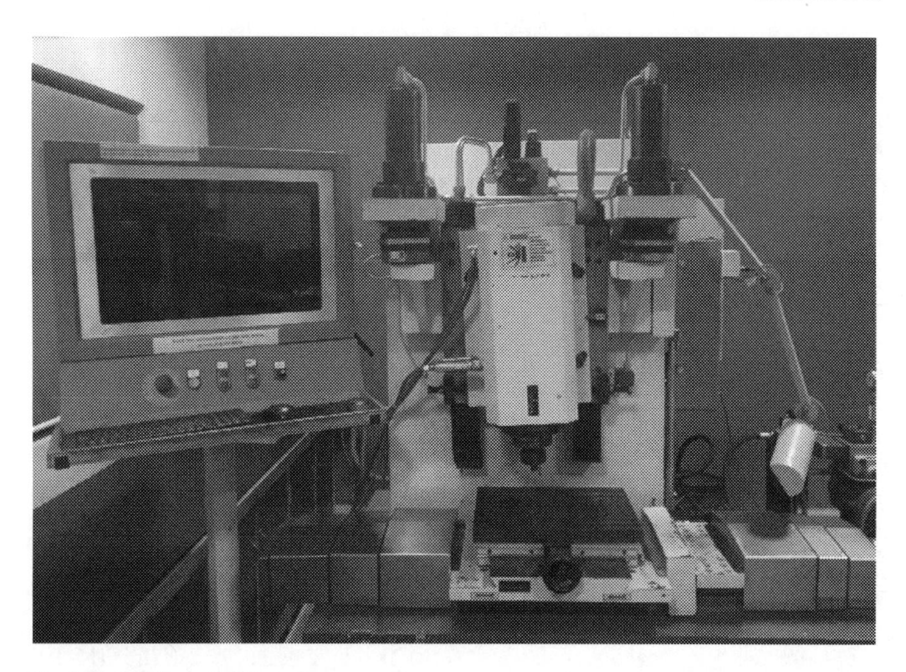

Fig. 4.12 NC linear FSW machine

The linear channels were successfully created with the FSC performed with a *C* of 0.5 mm and 0.8 mm. The FSC trial with zero value of *C*, however, was not successful. The cooling plates fabricated with the three values of *C* are shown in Fig. 4.14.

The force versus time data was captured during experiments for different values of *C*. It was found that the downward force was the highest in case of FSC with zero value of *C*, as compared to that of the other two trials.

This can be attributed to the fact that during *zero* value of *C*, the shoulder was in direct contact with the workpiece and experienced the maximum force during the first contact of the shoulder with the workpiece. Among the channels that were created with *C* of 0.5 mm and 0.8 mm, respectively, the latter has a lower F_z. As the pin used was the same thus, the variation in the clearance change the amount of pin plunge inside the plate, and hence, the amount of material extraction also. Material to be deposited by *C* of 0.8 mm will be less since pin plunge will be 3.7 mm in this case as compared to 4 mm pin plunge in case of *C* of 0.5 mm. In *C* of 0.8 mm, resistance force on the shoulder will be less.

Further, the flow rates were calculated with the help of flowmeters having a capacity of 500 ml/min. It was seen that the average flow rates through a single channel were 355 ml/min for the channels created with *C* of 0.5 mm, whereas, the flow rate for the channel with *C* of 0.8 mm was 450 ml/min. This result clearly shows that the channel created with *C* of 0.8 mm had a larger area than that of the channel created with *C* of 0.5 mm.

Fig. 4.13 FSC tool

Table 4.4 Process parameters for linear channels created by FSC

Process parameters	Values
ω (rpm)	800
v (mm/min)	120
C (mm)	0, 0.5, 0.8

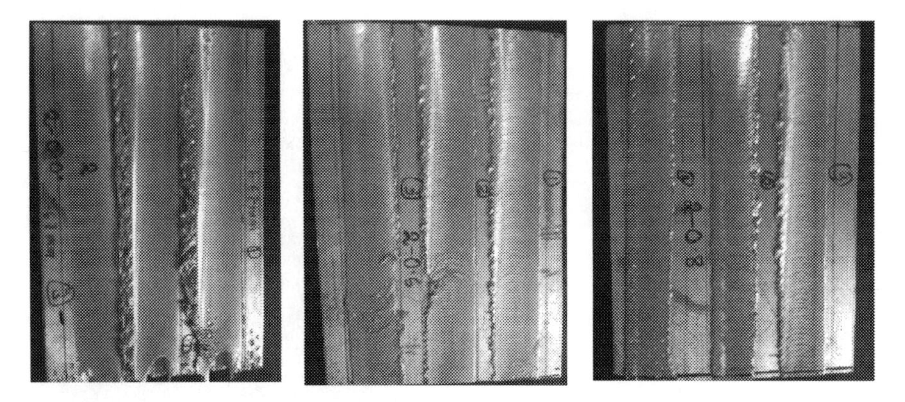

Fig. 4.14 Channel created by FSC with zero value of *C* (left), *C* of 0.5 mm (center), and *C* of 0.8 mm (right)

From the comparative plot in Fig. 4.15, it can be seen that with *C* of zero value, F_z is the highest amongst all three conditions. As the amount of material extracted did not get enough gap above the workpiece to be deposited and hence is further plunged back into the plate, hence no channel is created in this case. Again, at the highest clearance condition i.e. at *C* of 0.8 mm value the plunging force is the minimum. Moreover, enough gap is obtained by the material excavated by the pin to get deposited above the plate. The lowest value of F_z is also not sufficient to thrust back the complete material into the gap and close it. Hence, the height of the channel

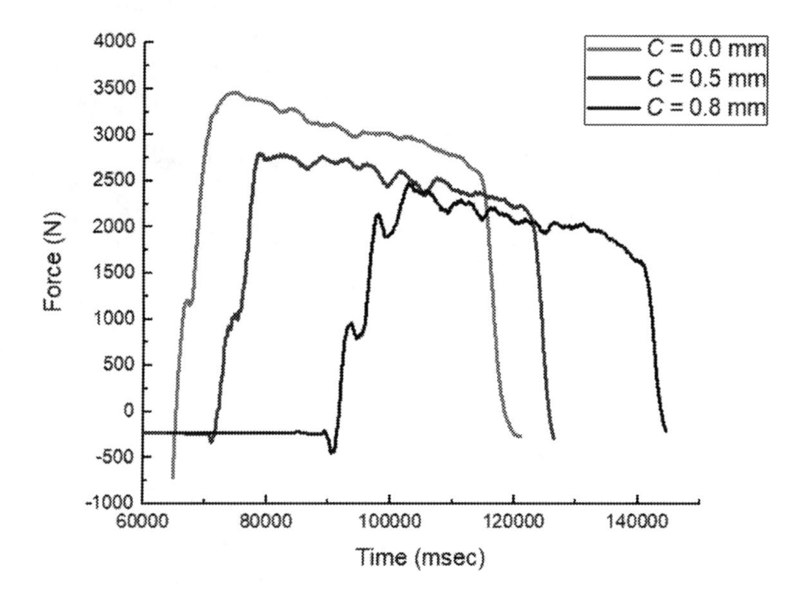

Fig. 4.15 Force versus time in FSC for different values of *C*

created in this case is the maximum which can be verified from the highest flow rate obtained by the channels created at C of 0.8 mm.

4.8 Future of FSC: Complex Heat Exchanger Used in Different Components

The above-mentioned disadvantages, associated with the existing processes, demand a requirement of a better manufacturing process that can satisfy the growing demand for heat exchanger fabrication processes.

With proper control of tool geometry and process parameters, the channel size and shape can be developed for different applications. Some of the potential applications of the cooling channels created by FSC are the cooling system used in automotive industries, car radiators where FSC made heat exchangers can be used.

The highly gazetted electronic systems in electric vehicles, powerful engines used in aerospace, lithium-ion battery packs, high-speed computing servers, etc. need proper thermal management for their enhanced product life and better functionality. The miniaturized heat exchanger created by FSC can prove to be a good solution for these in near future.

The conformal cooling channels used in moulds and the electrical wiring channels inside monolithic plates can also be created by FSC.

The water jackets used for cooling of pipes carrying hot fluids and also different cylindrical vessels cooling required for nuclear components can also be made by FSC because of its directional and dimensional independency, unlike other conventional channel fabrication processes.

4.9 Conclusion

Some of the greatest inventions in science were indeed achieved by accidents. The wormhole defect found in FSW is one such invention in manufacturing. The ability of the process to create closed cooling channels of complex shapes in one step makes it superior to the other conventional techniques. The FSC technique is very new, and the research in this field is in the nascent stage. To take this process to the industry ready level, further research is needed in this area. It is evident from the evolution of the FSC process and also from the parametric studies made by several researchers that tool geometry and process parameters like ω, v and C play a very significant role in controlling the shape, size, and quality of the channels created by FSC. Proper control of these parameters and studies correlating them and the thermal efficiency of the heat exchangers made by FSCare very much essential.

Acknowledgements and Funding This chapter is an outcome of the project funded by the Science and Engineering Research Board (SERB), Department of Science and Technology (DST), Government of India (Grant No-CRG/2019/005798).

References

1. https://www.millerwelds.com/resources/article-library/the-history-of-welding
2. Thomas WM, Nicholas ED, Needham JC, Murch MG, Templesmith P, Dawes CJ (1991) G.B. Patent 9125978.8
3. Yoshihiro K (2013) Honda develops robotized FSW technology to weld steel and aluminum and applied it to a mass production vehicle. Ind Robot Int J 40(3):208–212
4. https://stirweld.com/en/applications-2/
5. Huang YX et al (2011) New technique of filling friction stir welding. Sci Technol Weld Join 16(6):497–501. https://doi.org/10.1179/1362171811Y.000000003
6. Reynolds AP (2008) Flow visualization and simulation in FSW. Scripta Mater 58(5):338–342. https://doi.org/10.1016/j.scriptamat.2007.10.048
7. Hartl R et al (2019) IOP Conf Ser: Mater Sci Eng 480:01203
8. Zhang YN et al (2012) Review of tools for friction stir welding and processing. Can Metall Q 51(3):250–261. https://doi.org/10.1179/1879139512Y.0000000015
9. Thomas WM et al (2003) Friction stir welding tools and developments. Weld World 47:11/12
10. Thomas WM (2003) Materials science forum. Trans Tech Publ Switzerland 426–432:229–236. https://doi.org/10.4028/www.scientific.net/MSF.426-432.229
11. Thomas WM et al (2003) Friction stir welding-recent developments in tool and process technologies. Adv Eng Mater 5:7. https://doi.org/10.1002/adem.200300355
12. Dawes CJ et al (1995) Development of the new friction stir technique for welding aluminum—phase II. TWI member report 5651/35/95
13. Colligan KJ et al (2003) Welding tool and process parameter effects in friction stir welding of aluminum alloys. Friction stir welding and processing II, TMS, pp 181–190
14. Mishra RS, Mahoney MW (2007) Friction stir welding and processing, pp 7–35. https://doi.org/10.1361/fswp2007p007
15. Mahoney MW et al (2003) Microstructural modification and resultant properties of friction stir processed cast NiAl bronze. Mater Sci Forum 426–432:2843–2848
16. Ramasubramanian U et al (2005) Friction stir processing of class 40 grey cast iron. Friction stir welding and processing III, TMS, pp 115–122
17. Kumari K et al (2015) Friction stir welding by using counter-rotating twin tool. J Mater Process Technol 215(1):132–141
18. Pal SK et al (2014) Counter-rotating twin tool with variable gap for friction stir welding. Patent Filed Ref No: 1297/KOL/2014
19. Rai R et al (2011) Review: friction stir welding tools. Sci Technol Weld Join 16(4):325–342. https://doi.org/10.1179/1362171811Y.0000000023
20. Threadgill PL et al (2009) Friction stir welding of aluminium alloys. Int Mater Rev 54(2):49–93
21. Pacchione M, Lohwasser D (2004) Friction stir welding application to aircraft primary structures. Aircraft design principles PR0407463, Airbus
22. Christner B et al (2003) Development and testing of friction stir welding as a joining method for primary aircraft structure. In: 4th international FSW symposium, Park City, UT, USA, May 14–16
23. Talwar R (2000) Friction stir welding of airframe structures. In: 2nd international FSW symposium, Gothenburg, Sweden, June 26–28
24. Hattingh DG et al (2008) Characterization of the influences of FSW tool geometry on welding forces and weld tensile strength using an instrumented tool. J Mater Process Technol 203:46–57

25. Kallee SW (2010) Industrial applications of friction stir welding, Friction stir welding from basics to applications. Woodhead Publishing Limited, Sawston
26. Ferraris S, Volpone LM (2005) Aluminium alloys in third millennium shipbuilding: materials, technologies, perspectives. The Fifth International Forum on Aluminum Ships, Tokyo, Japan, pp 11–13
27. Dudzik K, Czechowski M (2009) Analysis of possible shipbuilding application of friction stir welding (FSW) method to joining elements made of AlZn5Mg1 alloy. Polish Maritime Res 4(62), 16:38–41
28. Kallee SW et al (2001) Industrialisation of friction stir welding for aerospace structures. In: Structures and technologies—challenges for future launchers, Third European conference, pp 11–14
29. Kallee SW et al (2002) Railway rolling stock manufacturers implement friction stir welding. Weld J
30. Thomas WM, Nicholas ED (1997) Friction stir welding for the transportation industries. Mater Des 18(4r6):269–273
31. Thomas WM et al (2006) Friction stir welding—process variants and developments in the Automotive Industry, 2006 SAE World Congress, 3–7. Cobo Center, Detroit, Michigan, USA
32. Zettler R et al (2010) Effects and defects of friction stir welds, Friction Stir Welding from Basics to Applications. Woodhead Publishing Limited, Sawston
33. Jain R et al (2015) Friction stir welding: scope and recent development, pp 179–229
34. Hartl R et al (2019) Automated surface inspection of friction stir welds by means of structured light projection, IOP Conf Ser: Mater Sci Eng 480:012035. https://doi.org/10.1088/1757-899x/480/1/012035
35. Kah P et al (2015) Investigation of weld defects in friction-stir welding and fusion welding of aluminium alloys. Int J Mechan Mater Eng 10:26. https://doi.org/10.1186/s40712-015-0053-8
36. Mahto RP et al (2020) Characterizations of weld defects, intermetallic compounds and mechanical properties of friction stir lap welded dissimilar alloys. Mater Charact 160:110115
37. Khan NZ et al (2015) Investigation on tunneling and kissing bond defects in FSW joints of dissimilar aluminium alloys. J Alloy Compd 648:360–367
38. Threadgill PL (2010) The future of friction stir welding. Friction stir welding from basics to applications. Woodhead Publishing Limited, Sawston
39. Global friction stir welding market to exceed US$ 1.04 Bn by 2026, Press Release, Credence Research
40. Friction stir welding market by equipment (fixed friction stir welding equipment, adjustable friction stir welding equipment, self-reacting friction stir welding equipment), by applications (shipbuilding and off-shore, automotive, railways, aerospace, fabrication)-growth, future prospects and competitive analysis, 2018–2026, Credence Research, 2017
41. Arbegast WJ (2008) A flow-partitioned deformation zone model for defect formation during friction stir welding. Scripta Mater 58:372–376
42. Mishra RS (2005) Integral channels in metal components and fabrication thereof. USA Patent US 6923362 B2, 2 August 2005
43. Kakaç K, Liu H, Heat exchangers: selection, rating, and thermal design, 2nd edn. CRC Press, Boca Raton
44. Dixit T, Ghosh I (2015) Review of micro- and mini-channel heat sinks and heat exchangers for single phase fluids. Renew Sustain Energy Rev 41:1298–1311
45. Tuckerman DB, Pease RFW (1981) High-performance heat sinking for VLSI. IEEE Electr Dev Lett Edl-2(5)
46. Mehendale SS et al (2000) Fluid flow and heat transfer at micro- and meso-scales with application to heat exchanger design. Appl Mech Rev 53(7)
47. Kandlikar SG, Grande WJ (2002) Evolution of microchannel flow passages-thermohydraulic performance and fabrication technology. Heat Transfer Eng 25(1):3–17
48. Vilaça P et al, Linear friction based processing technologies for aluminum alloys. Surf Stir Weld Stir Channel. http://dx.doi.org/10.5772/52026

49. Nordal D (2017) Development and analysis of tools for hybrid friction stir channeling. Dissertation for the Degree of Master of Science in Mechanical Engineering, Espoo
50. Vilaça P, Vidal C (2011) Modular adjustable tool and correspondent process for opening continuous internal channels in solid components. National patent pending Nº 105628T
51. Balasubramanian N et al (2007) Preliminary study of pressure drop and heat transfer through a friction stir channel. In: Proceedings of IMECE2007, 2007 ASME international mechanical engineering congress and exposition, November 11–15, Seattle, Washington, USA
52. Balasubramanian N et al (2009) Friction stir channeling: characterization of the channels. J Mater Proces Technol 209(2009):3696–3704
53. Balasubramanian N et al (2011) Process forces during friction stir channeling in an aluminum alloy. J Mater Process Technol 211:305–311
54. Balasubramanian N et al (2010) Development of a mechanistic model for friction stir channeling. J Manuf Sci Eng 132:054504-1-4
55. Pandya S et al (2019) Channel formation during friction stir channeling process—a material flow study using X-ray micro-computed tomography and optical microscopy. J Manuf Process 41:48–55
56. Pandya S et al (2020) Effect of axial conduction in integral rough friction stir channels: experimental thermo-hydraulic characteristics analyses, Heat and Mass Transfer. Springer-Verlag GmbH Germany, Part of Springer Nature, Berlin
57. Vidal C et al (2020) Monitoring of the mechanical load and thermal history during friction stir channelling under constant position and constant force control modes. J Manuf Proces 49:323–334
58. Vidal C et al (2012) Mechanical characterization of friction stir channels under internal pressure and in-plane bending. Key Eng Mater 488–489:105–108
59. Vidal C et al (2012) Metallographic characterization of friction stir channels. Material Science Forum 730–732:817–822
60. Rashidi A et al (2013) Channel formation in modified friction stir channeling. Appl Mech Mater 302:371–376
61. Rashidi A et al (2013) Modified friction stir channeling: a novel technique for fabrication of friction stir channel. Appl Mech Mater 302:365–370
62. Rashidi A, Mostafapour A (2014) Influence of machine parameters on material flow behaviour during channeling in modified friction stir channeling. Int J Mater Form
63. Rashidi A, Mostafapour A (2015) Influence of tool pin geometry and moving paths of tool on channel formation mechanism in modified friction stir channeling technique. Int J Adv Manuf Technol 80(5):1087–1096
64. Karvinen H et al (2017) New development and testing of FS channeling technique: hybrid FSC. In: 10th international conference on trends in welding research and 9th international welding symposium of Japan Welding Society (9WS), pp 376–379
65. Karvinen H et al, Application of hybrid friction stir channeling technique to improve the cooling efficiency of electronic components. Weld World. https://doi.org/10.1007/s40194-018-0576-8

Chapter 5
Welding of Dissimilar Metals—Challenges and a Way Forward with Friction Stir Welding

Suryakanta Sahu, Surjya Kanta Pal, Mahadev Shome, and Prakash Srirangam

Abstract The joining of dissimilar materials is proving beneficial for several industrial applications to achieve desired properties and improve the production economy. However, producing a satisfactory joint in between dissimilar metals is always being a challenge due to their distinct physical and chemical properties. This chapter will highlight those challenges, and also discuss about available welding methods for joining of dissimilar metals. Recently, the use of metals, for example, aluminium and magnesium, has been at the forefront in the automotive sectors because of their lightweight characteristics. These metals are being combined with steel to produce joints that are proving to be appropriate counterparts to the existing steel-based products. Thus, a detailed study regarding the importance, difficulties, and different welding methods for joining dissimilar metals for automobile industries has been outlined. Further, a case study has been presented, where aluminium-steel combination of dissimilar metals is joined using friction stir welding.

Keywords Dissimilar metals · Welding methods · Friction stir welding · Mechanism of joining · Intermetallic compound

---

S. Sahu
Advanced Technology Development Centre, Indian Institute of Technology Kharagpur, Kharagpur, West Bengal 721302, India
e-mail: suryakanta@iitkgp.ac.in

S. K. Pal (✉)
Department of Mechanical Engineering, Indian Institute of Technology Kharagpur, Kharagpur, West Bengal 721302, India
e-mail: skpal@mech.iitkgp.ac.in

M. Shome
Materials Welding and Joining, R&D, TATA Steel, Jamshedpur, Jharkhand 831001, India
e-mail: mshome@tatasteel.com

P. Srirangam
Warwick Manufacturing Group, The University of Warwick, Coventry CV4 7AL, UK
e-mail: P.Srirangam@warwick.ac.uk

J. P. Davim (ed.), *Welding Technology*, Materials Forming, Machining and Tribology, https://doi.org/10.1007/978-3-030-63986-0_5

5.1 Introduction to Welding of Dissimilar Metals

Dissimilar metal joints are becoming increasingly important in automotive, aerospace, electronics, and various other industrial applications. The reasons for using a combination of dissimilar materials in a single product are based on several considerations. Firstly, it is imperative to meet different technical requirements of the product, such as (i) structural requirements like a combination of strength and rigidity for static and fatigue loadings along with the reduction of weight, and (ii) operating conditions like local wear, corrosion and temperature resistance, etc. There are many cases arising from the above requirements where limited choices exist to join dissimilar materials. Second, combination of dissimilar metals provides designers and manufacturers of components and sub-assemblies more flexibility in fulfilling the functional requirement of the product or the component. Thirdly, the fabrication of products, containing dissimilar metals can yield economic benefits. Therefore, this strategy has found wide-scale applications in important sectors of manufacturing [1], and it has been explained in detail in the subsequent sections.

5.1.1 Applications of the Structures of Dissimilar Metals

5.1.1.1 Automotive

As mentioned in the previous section, joints of dissimilar metals have found major applications in the automotive industry. Light weighting is the main subject of automotive and transport manufacturing industries with the aim to improve the fuel efficiency. Worldwide, stringent regulations are being imposed to control the carbon footprint. In this regard, it is essential to implement multi-material concept, using different materials with specific properties for different parts of an automotive body. Automotive industry is thus attempting to come up with innovative solutions which would make the automotive lighter in weight, without compromising with safety of the passengers. A potential solution to this has been the tailor-welded blank (TWB) technology, combining different metals to be welded together. Such combinations aim at considering the most suitable properties of the materials which are then placed within the part depending on the application. This tailoring is done with respect to thickness, grades, strength, etc. [2].

5.1.1.2 Nuclear Reactor

Another area where the use of dissimilar materials is increasingly being explored is the nuclear reactor. The reactor consists of different high temperature-resistant materials present in the primary and secondary circuits. The performance of the reactor is mainly determined by the properties of the materials. This further considers devising proper development of required metals, and use of advanced integration technologies [3].

5.1.1.3 Aerospace

Aviation has been one of the major factors of global warming and environmental pollution. This is because of the fuel being consumed. Light-weight structures have been an effective way to increase energy efficiency and decrease the rate of fuel consumption. An example in this regard is the Boeing 787 airline, which has mentioned a 20% increase in fuel efficiency with a weight saving of 20% [4].

5.2 Welding—A Permanent Joining Approach

Welding is a permanent joining process performed either in a fusion state or a solid-state. Both these approaches have several different techniques for joining. Broadly, fusion welding produces a bond in which the base materials are chemically bonded because of the melting phenomenon. The process may use filler material as well. Some of the well-known techniques of fusion welding are: (a) gas metal arc gas (GMAW) welding, (b) gas tungsten arc (GTAW) welding, (c) laser beam welding (LBW), and (d) electron beam welding (EBW). With regard to the joining of dissimilar metals, the major demerits of these techniques are the melting and solidification of the base materials. Another important challenge with fusion welding is that it results in defects, such as porosity, oxidation of the molten metal, micro-segregation, hot cracking, etc., deteriorating the performance of the joint. Larger heat input in the case of fusion welding techniques leads to a wider heat-affected zone (HAZ), which degrades the mechanical properties of the joints. Some of these challenges and limitations are avoided in solid-state welding processes.

As the name suggests, in a solid-state welding method no melting happens; thus the protection of weld pool is not a concern here. The techniques in this approach are: (a) friction welding, (b) explosion welding, (c) friction stir welding (FSW), and (d) ultrasonic welding. Mostly, joints produced with these techniques retain the original properties of the substrate. In this case, the HAZ is very narrow as compared to the one produced in any of the fusion welding techniques because of the lower heat input. The major advantage comes with their ability to join dissimilar metals together. The thermal expansion is not severe in these processes as the joining phenomenon occurs below the melting temperature of the materials [5].

5.3 Influencing Factors for Dissimilar Welding

The development of various alloys has served many applications but welding them together perfectly and repeatedly continues to be a difficult process. The factors influencing a dissimilar joint have been discussed below:

5.3.1 Solubility

The most fundamental enabler of a dissimilar joint is the mutual solubility of the two metals. This ensures that the materials can dissolve in each other. With a very less solubility of both the mating materials, it becomes very difficult to join them. In many cases, a third material, soluble with the two base materials is often employed [6].

5.3.2 Intermetallic Compound

Intermetallic compounds (IMCs) are formed because of the limited solubility of alloy components. The microstructural composition of the resultant IMC, depends upon the mutual solubility, and it governs joint properties such as ductility, crack sensitivity, and corrosion [6].

5.3.3 Weldability

The solubility and type of intermetallic compound, together decide the level of weldability of that dissimilar combination. Post the determination, the appropriate selection of external filling metal or buttering layers can be done in order to ensure a smooth transition between the metals [6].

5.3.4 Melting Temperature

Large heat input in case of fusion welding may result in the melting of one metal before another in dissimilar welding. The solidification and shrinkage of the metal having a relatively higher melting temperature will induce stress in the other metal. In this regard, the buttering of the filler metal is often carried out with a material having an intermediate melting temperature. This helps in dealing with the problem of welding two metals having different coefficients of thermal expansions, and also restricts the detrimental elements moving to the weld pool at higher temperature [7].

5.3.5 *Thermal Conductivity*

A proper heat balance is required to be maintained while joining the dissimilar metalshaving significant dissimilarity in thermal conductivities. Thus, directional heat flow occurs towards the metal with higher thermal conductivitydue to the rapid conduction of heat. Often, preheating of the metal with greater thermal conductivity is performed to control the loss of heat and also to reduce the cooling rate [7].

5.3.6 *Thermal Expansion*

Large dissimilarity in thermal expansion coefficients of the two metals induces tensile and compressive stresses during solidification. This, in turn, leads to hot/cold cracking. This becomes critical for the weld joints deployed at high-temperature applications and in cyclic mode i.e. in expansion and contraction [7].

5.3.7 *Corrosion*

The base materials and the weld joint produced have different corrosion behavior which is another concern for welding of dissimilar metals. Galvanic corrosion is the most usual problem. To deal with this problem, cathodic protection is enabled on the base metals by varying the composition of the weld metal [8].

5.4 Welding Methods—Difficulties and Opportunities for Dissimilar Metals

The use of welding as a manufacturing process is believed to have been initiated in Egypt, somewhere around 4000 B.C. Since then, there have been various developments in this technology, which has brought several different techniques for welding of materials. At present, with the emergence of robots, welding is getting more popularity as it uses computer-controlled programs to weld metal more quickly and accurately.

The available welding techniques for joining dissimilar materials have been categorized in Fig. 5.1. Amongst these techniques, laser beam welding (LBW), electron

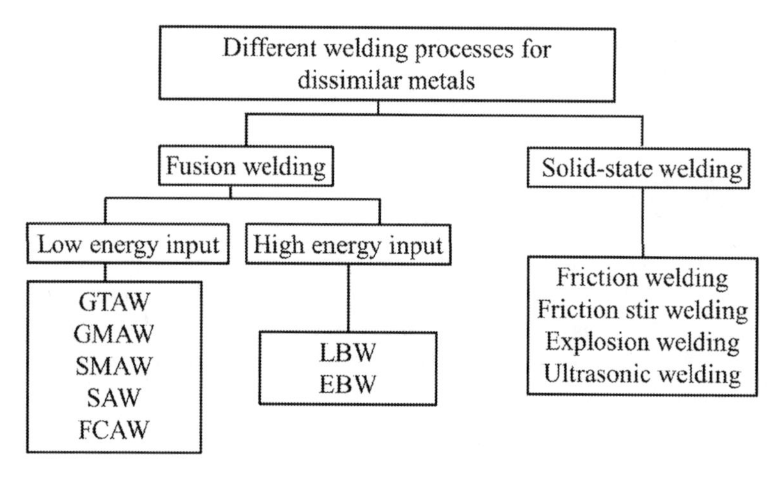

Fig. 5.1 Different welding processes for dissimilar metals

beam welding (EBW) which are advanced fusion welding techniques use the high energy beam during the joining of materials and non-fusion or solid-state welding techniques are being majorly preferred because of their higher production rate and efficiency to join the dissimilar materials. This is so because, with conventional fusion welding, the miscibility of the filler material with the base materials is a major challenge. These methods usually fail because of two broad reasons, (a) physical differences, which takes into account the dissimilarity in melting temperatures and the thermal expansion coefficients of base materials, and (b) faulty design parameters, which probably leads to the development of the brittle IMCand improper mixing. Furthermore, the relatively larger size of HAZ and the molten pool is one of the major challenges with the conventional low energy input fusion welding methods in comparison to the other welding methods.

The benefits and difficulties of each of those fusion welding methods have been discussed, to understand the associated problems, and later, the high energy input methods and solid-state welding methods have been discussed whose characteristics can help in overcoming the difficulties associated with the conventional low energy input fusion welding methods.

5.4.1 Gas Tungsten Arc Welding (GTAW)

GTAW (as shown in Fig. 5.2) is a fusion welding process in which the heat is developed by an electric arc. This arc is generated between the tungsten electrode and the workpiece [9, 10]. The process requires a filler material, and the features of this filler material depend upon the base materials [11]. This method is used for both similar and dissimilar welding [9, 10, 12–16]. However, special care is required in the case of

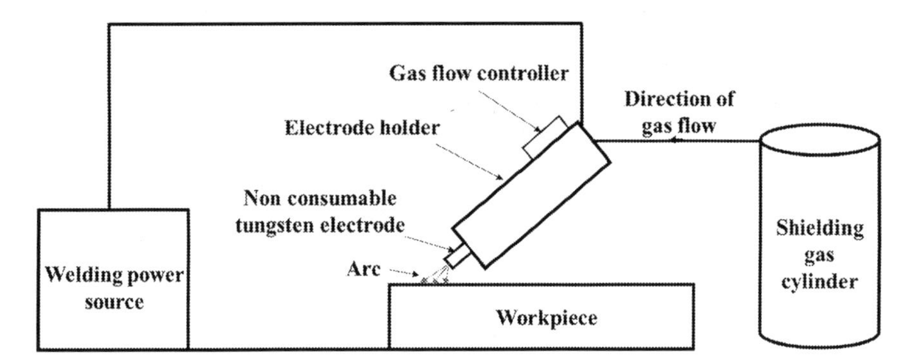

Fig. 5.2 Schematic of GTAW process

the dissimilar metals because of their compositional differences. The thermal expansion is one of the important concerns for welding dissimilar metals as it leads to the generation of internal stress. This thereby reduces the joint strength. In addition, a larger difference in their melting temperatures also reduces the joint strength. In GTAW process, melting of base metals and solidification of weld pool control the microstructures, that influence the joint properties. Furthermore, precipitation reactions, phase transformations, recrystallization, and growth of grain significantly alter the microstructure and weld properties [16].

5.4.2 Gas Metal Arc Welding (GMAW)

GMAW (as shown in Fig. 5.3) is another fusion welding method which also utilizes an arc generated between the electrode and the workpiece. A shielding gas is utilized to safeguard the molten pool from getting contaminated by the atmosphere [11]. It is advantageous than GTAW, as it consists of a spool containing the electrode which is constantly fed into the metals to be welded. This process can also be used for joining dissimilar metals [17]. But the concern remains with the differences in thermal and physical properties, such as thermal expansion of the materials to be combined. This is because larger differences may lead to the generation of post-welding residual stress. The generation of the residual stress at the joint indicates the development of crack within the joint, resulting in brittle failure of the component during the service period [18].

5.4.3 Shielded Metal Arc Welding (SMAW)

SMAW (as shown in Fig. 5.4), utilizes an arc delivered through an electrode leading to the coalescence of the base metals. A shielding gas protects the arc and the hot

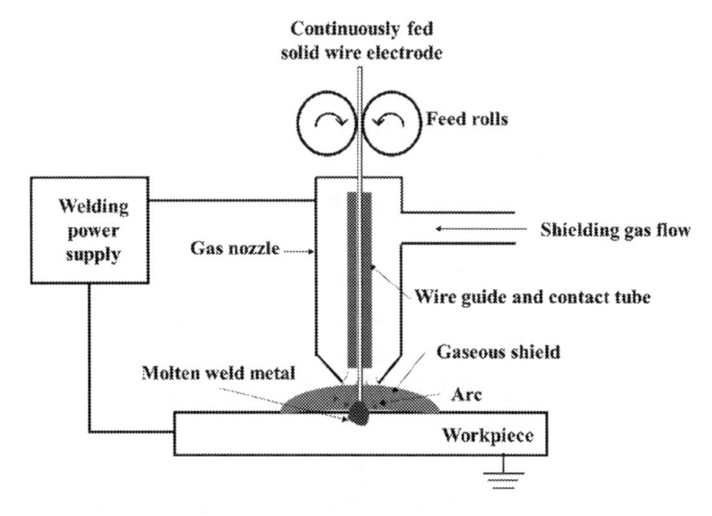

Fig. 5.3 Schematic of GMAW process

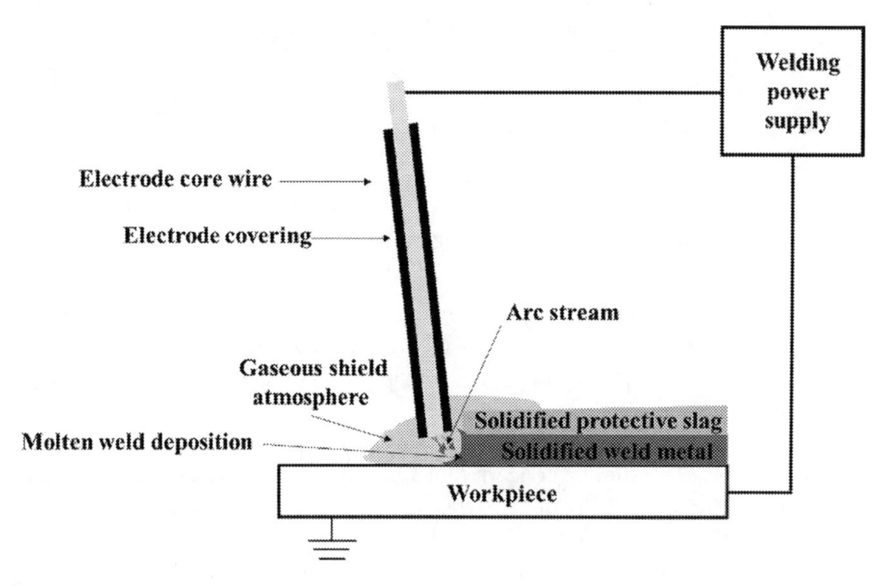

Fig. 5.4 Schematic of SMAW process

metal from getting contaminated by the atmosphere [11]. The parameters governing the weld quality are edge preparation, angle of electrode oscillation and movement, direction and position of the welding, etc. It is employed to weld various grades of steel, cast iron, copper, nickel, and aluminium alloys. This method has been also utilized for welding dissimilar metals [19, 20]. The factors posing challenges in joining dissimilar materials are the formation of residual stresses in the weldment because of the dissimilarities in the thermal expansion coefficients of the metals, precipitation of either of the materials at elevated temperature during heat treatment, and corrosion of either of materials because of environmental concerns. The other challenges in this process include porosity, weld spatter, shallow penetration, poor fusion, and cracking. Excessive heat input into the process causes spattering and may deflect the arc from the weld pool. This results in welding defects for example porosity and induces the residual stresses in the weld leading to cracking during solidification. On the other hand, lower heat input causes improper fusion of the metals.

5.4.4 Submerged Arc Welding (SAW)

In SAW (as shown in Fig. 5.5), the arc is generated underneath a layer of flux. This flux shields the molten metal from getting contaminated from the atmosphere [11]. It has been observed that the properties of the surfacing layer are directly affected by the arc generated during the process. Hence, it is essential to study the arc characteristics, especially the arc stability for improving the quality of the surfacing layer to ensure weld quality. SAW is many times preferred because of its automated nature which results in superior welds. This is because of the high heat availability which leads to full penetration. As such, this method is employed for joining different alloys of steel in both similar and dissimilar configurations [21]. However, joining of low melting point material like aluminium by this method is challenging as it has high thermal conductivity leading to loss of heat generated. Thus, the required amount of

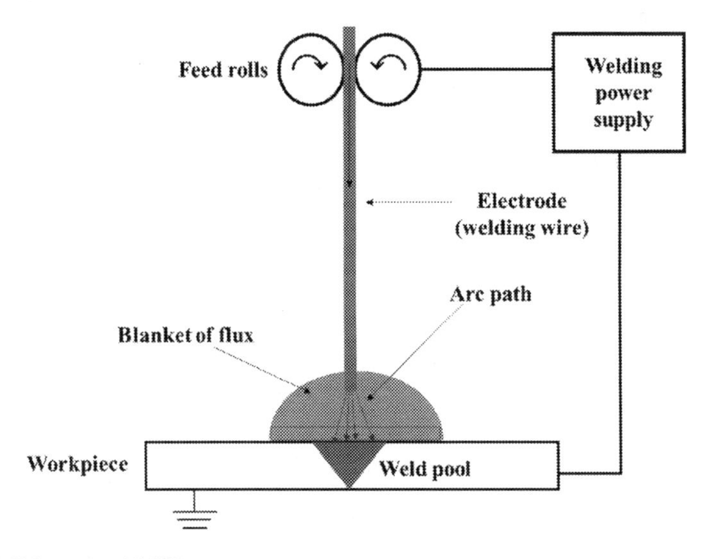

Fig. 5.5 Schematic of SAW process

penetration is not achieved. Moreover, joining of alloys of aluminium by using this method becomes expensive owing to the presence of oxide layer over the surface of aluminium. This oxide layer has high melting temperature. In order to remove the oxide layer, wire brushing, or chemical cleaning are performed which adds cost to the welding [21].

5.4.5 Flux-Cored Arc Welding (FCAW)

FCAW (as shown in Fig. 5.6) also utilizes an arc to produce coalescence of the base metals. An electrode containing flux is used for shielding [11]. In this welding process, different parameters such as the dilution, weld bead width, and penetration with respect to the current, nozzle to plate distance, and torch angle during welding are the key features affecting the quality of weld. The FCAW provides several advantages over other methods. This includes welding of thicker metals, endurance to strong breezes by the built-in shielding provided by the filler wire, flexibility with several alloys, and portability. It also offers high wire deposition rates and improved arc stability, which makes it suitable for applications requiring high speed. The disadvantages are the noxious smoke generated during welding which hinders the visibility of the weld pool, and formation of porosity with entrapping of the gases. FCAW has been also utilized for joining dissimilar metals [22]. It takes the advantage of the ability of the flux to braze and the rapid movement. The action of arc, with higher deposition rate and deeper penetration, results in proper mixing of the materials. However, a zone exists at the weld interface, where the materials remain un-fused. This zone is susceptible to corrosion, which is a concern for the joining of dissimilar metals. It has been noticed, that melting of the materials occursspeedily than that of the flux. This outcomes in a flux pole near the electrode tip. The occurrence of flux

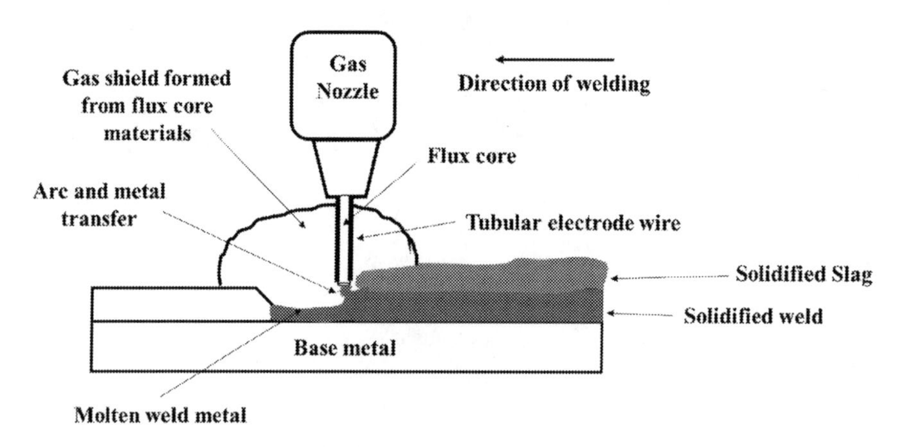

Fig. 5.6 Schematic of FCAW process

pole is unfavourable to the metal transfer, as the moving droplets have a tendency to grow in size which thereby increases the risk of spattering [22–24].

5.4.6 *Laser Beam Welding (LBW)*

LBW (as shown in Fig. 5.7) is an advanced fusion welding technique, uses a laser which is focused on the gap (cavity) formed in between the base metals. Laser upon striking the metal piece generates a high amount of heat leading to the melting of the metals, and this molten metal fills the cavity [25]. Dissimilar metal combinations can also be welded by the high energy input welding method involving the fusion process. However, it involves a relatively lower melting proportion of the base metals in the weld, with no filler material being added, and a better quality of weld in comparison to the conventionally used fusion welding technique. The use of LBW is favored for dissimilar metals because of its capability to control the amount of the brittle IMCs by using a high energy input being directed to a smaller area, higher rate of cooling, and shorter welding time. The high energy input retains a deeply penetrating weld pool, and as a result through thickness welds can be attained in a single-pass. Rapid cooling rates in LBW result in the development of fine solidified microstructures by limiting the grain growth in the HAZ. Metals such as aluminium and copper having high thermal conductivity pose difficulty to be welded by LBW as they disperse the energy more quickly posing a difficulty in maintaining the weld beam in molten condition. The irregular dissipation of heat in the joint of two dissimilar metals with inconsistent thermal conductivities might direct to the generation of an unequal weld bead with several root defects. These problems limit the application of LBW. On

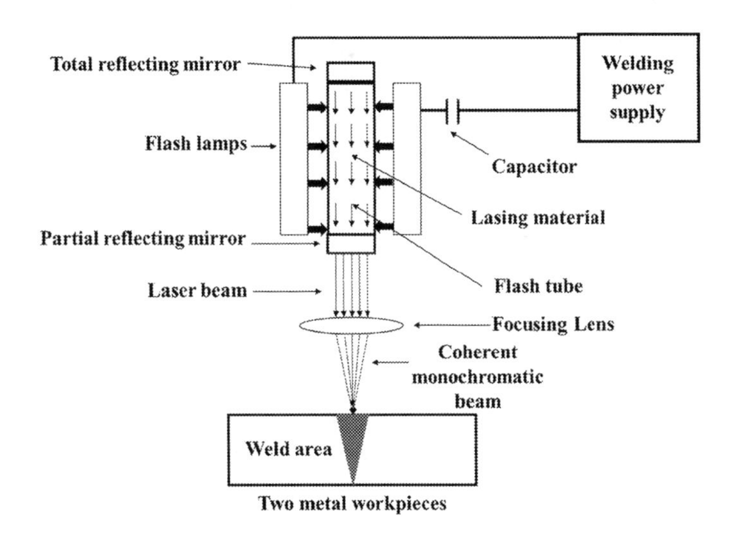

Fig. 5.7 Schematic of LBW process

the contrary, metals possessing higher degree of thermal conductivity generally have low absorptivity. Porosity is a commonly associated drawback in laser welds, mainly those involving partial penetration. The higher rate of solidification in the weld bead and the complexity in the formation of the keyhole become the important traits that exacerbate the problem compared to the conventional fusion welding methods.

5.4.7 Electron Beam Welding (EBW)

EBW (as shown in Fig. 5.8) which comes under advanced fusion welding methods employs electrons bearing high velocity, and bombards them on to the workpieces. These electrons are generated by an electron gun consisting of cathode and anode and is placed inside a high-vacuum chamber. The bombardment produces heat which melts the workpieces. The advantages of EBW include proper fusion of the metals because of the ability to precisely locate the heat input, and formation of smaller HAZ because of the high depth to width ratio. It is advantageous than the conventional fusion welding techniques, in the case of dissimilar metals joining, where the involvement of high temperature in case of conventional fusion welding deteriorates the weld quality by forming brittle and thick IMCs [26, 27]. The limitations of EBW with respect to joining of dissimilar materials are deflection of beam because of the variation in electrostatic and magnetic fields of the two materials, oscillation of beam

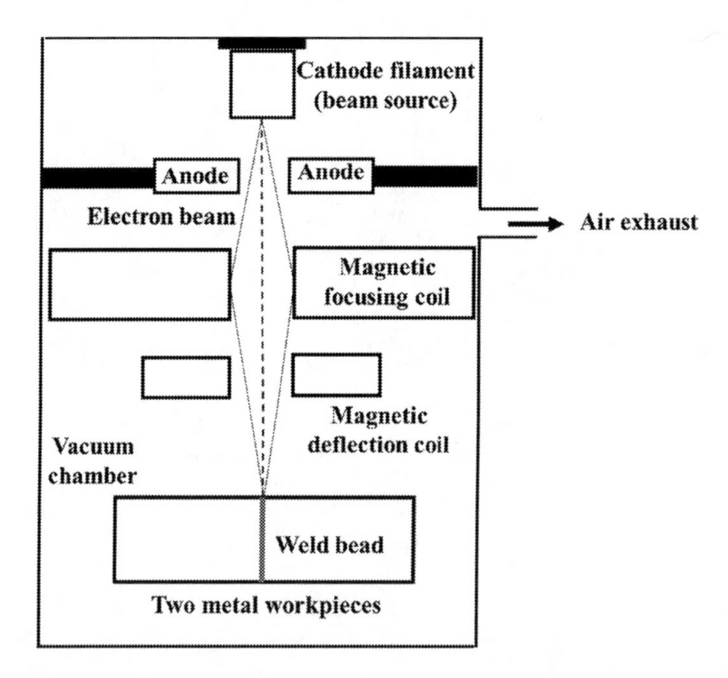

Fig. 5.8 Schematic of EBW process

resulting in uncontrollable fusion of the two metals, formation of welding defects such as porosity because of rapid solidification, and high cost of the equipment [28].

The welding of dissimilar metals by using the aforementioned fusion welding methods may be challenging because of: large variance in the physical properties of the dissimilar metals with respect to the heat available, and cooling rate of the welds. The formation of more number of brittle IMCs and thicker IMC layer as a result of the greater heat input in case of the fusion welding methods, further reduces the end quality of the weld. Thus, to overcome the difficulties associated in the above-discussed fusion welding techniques, an alternate to this is a solid-state welding method.

5.4.8 Friction Welding

The friction welding (as shown in Fig. 5.9), joins two materials by friction where one of the materials is forced to rub against the other. The rubbing action between the constantly rotating tool under pressure with the workpiece leads to the generation of the heat due to friction at the metallic interface. This leads to softening of the materials. Once sufficient softening is achieved, the materials (with mating edges) are maintained at the desired force for a period of time which leads to joint formation. Thus, the process involves conversion of mechanical energy into frictional heat near the joint interface by using relative movement between the materials [29]. This process has several benefits over fusion welding processes for joining the

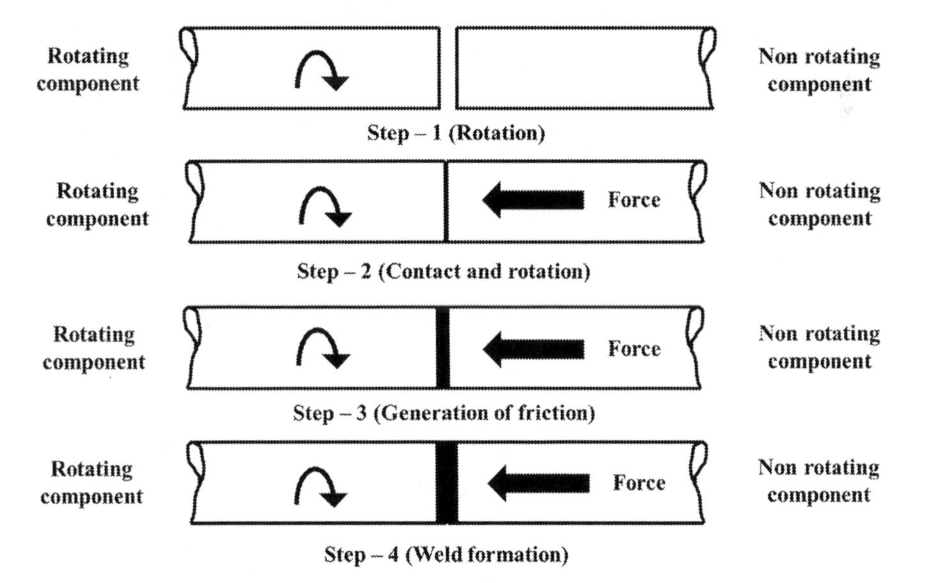

Fig. 5.9 Schematic of friction welding process

dissimilar materials. These benefits include the absence of porosity, hot cracking, and segregation. The other advantages of this process are the formation of refined microstructure and low distortion of workpiece. But the method has limitations with respect to joint configuration. It requires at least one of the workpieces to be of circular in cross-section [30]. These parameters majorly govern the formation of IMCs which is directly influenced by heat generation. The variations in the formation of the IMCs affect different mechanical properties specifically being the tensile strength and the hardness. The excessive friction at high rotational speed generates a greater amount of heat which ultimately substantiates the formation of the IMCs thereby deteriorating different mechanical properties [31, 32].

5.4.9 Explosive Welding

Explosive welding (as shown in Fig. 5.10) involves explosive which are detonated upon the surface of the workpieces. A stand-off distance between the flyer and the substrate is maintained during the joining process for two reasons. The first reason is that the stand-off offers the distance through which the flyer plate can be accelerated in the required impact velocity to produce welding. Another reason is that the air between the plates can easily come out through this stand-off gap. A buffer plate is kept in between flyer plate and explosive layer so as to prevent burning of flyer plate. The joining takes place under the influence of high pressure resulting in substantial plastic deformation. The bonding occurs metallurgically and the joint is even stronger than the base metals. Welding of both similar and dissimilar configurations is possible by this process [33]. This process uses detonation of explosives in a controlled manner to increase speed of one metal or both of the metals into each other to fuse them together [34]. Though in this process, solubility of hydrogen, which causes hydrogen embrittlement, is not a major issue during dissimilar metal welding but the formation of very hard metal oxide layer surface, having very high melting point is a major drawback for welding of dissimilar materials.The principle of joining in this technique is related to solid-state welding techniques, for example;

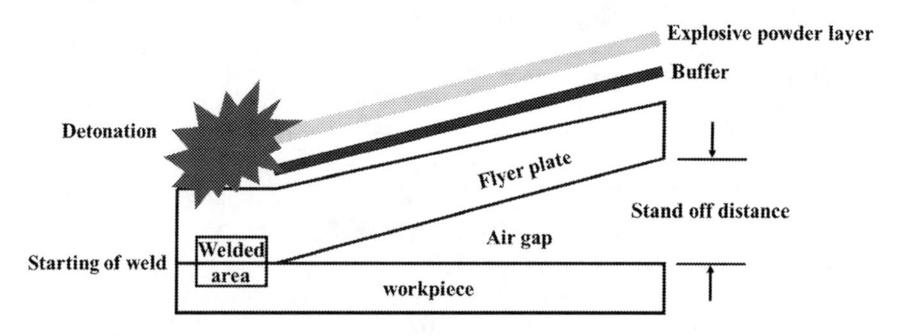

Fig. 5.10 Schematic of explosive welding process

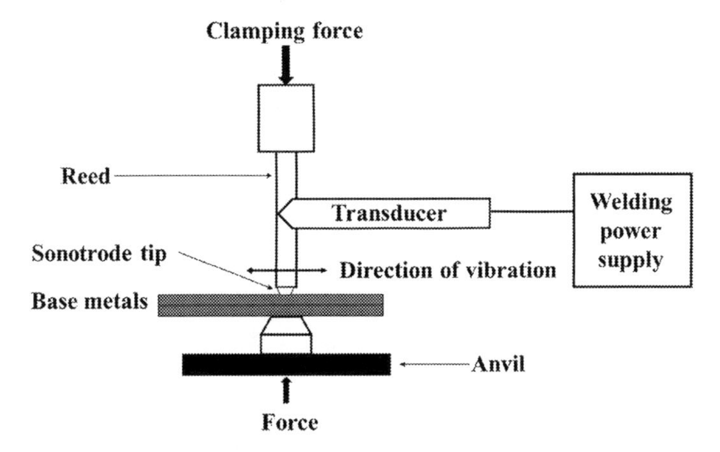

Fig. 5.11 Schematic of ultrasonic welding process

friction welding. The benefit of this process over the friction welding is that large areas can be welded in a short time, and a clean weld can be produced, because of the surface of both metals is aggressively expelled during the reaction.

5.4.10 Ultrasonic Welding

Ultrasonic welding (as shown in Fig. 5.11) is another technique in the solid-state joining approach that initiates coalescence of metals via application of high-frequency ultrasonic energy with a moderate clamping force [29]. The former helps in generating mechanical vibration along with the interface of the weld. The vibrations generate heat near the joint interface of the parts being welded and the latter helps in plastic deformation. This leads to the breakage of oxide layer and consolidation of the weld joint. This process is also implemented to weld a broad range of metallic and non-metallic metals, and dissimilar metalsas well, because of the lower temperature involved in the process [29]. The other factors such as the clamping force, ultrasonic frequency, and time required for welding are the key process variables that control the joint quality [35]. However, joining of two different metals using this process is generally limited to thin sheets.

5.4.11 Friction Stir Welding (FSW)

FSW uses a non-consumable rotating tool comprising of a pin, extending underneath shoulder that is plunged into the interface between two workpieces, as shown in Fig. 5.12. The process consists of four stages like plunging, dwelling, welding and

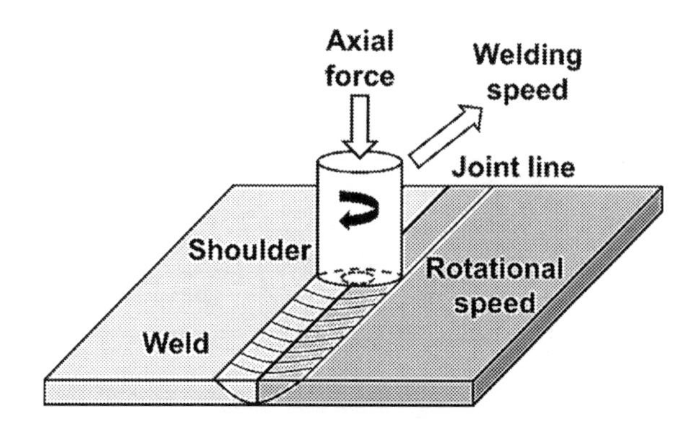

Fig. 5.12 Schematic of FSW

retraction. In the first stage, the workpieces are clamped to the machine base, and the rotating tool plunges at the mating surfaces of the workpieces. A substantial amount of heat is produced due to friction acting between the tool and the workpiece. The generated heat near the contact area softens both the workpiece further leading to the plastic deformation of the same. The dwelling operation follows this stage where the tool keeps stirring at the same position to take the workpieces to a viscous state. The third stage refers to the traversal of the tool along the joint line which stirs the viscous material and produces the weld joint. In the last stage, after completion of welding, the tool retracts and returns to the initial position [36]. In FSW process, because of the frictional heat, and stirring and forging action of the tool, plastic flow of the materials occurs which consequently leads to the development of the weld joint. There is no occurrence of melting, thus welding is done at the solid-state. This method was first established to join the low melting point metals like aluminium (Al) and magnesium (Mg) in similar and dissimilar combinations. However, with time this welding method is found to be efficient to join materials of high melting points, like steel, titanium, and copper in several joint configurations [37]. As in case of FSW, no macroscopic melting of metals involved, the control required to avoid fusion welding defects are absent. These recognize the benefits of solid-state joining which makes the FSW a suitable candidate for welding of a wide range of alloys, specifically in dissimilar metals [38]. In FSW, the thickness of the intermetallic layer, the determining factor of weld quality, is affected by the welding temperature. Since the process temperature does not exceed the solidus line, the amount of IMC formation can be reduced in comparison with fusion welding techniques [39].

As mentioned in the preceding sections, several problems do exist in case of the fusion weldingapproaches, for which the welding of dissimilar metals is not properly feasible. The problems include significant dissimilarity in the melting temperature of the two materials to be combined, high heat input during welding, solidification cracking, large HAZ, formation of residual stresses, etc. In addition, formation of more amount of brittle IMCs in the fusion welding methods adversely affecting joint properties. In order to overcome these challenges, ultrasonic welding and friction welding methods have been utilized to some extent. However, the limitation of joint configuration, and inability to weld metals with low ductility and toughness, in ultrasonic welding and friction welding methods, respectively, reduce their applications. FSW has been a potential alternative to these methods, in case of joining dissimilar materials. This is because of the advantages which have been discussed previously.

5.5 Mechanism of Joining Dissimilar Materials in FSW

IMC formation and mechanical interlocking are the two mechanisms governing FSW. The IMC formation refers to diffusion-controlled process. In FSW, the IMCs is formed below the melting temperature of the base metals. During the joining of dissimilar combination of metals in both configurations such as lap and butt, the number of IMCs are directly influenced by the kinetics and rate of diffusion of atom. Both types of IMCs and their thickness influence the joint quality [40]. The diffusion-controlled process involves three stages: (a) diffusion of atoms in between the workpieces, (b) formation of solid solutions at the IMC, and (c) growth in its thickness. The first stage depends upon the diffusion coefficients of the workpieces [41]. The second stage refers to the formation of solid solutions which is the mixture of the two workpieces. The growth in the thickness of IMC initiates as the solid solution in second stage reaches a saturated level in third stage. This thickness S varies as depicted in Eqs. 5.1 and 5.2 [41, 42].

$$S = Kt^{x} \tag{5.1}$$

$$K = K_0 \exp^{\frac{-Q}{RT}} \tag{5.2}$$

where t is the time of reaction at a specific temperature and pressure which does change as per the Eqs. (5.1) and (5.2) [41, 42]. Similarly, Q, R, and T refer to the activation energy in J/mol, universal gas constant in J/K mol, and temperature in K, respectively. K and K_0 are a constant and pre-exponential factor.

The mechanical interlocking phenomenon in FSWwhich is observed in case of lap configuration occurs because of the formation of sawtooth and it can be explained as shown in Fig. 5.13. In the process, metal-1 is considered as the softer material as compared to metal-2.

It begins with the plunging of the pin into the workpieces (metal-1 and metal-2), deforming them plastically. This is followed by rubbing of metal-2 leading to formation of metal debris which gets projected in the metal-1. This occurs because of the rotating pin. These debris have the sawtooth shape, as shown in Fig. 5.13. With further movement of the tool, bonding takes place. The third step: The direction of metal-1 fills the gap between an activated surface and the rotating pin is shown in Fig. 5.13. The activated metal-2 surface draws the metal-1 that travels over it and develops a metallic bond between both the metals. One of such profiles developed during welding of Al-steel using FSW is shown in Fig. 5.14, where the sawtooth

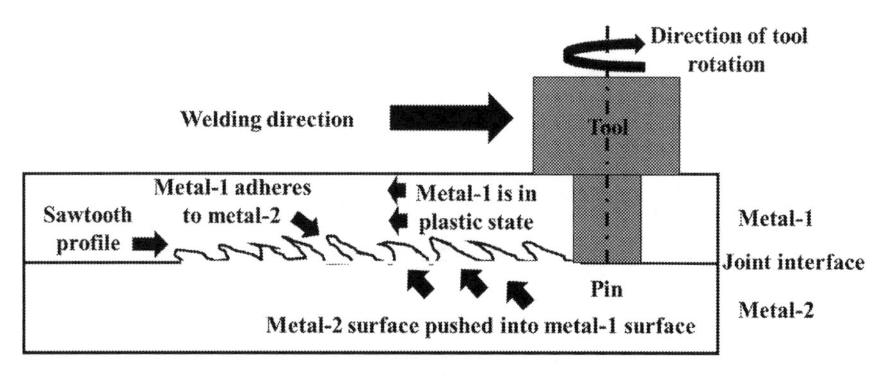

Fig. 5.13 Mechanical interlocking phenomenon; formation of sawtooth structure in FSW in lap configuration

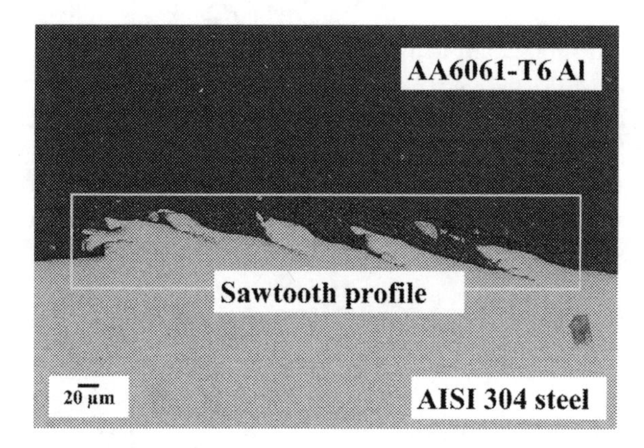

Fig. 5.14 Sawtooth structure in Al-steel FSW joint

interlocking has been observed as the hard steel is pushed upward into the softer Al surface by the action of the rotating tool.

5.6 Case Study on Dissimilar Material Welding by FSW

The environmental concerns such as controlling CO_2 emissions have commanded the transportation sector to make use of dissimilar material to manufacture components. Thus, it is crucial to identify an arrangement of metals that will optimize the weight and at the same time provide appreciable strength during crash situations. Light-weight structures with aluminium (Al) alloys and steel have shown a good combination in this regard. Al alloys deliver lower density and higher strength, and steel increases the stiffness [15]. Therefore, to attain the benefits out of this combination, a proper welding method is required to be adopted. One of the uses of Al-steel hybrid structure in automotive sector is the Honda manufactured sub-frames [43].

The significant dissimilarity in the melting point of these metals makes the combination very challenging for joining by using fusion welding [44, 45]. Moreover, the dissimilarity in different properties of the metals adds to the challenge in the joining process. Additionally, the formidable norms of the industries in terms of higher productivity and lower cost without compromising the product quality becomes a very difficult task for the manufacturers to produce a component from dissimilar materials. Because of the advantages mentioned above, FSW is one of the potent solutions for welding dissimilar metals. In FSW, the rotational speed of the tool (ω), the speed with which the tool traverses (v), the tilt angle of the tool (α), depth of the tool shoulder inserted into workpiece (*pd*), and the profile of the tool are the major variables that greatly affect the weld quality. Amongst these variables, ω and v have been identified as the key parameters affecting the joint quality [18].

The following subsections discuss the details of the case study where AA6061 is welded with AISI 304 steel using FSW. The FSW experiments were carried out in lap configuration using a 3-axis NC FSW machine (ETA Pvt. Ltd., WS004). A cylindrical tool made of tungsten carbide of 16 mm of shoulder diameter, 5 mm of pin diameter, and 1 mm of pin length was used for experimentation. To determine the lap shear load, rectangular-shaped samples of 172.6 mm length and 30 mm width were fabricated. Scanning electron microscopy (SEM) observation was carried out to study the fracture surface of samples. The specimens for microstructural studies of the welded samples were prepared by considering ASTM E3 standard. The samples were prepared using a hot mounting process; and these were then polished using different grades of polishing papers with varying grit size varying within the range of 120–2000 μm. The polished samples were then etched using Keller's reagent for AA6061. The weld cross-section microstructure was captured with the help of optical microscopy (Leica, DMILM). Furthermore, to identify IMCs near the joint interface, the weld line was characterized under the X-ray diffraction (Panalytical, Empyrean).

The discussed attributes include joint strength, IMCs, and interface-microstructure. The obtained results have also been qualitatively compared with the fusion welding methods. This helps in understanding the differences among the two. A process parametric combinations of ω of 1600 rpm and v of 50, 100, and 200 mm/min was selected by considering constant α and *pd* of 2° and 0.2 mm, respectively.

5.6.1 Joint Strength of Welds

The highest joint strength values of the sample welded using FSW is found to be 5950 N which is approximately 87.5% of base AA6061-T6 Al alloy, as shown in Fig. 5.15. A similar work [46], has reported the joint strength as 64.35%. This difference accounts for the selection of optimum parameters. In case of the fusion weldingof Al alloy and steel, the strength has been found to be varying in the range of 45–75% [9, 10]. This is significantly lower than the strength obtained by FSW.

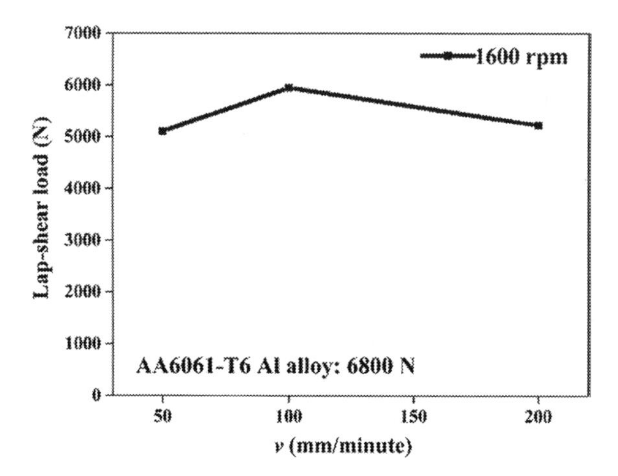

Fig. 5.15 Lap shear strength variations

As discussed earlier, the demerits of fusion welding technique, i.e., larger heat input, formation of brittle Fe_xAl_y IMCs and larger HAZ may lead to lower joint strength. The lap shear samples failed between SZ and TMAZ. Figure 5.16 shows the SEM of fractured surface. It was observed that the fractured surfaces comprised of small size

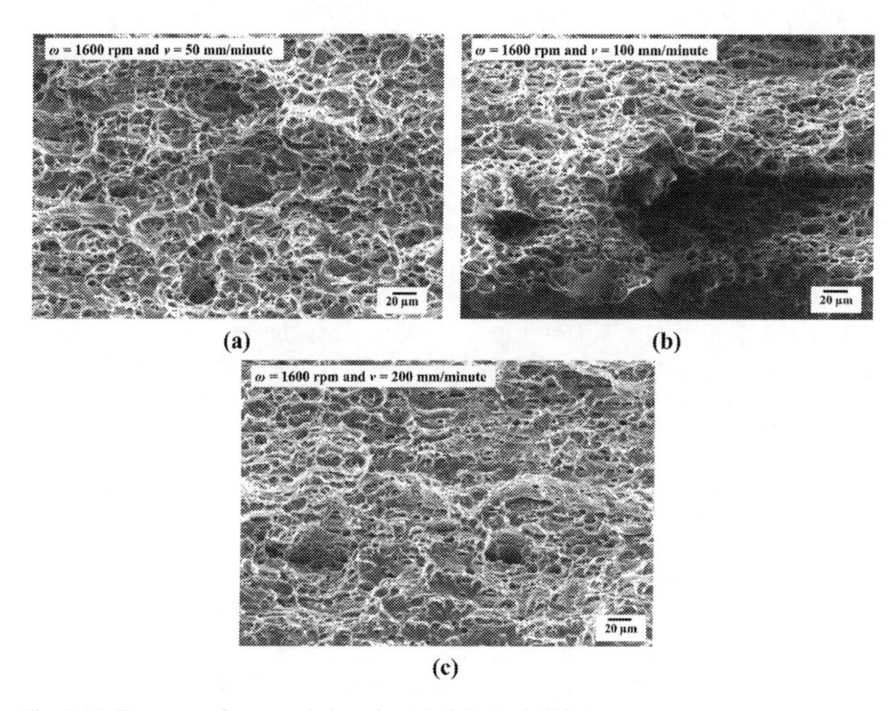

Fig. 5.16 Fracture surface morphology for AA6061-T6-AISI304 joints

dimples, microvoids, and tearing edges, indicating ductile fracture characteristics which is significantly lower as compared to fusion welding processes [9, 10].

5.6.2 IMCs

The formation of IMCs depends on the temperature during welding. Process parameters highly affect the heat generation, which leads to formation of IMCs. With proper metal flow, the chance of IMC formation reduces. In all the XRD results, as shown in Fig. 5.17, the IMCs like $Al_{13}Fe_4$, AlFe, $FeAl_3$, Al_5Fe_2, Fe_2Si, $Al_{0.4}Fe_{0.6}$, $Al_{0.5}Fe_{0.5}$, $Al_{0.42}Fe_{0.58}$, (Fe Ni), $Fe_{0.64}Ni_{0.36}$, FeSi, $Al_{0.42}Ni_{0.58}$, Mg_2Si, $Al_{1.1}Ni_{0.9}$, $Al_{0.9}Ni_{1.1}$ were formed for ω of 1600 rpm and at different values of v (50, 100, and 200 mm/min). The intensities of $Al_{13}Fe_4$, $FeAl_3$, Al_5Fe_2, and Mg_2Si were high, which resulted in weld strength reduction, as shown in Fig. 5.17. The IMCs like AlFe, $AlFe_3$, $Al_{0.4}Fe_{0.6}$, $Al_{0.5}Fe_{0.5}$, $Al_{0.42}Fe_{0.58}$, (Fe Ni) increased the joint strength due to high composition of Fe, as shown in Fig. 5.17. Similar observations with respect to FSW of Al alloy to steel lap welds were also found by other researchers [47–49]. Most of the fusion welding research investigating on joining of Al to steel

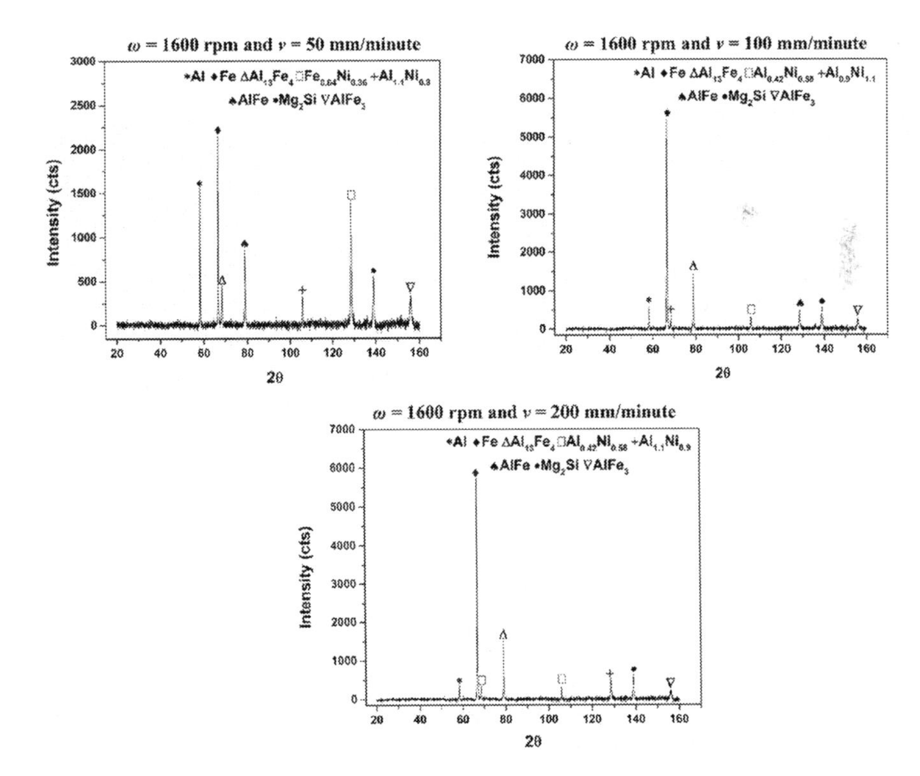

Fig. 5.17 XRD results

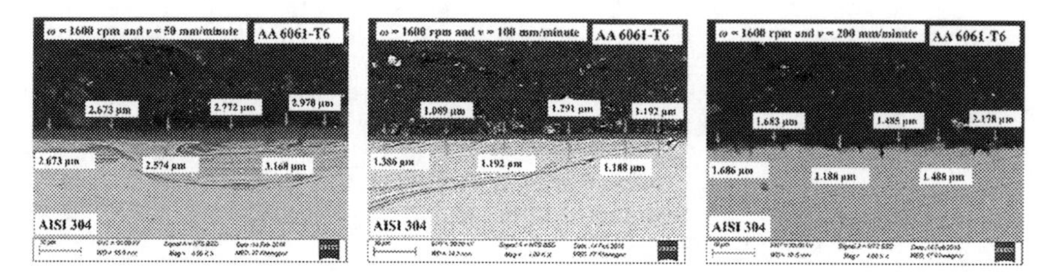

Fig. 5.18 Thickness of IMC layers

Table 5.1 Average thickness of IMCs

Process parameters		IMCs	Average thickness of IMCs (μm)
ω (rpm)	*v* (mm/min)		
1600	50	$Al_{13}Fe_4$, $Al_{0.5}Fe_{0.5}$, $Al_{0.42}Ni_{0.58}$, Mg_2Si, $Al_{1.1}Ni_{0.9}$	2.80
	100	$Al_{13}Fe_4$, $Al_{0.5}Fe_{0.5}$, $Fe_{0.64}Ni_{0.36}$, $Al_{0.42}Ni_{0.58}$, Mg_2Si, $Al_{0.9}Ni_{1.1}$	1.22
	200	$Al_{13}Fe_4$, $Al_{0.42}Ni_{0.58}$, Mg_2Si, $Al_{1.1}Ni_{0.9}$	1.61

have identified three distinct Al-rich IMCs such as, (a) Fe_2Al_5 which is close to the steel substrate, (b) $FeAl_3$, and (c) Fe_4Al_{13}, as the minor phases close to the solidified aluminium [50–54].

Figure 5.18 shows the variation of the thickness of IMCs for ω of 1600 rpm and different values of v (50, 100, and 200 mm/min). The average thickness of the IMCs near the joint interface for ω of 1600 rpm and at different values of v is found as 2.80 μm. The lower thickness of 1.22 μm was observed for ω of 1600 rpm and at a v of 100 mm/min. The average IMC thickness values are mentioned in Table 5.1. Comparable observations have been reported by other researchers [40, 55]. But in case of fusion welding, the IMC thickness varies between 4–40 μm which is significantly higher as compared to FSW [50–54]. Larger thickness of IMC reduces the joint strength, and thus, this may be the probable reason for inferior welds in dissimilar material joining while using fusion welding methods.

5.6.3 Weld Microstructure at the Interface

The microstructures of weld shown the development of coarser grains near the interface of the weld which was due to slower cooling rate. The weld formed underneath the tool shoulder area was exposed to heat and deformation which steered the development of finer grains. The micrographs for different welding conditions are shown

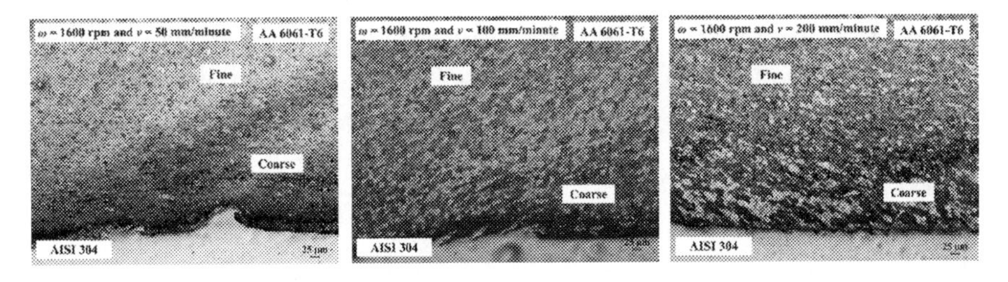

Fig. 5.19 Microstructure at the weld interface

in Fig. 5.19. At lower *v* (50 mm/min), coarser grains and the deformed microstructure were observed. This may be because the extra amount of heat that steered the severe deformation of the materials below the shoulder surface. At a *v* of 100 mm/min, coarser grains with lesser deformation and size compared to those obtained at lower speed. Thus, it was concluded that the *v* controlled the heat input which controls the size of grain during welding. Comparable microstructural observations have also been reported in other research works by using FSW to join Al-steel [47, 55–57]. In the case of fusion welding, due to the higher heat input as compared to FSW, larger grain size, porosity, and partially molten Al are observed at the weld interface [50–54].

5.7 Summary

In this chapter, the major applications of the dissimilar welded blanks in different sectors have been addressed. The complexities during the welding of the dissimilar materials with the difference in their mechanical and metallurgical properties have been critically analyzed, and the factors influencing the weld quality have also been stated. Considering the benefits of FSW technique, a case study with regard to the automotive requirement, i.e. AA6061-T6 and AISI 304 steel has been presented. It can be concluded from the joint interface study that the joints produced out of the FSW process are of good quality. A joint efficiency of approximately 87.5% was achieved in this case study.

Acknowledgement and Funding The necessary funding of Department of Science & Technology (DST), Science and Engineering Research Board (SERB), New Delhi to carry out the experimental work are gratefully acknowledged by the authors (Grant No: EMR/2015/001588).

References

1. Martinsen K, Hu SJ, Carlson BE (2015) Joining of dissimilar materials. CIRP Ann Manuf Technol 64:679–699. https://doi.org/10.1016/j.cirp.2015.05.006

2. Anand D, Chen DL, Bhole SD et al (2006) Fatigue behavior of tailor (laser)-welded blanks for automotive applications. Mater Sci Eng A 420:199–207. https://doi.org/10.1016/j.msea.2006.01.075
3. Zinkle SJ, Busby JT (2009) Structural materials for fission and fusion energy. Mater Today 12:12–19. https://doi.org/10.1016/S1369-7021(09)70294-9
4. Zhu L, Li N, Childs PRN (2018) Light-weighting in aerospace component and system design. Propuls Power Res 7:103–119. https://doi.org/10.1016/j.jppr.2018.04.001
5. Advances in Friction Stir Welding and Processing Related titles
6. No Title. https://www.spilasers.com/application-welding/dissimilar-metal-welding/. Accessed 11 May 2020
7. Kah P, Jukka Martikainen MS (2013) Trends in joining dissimilar metals by welding. Appl Mech Mater 440:269–276. https://doi.org/10.4028/www.scientific.net/AMM.440.269
8. Staple B (1997) Welding handbook volume 4 metals and their weldability
9. Sierra G, Peyre P, Deschaux Beaume F et al (2008) Galvanised steel to aluminium joining by laser and GTAW processes. Mater Char 59:1705–1715. https://doi.org/10.1016/j.matchar.2008.03.016
10. Borrisutth R, Mitsomwang P, Rattanacha S, Mutoh Y (2010) Feasibility of using TIG welding in dissimilar metals between steel/aluminum alloy. Energy Res J 1:82–86. https://doi.org/10.3844/erjsp.2010.82.86
11. Antonini JM (2014) Health effects associated with welding. Elsevier
12. Zhou W, Long TZ, Mark CK et al (2013) Hot cracking in tungsten inert gas welding of magnesium alloy AZ91D Hot cracking in tungsten inert gas welding of magnesium alloy AZ91D. pp 0836. https://doi.org/10.1179/174328407X213026
13. Ding J, Wang D, Wang Y, Du H (2014) Effect of post weld heat treatment on properties of variable polarity TIG welded AA2219 aluminium alloy joints. Trans Nonferrous Met Soc China 24:1307–1316. https://doi.org/10.1016/S1003-6326(14)63193-9
14. Sun DQ, Sun DX, Yin SQ, Li JB (2006) Microstructures and mechanical properties of tungsten inert gas welded magnesium alloy AZ91D Joints 46:1200–1204
15. Singh L, Singh R, Singh NK et al (2013) An evaluation of TIG welding parametric influence on tensile strength of 5083 aluminium alloy 7:2326–2329
16. Chaudhari R, Ingle A (2019) Experimental investigation of dissimilar metal weld of SA335 P11 and SA312 TP304 formed by gas tungsten arc welding (GTAW). Trans Indian Inst Met 72:1145–1152. https://doi.org/10.1007/s12666-019-01587-2
17. Zhang H, Liu J (2011) Microstructure characteristics and mechanical property of aluminum alloy/stainless steel lap joints fabricated by MIG welding-brazing process. Mater Sci Eng A 528:6179–6185. https://doi.org/10.1016/j.msea.2011.04.039
18. Mvola B, Kah P, Martikainen J (2014) Welding of dissimilar non-ferrous metals by GMAW processes. Int J Mech Mater Eng 9:1–11. https://doi.org/10.1186/s40712-014-0021-8
19. Srinivasan PB, Muthupandi V, Dietzel W, Sivan V (2006) An assessment of impact strength and corrosion behaviour of shielded metal arc welded dissimilar weldments between UNS 31803 and IS 2062 steels. Mater Des 27:182–191. https://doi.org/10.1016/j.matdes.2004.10.019
20. Kchaou Y, Haddar N, Hénaff G et al (2014) Microstructural, compositional and mechanical investigation of shielded metal arc welding (SMAW) welded superaustenitic UNS N08028 (Alloy 28) stainless steel. Mater Des 63:278–285. https://doi.org/10.1016/j.matdes.2014.06.014
21. McPherson NA, Millar DW (1997) Factors affecting quality of dissimilar plate submerged arc welds. Sci Technol Weld Join 2:149–154. https://doi.org/10.1179/stw.1997.2.4.149
22. Dong H, Yang L, Dong C, Kou S (2010) Arc joining of aluminum alloy to stainless steel with flux-cored Zn-based filler metal. Mater Sci Eng A 527:7151–7154. https://doi.org/10.1016/j.msea.2010.07.092
23. Katherasan D, Sathiya P, Raja A (2013) Shielding gas effects on flux cored arc welding of AISI 316L (N) austenitic stainless steel joints. Mater Des 45:43–51. https://doi.org/10.1016/j.matdes.2012.09.012

24. Bauné E, Bonnet C, Liu S (2001) Assessing metal transfer stability and spatter severity in flux cored arc welding. Sci Technol Weld Join 6:139–148. https://doi.org/10.1179/136217101101538677
25. Nee AYC (2015) Handbook of manufacturing engineering and technology. Handb Manuf Eng Technol 1–3487. https://doi.org/10.1007/978-1-4471-4670-4
26. Sufizadeh AR, Mousavi SAAA (2016) Metallurgical and mechanical research on dissimilar electron beam welding of AISI 316L and AISI 4340. Adv Mater Sci Eng. https://doi.org/10.1155/2016/2509734
27. Metzger G, Lison R (1976) Electron beam welding of dissimilar metals. Weld J (Miami, Fla) 55
28. Sun Z, Karppi R (1996) The application of electron beam welding for the joining of dissimilar metals: an overview. J Mater Process Technol 59:257–267. https://doi.org/10.1016/0924-0136(95)02150-7
29. Vural M (2014) Welding processes and technologies. Elsevier
30. Taban E, Gould JE, Lippold JC (2010) Dissimilar friction welding of 6061–T6 aluminum and AISI 1018 steel: properties and microstructural characterization. Mater Des 31:2305–2311. https://doi.org/10.1016/j.matdes.2009.12.010
31. Mehta KP (2019) A review on friction-based joining of dissimilar aluminum-steel joints. J Mater Res 34:78–96. https://doi.org/10.1557/jmr.2018.332
32. Rn S, Surendran S (2012) Friction welding to join dissimilar metals. Int J Emerg Technol Adv Eng 2:200–210
33. Mahi F, Dilthey U (2016) Joining of metals. Ref Modul Mater Sci Mater Eng 1–6. https://doi.org/10.1016/b978-0-12-803581-8.03785-1
34. Kovacs-Coskun T, Volgyi B, Sikari-Nagl I (2015) Investigation of aluminum-steel joint formed by explosion welding. J Phys Conf Ser 602. https://doi.org/10.1088/1742-6596/602/1/012026
35. Watanabe T, Sakuyama H, Yanagisawa A (2009) Journal of materials processing technology ultrasonic welding between mild steel sheet and Al–Mg alloy sheet 209:5475–5480. https://doi.org/10.1016/j.jmatprotec.2009.05.006
36. Mishra D, Roy RB, Dutta S et al (2018) A review on sensor based monitoring and control of friction stir welding process and a roadmap to Industry 4.0. J Manuf Process 36:373–397. https://doi.org/10.1016/j.jmapro.2018.10.016
37. Thomas W, Nicholas E (1997) Friction stir welding for the transportation industries. Mater Des 18:269–273. https://doi.org/10.1016/S0261-3069(97)00062-9
38. DebRoy T, Bhadeshia HKDH (2010) Friction stir welding of dissimilar alloys—a perspective. Sci Technol Weld Join 15:266–270. https://doi.org/10.1179/174329310X12726496072400
39. Zens A, Zaeh MF, Marstatt R, Haider F (2019) Friction stir welding of dissimilar metal joints. Materwiss Werksttech 50:949–957. https://doi.org/10.1002/mawe.201900023
40. Mahto RP, Kumar R, Pal SK (2020) Characterizations of weld defects, intermetallic compounds and mechanical properties of friction stir lap welded dissimilar alloys. Mater Charact 160:110115. https://doi.org/10.1016/j.matchar.2019.110115
41. Naoi D, Kajihara M (2007) Growth behavior of Fe2Al5 during reactive diffusion between Fe and Al at solid-state temperatures. Mater Sci Eng A 459:375–382. https://doi.org/10.1016/j.msea.2007.01.099
42. Xu L, Robson JD, Wang L, Prangnell PB (2018) The influence of grain structure on intermetallic compound layer growth rates in Fe-Al dissimilar welds. Metall Mater Trans A Phys Metall Mater Sci 49:515–526. https://doi.org/10.1007/s11661-017-4352-y
43. Kusuda Y (2013) Honda develops robotized FSW technology to weld steel and aluminum and applied it to a mass-production vehicle. Ind Robot Int J 40:208–212
44. Chen YC, Nakata K (2009a) Effect of tool geometry on microstructure and mechanical properties of friction stir lap welded magnesium alloy and steel. Mater Des 30:3913–3919. https://doi.org/10.1016/j.matdes.2009.03.007
45. Chen YC, Nakata K (2009b) Friction stir lap welding of magnesium alloy and zinc-coated steel. Mater Trans 50:2598–2603. https://doi.org/10.2320/matertrans.M2009022

46. Mahto RP, Bhoje R, Pal SK et al (2016) A study on mechanical properties in friction stir lap welding of AA 6061–T6 and AISI 304. Mater Sci Eng A 652:136–144. https://doi.org/10.1016/j.msea.2015.11.064
47. Chen ZW, Yazdanian S (2012) friction stir lap welding : material flow, joint structure and strength. J Achiev Mater Manuf Eng 55:629–637
48. Ogura T, Saito Y, Nishida T et al (2012) Partitioning evaluation of mechanical properties and the interfacial microstructure in a friction stir welded aluminum alloy/stainless steel lap joint. Scr Mater 66:531–534. https://doi.org/10.1016/j.scriptamat.2011.12.035
49. Das H, Basak S, Das G, Pal TK (2013) Influence of energy induced from processing parameters on the mechanical properties of friction stir welded lap joint of aluminum to coated steel sheet. Int J Adv Manuf Technol 64:1653–1661. https://doi.org/10.1007/s00170-012-4130-3
50. Torkamany MJ, Tahamtan S, Sabbaghzadeh J (2010) Dissimilar welding of carbon steel to 5754 aluminum alloy by Nd:YAG pulsed laser. Mater Des 31:458–465. https://doi.org/10.1016/j.matdes.2009.05.046
51. Meco S, Pardal G, Ganguly S et al (2015) Application of laser in seam welding of dissimilar steel to aluminium joints for thick structural components. Opt Lasers Eng 67:22–30. https://doi.org/10.1016/j.optlaseng.2014.10.006
52. Zhou D, Xu S, Peng L, Liu J (2016) Laser lap welding quality of steel/aluminum dissimilar metal joint and its electronic simulations. Int J Adv Manuf Technol 86:2231–2242. https://doi.org/10.1007/s00170-015-8254-0
53. Bach FW, Beniyash A, Lau K, Versemann R (2005) Joining of steel-aluminium hybrid structures with electron beam on atmosphere. Adv Mater Res 6–8:143–150. https://doi.org/10.4028/www.scientific.net/amr.6-8.143
54. Dinda SK, kar J, Jana S, et al (2019) Effect of beam oscillation on porosity and intermetallics of electron beam welded DP600-steel to Al 5754-alloy. J Mater Process Technol 265:191–200. https://doi.org/10.1016/j.jmatprotec.2018.10.026
55. Movahedi M, Kokabi AH, Seyed Reihani SM et al (2013) Effect of annealing treatment on joint strength of aluminum/steel friction stir lap weld. Mater Des 44:487–492. https://doi.org/10.1016/j.matdes.2012.08.028
56. Mahto RP, Kumar R, Pal SK, Panda SK (2018) A comprehensive study on force, temperature, mechanical properties and micro-structural characterizations in friction stir lap welding of dissimilar materials (AA6061-T6 & AISI304). J Manuf Process 31:624–639. https://doi.org/10.1016/j.jmapro.2017.12.017
57. Zheng Q, Feng X, Shen Y et al (2016) Dissimilar friction stir welding of 6061 Al to 316 stainless steel using Zn as a filler metal. J Alloys Compd 686:693–701. https://doi.org/10.1016/j.jallcom.2016.06.092

Chapter 6
Microstructure and Texture in Welding: A Case Study on Friction Stir Welding

Soumya Sangita Nayak, Raju Prasad Mahto, Surjya Kanta Pal, and Prakash Srirangam

Abstract This chapter provides a detailed discussion on the evolution of microstructure and texture during welding. The major focus of the chapter has been on various metallurgical reactions and events occurring in the welded zone of both fusion and solid-state welding processes. Along with this, the chapter also discusses evolution of texture in various welding techniques. For more insights, a case study on evolution of microstructure and texture in an innovative welding technique, friction stir welding (FSW) has been provided at the end of this chapter. The case study also shows the importance of microstructure and texture on the mechanical properties of the weld joint.

Keywords Microstructure · Texture · Fusion welding · Solid-state welding · Friction stir welding

6.1 Introduction

The final performance of a material in any application depends on three aspects, i.e. processing, microstructure and properties. Metal processing (casting, forming and welding) is the initial step in manufacturing a component, and the evolved microstructure during processing influences the mechanical and other properties

S. S. Nayak
Advanced Technology Development Centre, Indian Institute of Technology Kharagpur, Kharagpur, West Bengal 721302, India
e-mail: soumyasangitanayak@iitkgp.ac.in

R. P. Mahto · S. K. Pal (✉)
Friction Stir Welding Laboratory, Department of Mechanical Engineering, Indian Institute of Technology Kharagpur, Kharagpur, West Bengal 721302, India
e-mail: skpal@mech.iitkgp.ac.in

R. P. Mahto
e-mail: rajukec1@gmail.com

P. Srirangam
Warwick Manufacturing Group, University of Warwick, Coventry CV4 7AL, UK
e-mail: P.Srirangam@warwick.ac.uk

J. P. Davim (ed.), *Welding Technology*, Materials Forming, Machining and Tribology, https://doi.org/10.1007/978-3-030-63986-0_6

(e.g. electrical, chemical, etc.) of the material. Complete knowledge of microstructure conveys the idea of the arrangement and orientation of grains in a material, their size and shape, phases present, grain boundaries, dislocations, etc. The components of the microstructure are large groups of atomic arrangements which mean the position of all atoms present in the unit cell. It is described by the fractional coordinate of the atoms and the type of lattice. Single-crystal/grain consists of no grain boundary, and its atomic structure repeats periodically across its whole volume. But most of the materials are polycrystalline in nature. This means the material is made up of more than one grain or crystal, and the orientation of each grain (texture) differs from its neighbouring one. In a real polycrystalline material, some of the individual grains or crystals have a particular kind of orientation, whereas others do have a set of different orientations. Here, the polycrystalline material is called a "textured" one. The texture is known as the preferred orientation of grains. Different textures form anisotropy, which means material properties, are different in different crystallographic planes and directions. This anisotropic property may or may not be desirable from a realistic viewpoint. A careful control of texture will enable improving many properties such as yield strength, elastic constants, magnetic susceptibility, electrical conductivity, piezoelectricity, wave propagation and refraction of light. This will enhance the performance of a material for various applications. An example can drag readers' attention towards the importance of reading this chapter and understand the importance of microstructure and texture.

Nowadays, in the automotive industry, the main aim is to increase the application of lightweight materials so as to decrease the fuel consumption and the greenhouse gas emissions [1]. For example, magnesium is a widely used material among all light weight automotive materials, but it has poor elongation and low strength as compared to steel and aluminium. However, by careful control of texture in magnesium, it is possible to improve its elongation. This is possible by the application of multiaxial forging operation on the magnesium material which leads to refinement of its grain microstructure thereby improving its strength and elongation [2], enabling it to be used in automotive and aerospace applications.

Texture plays an important role in rolling, forming, machining, and welding. In the present chapter, evolution of microstructures and their preferred orientation have been discussed in the welding. Most of the industries (i.e. mechanical construction including naval, automotive, locomotive, nuclear and aviation industries) utilize welding to assemble the metallic materials. Welding is a manufacturing process used to join similar or dissimilar materials permanently. At the intersection of two metallic materials, a weld-bead is formed which joins them. It may be divided into two broad categories, depending on whether or not the parent materials are fused for weld-bead formation. The categories are fusion and solid-state welding processes. In the former, both the base materials along with the filler material are fused by the application of heat which is applied externally for the formation of the weld-bead. In this, phase transformation occurs from solid to liquid and then again from liquid to solid. Examples of a few fusion welding processes are: arc welding, gas welding, resistance welding, laser and electron beam welding processes. In solid-state welding processes, the melting of materials does not take place, rather the joining occurs in the solid-state itself. Phase change does not occur here, but the parent material is heated up below its melting point temperature. Pressure is usually applied instead of external heat for this category of welding process. A few examples in this category are roll welding, ultrasonic welding, friction welding, diffusion welding and friction stir welding (FSW). A sufficient knowledge of physical metallurgy is crucial to understand the welding metallurgy which includes the welding operation and the effects of welding parameters on the properties of the metals to be joined [3].

Over the last few decades in metal industries, especially in aluminium industry, FSW seems to be an important joining process. In FSW, the metals to be joined undergo deformation, and the texture distribution in the stir zone (SZ) is very complex. It may deteriorate the joint performance, if not controlled. Therefore, the control of texture, either by controlling the welding parameters or by any post-weld analysis in FSW, is crucial to optimize the mechanical properties of the joint. The current chapter primarily highlights the fundamental principles which govern the evolution of microstructure and texture in the weld zones of fusion and solid-state welding processes. Towards the end, a case study has been presented which discusses on the formation of microstructure and texture in FSW of aluminium to steel.

The success or failure of a welded joint depends upon several aspects such as process parameters, type of filler material, heating and cooling cycles, compositional variation with respect to percentage of alloying elements such as carbon, sulphur, nitrogen, stress patterns around the weld because of thermal expansion during heating and contraction during welding, environmental factors, formation of intermetallic compounds, etc. These factors govern the final microstructure and texture of a welded structure. Thus, it is highly essential to keep a track of the change in the microstructure of the weld to preserve its properties by taking care of the aforementioned points.

6.2 Microstructure in Welding

The microstructure is affected by the way a welding process is performed, and the details in this regard are discussed in the following subsections. The discussions have been carried out for fusion and solid-state welding processes.

6.2.1 Microstructure in Fusion Welding

This section will give a brief overview of microstructural regions of fusion welding processes, followed by various metallurgical processes that influence the welding.

6.2.1.1 Regions of a Fusion Weld

There are three distinct microstructural regions in a fusion weld. Those are fusion zone (FZ), heat affected zone (HAZ) and base metal (BM). FZ comprises of a mixture of BM and filler metal, which is similar to BM in homogeneous welding, different in case of heterogeneous welding. Sometimes FZ is a mixture of the two BMs only in case of autogeneous welds. In HAZ, melting does not occur. Heat generated out of welding itself or any post-weld heat treatment (PWHT) affects HAZ. BM, which is beyond the HAZ, mostly stays unaffected. Fusion weld microstructural regions are shown in Fig. 6.1a. Weld microstructure depends on the cooling rates. The temperature variation plots along the fusion boundary and the outer boundary of HAZ are depicted in Fig. 6.1b, c, respectively. Along the centreline of the weld, the temperature is uniform. But along the fusion boundary, there is a sharp temperature gradient shown as a straight line in Fig. 6.1b. The centre has high cooling rate condition than the fusion boundary. In FZ, the temperature crosses the melting point temperature of BMs and then it cools. Similarly, in HAZ, the temperature does not cross the melting point temperature of BMs but it is sufficient enough for a metallurgical change. After the curved part in the cooling curve shown in Fig. 6.1c, no metallurgical changes happen in HAZ boundary.

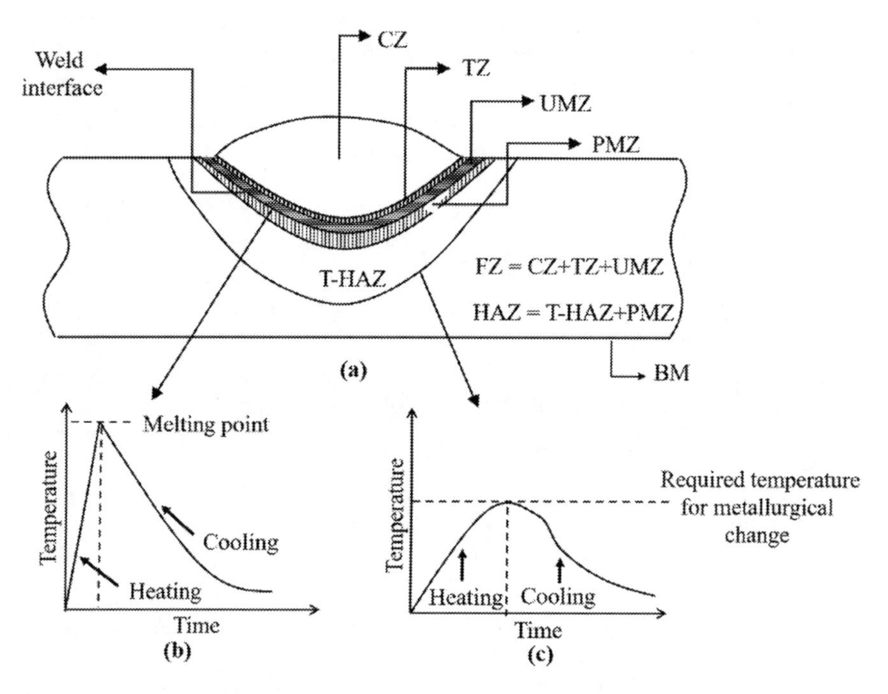

Fig. 6.1 **a** Zones of a fusion weld and temperature variation plot with time at **b** fusion boundary **c** outer boundary of HAZ

The FZ consists of two regions. The first zone is the composite zone (CZ), where the parent metal and filler metal are mixed in a composite composition. Along the fusion boundary, surrounding CZ, the second region is present, which is known as the unmixed zone (UMZ). This zone comprises of melted and resolidified parent material which never mixes with the filler material. Between CZ and UMZ, another zone named transition zone (TZ) exists, and it shows a microstructure different from its surrounding regions, especially in case of heterogeneous welds. The HAZ is subdivided into two zones, named as true HAZ (T-HAZ) and partially melted zone (PMZ). PMZ is present in most of the fusion welding processes because of the transition from 100% liquid to 100% solid along the FZ boundary [4]. T-HAZ is a part of the BM and is differentiated from HAZ, where no melting occurs, and all the metallurgical events happen in the solid state.

6.2.1.2 Metallurgical Processes Influencing Microstructure in Fusion Welding

The BM is usually manufactured after casting, followed by rolling, heat treatment and other secondary manufacturing processes. Heat in the fusion welding modifies the microstructure of material. The difference in microstructures of the base and the weld material is dictated by the differences in the mechanical and thermal histories. The following sections will give a brief knowledge of the metallurgical events that occur in FZ and HAZ of a fusion weld responsible for change in microstructure.

Metallurgical Events in FZ

Three most important metallurgical events occur in FZ such as gas–metal reactions, i.e. reaction of weld metal with gases, liquid–metal reactions, i.e. reaction of weld metal with any liquid phase and solid-state reaction, which occurs during and after solidification in FZ.

Melting and Solidification

These are two important processes to achieve acceptable joints in fusion welding processes. Microstructure development in FZ is mostly dependent on solidification behaviour. The ideal substrate for solidification is the unmelted part of the grain in HAZ at solid–liquid (S–L) interface. Various solidification parameters are useful to describe microstructure development in fusion welding. Those are: (a) partition coefficient, (b) liquid temperature gradient, (c) solidification rate, and (d) cooling rate. Solidification parameter is generally defined by a ratio of temperature gradient and square root of solidification growth rate (how fast S–L interface moves). To start solidification, it is mandatory to have the nucleation (homogeneous/heterogeneous) of solids inside the liquid phase in the molten weld pool, and the S–L interface provides the ideal nucleation site. Few solidification modes that exist at the S–L interface are (a) planar growth (Fig. 6.2a), (b) cellular growth (Fig. 6.2c), (c) cellular dendritic growth (Fig. 6.2e), (d) columnar dendritic growth (Fig. 6.2g) and (e) equiaxed dendritic growth (Fig. 6.2i)[5]. The grain structure which can be seen in a typical weld is planar grains, cellular grains, dendritic grains and equiaxed grains. Factors that govern different types of grain structure are: (a) composition of weld metal which is solidifying (it basically affects the constitutional supercooling) and (b) heat transfer rate away from weld metal (cooling rate experienced by the weld metal and it actually affects the temperature gradient in weld metal). Cooling rate determines the rate of growth of S–L interface movement. In Fig. 6.2a, c, e, g, solidified regions as well as liquid region which is to be solidified can be seen. Heat is rejected to the BM from the fusion boundary. Due to heat extraction, the grains grow in opposite direction of the heat flow and perpendicular to the fusion boundary. First, it forms the planar grain structure next to the fusion boundary. Thereafter, the cellular grains form. Width of these zones is governed by the composition as well as the cooling condition experienced by the weld metal during the solidification. Thereafter, the dendritic structure forms and in the centre of the weld, equiaxed grain structure forms. Similar kind of things is repeated from both the sides and gets accommodated in centre of the weld, where solidification occurs at the end. In the alloy, solidification starts with the formation of pure metal or the metal with a minimum concentration of the alloying element and one having the higher liquidus temperature. The solid region is formed by consuming the liquid metals and rejecting the excess alloying elements which are beyond the solubility limit in the solid state. The region next to the S–L interface is enriched with alloying elements. Therefore, in this region, equilibrium temperature gradient is different from the remaining portion due to compositional variation. In S–L interface, if the equilibrium temperature gradient is below the actual temperature gradient as per the cooling condition, then planar grains are formed (Fig. 6.2b, dark line). If this condition is reversed, it results in the formation of zone having an effect of constitutional supercooling (Fig. 6.2d, f, h, j). When the zone is very limited, cellular grain results and plane interface are broken and cells start to grow. Further increase in constitutional supercooling leads to the formation of dendritic structure, which causes equiaxed grain structure.

In fusion welding, the molten weld pool is surrounded by the solid metal. The heat flow is always directed towards the cold metal. Therefore, the weld metal attains a column-like shape with long grains that are parallel to the heat flow direction. If solidification is extremely rapid, a cell-like microstructure [6] is obtained where dendrites are not fully developed.

The dendrites grow when the molten metal gets solidified. The tree-like shape is produced because of the faster growth of grains in favourable crystallographic direction. In the beginning, one solid nucleus grows in the melt. In metals, this solid tries to minimize the area of the surface having high surface energy. The dendrites, which are initially formed, solidify first. The solid rejects the solutes which are most soluble in liquid, and these, in turn, get diffused into the remaining liquid and become concentrated near the S–L interface. The growth of crystals is locked in that direction. Dendritic arm characteristics of as-solidified metals are produced

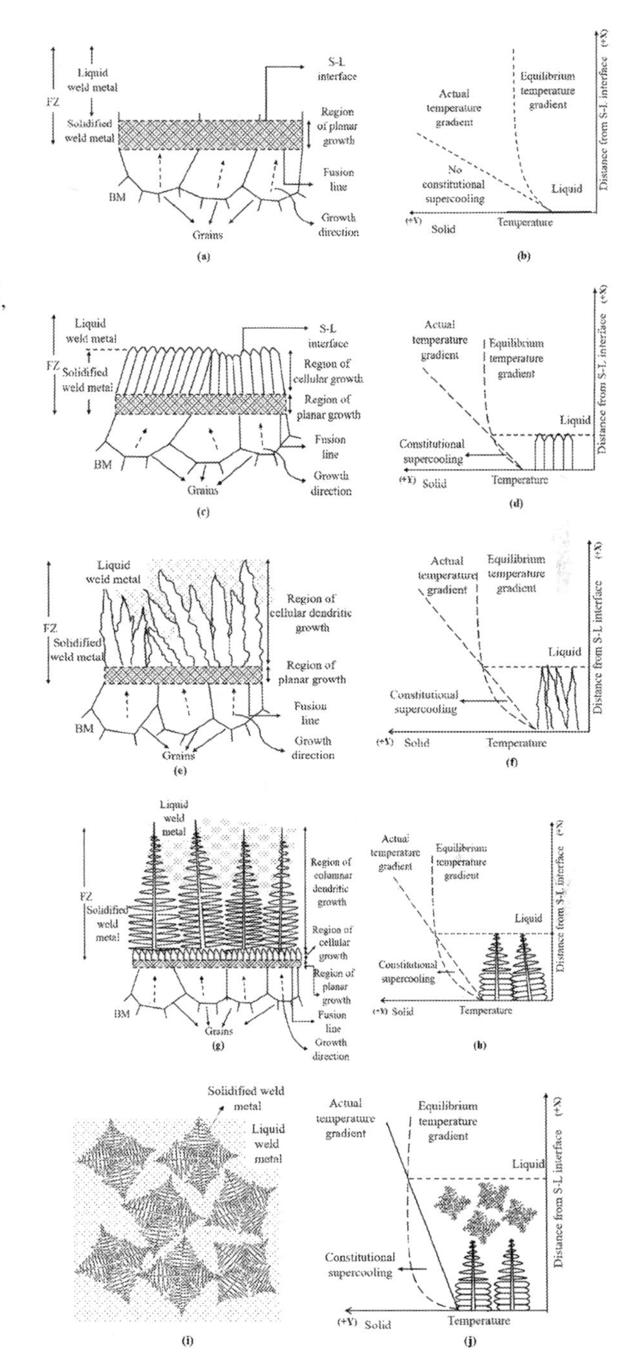

Fig. 6.2 **a** Planar growth, **b** influence of constitutional supercooling on planar growth, **c** cellular growth, **d** influence of constitutional supercooling on cellular growth, **e** cellular dendritic growth, **f** influence of constitutional supercooling on cellular dendritic growth, **g** columnar dendritic growth, **h** influence of constitutional supercooling on columnar dendritic growth, **i** equiaxed dendritic growth and **j** influence of constitutional supercooling on equiaxed dendritic growth

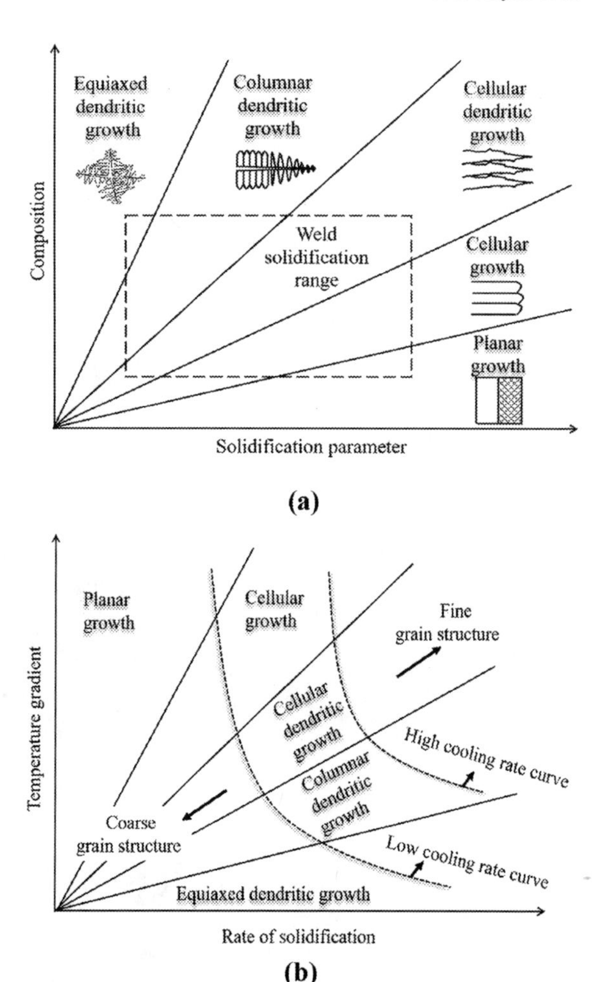

Fig. 6.3 Effect of solidification on microstructure: **a** solidification versus composition and **b** solidification rate versus temperature gradient

as and when the grains grow laterally. During solidification, many dendrites grow at a time from a single grain. All of them have nearly similar crystal orientation. Grain size is affected by heat input, nucleating agents, nucleation and growth and various other factors. The effect of the solidification parameter and composition and the effect of solidification rate and temperature gradient on the various modes of solidification have been depicted in Fig. 6.3a, b, respectively. Two curved lines are shown in Fig. 6.3b from which one is low cooling rate curve (curve at left side) and the other one is high cooling rate curve (curve at right side). Cellular grains and dendritic grains are fine when cooling rate is high. The cellular grains become wider and coarse dendritic structure with widely spaced arms form when the cooling rate is low. The dendritic arm spacing depends on solidification rate and defines the grain size. Fine equiaxed grains form when cooling rate is high, and when it is low, coarse equiaxed structure forms.

Gas–Metal Reactions

Gas–metal reactions happen because of absorption of the reactive gases like nitrogen, oxygen or hydrogen from arc/flame into the weld pool, and it has an impact on microstructure. In few cases, the entrapped gas dissolves in the liquid metal. In flux-cored arc welding (FCAW) and shielded metal arc welding (SMAW), a main source for hydrogen is chemically bonded with water/moisture in the electrode coating or core and it causes embrittlement [7]. To stabilize the arc, argon along with oxygen is used in gas metal arc welding (GMAW) [8]. The weld metal microstructure is observed to be changing due to variation in oxygen activity in case of high-strength low-alloy steels. Oxide inclusion acts as nucleation sites during cooling, which promotes decomposition of austenite [9].

In few cases, liquid metal and the gas reaction lead to formation of a chemical compound. This compound, if it is soluble, it causes embrittlement of the welded joint. However, if it is insoluble, then produces slag which obstructs the formation of the weld pool. Therefore, the use of excessive gas during welding is not desirable, and flux can be used to dissolve the so formed chemical compound, as found in submerged arc welding process (SAW) [10]. The microstructure in the weldment of ferritic stainless steel depends on types of shielding gas mixture used. If the amount of carbon dioxide (oxidizing gas) increases, it will increase the carbon content in the weld and also there will be an increase in the martensite content at the grain boundary. If a gas dissolves in the liquid weld pool, it tries to come out during cooling in the form of bubbles. Otherwise, the weld becomes porous in nature, if the gas gets entrapped. In order to avoid this situation, deoxidizers are utilized [11].

Nitrogen helps in grain refining and is a powerful austenite stabilizer with an ability to reduce the ferrite formation [12]. Nitrides and oxides help in reducing both notch toughness and ductility of steel weld metal. The availability of hydrogen in excess amount during solidification is not desirable because it causes porosity. It has been reported that nucleation gets affected by the shielding gas pressure and can be a reason to suppress the weld porosity [13]. The diffusion of gas into BM is one more important gas–metal reactions which helps in altering the microstructure.

Liquid–Metal Reactions

One of the most important liquid–metal reactions is the formation of non-metallic liquid phases (e.g. slag layers) which interacts with the molten weld pool produced in electro slag welding, SAW and SMAW [14]. The flux layers used in these processes absorb those contaminated and deoxidation products. These products are initially silicates of iron, manganese and aluminium floating on top surface of molten pool which get changed into slag. Some of them reside inside the weld metal, which are known as inclusions. The second type of liquid–metal reaction is hot cracking. If the interdendritic liquid has a low freezing temperature in comparison to formerly solidified metal, hot cracking occurs, which in turn produces shrinkage stress in the liquid portion, forming microcracks within the dendrite [15]. Both slag layer formation and hot cracking have significant impact on altering weld microstructure.

Solid-State Reactions

In fusion welding, the reactions helping to strengthen the weld metal are: solid-state phase transformations, delayed cracking [16], precipitation of secondary phases during cooling, etc. These reactions notably change the microstructure and properties of the weldment with respect to the BM.

Various Strengthening Mechanisms in Weld Metal

There are various strengthening mechanisms such as precipitation hardening, transformation hardening, solid solution strengthening and solidification grain structure. Solidification grain structure strengthening mechanism applies to all types of welding processes; the obtained microstructure is having fine dendrites because of rapid freezing of the weld metal which blocks the plastic flow in case of a tensile test. Therefore, in weld metals, the ratio of yield strength to tensile strength is high relative to BM. The weld metal is strengthened by the addition of various alloys (both interstitial and substitution alloying) in case of solid solution strengthening mechanism, and it can be applied to any type of alloy. The grains form in this mechanism is due to modulus effect and lattice distortion. A solid solution is formed due to the addition of alloying element and sometimes it leads to non-uniformity in crystal lattice. Precipitation hardening mechanism is accomplished by appropriate heat treatments and applicable for alloys of aluminium, magnesium, nickel, etc. In nickel, a vast range of precipitations is found and many phases present in the interior of grain. Precipitates act as obstacles to the dislocation motion, and stress increases to push the dislocations leading to increase in strength. Transformation hardening mechanism is applicable for alloys of steels [17]. Even if the austenite decomposition product is not martensitic, the transformation hardening mostly occurs in ferrous steels. In precipitation hardening, the weld metals are strengthened with the ageing phenomenon and different microstructures can be seen in it.

Metallurgical Events in HAZ

The microstructure of HAZ depends on the heat input, metals used, metal condition before welding, composition, etc. Various metallurgical reactions in HAZ occur such as recrystallization, recovery, grain growth, precipitate formation, over-ageing/dissolution of precipitates, phase transformations, stress relaxation and residual stress [18].

Recovery, Recrystallization and Grain Growth

During the recovery stage, rearrangement of dislocation occurs resulting in the formation of a cellular dislocation structure. This in turn produces strain-free regions due to the reduction of internal energy. These regions help in the formation of new grains by acting as nuclei. The strain-free nuclei grow in the recrystallization stage, and the

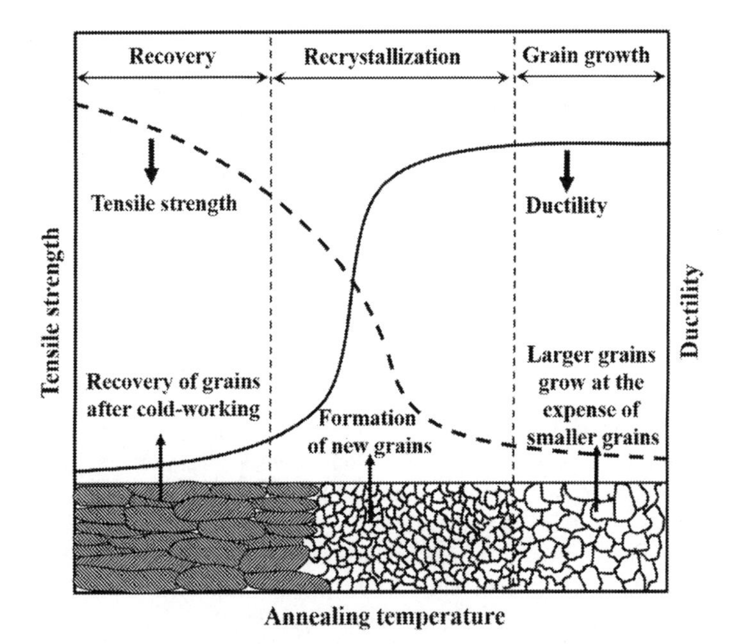

Fig. 6.4 Effect of recrystallization on ductility and strength for cold-worked materials

dislocations are annihilated. These nuclei develop into new grains and absorb the old ones, resulting in decreased strength and hardness and increased ductility. With the advancement of time or temperature, grain growth continues due to the reduction in the energy of overall grain boundaries because of reduction in boundary areas. The grains get larger, and this process stops. Beyond the recrystallization stage, if the grain size increases, it results in a decrease in the strength, which is known as the Hall–Petch effect [19]. In fusion welding of single-phase cold-worked material, the inner crystals/grains are having low ductility but high strength. However, within the melt boundary, in fusion welding, an abnormal grain growth occurs, yielding low strength. Due to annealing, the grain becomes coarser within HAZ, and it leads to partial recrystallization. Strength falls much below that of the BM in this case. The effect of temperature on strength and ductility in cold-worked materials is shown in Fig. 6.4. From the figure, it can be seen that the annealing temperature has an impact on the amount of the recrystallization. At low-annealing temperatures, the grains do not reform. But some of the internal stresses get relieved within the material. As those internal stresses relieve, some recovery in the material property is observed. At higher temperature, the grains begin to reform and larger grains are seen. This diagram is for brass material. At lower temperature, recovery is seen. Here, reduction in strain energy occurs, and some of the properties get restored. From Fig. 6.4, it can be observed that the tensile strength as well as ductility have not recovered a huge amount in this phase. At recrystallization stage, when the grains begin to reform, significant recovery of the properties can be seen. Tensile strength reduced significantly and ductility increased at this stage. Finally, when each of the grains begins to enlarge at high temperature, a small amount of additional recovery of the properties can be seen. The vast majority difference occurs during the recrystallization stage when each of the deformed grains is given the opportunity to reform its crystalline structure.

Allotropic Phase Transformation

Allotropy signifies the existence of different crystallographic forms of a metal at different temperature values. Due to allotropic phase transformation, various microstructures get evolved in steel. Austenite is formed at elevated temperature in HAZ; however with cooling, various other transformational products like ferrite, bainite and martensite are also evolved. Sometimes, the phase transformation is taken into consideration to optimize the strength of a material. In some of the welding processes, the temperature gradient is intentionally decreased by preheating the alloy steels to form martensite or bainite, and finally, to avail a microstructure consisting of both tempered martensite and bainite.

Precipitation reactions are another most important metallurgical event which cause microstructural change in HAZ.

6.2.2 Microstructure in Solid-State Welding

In solid-state processes, the formation of metallic bonds occurs at the atomic levels. As stated above, melting and solidification do not take place in these welding processes. Instead, heat and deformation play a vital role at weld interface. A few of the examples are mentioned in the introduction section. This section starts with the regions of solid-state welding processes, followed by the influence of metallurgical processes influencing these welding processes.

6.2.2.1 Regions of a Solid-State Weld

The BM, HAZ, and heat and deformation zone are the three regions found in these welding processes. Heat and deformation zone refers to the weld zone, where dynamic recrystallization occurs due to forging action by various means which in turn results in extremely fine grains, which thereby provides higher strength to this zone. The properties of HAZ are significantly different from BM (which as usual stays unaffected during these welding processes).

6.2.2.2 Metallurgical Processes Influencing Microstructure in Solid-State Welding

In friction welding, frictional heat and pressure are the main reasons for welding due to the relative movement of the workpieces. In diffusion welding, interdiffusion takes place due to heating. Here, the two metals to be joined are maintained at close vicinity with each other and are heated up to an elevated temperature, followed by application of pressure. A local deformation occurs at the interface of the two metals, followed by boundary migration, recrystallization and reduction in pore size, along with bulk diffusion phenomena [20]. Explosion welding is a kind of collision welding process which happens instantaneously. Due to collision and impact, the heat generation occurs and then huge plastic flow occurs at the interface. These two facilitate the development of the metallurgical bond between the materials to be joined. As it is an instantaneous process, many metallurgical reactions are suppressed, and a very narrow or undetectable HAZ is seen in this welding process [21]. Metals which are metallurgically compatible, for example, stainless steel–titanium, titanium–copper, aluminium–steel, stainless steel–carbon steel, etc., can be joined with this process. In ultrasonic welding, very less heat is generated at the interface (30–50% of melting point temperature of base material) by friction. By using ultrasonic vibratory energy (frequency more than 20 kHz), the joint is made through localized plastic deformation at the interface. The generated heat helps in softening of metal and facilitates plastic flow at the interface. This heat is high enough for recrystallization and the formation of the new grains. Refinement of grains at the interface occurs due to the elastic plastic flow along with work-hardening due to the plastic flow of the metal. No melting occurs as well as no heat related issues, i.e. solidification cracking, porosity, blow holes, etc., are there in this welding, heat being generated over a small volume of the metal, so very less HAZ is found. There is no evidence of grain growth in this kind of welding [22]. In all the cases, there exists a different bonding mechanism which in other ways affects the weld microstructure.

Since the authors have considered aluminium to stainless steel welding in the case study, therefore, a brief overview of the microstructure and weldability of steel, as well as aluminium, is stated below.

6.2.3 Microstructure of Steel and Its Weldability

Steel is primarily iron with few carbon and other alloying elements. Properties of steel get affected depending on these alloying elements. The main alloying element for the formation of microstructure in steel is carbon. Iron–carbon equilibrium diagram is a convenient way to understand the microstructure of steel at different temperatures and carbon content. Various structures are austenite, pearlite, cementite, ferrite, etc. The typical microstructure of steel can be understood in one of the literature [23]. Most commercial steels can be divided into low-alloy steel (ferrite + small grains of pearlite), high-alloy steel (pearlite at 0.8% c, above 0.8% c pearlite + cementite), medium-alloy steel (pearlite + small pearlite) and plain carbon steels. All these steels are specifically designed with several factors such as section thickness, preheat, hydrogen content and heat input and the formation of undesirable transformation products, such as martensite to have good weldability to be used in many applications. During welding of a steel component or structure, it is important to select a grade that has good weldability and meets the mechanical requirements of the design. One example of a specially designed steel is ASTM A 216, Grade WCB, which is used for pressure vessel, valve and pump castings. Its ability to be welded and repaired makes it as a popular material [24].

Stainless steels are one of the important commercial alloy steels that contain more than 12% chromium and other alloying elements like nickel, manganese, phosphorous and silicon. The four classes of stainless steel are duplex, ferritic, austenitic and martensitic. Because of chromium, steel forms a tiny layer of chromium oxide which is non-porus, coherent and protective to BM. The austenitic alloys are used in applications where good toughness, high temperature resistance, good corrosion resistance are needed. They do not harden in the HAZ and do have a good weldability. But few problems associated with welding these steels are solidification cracking of weld metal and other problems such as weld decay/sensitization, sigma phase formation, hair like cracking in HAZ. The chromium percentage in ferritic stainless steels is 12–27%. It also has a small amount of austenite-forming elements. These steels have good formability. Martensitic stainless steels have the lowest percentage of chromium. Due to this, it shows hardenability. Due to this property, it is used in material where sharp edges need to be maintained, e.g. cutlery, surgical equipment's, etc. Necessary care such as preheating and post-heating must be taken during welding of these steels, as the martensitic HAZ is inclined towards cracking. Duplex stainless steels are chosen for their improved corrosion resistance [25]. They do not lead to the formation of martensitic transformation product during welding. Most stainless steels are easily welded by welding method such as resistance spot (RSW), laser beam, electron beam, arc (plasma arc, SMAW, gas tungsten arc welding (GTAW), etc.) and friction welding processes. In the case study section, FSW is used to weld steel along with aluminium.

6.2.4 Microstructure of Aluminium and Its Weldability

The alloys of aluminium have face-centred cubic (FCC) crystal lattice structure. It is not possible to design aluminium's microstructure by using phase transformation. They have properties such as: (a) low density (b) good corrosion resistance (c) capability to form oxide film which is insoluble in the molten pool of aluminium and (d) high electrical and thermal conductivity. Aluminium has a very less melting point. Therefore, it is easy to produce the fusion weld joints of the aluminium alloys from the melting point of view. But the problem is thermal expansion coefficient of aluminium. It expands too much because of the high expansion coefficient (25E-6/°C) as compared to iron. It expands much and contracts leading to high residual stresses and creates residual stress cracking (similar to solidification cracking), partial melting zone cracking and distortion. Aluminium has a higher thermal conductivity than steels hence making the melting difficult because whenever heat is applied on abutting surfaces of the BMs to be joined, heat is dissipated towards BM. It produces wider HAZ. Another important concern related to aluminium is solidification temperature range. For this reason, aluminium shows cracking tendencies and softening in HAZ. To overcome these problems, it is always essential to use special processes rather than conventional welding processes (e.g. gas welding and SMAW) to weld aluminium because in both the cases, shielding is not that much effective to prevent oxidation. Gas-based welding methods like GMAW, GTAW and plasma arc welding are more effective to weld aluminium alloys. But these processes cause porosity, solidification cracks, liquation cracks, etc. Thermal softening is another major problem in aluminium. It can be possible to get rid of this problem of aluminium by various strengthening mechanisms applied to aluminium. To weld thick aluminium plates, preheating is required [26]. When compared with the alloys of steel, aluminium alloys will require high heat input in fusion welding, high current and small welding time in RSW. Alloying with various materials like lithium, zinc, silicon, manganese, magnesium and copper is performed to obtain the required properties and for grain refinement. To get optimum mechanical properties, heat treatment, in addition to welding, must be carried out for aluminium. The typical microstructure of the heat-treatable aluminium alloy is given in [27]. Solid-state welding such as FSW is proved to be one of the best joining technologies to weld aluminium.

6.3 Texture in Welding

As mentioned in the introduction section, texture in a polycrystalline material is the preferred orientation of grains in regard to a reference co-ordinate system. This reference system is of two types: (a) crystal reference system and (b) sample reference system. In the first case, the reference is three mutually perpendicular axes of a unit cell. For all the crystals present in the sample, this reference is considered the same. In the second one, three mutually perpendicular axes: [(a) rolling direction (RD), (b) transverse direction (TD) and (c) normal direction (ND)] of the sample are taken as reference.

There are two broad categories of texture present known as: (a) macrotexture and (b) microtexture. Macrotexture is a bulk measure of the orientations of all grains present in a sample. It does not give any idea about any particular grain within the sample. It is measured by X-ray diffraction (XRD). Microtexture gives information about the orientation of a particular grain along with the neighbouring ones as well. It is measured by electron back scattered diffraction (EBSD).

Texture changes in all stage of processing. Based on it, there are four kinds of texture present, i.e. (a) solidification texture, (b) recrystallization texture, (c) deformation texture and (d) transformation texture. The first one develops during melting and

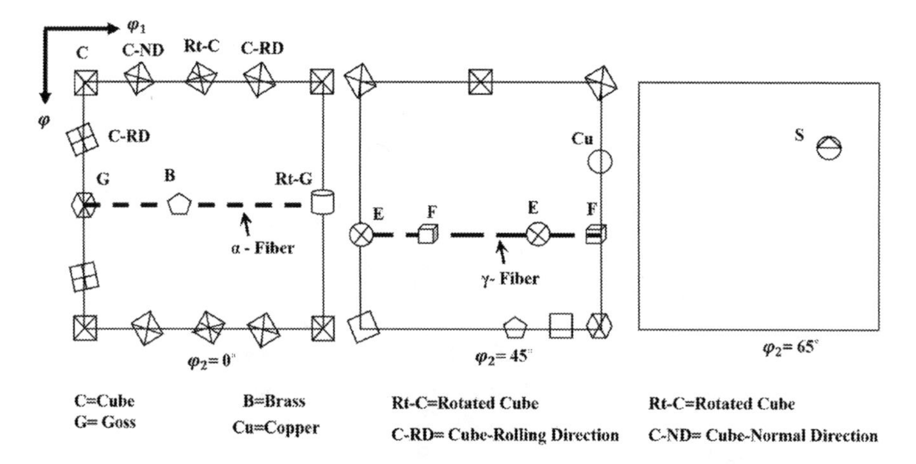

Fig. 6.5 Representation of different texture components for aluminium with different values of φ_2 Euler angle

solidification, e.g. casting. Third one develops due to different deformation processes, e.g. rolling, forging, drawing, etc. Second one develops when a deformed material is annealed at elevated temperature. Last one arises due to the crystallographic transformation from BM to final product but depends on the texture of the BM. There are three ways to represent texture, i.e. (a) pole figure method, (b) inverse pole figure (IPF) method and (c) orientation distribution function (ODF).

Figure 6.5 shows the schematic representations of the ODFs for FCC material. Here, the ideal location of texture components is given.

In welding, the weld zone is subjected to heat and load/pressure in general. Heat flow characteristics and the pressure dictate microstructure along with their orientations. Properties of the welded zone are not only determined by microstructure, but also by the orientation of microstructures.

This section will give an overview of texture in some of the fusion welding and solid-state welding processes.

6.3.1 Texture in Fusion Welding

The fusion welding techniques yield random textured microstructures in the FZ and highly textured grains at the HAZ because of the large heat input in the weld zone. Texture in these welding processes develops mostly due to crystallization/solidification (from a non-crystalline/liquid state), annealing (recrystallization/grain growth), phase transformation (due to orientation relationship) and thin-film growth (substrate orientation and strain energy).

The grain growth occurs in certain crystallographic directions because of heating and cooling cycles in welding processes as shown in Fig. 6.2a, c, e, g. Therefore, favourable-orientated grains grow for a considerable distance, while the growth of less favourable-oriented grains is blocked by fast-growing grains [28]. For this reason, the columnar microstructure is seen in weld metal where relatively longer grains are present, and these grains grow parallel with the direction of heat flow. These grains grow due to favourable crystal orientation and solidification. The initial grains to solidify are the ones that are close to the unmelted parent material. The unmelted parent metal nucleates these grains to maintain equal crystal orientation. In metals having body-centred cubic (BCC) and FCC crystal structures, solidification growth occurs in preference to <100> crystallographic direction. As solidification is most prominent in this crystallographic orientation, this direction is sometimes called as the direction of "easy growth". In metal having hexagonal closed pack (HCP) crystal structure, grain growth occurs in the <1010> direction parallel to the basal plane. The grain growth becomes prominent if this direction of easy growth comes parallel with the direction of heat flow over the S–L interface [29].

The crystallographic orientations in the weld regions obtained in different fusion welding techniques have been discussed in the following sections.

6.3.1.1 Resistance Spot Welding (RSW)

The values of current and voltage largely affect the grain size and the texture in RSW as found in [30]. RSW at higher values of the time and current (300 ms–9.5 kA), produces randomly textured microstructure in the weld zone. In reverse, strongly textured microstructures are obtained at a low value of time and current (i.e. 200 ms–8 kA). The major texture components in RSWed samples can be classified as deformation and recrystallization textures. Later one includes Q {013} <123> , Goss {110} <001> , Cube {001} <100> and RCND1{001} <310> , whereas the deformation texture components include S {123} <634> , Brass {110} <112> , Shear1{001} <110> and Shear2 {111} <110> . The volume fractions of these texture components depend on process parameters of the welding. Welding at a lower range of time and current (200 ms–8 kA) produces dominating volumetric percentage of cube texture in the weld zone and higher volume fraction of deformation texture components at an increased value of time and current (i.e. 300 ms–9.5 kA).

6.3.1.2 Gas Metal Arc Welding Process (GMAW)

In GMAW, directional solidification and preferred grain growth are largely affected by Lorenze and Marangoni force [31]. Magnitudes of the current and voltage determine the value of Lorenze and Marangoni force. In GMAW of austenitic stainless steel, the preferred grain growth occurs along <001> and <101> directions parallel to the RD in the weld at a higher value of current and welding speed. In contrary to this, at lower value of the current and welding speed, the direction of grain growth aligns to <101> , i.e. parallel to RD. Later, in the cooling stage of the welding process, the grain growth further changes, either along <001> or <111> . This change in the cooling direction is because of the temperature gradient change. In GMAW, higher values of current increase the Lorenze force and reduces the Marangoni force [32]. This promotes a strong downward fluid flow force along the thickness direction of the weld and thereby causes preferred grain growth along <001> and <101> directions. The low electromagnetic force and high heat for a long period causes grain growth along <101> and <111> directions.

6.3.1.3 Laser Beam Welding (LBW)

In LBW, the amount of heat input and time of heating largely affect the secondary phase precipitates, size of the microstructure and orientations [33]. The existence of the columnar and equiaxed grains is found in the LBW of aluminium. Columnar grains orient along <100> plane, i.e. parallel to RD. LBW produces strong cube {001} <100> textured microstructure in the welded regions which makes the material prone to slip, preferentially aligns to {100} plane. Cube-oriented microstructure has a low value of critically resolved shear stress. As a result, columnar grains with Cube {001} <100> microstructure reduces the joint strength.

6.3.2 Texture in Solid-State Welding

In solid-state welding processes, texture develops mostly due to plastic deformation (by glide or slip and twinning) and annealing (recrystallization/grain growth). As a result, the solid-state welding techniques produce moderately textured grains in the welded zone. Initially, due to dynamic recovery, the formation of low-angle boundaries occur. Then subgrain boundaries transform into grain boundaries followed by local migration of newly formed grain boundaries. The complete mechanism is linked with texture development. This section is limited to aluminium welding in all the solid-state processes as discussed below and the reason for taking only aluminium is found in one of the subsequent sections.

6.3.2.1 Ultrasonic Welding (UW)

It has been extensively applied in the welding of thin sheets of aluminium, copper, etc. Oscillation amplitude and the falling height are the two important process parameters of UW which affect the orientations of microstructure and the weldability. Large value of oscillation amplitude and falling height weakens the texture intensity but yields fine microstructures in UWed regions [34]. The UW at an increased value of ultrasonic amplitude causes large deformation and heat in the build. Deformation leads to the formation of strong deformation texture components such as Brass {011} <211> , Cube {001} <100> and Rotated-cube {001} <110> at the cost of recrystallized texture components. Similarly, UW at an increased value of the falling height (i.e.

high deformation reduction) leads to the formation of strong Rotated-cube and Brass textured grains. However, at a medium range of deformation (<50% deformation reduction), recrystallized textured grains form at the weld.

6.3.2.2 Friction Welding (FW)

It is a solid-state welding process and capable to weld similar and dissimilar materials. In this technique, at the joint interface, frictional heating and axial pressure help to establish the metallurgical bond. Heat and pressure modify the grain size and orientations of microstructures at the weld [35]. Material deformation mechanism in this process is shear and forging which lead to the formation of shear texture components such as B $\{1\,\bar{1}\,2\}\langle 110\rangle$, A$\{1\,\bar{1}\,1\}\langle 110\rangle$ and C$\{001\}\langle 110\rangle$ in the weld regions.

6.3.2.3 Explosive Welding (EW)

The EW is a widely adopted solid-state joining technique for high ductile and impact resistance materials such as aluminium and titanium. High pressure in this technique produces a significant amount of deformation on the sheets which help to establish metallurgical bonding. Deformation at a high colliding speed (0.6–3 mm/μs) causes low heat and severe deformation in the joint interface. This leads to formation of deformation textured grains in the weld regions. S$\{123\}\langle 634\rangle$, Brass$\{011\}\langle 211\rangle$ and Copper $\{112\}\langle 111\rangle$ are the major deformation texture components which are often obtained in EWed regions [36]. Low heat input in this process prevails the formation of recrystallized texture grains.

6.4 Case Study for Microstructure and Texture Evolution in Solid-State Welding of Dissimilar Materials

This case study section begins with an overview of the FSW process, and its various microstructural zones, followed by experimentally obtained microstructure and texture results of FSW of two dissimilar materials: AA6061-T6 aluminium alloy to AISI 304 stainless steel.

6.4.1 FSW Process and Microstructural Zones

FSW is an efficient, economic and eco-friendly joining process, which is accomplished in the solid-state without melting the base materials. A rotating and non-consumable tool is inserted in the mid-line of two neighbouring edges of the BMs to be joined. Heat is generated by the friction occurring between tool and BMs along with plastic deformation of BMs. The BM gets heated up to an elevated temperature and material flows in the vicinity of FSW tool by stirring action leading to joining. In FSW, various problems associated with the fusion welding are reduced and high integrity joints are formed in solid state. The details have been mentioned in a previous chapter.

Four microstructural zones are present in a FSWed sample. The first one is known as the SZ, sometimes also called as the nugget zone (NZ). In this zone, the material undergoes severe plastic deformation since this zone is directly under the influence of the tool action. The grains in this zone undergo DRX, and the grain size gets reduced in comparison to the parent material. The flow of the material starts with the movement of the tool, extruding from the advancing side (AS) in NZ, and flows around retreating side (RS). Surrounding the NZ, thermo-mechanically affected zone (TMAZ) exists. Very little recrystallization is noticed here. It is often difficult to observe the TMAZ in few materials (steels and titanium alloys) [37]. In FSW, the third zone is the HAZ, which sometimes exhibits the same metallurgical reactions as in fusion welding. The softening of HAZ in aluminium occurs because of over-ageing, both in fusion welds and FSW [38, 39]. In order to obtain a sound weld joint in FSW, sufficient deformation of the BMs is to be ensured so that and they can flow alongside the tool.

It is well understood that the material deformation occurs in FSW primarily due to extrusion, forging and shear created by the FSW tool. The deformation mechanism in the process can be understood through a detailed investigation of crystallographic texture. The detailed experimental procedure of FSW for AA6061-T6 aluminium alloy–AISI304 austenitic stainless steel has been given in Sect. 6.4.2. In the FSWed joint, the authors have studied the microstructure and texture in the aluminium alloy by using EBSD technique, and the results are shown in Sect. 6.4.3.

6.4.2 Experimental Details

The materials were austenitic stainless steel AISI304 and aluminium alloy AA6061-T6 of 1 mm thickness each and were welded in the lap configuration, keeping the aluminium sheet over the steel sheet. The tool used in this process was made with bimetallic materials: tungsten carbide (WC) and H13 die steel. The tool pin material was taken as WC to avoid tool wear and the rest of the tool parts were of H13 die steel. The WC pin was inserted into the H13 die steel and assembled through interference fit and supported through the grove screw. The tool has a flat shoulder, and cylindrical pin, having diameters of 18 mm and 5 mm, respectively. Welding was performed at a welding speed (v), rotational speed (ω), plunge depth *(PD)* and tilt angle (α) of 125 mm/min, 900 rpm, 0.2 mm and 1°, respectively. The experiment was performed in a computer numerical control (CNC) FSW machine (ETA Technology Bangalore, WS004).

To determine the tensile strength of the parent material (AA6061) and weld joint, tensile samples were cut in a rectangular shape in the CNC wire electrical discharge machining (EDM) machine (Electronica, Maxicut 5231) [40]. The cut samples were tested in a 5 kN universal testing machine (UTM), (Instron, 3325). SEM observations were performed to study the fracture surface of failed samples.

The cross sections of the BM and weld sample (from the welded region) were cut having a length, width and thickness of 20 mm, 7 mm and 1.8 mm, respectively, in a low-speed saw machine (Buehler, Isocut) and then polished mechanically with silicon carbide emery papers of various grit sizes followed by diamond cloth polishing. For EBSD characterization, the cut samples were electro-polished in a solution of 1000 ml (200 ml perchloric acid and 800 ml ethanol) for 15 V and at 10 s only. EBSD scan was performed on different weld regions at a scan rate of 0.5 step size in a field emission gun scanning electron microscope (ZEISS, AURIGA with oxford detector). Obtained data from EBSD were cleaned and analysed in a software (TSLOIM Analysis 7.0) by following the standard grains dilation method in a single iteration. OIM means orientation imaging microscopy which is a tool to analyse local texture in polycrystalline material. Spatial distribution of crystallographic orientation can be visualized through OIM measurements. It also provides a tool to link orientation as well as microstructure.

Different maps are used to allow colours to be assigned to points in an OIM scan based on some parameter associated with the point such as orientation or image quality (IQ). IQ defines the quality of an EBSD. IQ depends on the materials and their state. It is a function of parameters used and the technique to index the pattern. Authors have shown grain orientation spread (GOS) map, kernel average misorientation (KAM), ODF map and IPF map in the following sections to describe texture. In KAM, the misorientation is calculated between a grain at the kernel core and all points at kernel perimeter. The average of these misorientations is the locale misorientation value assigned to the centre point. In GOS, each point is shaded according to the orientation spread of the grain to which the point belongs. This can also be normalized to the area of the grain; however, this tends to make the large grains have smaller spreads than those observed in the small grains. Instead of plotting crystal orientations with regard to an external frame of reference, IPFs can be produced which shows the RD, TD and ND directions with reference to the crystallographic axes. These arc plotted in a standard stereographic triangle shown in Fig. 6.6a as well as in the bottom left portion of Fig. 6.7. ODFs are developed by the combination of data from various pole figures. ODFs define the texture of each crystal with regard to three Euler angles [defines the variation with three orientation axes (RD, ND, TD)].

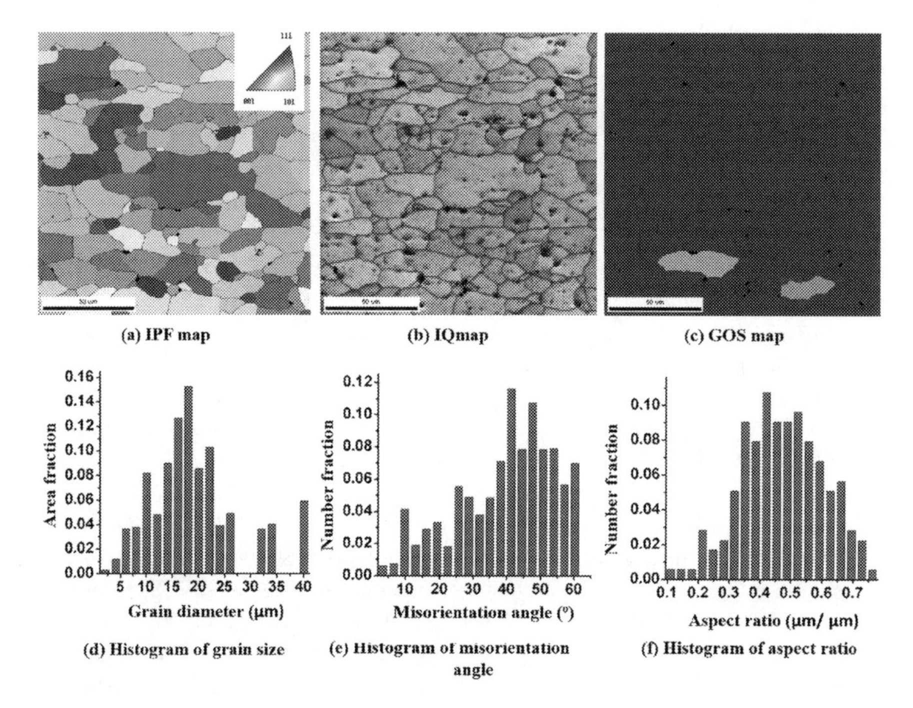

Fig. 6.6 Grain boundary characteristics of base AA6061-T6

6.4.3 Results and Discussion

FSW modifies the microstructures of the weld regions. As mentioned earlier, FSW is a solid-state joining process and from the experiments, it was found that the weld temperature was below the: (a) melting point temperature of base AA6061-T6 aluminium alloy (i.e. 660 °C) and (b) recrystallization temperature of AISI304 stainless steel. The lap FSW welded sample has failed from HAZ of aluminium in the tensile test. In general, any kind of study is carried out on the failed sample first. Therefore, in the present study, a detailed microstructural investigation of only aluminium substrate has been carried out.

6.4.3.1 Microstructure and Texture in Aluminium Substrate: AA6061-T6

Heat and deformation are the two causes for DRX, grain growth and recovery in the aluminium substrate. The microstructure of base AA6061-T6 has been depicted in Fig. 6.6a, b where the pancaked rolled microstructure can be seen. The fraction of recrystallized grains was found to be 96%, which can be attributed due to the solution heat treatment for obtaining tempered state (T6) of AA6061 (Fig. 6.6c). The base

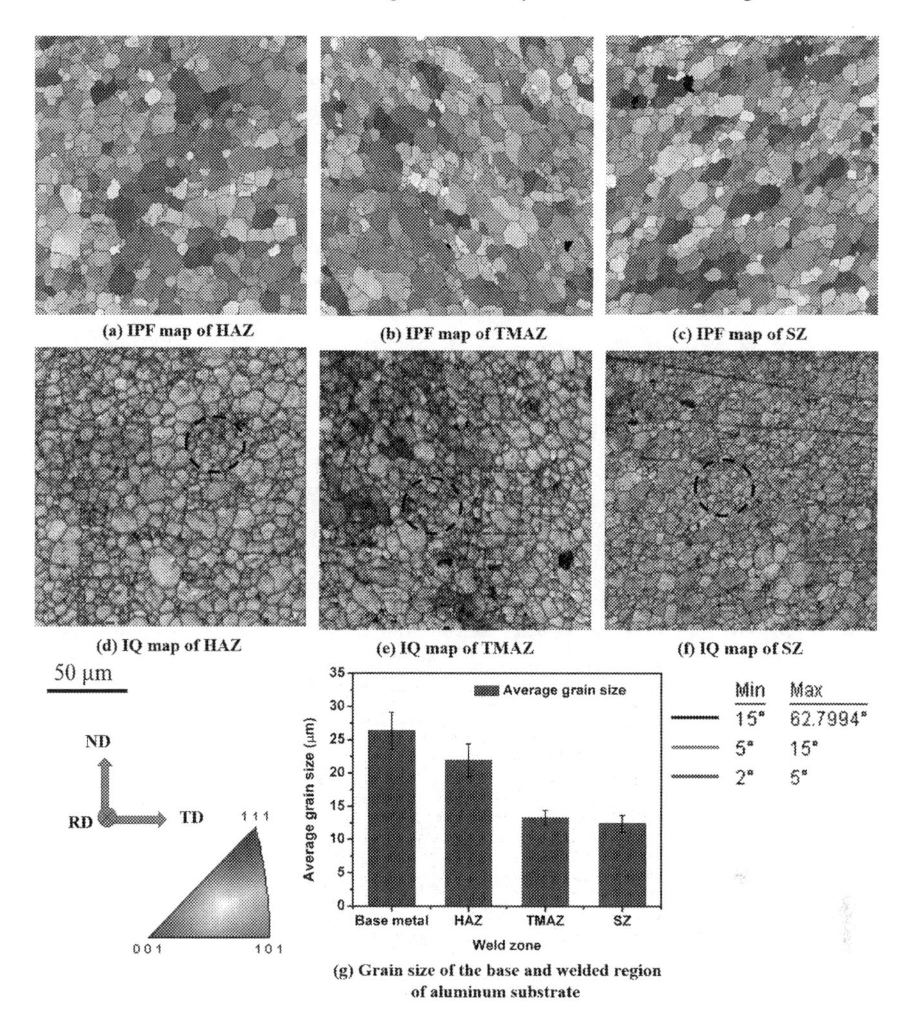

Fig. 6.7 Grain boundary characteristics of the aluminium substrate at the different FSWed regions. The red colour square box shown in IQ maps indicates recrystallized grains

AA6061-T6 material was found consisting of bimodel grains of size varied in the range of 2–42 μm (Fig. 6.6d). Also, the T6 state of base aluminium resulted in strain-free grains, and it has a negligible fraction of low-angle grain boundary (LAGBs) (Fig. 6.6e). The average grain diameter and grain aspect ratio (length/width) of base material were 26 μm and 0.47, respectively (Fig. 6.6f). Less the aspect ratio, more will be the strength and elongation. A good example to understand aspect ratio is rolling operation. The tensile strength of a 3 mm rolled aluminium comes around 240–245 MPa. But the tensile strength of a 1 mm rolled aluminium comes around 330 MPa [41, 42]. Because of more rolling operation, the grains get elongated and thin having less aspect ratio in the elongated direction. The material is more deformed here consisting of more subgrains and having more strain and dislocation defects. Therefore, it is having high strength.

Figure 6.7a–c shows the EBSD IPF maps of different weld zones of the aluminium substrate. It can be noticed that the grain boundary characteristics and their orientations have been changed in the FSWed regions. The HAZ has equiaxed and coarser microstructures. The average aspect ratio of the microstructures was 0.55. The grain size varied in a range of 2.5–36.30 μm and the average size was calculated as 19.6 μm. The SZ of the weld confined of elongated microstructures with an average aspect ratio of 0.49. The grain size varied in the range of 2.5–30 μm with an average value of 12.2 μm. Histogram plot of base and welded regions' grain size in aluminium substrate are shown in Fig. 6.7g.

Further, the EBSD map of the transition region (i.e. TMAZ) between SZ and HAZ revealed similar microstructures as the SZ. The average values of grain size and their aspect ratio in TMAZ are 13.8 μm and 0.52, respectively. Thermo-mechanical action during FSW causes the elongated microstructures at SZ and TMAZ. However, frictional heat is the cause for the formation of equiaxed grains at HAZ.

FSW-led recrystallization and grain growth of the microstructures have been explained in the subsequent section. Heat and deformation of the FSW modified the microstructure size [43] and their orientations. Both are responsible for the DRX and recovery in the aluminium substrate. Deformation due to the tool causes defects (dislocations, point defects) in the lattice structure of aluminium. Dynamic recovery occurs when an excessive amount of dislocations or point defects form in weld regions. Later, dislocations get tangled and form a cell wall, which is the early sign for the recovery. Further, as the welding proceeds, it adds dislocation and heat into the lattice structure which makes easy cross-climb and slips for the dislocations. It leads to the annihilations of dislocations into the existing cell walls, which later turn into LAGBs and subgrains (misorientation angle <15°). The driving force for the recovery is the reduction in strain energy (i.e. dislocations and point defects) of the weld material [44]. Recovery forms the serrated grain boundaries which later gets converted into strain-free grains after DRX. Recrystallization and grain growth take place after recovery. The available dislocations and LAGBs in the weld region drive both recrystallization and grain growth in the annealing stage of the weld.

Recrystallized and subgrains can be seen in IQ maps of weld regions depicted in Fig. 6.7d–f. Grains having misorientation angle 2°–15° and beyond have been taken as LAGBs and high angle grain boundaries (HAGBs), respectively. Grains having misorientation angle 2°–5° are known as very low-angle boundaries as shown in bottom right corner of Fig. 6.7. This division is based on the extent of misorientation between two grains. In simple words, LAGBs are comprised of an array of dislocations and their properties are a function of misorientations, while HAGBs are normally not independent of misorientation. The term min. denotes to minimum misorientation angle and the term max. denotes to maximum misorientation angle of the grain boundary as shown in bottom right corner of Fig. 6.7.

The recrystallized grains have been indicated by the square box and subgrains in red and blue colour boundaries are circled in HAZ, TMAZ and SZ. After welding, in aluminium alloys, grain growth and recrystallization occur through two mechanisms [45, 46]: (a) strain-induced boundary migration (SIBM) and (b) subgrain rotation and coalesce (SRC). SIBM is a recrystallization mechanism which occurs in polycrystalline materials deformed to a strain less than around 0.2. It occurs at the original grain boundary. In SIBM, the new grain boundary occurs which moves to next crystal and then recrystallized area expands. SIBM has been identified in the present study, as depicted by a circle in Fig. 6.8a, b, between grains 1 and 2. This mechanism is mostly observed between two grains which have different density of dislocations [47]. KAM map, as depicted in Fig. 6.8b, verifies the difference in strain between grains 1 and 2. SIBM has been further illustrated in Fig. 6.8c. The difference in dislocation density, subgrains and grain boundaries between two grains provides the driving force for SIBM [48].

The SRC is based on the rotation of subgrain with regard to adjacent grains. Reduction in the boundary energy forms the driving force for SRC. In annealing stage of the FSWed joint, as the difference of the misorientation angle among two adjacent subgrains is less than their average misorientation angles, then one of the subgrains may rotate till both reaches the same orientation. SRC eventually leads to the coalescence of subgrains formation of strain-free recrystallized grains (Fig. 6.9). This mechanism has been identified between two grains 3 and 4, depicted by an elliptical box as shown in Fig. 6.8a, b. Later, the recrystallized grains grow through coalesces of HAGBs at triple point junctions which can be understood through Burke and Turnbull mechanism [49]. It has been identified by a square box between two grains 5 and 6, in Fig. 6.8a, b. Burke and Turnbull mechanism of grain growth has been schematically represented in Fig. 6.10. The driving force for grain growth

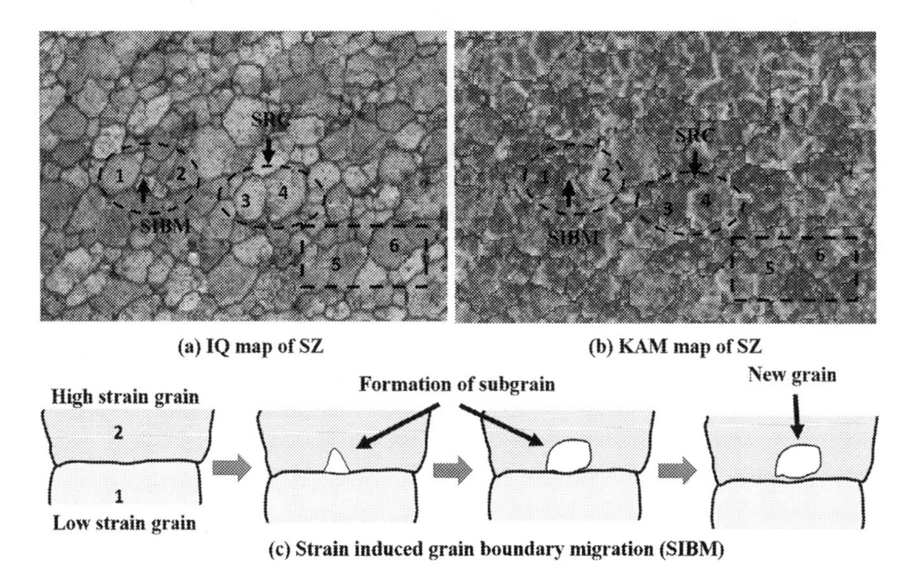

Fig. 6.8 Recrystallization of microstructure through SIBM, **a** IQ map of SZ, **b** KAM map of SZ and **c** SIBM mechanism

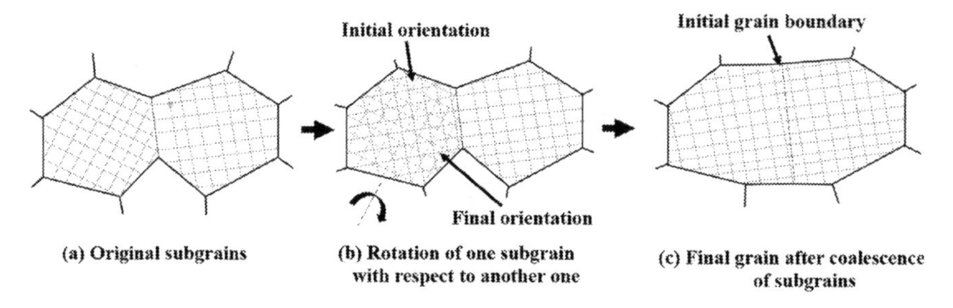

Fig. 6.9 Coalescence of two subgrains by rotation of one of the grains

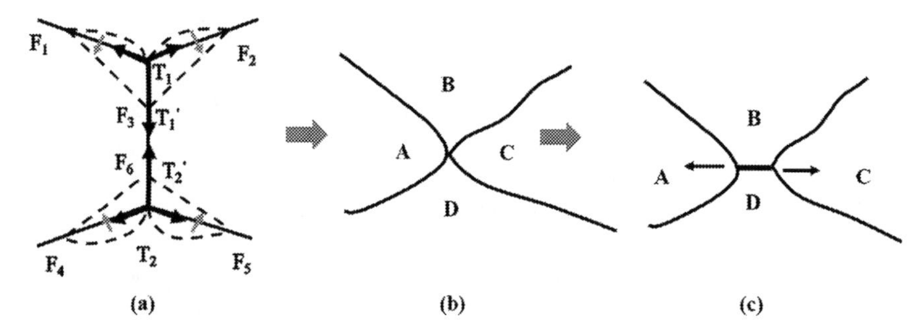

Fig. 6.10 Migration of grain boundaries at the triple point

comes from the reduction in boundary energy triple point junction where the grain boundary energy reduces to obtain stable dihedral configuration, i.e. angle between grain vertices is 120° [50]. At the triple points (T_1 and T_2) junction, where the three grain boundaries meet, it is subjected to the three forces (F_1, F_2 and F_3). If the grain boundaries attain equal specific energy at point T_1 or T_2, then boundaries form curve line (dotted line) which later tends to migrate along the direction of arrows in order to reduce their lengths. As a result, triple point will migrate from point A to A' and form straight grain boundaries (dotted line) which have an angle of 120°. It leads to the formation of stable grains. SZ has a comparatively larger number of triple point junctions, causing the formation of fine grains of nearly equal size. Later, as the grain growth continues in the annealing stage, lattice misfit took place introducing additional dislocations in the lattice structure of the aluminium substrate. It led to the formation of subgrains which have been verified by mapping of misorientation through point to point and point to an origin in a grain, as shown in Fig. 6.11b. Arrow marked on grain shows the direction of misorientation mapping as depicted in Fig. 6.11a. Mapping revealed the presence of LAGBs (i.e. subgrains) whose misorientations angle was below 15°. The formation of subgrains during grain growth leads to the development of strain-induced/deformed grains.

The fractions of recrystallized and deformed grains have been analysed at SZ, TMAZ and HAZ and are shown in Fig. 6.12. GOS approach has been used for the partition of recrystallized grains from deformed ones. Grains that have GOS < 1.5°

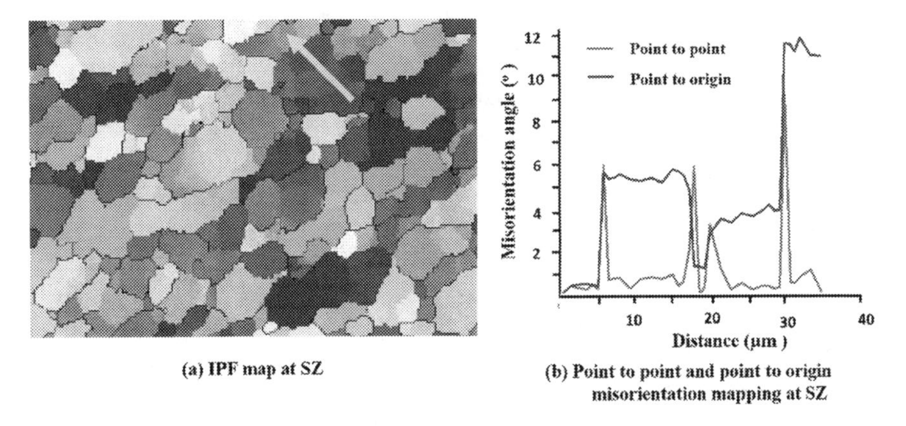

Fig. 6.11 **a** IPF map of SZ and **b** mapping of misorientation in a grain

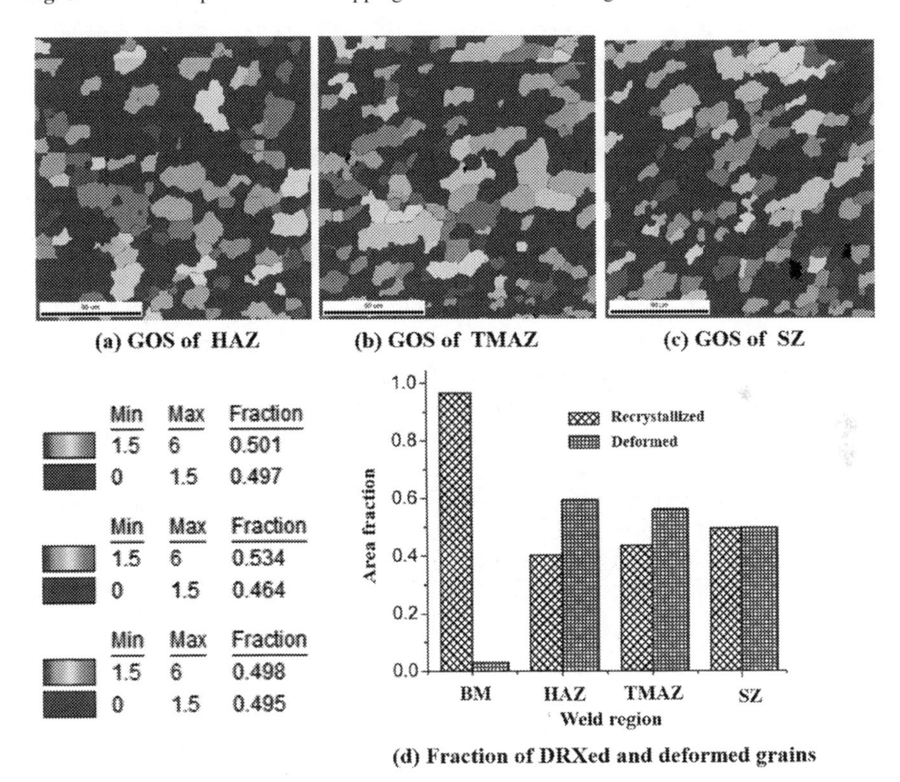

Fig. 6.12 GOS map of welded region: **a** HAZ, **b** TMAZ, **c** SZ and **d** area fraction of DRXed grains

have been taken as DRXed grains. Results showed that the highest fraction of DRXed grains is at SZ than in other regions. This is due to larger material deformation at the SZ generating a high density of dislocations. This provided an intense driving force for DRX and grain growth at SZ.

As discussed above, there was an existence of inhomogeneous microstructure along the weld cross section. This is due to the different degrees of frictional heat and plastic deformation in the weld cross section. Materials at SZ are mechanically stirred from AS to the trailing edge of the tool. TMAZ and HAZ are subjected to low plastic deformation, but are primarily affected by the heat. This causes different orientations of grains in the polycrystalline materials and anisotropic mechanical properties in the weld regions. Anisotropy is dependent on the crystallographic texture components of weld regions. EBSD ODF maps on the plane comprising of ND and TD are shown in Fig. 6.13, and it can be noticed that texture components of base aluminium were changed in the weld regions.

Referring to Fig. 6.13a, BM was dominated with Cube RD {013} <100>, Rotated-cube {001} <110> and Goss {011} <001> which are recrystallized and deformed textures, respectively. HAZ was dominated of recrystallized type texture component which includes Cube {001} <100>, whereas the TMAZ was dominated with Cube RD {013} <100> and Brass {011} <211> which are recrystallized and deformed texture component, respectively (Fig. 6.13b, c). The effect of frictional heating and material deformation at TMAZ produced both recrystallized and deformed textures whereas the effect of only frictional heating at HAZ led to the formation of recrystallized textured grains. The SZ was dominated with Goss {011} <100>, E {111} <110> and F {111} <112> textured grains which are known as deformed and shear textures (Fig. 6.13d). Material at SZ was extruded and sheared under the effect of the tool which is the likely explanation for shear and deformed textured grains. The texture components, their levels of intensity and the types obtained from ODF map are given in Table 6.1. It can be noticed that FSW reduced the texture intensity in the weld regions as compared to the base aluminium. In addition, FSW transformed recrystallized and deformed textured grains of BM to shear and deformation textured grains at the SZ. TMAZ has similar texture grains as base AA6061-T6 whereas HAZ has only recrystallized texture grains. It can be seen that various types of texture components have been found depending on the weld zones. Texture components, such as Cube {001} <100>, Rotated-cube texture {013} <100> and Goss texture {011} <100> components are known as recrystallized texture, whereas texture components such as Brass {011} <211>, Cu {112} <111> and S {123} <634> are known as deformed textures.

HAZ has both deformed and recrystallized texture components, whereas TMAZ and SZ have only recrystallized texture components. This is due to the synergetic effect of frictional heat and deformation at SZ and TMAZ, whereas HAZ was affected by the frictional heat only.

The major texture components in the FSW for aluminium are {110} <110> and {114} <221>, which are shear texture components with respect to the local reference plane. Further, the post-weld study of the FSW of aluminium revealed the presence of an alternate shear texture components of $B/\overline{B}$ and C in the weld SZ region. Similar

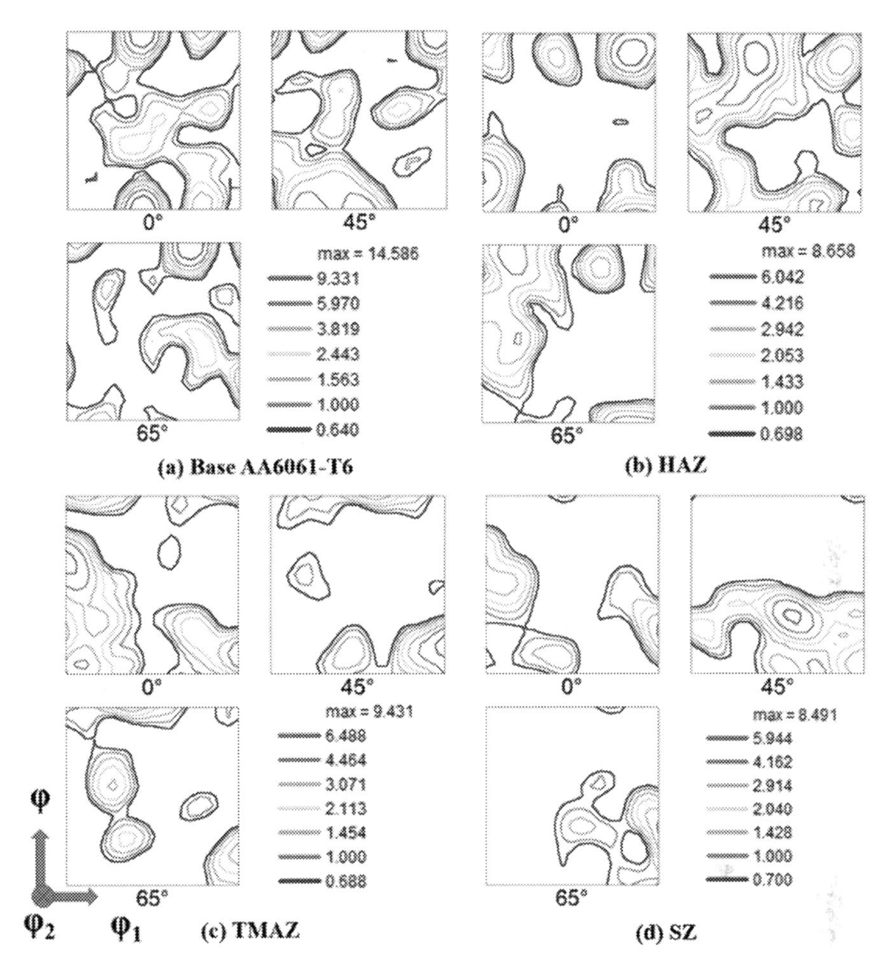

Fig. 6.13 EBSD ODF maps of HAZ, TMAZ and SZ of the welded sample

observations have been found for FSW of titanium Ti6Al4V [51]. However, for materials such as steel and other titanium alloys, major texture components are simple Shear {11 $\overline{2}$} <111> in pure iron and Rotated-cube {1 $\overline{11}$} <110> , {112} <110> and {001} <110> in AISI304 steel.

6.4.3.2 Microstructure and Tensile Strength in FSW

Tensile strength of BM and welded sample have been studied. One of the tensile samples has been shown in Fig. 6.14, i.e. test specimen dimension in Fig. 6.14a, and the physical appearance of weld in Fig. 6.14b.

Table 6.1 Texture components and their intensity at the different FSWed zones

Texture components	Relative intensity at HAZ	Relative intensity at TMAZ	Relative intensity at SZ	Texture types
Cube {001} <100>	Significant	Weak	Not found	Recrystallized
Cube ND {001} <310>	Significant	Significant	Not found	Recrystallized
Cube RD {013} <100>	Significant	Not found	Not found	Recrystallized
Goss {011} <100>	Significant	Significant	Significant	Recrystallized
Brass {011} <211>	Significant	Weak	Not found	Deformed
Cu {112} <111>	Weak	Weak	Weak	Deformed
S {123} <634>	Weak	Weak	Not found	Deformed

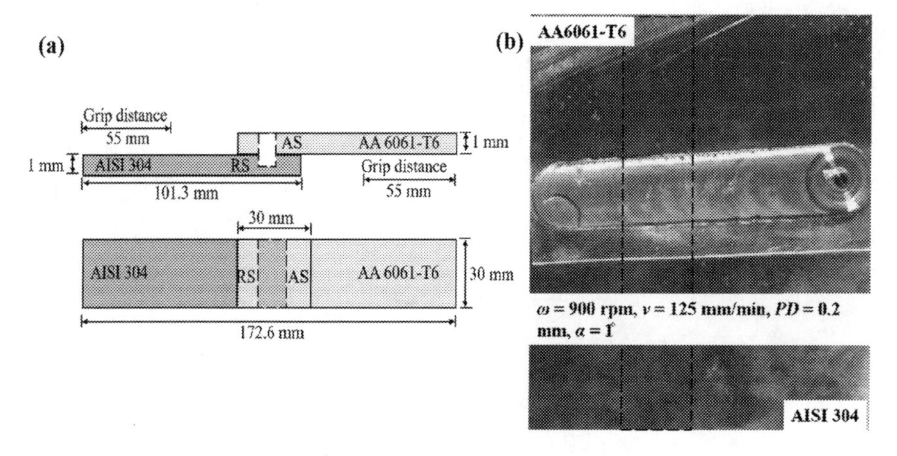

Fig. 6.14 **a** Tensile test specimen dimension and **b** physical appearance of weld

The coarser microstructure produced due to the softening effect at HAZ made it the weakest zone. Thus, the welded sample failed from HAZ. This sample has fewer welding defects and a coarser microstructure at HAZ. Consequently, during the tensile test, the sample has a larger deformation before failure as shown in Fig. 6.15. The tensile strength of the parent aluminium AA6061-T6 alloy and of the weld sample was found to be 328 MPa and 180.6 MPa, respectively.

Grain size of a weld is determined by temperature and strain experienced during welding [52]. In general, a material having fine grains is stronger and harder than a material having coarse grains. This may be well understood with the relationship between the grain size and yield strength by Hall–Petch equation shown in Eq. (6.1)

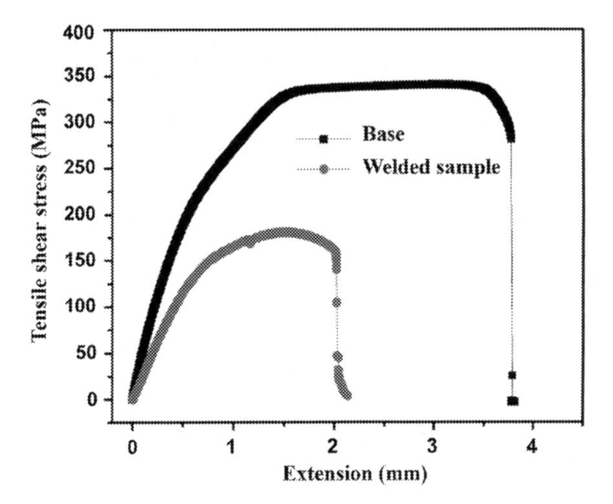

Fig. 6.15 Tensile plot of base material and welded sample

[53, 54]:

$$\sigma_y = \sigma_0 + kd^{-\frac{1}{2}}, \tag{6.1}$$

k is material correlation coefficient, d is average grain size, σ_0 is lattice resistance and σ_y is yield strength. Hall–Petch equation is majorly valid for polycrystalline materials with grain size greater than 1 μm [55]. It is true from this relation that, the joint strength decreases with increase in grain size [56]. As mentioned above, as strength depends on the average grain size, and therefore, with the change in average grain size for BM as well as welded sample, the tensile strength also varied. This justification is in a good argument with the literature [57], and one can conclude that microstructure has a strong correlation with the property. In addition to that, the preferred orientation of microstructure also affects the mechanical and metallurgical properties [58].

6.4.3.3 Texture and Tensile Strength in FSW

In the present chapter, a glimpse of the effect of texture on tensile strength is given. Mechanical testing shows that the BM has a higher tensile strength than the welded joint. The reason can be attributed to the presence of different texture components. Texture components, such as Brass {110} <112> and Goss {110} <001> , have been found in the base aluminium (i.e. AA6061-T6), whereas welded samples have different texture components at the different weld regions. As mentioned, major texture components were Cube ND {001} <310> , Cube RD {013} <100> and Cube {001} <100> at HAZ, whereas Goss {110} <001> and Shear texture E {111} <110> at TMAZ and SZ, respectively. In literature, the relationship between texture and tensile strength is shown [59]. During tensile testing, material deformation occurs by the activation of a slip-plane, which is determined by the preferred orientation of crystals [58, 59]. The stability of the slip systems of the materials is dependent on the Schmid factor. Texture components having high Schmid factor are unstable under tensile load. Crystals of lower Schmid factor show higher yield strength.

In a single crystal, for a particular slip-plane and slip-direction to applied stress in a given direction, Schmid factor ($\cos\phi \times \cos\lambda$) is related to the critically resolved shear stress (CRSS), where ϕ and λ are angle made by shear-plane normal and slip-direction with stress axis, respectively. It is useful for FCC polycrystalline materials to analyse the relationship between tensile strength and texture where lower the Schmid factor gives both higher yield as well as ultimate tensile strength [60]. Under the action of stress, materials show lowest value of resistance (i.e. yield strength) when the slip-plane normal and slip-directions are inclined at an angle of 45° with direction of an applied force. The value of Schmid factor is 0.5 when the slip-plane normal and slip-direction are inclined at angle of 45° with stress axis. Schmid factor of the crystals affects the activation of slip-planes. The Schmid factor values of Brass, Goss and Rt-Cube textures are 0.2747, 0.4082 and 0.4079, respectively. For aluminium material, the most favourable grain growth direction is <100> under the effect of the heat. Another aspect that can describe the effect of texture on various mechanical properties is the Taylor factor [61].

Taylor factor gives an indication of the resistance to deformation of a particular point for a given stress state. It is a geometric factor based on the texture of grain with respect to the applied strain gradient. An FCC material having 100% <111> texture component would have the highest possible Taylor factor, which would create a material with the highest tensile strength and with lower Taylor factor would create a material with lower yield strength. Another aspect is the volume fraction of grains. The higher the volume fraction of <111> texture in an FCC material, the higher is the yield strength [62]. Texture has a strong relation with elongation stated in the introduction. It is also a fact that when a polycrystalline material is subjected to plastic deformation, hardness increases due to strain hardening. FSW is a process where plastic deformation takes place. Strength and hardness have a linear relationship. This provides a way to comment that texture affects hardness and tensile strength.

FSW produces various texture components in the FSWed cross section. For almost all the metallic alloys, shear textured grains dominate at the SZ and recrystallized and deformation textured grains at the other weld regions in this process. Intensity of texture components slightly vary in FSWed regions. FSW also produces shear texture at the joint line. Similar to FSW, other solid-state welding techniques such as UW and EW also produce strong shear textured grains at the weld interface for aluminium alloys [34, 36]. Comparing with FSW, fusion welding techniques produce weak textured grains. GTAW and LBW produce weak textured grains at NZ as compared to HAZ [33]. Cube {001} <100> is the dominating recrystallized texture component in the mentioned fusion welding techniques. The deformation (i.e. deformation direction and deformation rate) and heat are the two important parameters which determines the texture components in the weld regions for a given material.

6.5 Conclusion

This chapter explains the basic principle of the formation of microstructure and texture in fusion welding, as well as in solid-state welding processes. At the end, the evolution of microstructure and texture in the FSW process is discussed. A detailed investigation of crystallographic texture has also been shown to understand the deformation mechanism in the process. Strong correlation does exist between property and microstructure, as well as texture.

References

1. Kulkarni S, Edwards DJ, Parn EA et al (2018) Evaluation of vehicle lightweighting to reduce greenhouse gas emissions with focus on magnesium substitution. J Eng Des Technol 16:869–888. https://doi.org/10.1108/JEDT-03-2018-0042
2. Biswas S, Suwas S (2012) Evolution of sub-micron grain size and weak texture in magnesium alloy Mg-3Al-0.4Mn by a modified multi-axial forging process. Scr Mater 66:89–92. https://doi.org/10.1016/j.scriptamat.2011.10.008
3. Easterling K (1992) Introduction to physical metallurgy of welding. Butterworth Heinemann
4. Savage WF, Nippes EF, Szekeres ES (1976) Study of weld interface phenomena in a low alloy steel. Weld Res Suppliment, 260–268
5. Bradley G, James MN (2000) Geometry and microstructure of metal inert gas and friction stir welded aluminium alloy 5383-H321
6. Callister WD, Rethwisch DG (2010) Materials science and engineering—an introduction, 8th ed. Wiley, New York
7. Padhy GK, Komizo Y (2013) Diffusible hydrogen in steel weldments-a status review. Trans JWRI 42:39–62
8. Soonrach R, Poopat B (2016) Effect of oxygen addition in argon/carbon dioxide gas mixture on metal transfer behavior in gas metal arc welding. Int J Mech Prod Eng 4:2320–2392
9. Francis R, Jones J, Olson D (1990) Effect of shielding gas oxygen activity on weld metal microstructure of GMA welded microalloyed HSLA steel. Weld J 69:408–415
10. Rehal A, Randhawa JS (2014) Submerged arc welding fluxes-a review. Int J Sci Res 3:230–234. https://doi.org/10.21275/02014158
11. Yamamoto H, Danno Y, Ito K et al (2018) Weld toe modification using spherical-tip WC tool FSP in fatigue strength improvement of high-strength low-alloy steel joints. Mater Des 160:1019–1028. https://doi.org/10.1016/j.matdes.2018.10.036
12. Sathiya P, Mishra MK, Shanmugarajan B (2012) Effect of shielding gases on microstructure and mechanical properties of super austenitic stainless steel by hybrid welding. Mater Des 33:203–212. https://doi.org/10.1016/j.matdes.2011.06.065
13. Sugizaki Y, Nakagawa T, Inoue H (2013) Kobelco technology review. Secretariat & Publicity Dept, Kobe Steel Ltd.
14. Mitra U, Eagar TW (1991) Slag-metal reactions during welding: Part II. Theory. Metall Trans B 22:73–81
15. Zhong H, Li X, Wang B et al (2019) Hot tearing of 9Cr3Co3W heat-resistant steel during solidification. Metals (Basel) 9:1–13. https://doi.org/10.3390/met9010025
16. Guo X, Bleck W (2008) Delayed cracking in high strength steels. In: Aachener Stahlkolloquium
17. Takechi H (2008) Transformation hardening of steel sheet for automotive applications. JOM 60:22–26. https://doi.org/10.1007/s11837-008-0160-6
18. Messler RW (1999) Principles of welding. Wiley, Second Col

19. Dunstan DJ, Bushby AJ (2014) Grain size dependence of the strength of metals: the Hall—Petch effect does not scale as the inverse square root of grain size. Int J Plast 53:56–65. https://doi.org/10.1016/j.ijplas.2013.07.004
20. Owczarski WA (1965) Diffusion welding. SAE Trans 73:537–548
21. Crossland B, Williams JD (2013) Explosive welding. Metall Rev 15:79–100. https://doi.org/10.1179/mtlr.1970.15.1.79
22. Kicukov E, Gursel A (2015) Ultrasonic welding of dissimilar materials : a review. Period Eng Nat Sci 3. https://doi.org/10.21533/pen.v3i1.44
23. Karkhin VA (2015) Thermal processes in welding. Springer Nature Singapore Pte Ltd
24. Nesbitt B (2006) Handbook of pumps and pumping. Elsevier Science & Technology Books
25. Walker RA (1988) Duplex and high alloy stainless steels - Corrosion resistance and weldability. Mater Sci Technol (United Kingdom) 4:78–84. https://doi.org/10.1179/mst.1988.4.1.78
26. Poongkundran R, Senthilkumar K (2016) Effect of preheating on microstructure and tensile properties of friction stir welded AA7075 aluminium alloy joints. Brazilian Arch Biol Technol 59:1–15. https://doi.org/10.1590/1678-4324-2016161056
27. Ambriz RR, Jaramillo D (2014) Mechanical behavior of precipitation hardened aluminum alloys welds. In: Light metal alloys applications, pp 39–59
28. Wu Z, David SA, Leonard DN et al (2018) Microstructures and mechanical properties of a welded CoCrFeMnNi high-entropy alloy. Sci Technol Weld Join. https://doi.org/10.1080/13621718.2018.1430114
29. Wang Y, Huang J (2003) Texture analysis in hexagonal materials. Mater Chem Phys 81:11–26. https://doi.org/10.1016/S0254-0584(03)00168-8
30. Xiaoqing J, Shujun C, Jinlong G, Zhenyang L (2019) Effect of microstructure and texture on mechanical properties of resistance spot welded high strength steel 22MnB5 and 5A06 aluminium alloy. Met Artic MDPI 9:1–17. https://doi.org/10.3390/met9060685
31. Soderstrom E, Mendez P (2006) Humping mechanisms present in high speed welding. Sci Technol Weld Join 11:572–580. https://doi.org/10.1179/174329306X120787
32. Saha S, Mukherjee M, Pal TK (2015) Microstructure, texture, and mechanical property analysis of gas metal arc welded AISI 304 austenitic stainless steel. J Mater Eng Perform 24:1125–1139. https://doi.org/10.1007/s11665-014-1374-0
33. Chu Q, Bai R, Jian H et al (2018) Microstructure, texture and mechanical properties of 6061 aluminum laser beam welded joints. Mater Charact. https://doi.org/10.1016/j.matchar.2018.01.030
34. Ji H, Wang J, Li M (2014) Evolution of the bulk microstructure in 1100 aluminum builds fabricated by ultrasonic metal welding. J Mater Process Technol 214:175–182. https://doi.org/10.1016/j.jmatprotec.2013.09.005
35. Avettand-Fenoel M-N, Racineux G, Debeugny L, Taillard R (2016) Microstructural characterization and mechanical performance of an AA2024 aluminium alloy—Pure copper joint obtained by linear friction welding. Mater Des 98:305–318. https://doi.org/10.1016/j.matdes.2016.03.029
36. Fronczek DM, Wojewoda-Budka J, Chulist R et al (2016) Structural properties of Ti/Al clads manufactured by explosive welding and annealing. Mater Des 91:80–89. https://doi.org/10.1016/j.matdes.2015.11.087
37. Edwards PD, Ramulu M (2009) Investigation of microstructure, surface and subsurface characteristics in titanium alloy friction stir welds of varied thicknesses. Sci Technol Weld Join 14:476–483. https://doi.org/10.1179/136217109X425838
38. Kamp N, Sullivan A, Robson JD (2007) Modelling of friction stir welding of 7xxx aluminium alloys. Mater Sci Eng a 466:246–255. https://doi.org/10.1016/j.msea.2007.02.070
39. Wang B, Xue S, Ma C et al (2017) Effects of porosity, heat input and post-weld heat treatment on the microstructure and mechanical properties of TIG welded joints of AA6082-T6 Met Artic MDPI 7 https://doi.org/10.3390/met7110463
40. Lan S, Liu X, Ni J (2015) Microstructural evolution during friction stir welding of dissimilar aluminum alloy to advanced high-strength steel. Int J Adv Manuf. https://doi.org/10.1007/s00170-015-7531-2

41. Jain R, Pal SK, Singh SB (2019) Investigation on effect of pin shapes on temperature, material flow and forces during friction stir welding: a simulation study. Proc Inst Mech Eng Part B J Eng Manuf 233:1980–1992. https://doi.org/10.1177/0954405418805615
42. Mahto RP, Kumar R, Pal SK (2020) Characterizations of weld defects, intermetallic compounds and mechanical properties of friction stir lap welded dissimilar alloys. Mater Charact 160:110115. https://doi.org/10.1016/j.matchar.2019.110115
43. Iqbal MP, Tripathi A, Jain R et al (2020) Numerical modelling of microstructure in friction stir welding of aluminium alloys. Int J Mech Sci 185. https://doi.org/10.1016/j.ijmecsci.2020.105882
44. Miller VM, Johnson AE, Torbet CJ, Pollock TM (2016) Recrystallization and the development of abnormally large grains after small strain deformation in a polycrystalline nickel-based superalloy. Metall Mater Trans a Phys Metall Mater Sci 47:1566–1574. https://doi.org/10.1007/s11661-016-3329-6
45. Nadammal N, Kailas SV, Szpunar J, Suwas S (2015) Restoration mechanisms during the friction stir processing of aluminum alloys. Metall Mater Trans a 46:2823–2828. https://doi.org/10.1007/s11661-015-2902-8
46. Trautt ZT, Mishin Y (2012) Grain boundary migration and grain rotation studied by molecular dynamics. Acta Mater 60:2407–2424. https://doi.org/10.1016/j.actamat.2012.01.008
47. Halfpenny A, Prior DJ, Wheeler J (2006) Analysis of dynamic recrystallization and nucleation in a quartzite mylonite. Tectonophysics 427:3–14. https://doi.org/10.1016/j.tecto.2006.05.016
48. Gutierrez Castaneda EJ, Hernandez Miranda MG, Salinas Rodriguez A et al (2019) An EBSD investigation on the columnar grain growth in non-oriented electrical steels assisted by strain induced boundary migration. Mater Lett 252:42–46. https://doi.org/10.1016/j.matlet.2019.05.073
49. Humphreys MH (2004) Recrystallization and related annealing phenomena, 2nd edn. Elsevier, Oxford
50. Huda Z (2020) Recrystallization and grain growth. In: Metallurgy for physicists and engineers. CRC Press, pp 311–320
51. Knipling KE, Fonda RW (2009) Texture development in the stir zone of near- a titanium friction stir welds. Scr Mater 60:1097–1100. https://doi.org/10.1016/j.scriptamat.2009.02.050
52. Iqbal MP, Jain R, Pal SK (2019) Numerical and experimental study on friction stir welding of aluminum alloy pipe. J Mater Process Technol 274:116258. https://doi.org/10.1016/j.jmatprotec.2019.116258
53. Weertman JR (1993) Hall-Petch strengthening in nanocrystalline metals. Mater Sci Eng a 166:161–167
54. Armstrong RW (1987) The (cleavage) strength of pre-cracked polycrystals. Eng Fract Mech 28:529–538
55. Yang G, Park S (2019) Deformation of single crystals, polycrystalline materials, and thin films: a review. Mater Rev MDPI 12:1–18. https://doi.org/10.3390/ma12122003
56. Fujita H, Tabata T (1973) The effect of grain size and deformation sub-structure on mechanical properties of polycrystalline alumunium. Acta Metall 21:355–365
57. Mahto RP, Kumar R, Pal SK, Panda SK (2018) A comprehensive study on force, temperature, mechanical properties and micro-structural characterizations in friction stir lap welding of dissimilar materials (AA6061-T6 & AISI304). J Manuf Process 31:624–639. https://doi.org/10.1016/j.jmapro.2017.12.017
58. Hasan SM, Ghosh A, Chakrabarti D, Singh SB (2020) Orientation dependence of deformation-induced martensite transformation during uniaxial tensile deformation of carbide-free Bainitic steel. Metall Mater Trans a 51:2053–2063. https://doi.org/10.1007/s11661-020-05694-4
59. Gustafsson M, Thuvander M, Bergqvist E et al (2007) Effect of welding procedure on texture and strength of nickel based weld metal. Sci Technol Weld Join 12:549–555. https://doi.org/10.1179/174329307X213800
60. Lancaster JF (1999) Metallurgy of welding. Abington Publishing, Woodhead Publishing Limited, Abington Hall

61. Shen JH, Li YL, Wei Q (2013) Statistic derivation of Taylor factors for polycrystalline metals with application to pure magnesium. Mater Sci Eng a 582:270–275. https://doi.org/10.1016/j.msea.2013.06.025
62. Ceramic Transactions V 201 (2012) Applications of texture analysis. A Wiley 81 Sons, Inc., Publication

Chapter 7
Tubular Structures: Welding Difficulty and Potential of Friction Stir Welding

Debolina Sen, Surjya Kanta Pal, and Sushanta Kumar Panda

Abstract Tubular structures find usage in various industries such as aerospace and automobiles. Different techniques involving casting, forming and welding are used to fabricate tubes; among all, welding is the most utilized manufacturing route because of its versatility and cost-effectiveness. This chapter provides an exhaustive understanding of various welding techniques utilized for manufacturing of tubes, stating their advantages and disadvantages. In addition, the production of tubular structures with a newer welding technique known as "Friction Stir Welding" has been discussed in detail. A case study has been presented which shows the potential of this process over other welding techniques.

Keywords Tubular structure · Seamless tube · Welded tube · Fusion welding · Solid-state welding · Friction stir welding

7.1 Introduction

Production of tubular components was introduced in the early 1800s, and since then these components have found their applications in several industries, such as in automobile, aerospace, nuclear, oil and gas and food processing, either as a structural component or for conveying fluids [1, 2]. In aerospace and aviation industries, tubular structure forms the core part of the aircraft which can operate flawlessly under high pressure and temperature. The rocket fuel tanks, webbed fuselage structures, steering columns, landing gears for an aircraft, etc., are some of the tubular components which are being used by these industries [3]. Automobile industries also find

D. Sen · S. K. Pal (✉) · S. K. Panda
Department of Mechanical Engineering, Indian Institute of Technology Kharagpur, Kharagpur, West Bengal 721302, India
e-mail: skpal@mech.iitkgp.ac.in

D. Sen
e-mail: debolinasen90@gmail.com

S. K. Panda
e-mail: sushanta.panda@mech.iitkgp.ac.in

J. P. Davim (ed.), *Welding Technology*, Materials Forming, Machining and Tribology, https://doi.org/10.1007/978-3-030-63986-0_7

a wide application of tubular components, such as the crash box, shock absorbers, suspension arms, seats and instrument panel beams. According to a report published by Arcelormittal, 40 kg of the total mass of a car comprises of the tubular components [4, 5]. Moreover, in automobile industries, there is a huge demand for light weighting of automotive body structures so as to meet the environmental norms, and simultaneously fulfilling the economic and safety concerns for the end-users. Tubular components of various materials such as aluminium, magnesium, extra deep drawing steel and advanced high strength steels are extensively used to meet these conflicting demands. Moreover, in nuclear reprocessing plants, various important equipment and process vessels such as vent pots, cylindrical tanks, annular tanks and high-level waste storage tanks make use of tubular structures with high service life [6]. Similarly, in oil and gas industries, tubes with high strengths are used to endure the high pressure during transportation of natural gas or crude oil over long distance [7].

These tubular components are produced by utilizing various manufacturing processes such as centrifugal casting, mandrel mill process, tube drawing process, extrusion and welding. The tube manufacturing industries face a lot of challenges due to the versatility in application of tubular components in various industries. The demand for a wide variety of shapes and sizes of the tubular products are increasing day by day. The manufacturers face difficulty when the required shape is not the regular one, and the size of the tube is either very small or very large. Additionally, the application of tubular components of advanced high strength steels and titanium alloys requires special methods and manufacturing techniques to fabricate them [8]. Finally, enhanced product quality in terms of durability, surface finish and uniform thickness of tube with a reasonable cost is another aspect worth considering during manufacturing of the tubular components. Therefore, to meet these challenges, manufacturing industries are trying to utilize advanced techniques to produce quality tubular components.

Over the past decades, a huge development has been made to master the manufacturing process of tubular components in terms of product quality, process quality, automation, etc. Several techniques of tube manufacturing have evolved among which two distinct processes, namely extrusion and welding are largely being utilized by the manufacturers to produce tubular structures, as shown in Fig. 7.1 [9]. Extrusion process results in seamless tube, whereas the welding process yields in seamed tubes. Extrusion process is a bulk metal forming process where a round cylindrical billet is forced to move through a die and a mandrel. The die opening is of a smaller cross-sectional area which is kept in accordance to the desired diameter of the tube to be produced. A small clearance between the die and the mandrel is present which is equal to the desired thickness of the tube to be formed. The interior profile of the tube is shaped by the mandrel, while the exterior profile is shaped by the extruding die. In extrusion process, the seamless tubes formed are homogeneous in nature, and hence have good mechanical properties in terms of strength and corrosion resistance. Also, extruded tubes have better pressure resistance capacity, i.e. they can withstand high pressure, and hence are preferred in oil and gas industries. However, seamless tubes are expensive to produce and there is limitation in production of larger

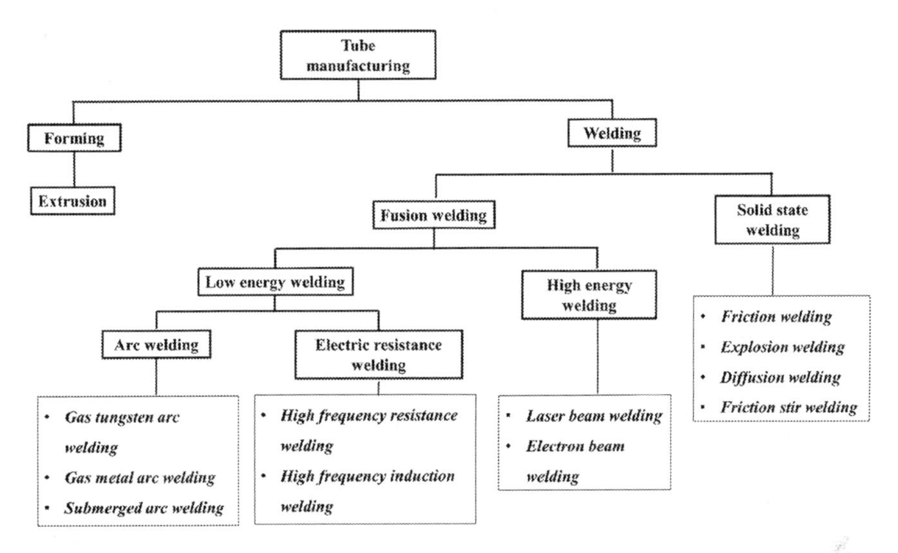

Fig. 7.1 Various processes involved in manufacturing of tubes

diameter and higher thickness tubes. It is difficult to maintain the roundness of tubes having lager diameter. In addition, the force required to extrude such larger diameter tube is very high. Moreover, seamless tubes have an inconsistent wall thickness across their length due to various conditions such as uneven temperature distribution and improper centering of the mandrel which restricts its usage in areas where dimensional accuracy is of importance.

In tubular welding process, metal sheets or plates are rolled or bent in the shape of a cylindrical open tube of the desired diameter. The edges of these open tubes are then aligned and longitudinally welded with or without a filler material to form a closed tubular structure. Seamed tubes do not have a good finish as there is a thin line of welding on it, and the welded region is considered as the weakest part which is vulnerable to failure and corrosion [7, 10]. On the other hand, these tubes are generally cost effective than seamless tubes [11]. In welding process, tubes having consistent wall thickness throughout can be produced for long continuous length. Welded tubes can be manufactured for large diameter to thickness ratio without any restriction. It has been reported that the diameter and the wall thickness of welded tubular structure generally range from 6 mm to 2500 mm, and from 0.5 mm to 40 mm, respectively [2]. Presently, out of the total steel tube fabrication in the world, about two-third of it is being manufactured by welding processes [12]. Various welding techniques are used in industries for manufacturing seamed tubes, such as arc welding, electrical resistance welding, high energy beam welding and solid-state welding. Each of these welding techniques has got its advantage as well as disadvantage on the tube production. Hence, selection of the welding process is very crucial in manufacturing of tubes and it depends on various aspects, such as the tube material, thickness of the tube and application area.

In arc welding processes, the tube welding is carried out mostly by using a filler material. Filler material, even if is of the same material as the tube, forms an in homogenity in the microstructural characteristics. Also, in arc welding, always a weld reinforcement is present, both internally and externally, in the surface of the tube which deteriorates its surface quality. Welding of thicker tubular structures needs multiple passes in arc welding. Due to these multi-passes, the residual stresses in the joint increase, which in turn reduce the life of the component. To avoid these problems, manufacturers have started joining tubes with resistance welding technique. In this method, no filler material is required resulting in microstructural homogeneity and good surface finish. However, in this technique, the edges of the tube are melted for joining, thus in some materials, the alloying elements tend to combine with atmospheric oxygen and form high melting point oxides. These oxides persist in the seam and generates welding defects such as inclusions, incomplete joining, lack of fusion and hook crack, and reduces the long sustainability of the tubes [10]. In arc welding, as well as resistance welding, a high heat input is required since the melting of the tube material takes place for joining. Due to this, a wider heat-affected zone (HAZ) is created in the welded tubes. This changes the microstructure of this zone and results in softening which reduces its strength and leads to failure. In some industries, tubular structures are used in critical places, like in nuclear reactors, where no compromise with the desired properties is allowed to avoid a miserable disaster. Therefore, very precision welding of tubes is required with minimum defects and lower HAZ to minimize the risk. In such cases, high energy density welding technique is used where the energy required for welding is focused into a very small area which reduces the overall heat input for welding. This results in minimal degradation of the base material with narrow HAZ and low residual stress [13]. There are a variety of materials used for tubing which include stainless steels, ferritic steels and nickel base alloys to name a few. These materials are generally welded by using conventional welding techniques. However, there are many applications where welding cannot be used owing to environmental concerns or where non-weldable materials must be used due to the necessity of specific material properties. Also, as mentioned earlier, lightweight structures are preferred in automotive and aerospace industries; thus, materials like aluminium alloys having high strength-to-weight ratio are generally preferred, but cannot be welded by the above-mentioned welding techniques. Aluminium metal has high thermal conductivity and high reflectivity, causing it difficult to be welded. Also, they have high hydrogen solubility at higher temperature which leads to defects, such as porosity and blow holes during fusion welding [14]. Thus, to overcome these difficulties, and to manufacture tubes of aluminium alloys, solid-state welding technique is used. In this welding technique, no melting of the tubular material takes place, only softening occurs due to temperature and pressure, which on cooling creates a solid phase bond. Since solid-state welding occurs below the melting point of the tube material, a very low heat input is required in welding. This reduces the solidification defects and distortion of the weld and improves the dimensional accuracy of the tubes. In solid-state welding, the weld zone comprises of fine recrystallized grains which improves the mechanical as well as microstructural properties of the tubes. The heat input in solid-state welding is lesser in comparison to the fusion welding techniques, thus a narrow or sometimes negligible HAZ is formed which further enhances its durability [15]. Solid-state welding has got various advantages in tube welding, such as better surface finish of the welded tube, no residual stress generation and enhancement of corrosion resistance. These advantages make solid-state joining process effective to fabricate tubes with higher precision and accuracy. Thus, with the availability of these many variety of welding techniques, selection of suitable one to produce seamed tubes depends on the application, material and the quantity of the tube to be produced. A detailed explanation of all these welding techniques, for tubular structure fabrication, is explained below.

7.2 Welding Techniques Used to Manufacture Tubes

7.2.1 Arc Welding

Various arc welding processes, such as gas metal arc welding (GMAW), gas tungsten arc welding (GTAW) and submerged arc welding (SAW) are used in manufacturing tubular components [6, 16, 17]. Among these processes, the most commonly used by the tube manufacturing industry is the SAW process. SAWed tubes are generally the most economical solution where application requires high wall thickness to sustain high internal or external pressure. In SAW, the welding is done with the help of an arc which is submerged inside the blanket of granular fusible flux. It reduces the heat loss during welding without any atmospheric contamination of the weld pool. In this welding, the edges of the rolled or bent plates are tack welded at the ends to hold the open tube in proper alignment. A continuous solid filler wire is fed from outside, and welding is done both on internal and external surfaces of the tubular material, as shown in Fig. 7.2. This creates a weld bead in both the inner as well as the outer surfaces of the tube and deteriorates its surface finish. The weld bead geometry is a significant parameter in deciding the quality of the tubular structure.

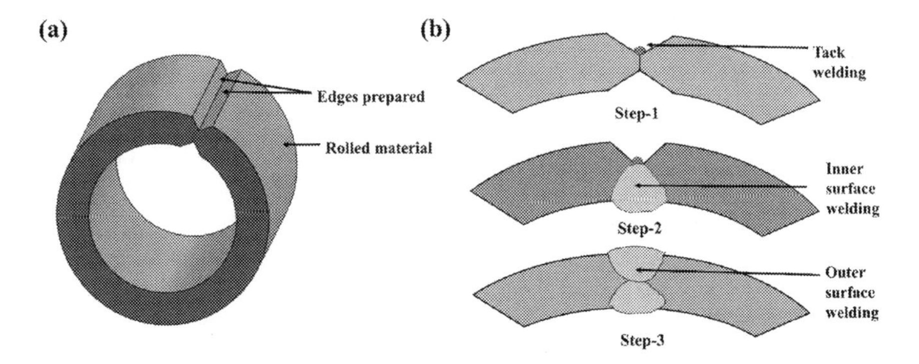

Fig. 7.2 **a** Joint preparation for SAW of tubes and **b** various steps involved during SAW of tubes

Process parameters such as arc voltage, welding speed, line power, wire feed rate and nozzle to plate distance influence the weld bead parameters, particularly the bead penetration, width of the bead, reinforcement and dilution. Hence, it is important to find out an optimum relationship among the above-mentioned bead parameters and process parameters to control the bead shape to produce a quality welded tubes [18, 19].

In a SAW process, higher heat input is induced in the welded zone because the heat losses are less as the welding is being done inside the flux. This allows welding of thicker tubes without any contamination in the welded zone. High heat generation affects the microstructure and results in coarser grain structures in the welded zone, which in turn modifies the properties of the tubular structure. Properties, such as the strength and toughness determining the suitability of tubular structures in various applications, get deteriorated in the nugget zone (NZ) and the HAZ due to these microstructural changes. Hence, these zones form the weakest link in the tube where failure can occur easily [17, 20].

Further, in SAW, one of the unavoidable limitations is the generation of residual stresses in the weldments because of the complex thermal cycle produced. Heat transfer feature of the fusion zone and the HAZ which leads in contraction of the weld bead during solidification process induces residual stress and strain near the weld bead [21]. This residual stress affects the mechanical performance of the welded tubes by increasing its susceptibility to fracture and fatigue. Moreover, the presence of inclusions in the welded zone leads to crack growth. This crack gets accelerated with the existence of residual stresses in the welded zone along the weld line which leads to an early failure of the tube [22]. Similarly, the presence of residual stresses in SAWed joint accelerates the occurrence of corrosion cracks and notably increases the stress corrosion cracking susceptibility of the tube [23]. Thus, even though SAWed tubes are easily manufactured with very less cost, these cannot be used in places where failure of tubes can lead to a catastrophic disaster. Therefore, to fabricate tubes with minimum residual stress and enhanced properties, another welding technique, known as the electrical resistance welding (ERW), is being used by the manufacturers widely.

7.2.2 Electric Resistance Welding (ERW)

In ERW of tubular structures, the edges of the rolled sheet or plate are held together and high current is passed through them. Due to skin and proximity effect, the current gets localized at the edges and heats them up to the melting point by Joule heating. The current feeding in ERW can be done either by contact method or by the induction method. Contact method is preferred to weld large diameter tubes where copper based shoes are applied close to the strip edges. It allows the current to flow into the edges and melt. Further, to weld small diameter tubes, induction method is preferred where high-frequency induction coils are used to concentrate the current in the edges and heat it for welding. In order to complete the welding of the tubes,

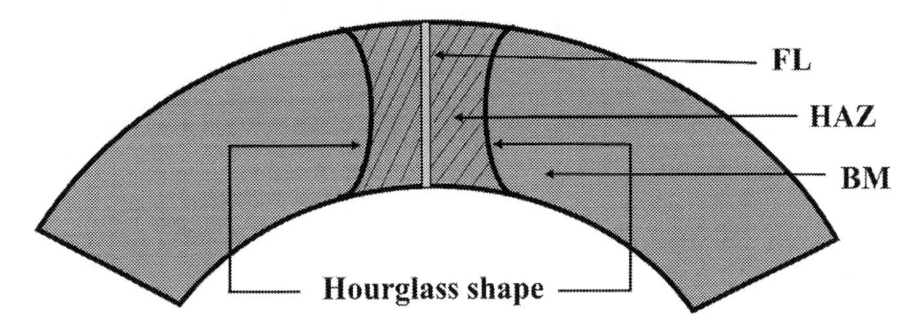

Fig. 7.3 Schematic diagram of various zones in ERWed tubular structure

pressure is applied, and this extrudes the molten metal having oxide particles from the edges. This extruded metal deposited on to the inside and outside surfaces, and the welded tube can be easily removed by scarfing process. The weld seam produced is small, flat and smooth with no pits. The welding of the tubes by ERW process is autogenous in nature as no filler material is applied. Various zones are formed in the ERWed tubular samples, as shown in Fig. 7.3, namely the fusion line (FL), which is formed in the joining area; next to the FL is the HAZ which forms like an hourglass shape due to the heat produced by the high-frequency current, and lastly the base material (BM).

In ERW, current, voltage, welding power, welding frequency, weld speed, forge pressure, etc., are the typical process parameters which are the interdependent factors significantly affecting the weld quality. These factors control the heating and cooling of the welded zone which influence the grain structures and mechanical properties of the tube. It has been found that in order to ensure no defect, the heat generation during welding and upset amount must be kept within the limit. Optimization of aforementioned parameters is the key to achieve proper heat generation and upset amount in order to achieve quality welded tubes [24]. Also, the heat is generated in a small volume of metal along the edges of the tubes to be joined, which results in a narrow HAZ. Moreover, the heat generated results in metallurgical changes in the welded zone, and these changes deteriorate properties such as toughness, hardness and residual stresses. These properties can be recovered by performing post-weld heat treatments to the welded tube [25]. Further, in ERW, defects are formed in welded joint due to inherent properties of the base material or improper selection of process parameters. The most common defects formed are hot cracks, hook cracks, cold weld and penetrators. The hot cracks and hook cracks are material defects, produced by the occurrence of inclusions in the tube material. The cold weld and penetrators are process defects which are formed due to the variation in heat input. Cold weld defect is a thin film of oxide inclusions formed because of improper squeezing of the surface oxides during upsetting due to insufficient heat input. Penetrator defect is a pancake-type oxide inclusion formed due to high heat input which is generally discharged in the gap because of short circuiting at the apex point. Various researchers studied the consequence of these defects on the mechanical properties of the welded tube

[26, 27]. It was observed that these defects were unavoidable because of the alloying content present in the tube material which significantly decreased the properties of the welded tube, such as ductility and corrosion resistance. In this context, it was also reported that the weld zone exhibits severe localized corrosion due to microstructural changes. These microstructural changes were because of faster cooling in the weld zone. Thus, change in microstructure and the variance of alloying elements between the welded metal and the parent metal act as a potential corrosion sites in the tubular structure. To mitigate these potential corrosion sites, heat treatment of the tubes is suggested by the researchers [28].

7.2.3 High Energy Beam Welding

High energy beam welding of tubular components are used by industries where very precise welded tubes are required with high performance. This welding technique have small beam size which helps in getting deep penetration, thus making it possible for welding thicker tubular components with less heat input. Welding processes, such as laser welding and electron beam welding (EBW), fall in the category. Among all, laser beam welding (LBW) is widely used for manufacturing of tubes.

LBW is an advanced technique which provides high precision and flexibility in welding of tubular components. In LBW for tubular components, the welded joint is formed by melting the edges of the rolled sheet or plate by a laser beam. Energy required for welding tube is focused into a very small area allowing it to weld with a high energy density and a very less heat input, resulting in a narrow fusion zone with small HAZ. Due to less heat input, the cooling rate becomes high resulting in low residual stress, fine grain structure, low thermal distortion maintaining good dimensional accuracy and good surface quality. LBW results in high productivity and produces deep penetration and narrow weld bead with fine grain structure [29]. In laser welding, owing to high density and reduced heat input, rapid solidification of the molten metal occurs at the welded zone. This allows for microstructural grain refinement and chemical homogeneity of the joint, which enhances the mechanical properties of the overall LBWed tube [8]. Researchers compared the performance of LBWed and conventionally welded (GTAWed) austenitic steel tubes and observed that LBWed tube offers better metallurgical properties and exhibits near-seamless quality [30].

LBW, for fabrication of tubular structures, mainly depends on various factors which can be classified into three categories, namely material properties, process parameters and system parameters, as shown in Fig. 7.4. Even though LBW has got many advantages, the efficiency and the quality of LBWed tubular structure depend on the material of the tube to be welded and its thickness. For a thick tube, multi-pass welding is used with an addition of filler material. However, with an addition of the filler metal, many of the advantages associated with LBW either get reduced largely, or substantially get offset by other complications, such as residual stress generation, formation of defects and intermetallics [31]. Moreover, properties of the material of

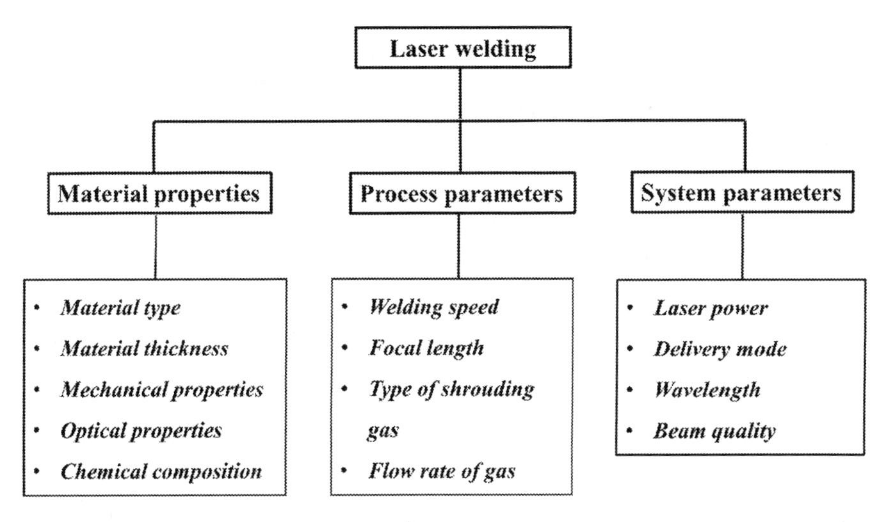

Fig. 7.4 Factors influencing LBW in tubular structures

the tube such as thermal conductivity, absorption, diffusivity and reflectivity of the laser beam affect the efficiency of the welding. Materials, such as aluminium alloys and copper alloys, are considered to have lower absorption of laser beam, and hence tubes made up of these materials are not welded by laser technique [32]. Hence, for fabrication of tubes of such materials, welding technique known as the solid-state welding can be used as a primary choice.

7.2.4 Solid-State Welding

Solid-state welding process is employed to weld tubes at a temperature essentially lower than the melting point of the parent materials. Welding occurs without usage of any filler material and with or without the application of pressure. It eliminates all the complexities as well as defects which are associated with the solubility of gases and solidification of the molten material in conventional welding processes such as porosity, blow holes and inclusions [33]. Further, manufacturing of tubular components of difficult-to-weld materials like aluminum alloys can be easily done by solid-state welding techniques. This approach might help in encouraging various industries with the increase usage of the tubular structures made of aluminum alloys. Solid-state welding of tubular components is economic, as well as an easy process as compared to other known fusion welding processes. Various solid-state welding processes that are used in welding of tubes are friction welding, friction stir welding (FSW), diffusion welding, explosion welding, etc. [34–38]. Among all these processes, FSW has emerged recently and has drastically changed the scenario of the tube manufacturing industries by increasing their productivity, as well as the product variety. Recently, it is being widely used because of yielding near-seamless finish, excellent properties, and most importantly of an energy-efficient process in tube fabrication [39]. A brief explanation of this process is mentioned in the section given below. The detailed description of FSW, in general, has already been mentioned in the previous chapter.

7.3 Potential of FSW in Fabrication of Tubes

FSW was invented by TWI with the advent of joining materials having low melting temperature, such as copper, brass and aluminium in flat plates or sheets as described in previous chapters. This process offers good mechanical properties, flexibility and ease for automation; hence, it can be applied in manufacturing high-quality welded tubular structures. Not only metals, these days even polymeric pipes are being produced by FSW process [40]. Unlike conventional welding, FSW can be used to weld thicker tubes in a single pass. Similarly, thin-walled tubes can also be welded easily by FSW process. The thickness of the tube which can be welded in a single pass ranges from 1 to 12 mm [41, 42].

The FSW can be done circumferentially, as well as longitudinally, in tubular structures. The circumferential FSW tubes are used to join two seamless cylindrical tubes, whereas longitudinal FSW process is used to fabricate the seamed tube. The circumferential FSW process has not been discussed in details, as this book chapter is primarily focused on the manufacturing of seamed tubes. In longitudinal FSW process, sheets or plates are rolled to form a cylindrical open tube like structure of the desired diameter, and edges are then welded as shown in Fig. 7.5. In case of FSW of tubes, the foremost important thing is the joint preparation so as to achieve a sound weld joint. However, one of the challenges in tubular structure is its curvature, due to which a gap is always present between the edges at the root. This gap cannot be completely exempted, but it is minimized by polishing or filing the edges. After the joint preparation, open ends of the bent sheet are properly clamped in order to restrict

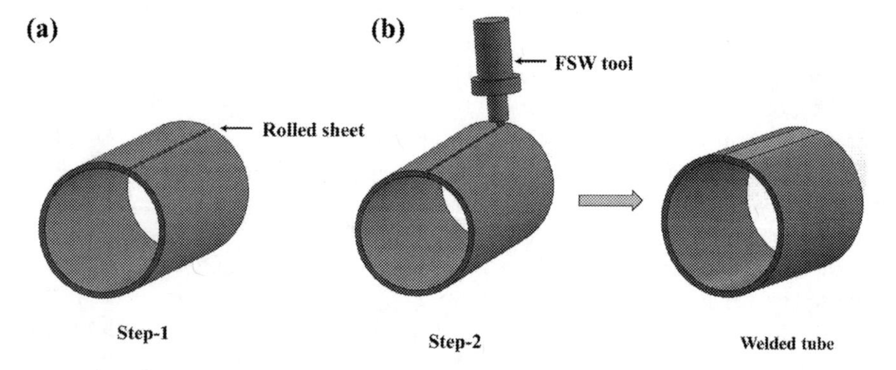

Fig. 7.5 Steps involved in FSW of tubular structure: **a** rolling of sheet and **b** welding using FSW tool

all the movements during welding. A non-consumable tool is used for welding which comprises of a shoulder and a pin, designed according to the thickness and the type of the tube material to be joined. The tool is rotated with a certain speed and plunged into the abutting edges of the tube and is subsequently traversed along the longitudinal joint line. High amount of heat generation takes place because of frictional heating among the tool shoulder and the tube material. The material undergoes severe plastic deformation which also contributes to the total heat generation. This localized heating softens the material and because of the rotation of the tool the material gets stirred and mixed by the pin, which leads to the joint formation. FSW typically exhibits four main types of microstructural region, namely NZ, the thermo-mechanically affected zone (TMAZ), the HAZ and the BM. One of the significant benefits of the FSW process is that fully recrystallized, equiaxed and fine grain microstructure is formed in the NZ [43]. The properties of the welded tubular structure possess excellent mechanical properties, good corrosion resistance, better fatigue properties, enhanced formability, etc., than its BM [44].

There are various factors involved in obtaining a good joint by FSW process, but in case of tubular structures those factors are even more challenging. Some of those key factors are mentioned below.

7.3.1 Fixture

In FSW, a fixture is required for holding and clamping the workpiece. It plays an important role in getting a sound weld, as it bears the forces during the plunging phase and maintains the position of the workpiece. It also absorbs all the vibrations generated during welding. In FSW of tubular structures, the design of the fixture is a bit complex than that required for flat sheets. Due to the curvature of the tubular structure, the clamping system required should be such that it can hold the rolled material firmly with the outer surface of the tube. Any gap between the tube and the clamp can lead to distortion or buckling of the tube because of higher forces generated during welding. Various researchers developed different types of fixtures in order to weld tubular structures of different diameter and thickness. Researchers developed a fixture where a die of semicircular shape was placed at the bottom as a basic support for the bent sheets to be positioned. Two clamps were utilized to grip the bent sheets from sideways to restrict the movement during welding [45]. Similarly, another fixture was developed where the sheets can be rolled and then longitudinally welded by FSW so as to obtain a tubular structure. The fixture consisted of three main systems: the clamping system which could form the sheet into an open tube; the fixing system which restricted the movement of the open tube; and the supporting system which provided proper support from inside of the hollow tube [46]. Thus, designing a fixture in FSW of tube is a complex yet an essential process for obtaining a flaw-free tube. Another complexity arises in FSW while welding tubes with smaller diameter, because of less contact area available to hold it and resist the forces applied by the tool. Special holding devices are required during welding of tubular structures which needs to be designed according to the diameter of the tubular structure [41, 47].

In FSW, large vertical force is exerted by the tool during the welding process. This can damage the surface of the tubular structure, and hence compromise its circularity. Thus, during FSW of tubular components, a mandrel is provided inside the tube which provides a counter balancing force to the force employed by the tool on the tube. In other words, the mandrel bears the vertical force applied by the tool and prevents distortion of the tube. Thus, the mandrel should be designed in such a way that it fits properly with the inner diameter of the tubular structure to guarantee a good welded joint. Also, the mandrel should easily mantle and dismantle without damaging the tube before and after the welding. In this context, many researchers tried to design and fabricate an expandable mandrel which could be easily expanded by fastening screws to be used for various tubes with varying inner diameter. Again, on unscrewing it, the mandrel can easily come out from the tube without any damage to the interior surface of the tube. Moreover, this mandrel could be adjusted to apply the desired degree of support to the tube [41, 48, 49].

7.3.2 Tool Design

The tool design is a significant parameter in attaining a flawless joint in welding tubular components by FSW. In FSW, the heat required to plasticize the material is produced because of friction among the tool and the material. The shoulder of the tool rubs against the curved surface of the tube and the generated heat softens the material enabling the material to flow. Unlike in flat butted plates, in tubular structures, the tool shoulder makes contact with the curved outer surface of the tube. Because of the curvature of the tube, the tool shoulder does not have full contact with the tube surface, as shown in Fig. 7.6. As a result of this, the heat generated is less in case of tube welding. This, in turn, results in less softening of the material and improper material flow leading to a defective joint. It has been reported that welding tubes using flat shoulder tool decreased the tensile strength of the joint by 65% with respect to the tube material due to ineffective material flow [37]. Thus, researchers suggested that joining tube with a double pass can lead to a defect-free weld where the first pass can improve the shoulder contact with the tube by flattening the curved surface during welding, and the second pass can improve the surface quality of the welded tube [50].

Tool shoulder diameter is an important parameter in FSW of tubular components. Tools having smaller shoulder diameter will result in a defective welded tubular joint since it will not generate enough frictional heat required for the material softening. It may result in defects such as tunnels or wormholes in the welded tube. Conversely, tools with large shoulder diameter cannot be used because it will not increase the heat generation further. This is because the contact of the shoulder of the tool will not increase due to the curvature of the tube. Thus, selection of an optimum shoulder

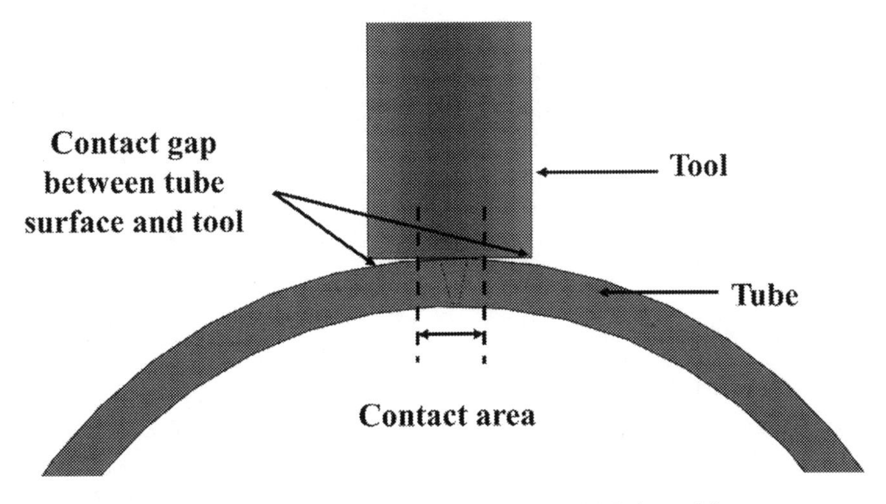

Fig. 7.6 Contact area between the FSW tool and the tube material during welding process

diameter of the tool is necessary in FSW of tubular components to produce a defect-free joint which is depicted graphically in Fig. 7.7 [50].

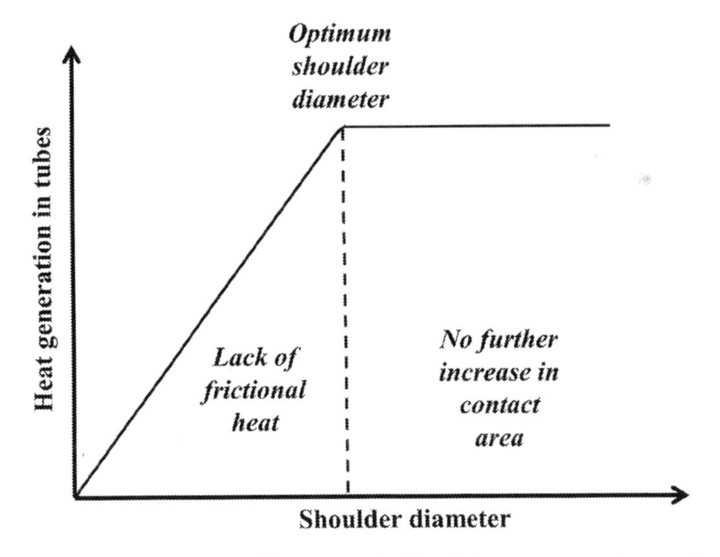

Fig. 7.7 Graph showing the effect of heat generation in tubular structures on increase in shoulder diameter

7.3.3 *Process Parameters*

Process parameters play a very important part in obtaining a defect-free FSWed tube. In FSW of tubular components, the tool does not sit flushed with the tube surface due to its curvature, thus inadequate heat generation takes place. This short-fall in heat requirement is compensated suitably by changing the welding process parameters for a defect-free weld. Parameters such as the rotational speed, transverse speed, plunge depth and tilt angle are some of the important parameters by which this heat generation can be controlled.

Researchers studied the contact characteristics and the heat generation between the tool and the tube by changing the plunge depth during FSW. According to them at a low plunge depth, adequate heat generation is not produced because of the smaller interface between the tube surface and the tool shoulder leading to tunnel formation. By increasing the plunge depth, contact surface between the tube outer surface and the tool shoulder increases, and as a result more heat generation takes place to soften the material. Alternatively, excessive increase in plunge depth consequences in thinning of the tube thickness along the weld line leading to degradation in the mechanical properties of the welded joint [51].

7.4 Limitations of FSW of Tubes Over Other Welding Techniques

Even though FSW has got many benefits in manufacturing of tubular components, still there are some limitations associated with this process. During manufacturing of tubular structures by FSW process, as the shoulder of the tool comes in contact with the tube surface for joining, it forms a flat surface in the welded zone. Thus, the tube formed is not perfectly curved in shape, and hence dimensionally it is not perfect. Hence, it requires some post weld forming operations to get a perfectly circular tube. This flatness is more prominent in small diameter tubes, but as the diameter increases this flatness does not matter much. Another non-avoidable limitation of FSW process in production of tubular components is keyhole, which gets generated at the end of the tube resulting from the extraction of the tool pin after welding. This leads to removal of some portion of the tube material at the end causing material loss.

7.5 Case Study: Longitudinal FSW of Tubes Fabricated from AA5083

Aluminium alloys are extremely corrosion resistive and have high strength-to-weight ratio which makes it suitable for application as tubular structures in automotive bodies. Among various aluminium alloys, non-heat treatable alloy AA5083 is extensively used by the automobile industries in making of inner body panels, tanks, gas cylinders, etc., due to their high formability and high corrosion resistance property [52]. Hence, it is intended to manufacture AA5083 tubes by longitudinal FSW process and show the suitability of the process by studying its microstructural as well as mechanical properties.

7.5.1 Experimental Procedure

AA5083-O sheet of 4.6% (weight percentage) of magnesium content has been selected for experimentation. The uniaxial tensile properties were evaluated by carrying out tensile test according to ASTM E8M standard along the rolling direction of the sheet with a repeatability of three times. Properties such as yield strength (YS), ultimate tensile strength (UTS), percentage elongation (% elongation) and strain hardening exponent (n-value) were evaluated for AA5083 material, and the obtained average values of these properties from repeated tensile tests are presented in Table 7.1.

FSW was carried out on the rolled sheets of 2.5 mm thickness and 80 mm outer diameter. The welding was accomplished on a 6-axis FSW machine (BISS-ITW, SP-02-05-10). The tool utilized in this welding was made of H13 steel. The tool design parameters and the process parameters used for these study are given in Table 7.2. Improper choice of parameters may lead to a defective joint, as shown in Fig. 7.8. After welding, a non-destructive method was used to visualize the presence of any defect in the welded tube by using a 3D X-ray microcomputer tomography (CT) (General Electric, Phoenix v/tome/xs) scanning system. Uniaxial tensile tests were carried out to determine the strength and ductility of the welded tube. Tensile

Table 7.1 Mechanical properties of AA5083-O sheets of 2.5 mm thickness

Orientation	0.2% *YS* (MPa)	*UTS* (MPa)	% Elongation	n-value	K-value
Rolling direction (RD)	159.2	330	31.7	0.3	0.76

Table 7.2 Friction stir welding parameters for AA5083-O

Tool design parameters		Process parameters	
Tool pin geometry	Conical	Rotational speed (rpm)	1000
Shoulder diameter (mm)	14	Traverse speed (mm/min)	30
Pin diameter (mm)	5	Tilt angle (°)	2°
Pin length (mm)	2.2	Plunge depth (mm)	0.1

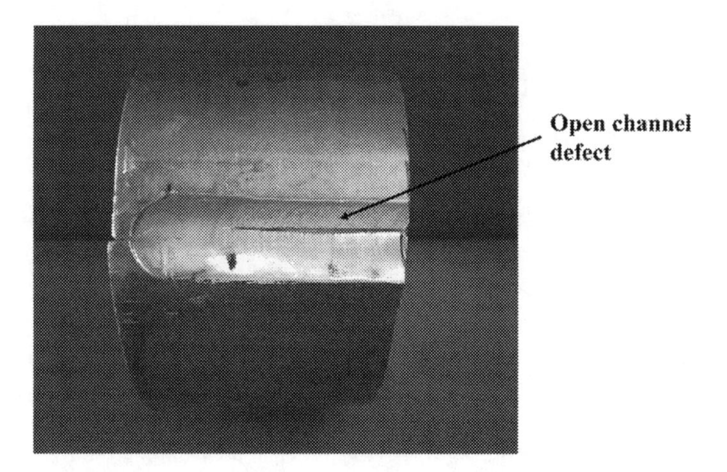

Fig. 7.8 Defective welded tube due to improper parameter selection

test specimens having 50 mm gauge length was cut by using CNC wire electro-discharge machine (Elecktra, Maxicut 523) in the direction of the welding and also in the perpendicular direction. The tests were performed with a crosshead speed of 2 mm/min in a Universal Testing Machine (UTM) (BISS, Elektra 50 kN) of 50 kN capacity at room temperature. For metallographic studies and hardness measurement, samples were cut in the transverse direction to the welding. Polishing of the samples was done (Struers, Laboforce 50) as per standard metallographic procedure and polished samples were etched using Poulton's reagent. Macroscopic and microscopic images were observed in stereo zoom microscope (Leica, S8APO) and optical microscope (Leica, DMi8A), respectively. Hardness test was carried out on a Vickers microhardness tester (UHL, VMHT 001) with an indentation load of 50 gf and a dwell time of 15 s. Further, fracture analysis of the tensile failure specimens was carried out under scanning electron microscopy (SEM) (ZEISS, EVO 50) in order to get an insight into the failure modes in the welded joint.

7.5.2 Results and Discussion

7.5.2.1 Macrographic Analysis

The macroscopic image of the welded AA5083-O tube is shown in Fig. 7.9a with different zones marked on it. The NZ consists of the FSWed tool pin-affected area wherein the joining of the material takes place because of plastic deformation and recrystallization. Some ring-like structures are also visible in the NZ which are known as onion rings. Adjacent to the NZ is the TMAZ, which includes the region under shoulder undergoing plastic deformation without any recrystallization. Next to the TMAZ is the HAZ, which are formed because of thermal cycle of the welding. To

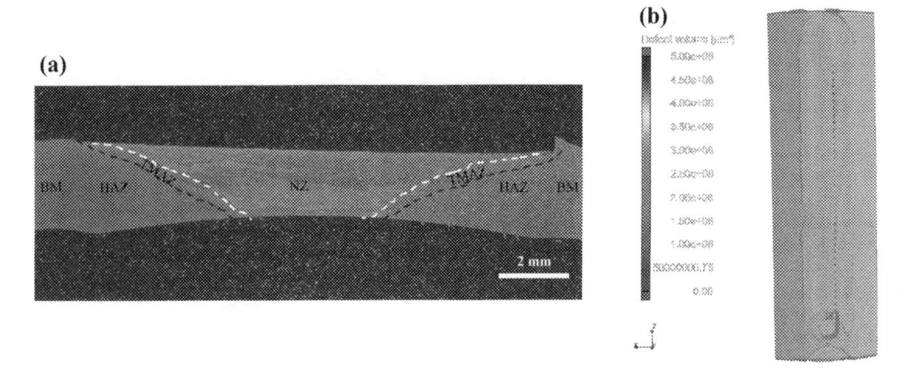

Fig. 7.9 **a** Macroscopic view of FSWed AA5083-O sheets depicting various weld zones and **b** CT scan image showing defect-free weld

investigate the soundness of the joint, micro-CT scan was done on the welded tube, as presented in Fig. 7.9b. It can be observed that there is no occurrence of any void or flaw in the weld zone which confirms it to be a defect-free welded tube. Hence, it can be concluded that the chosen parameters for welding produces a defect-free welded tube.

7.5.2.2 Micrographic Analysis

The microstructures of the as-received AA5083-O alloy and its rolled sheet along RD are presented in Fig. 10a, b, respectively. The structure of the grains of the rolled metal (RM) was found to be more elongated in the RD than the as-received base material (BM). This elongation was caused due to the rolling operation in due course of forming the sheet. There were no other microstructural changes observed in rolled AA5083-O sheets due to rolling. The average grain size was found to be

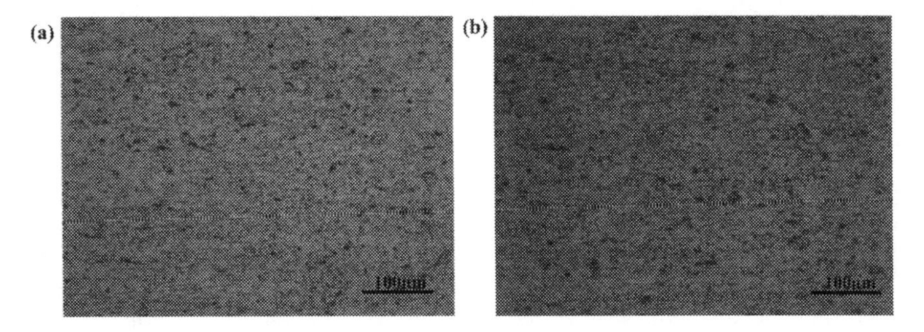

Fig. 7.10 Optical microscopic images of **a** as received BM and **b** RM

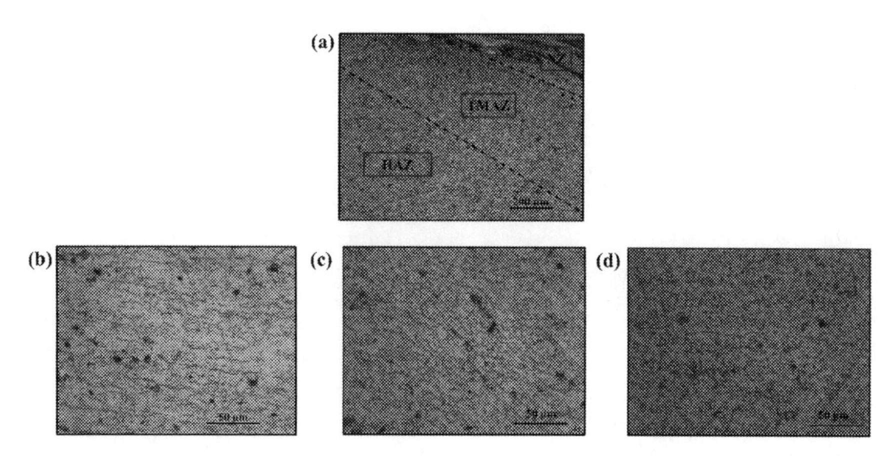

Fig. 7.11 Optical microscopic images of FSWed AA5083-O **a** showing all the zones, **b** magnified HAZ, **c** magnified TMAZ and **d** magnified NZ

approximately 16 μm for the rolled sheet and 11 μm for the as-received material, respectively.

Microstructures of the weld metal (WM) is presented in Fig. 7.11a showing all the three regions formed during welding. The magnified optical image of the NZ is shown in Fig. 7.11d which comprises of fine equiaxed grains, formed because of dynamic recrystallization produced by plastic deformation and large heat generation during welding. The average grain size of the NZ was found to be nearly equal to 3 μm. These equiaxed fine grain structures lead to improvement in the mechanical properties of the joint [53]. The grain size of the TMAZ was little larger than the NZ which was close to 6 μm, and that the HAZ region was much coarser which was approximately 9.5 μm, as shown in Fig. 11b, c, respectively. The coarser nature of grains formed in HAZ was due to the slower rate of cooling taking place than the NZ. It was found that the width of the HAZ was approximately 1.8 mm which is pretty less than that in other fusion welding techniques [54]. These changes in microstructure influence the mechanical properties of the welded tube and are discussed in the below section.

7.5.2.3 Mechanical Properties

Mechanical properties such as microhardness and uniaxial tensile strength of the welded tube have been evaluated. The microhardness across the as-received material and along the cross section of the welded region was measured and is presented in Fig. 7.12. The average hardness of the as-received material was found to be 77 HV. The hardness graph of the welded sample followed the typical "*W*" pattern although not much difference was observed among the zones. On the contrary, with other fusion welding techniques, a huge drop in the hardness value in the HAZ has been observed which acts as potential site for failure in the pipe [29]. The average hardness

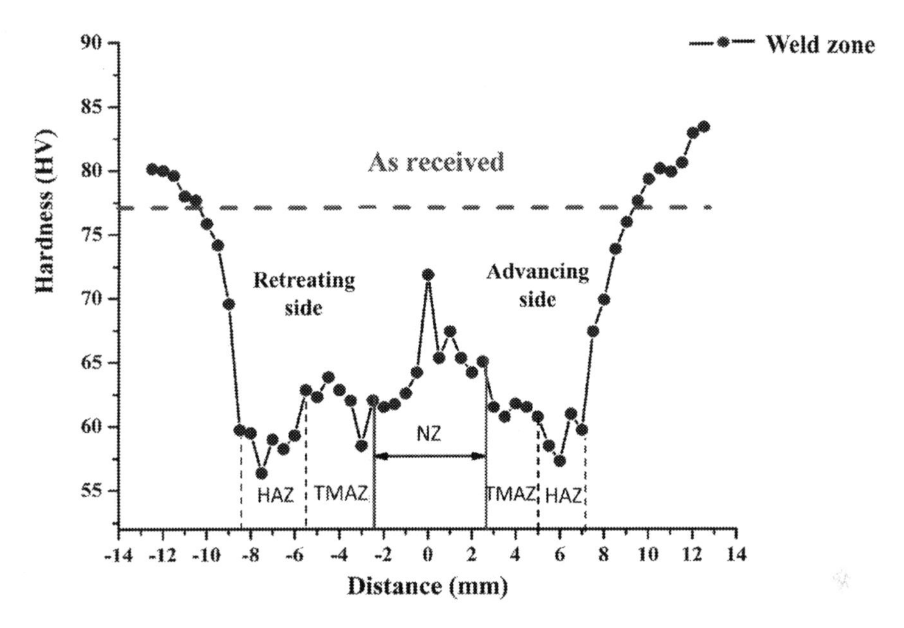

Fig. 7.12 Hardness distribution of FSW AA5083-O along the cross-sectional area

of the rolled base material was observed to be 83.4 HV. This increase in the hardness value with respect to the as-received material was due to the strain hardening caused during the rolling of the sheet to form the open tubular structure. The hardness of the welded region decreased with respect to both the rolled base material, as well as the as-received material. This decrease in hardness value was due to the reason that the precipitates might have dissolved due to the heat generated during welding. The HAZ region had the lowest hardness of 57.75 HV among all the zones because of coarsening of the grains. The NZ had the highest hardness among all the zones, which was around 71 HV. This was due to the refinement of the grains.

The engineering stress–strain response for the BM, RM and the welded tube in both longitudinal and transverse directions have been plotted as shown in Fig. 7.13a, b, respectively. Along longitudinal direction, the BM had the highest elongation of about 36.9% as compared to the RM and the WM, as shown in Fig. 7.13a. The elongation decreased to 27% in the WM with respect to the BM, because of the presence of finer grains in the NZ. Similarly, the strength of the welded joint along the longitudinal direction was found to be 327 MPa, i.e. it increased by 3.15% with respect to the RM and 1.8% with respect to the BM. Along the transverse direction, the percentage elongation of the BM and the RM remained same but it drastically reduced to 12.8% in the WM. Similarly, the strength of the BM and the RM was comparable but it decreased by approximately 9% with respect to the BM, as presented in Fig. 7.13b. The fracture location of the tensile specimen was from the HAZ, and hence in agreement with microhardness profile. However, it was noticed that the tensile strength of the welded transverse specimens was similar to the BM and

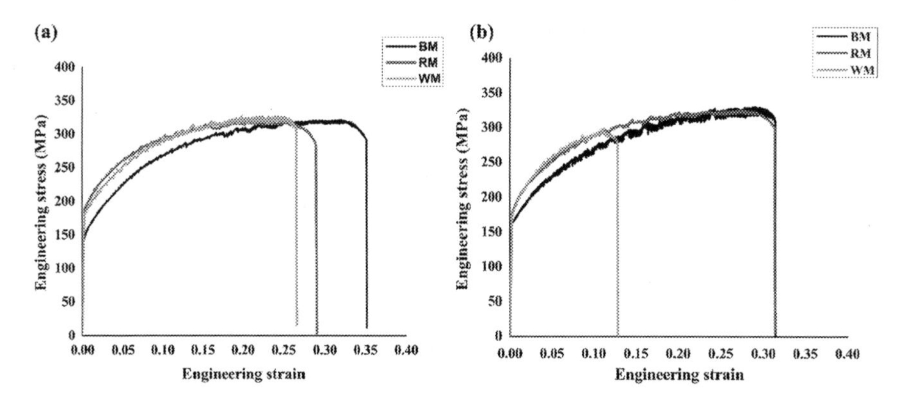

Fig. 7.13 Engineering stress–strain curve of BM, RM, and WM **a** along longitudinal direction and **b** along transverse direction

RM which shows that the presence of joint does not significantly affect the strength of the tubes.

7.5.2.4 Fractography

The fractured surfaces of the BM, RM and FSWed along both the longitudinal and the transverse regions, subjected to tensile test, were explored in order to get insight of the failure mode, as shown in Fig. 7.14. All the fractographs show that dimples were present which indicate ductile fracture of the specimens. The fracture surface of the base metal shows macrodimples having average size of 22.38 μm. The size of the dimples decreased in the rolled sample. Also, some cleavage facets were observed which showed brittle nature of the fracture surface. This might have occurred during the rolling process. In the longitudinal welded samples, microdimples were visible with no other mode of fracture. These microdimples were formed because of the fracture along the refined NZ which decreased the elongation of the sample. The welded sample along the transverse direction which fractured along the HAZ showed mixed mode of fracture. Both intergranular fracture and cleavage facets along with microdimples were present. The intergranular fracture mode occurred because of the coarser grains present in the HAZ zone. These observations were in agreement with the tensile results obtained. Hence, it can be concluded that the failure in the welded zone is predominantly occurred by dimples which means that it was a ductile fracture mode. Thus, the existence of the weld might not significantly affect the subsequent forming operations due to lesser reduction in ductility of the welded zone.

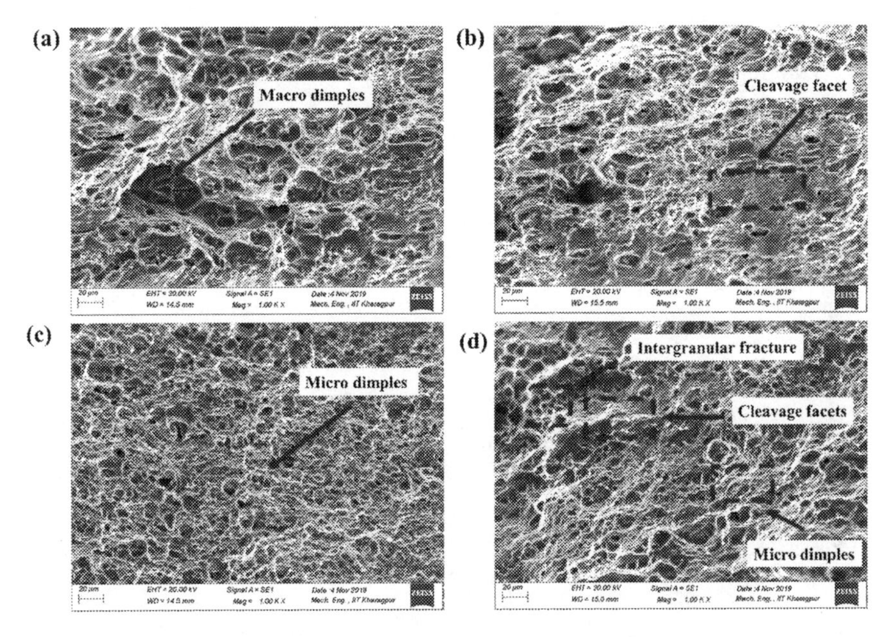

Fig. 7.14 Fractured surface of: **a** BM, **b** RM **c**, **d** WM along longitudinal and transverse directions, respectively

7.6 Conclusion

In the present chapter, the fabrication of tubular components by welding process has been discussed. Various commonly used welding techniques, like the fusion welding (arc welding, ERW and laser welding) and solid-state welding for tube fabrication are discussed. Among these welding techniques, FSW, a solid-state welding process, has got numerous advantages over other techniques. Thus, the potential of FSW process in manufacturing tubular components has been discussed broadly. The defects related to the fusion welding process diminishes by implementing solid-state welding technique, and a better quality tubular structure is produced. Also, this process can be used for producing tubes of different lightweight materials such as aluminium and magnesium which are difficult to be welded by other techniques. Overall, FSW, a newer welding technique, can be used for welding tubular structures with good quality, consistency and cost effectiveness. To justify it, a case study has also been shown where a defect-free and good quality welded tube of aluminium alloy has been produced. It has been observed that with proper selection of process parameters, the weld zone obtained has finer microstructure with improved mechanical properties. The tensile strength and hardness of the welded zone are comparable with that of the BM which makes the welded tubular structure with seamless product quality. However, fusion welded tubular structures are observed to have coarser microstructure and lesser strength as well as lower hardness because of the involvement of higher heat input. This deteriorates the mechanical properties of the welded tube leading to limited usage of the tube and higher risk factor.

References

1. Mei Z, Kun G, He Y (2016) Advances and trends in plastic forming technologies for welded tubes. Chin J Aeronaut 29:305–315. https://doi.org/10.1016/j.cja.2015.10.011
2. Hashmi MSJ (2006) Aspects of tube and pipe manufacturing processes: meter to nanometer diameter. J Mater Process Technol 179:5–10. https://doi.org/10.1016/j.jmatprotec.2006.03.104
3. Wang G, Zhao Y, Hao Y (2018) Friction stir welding of high-strength aerospace aluminum alloy and application in rocket tank manufacturing. J Mater Sci Technol 34:73–91. https://doi.org/10.1016/j.jmst.2017.11.041
4. Hirsch J (2011) Aluminium in innovative light-weight car design. Mater Trans 52:818–824. https://doi.org/10.2320/matertrans.L-MZ201132
5. Europe Communications AM (2020) Automotive business unit. https://tubular.arcelormittal.com/business_units/67/language/EN. Accessed 27 Jun 2020
6. Velaga SK, Rajput G, Murugan S et al (2015) Comparison of weld characteristics between longitudinal seam and circumferential butt weld joints of cylindrical components. J Manuf Process 18:1–11. https://doi.org/10.1016/j.jmapro.2014.11.002
7. Musraty W, Medjo B, Gubeljak N et al (2018) Seam pipes for process industry—fracture analysis by using ring-shaped specimens. Hem Ind 72:39–46. https://doi.org/10.2298/HEMIND170530014M
8. Wang K, Liu G, Zhao J et al (2016) Formability and microstructure evolution for hot gas forming of laser-welded TA15 titanium alloy tubes. Mater Des 91:269–277. https://doi.org/10.1016/j.matdes.2015.11.100
9. Yuan SJ, Hu ZL, Wang XS (2012) Evaluation of formability and material characteristics of aluminum alloy friction stir welded tube produced by a novel process. Mater Sci Eng A 543:210–216. https://doi.org/10.1016/j.msea.2012.02.076
10. Fazzini PG, Ã JLO (2007) Experimental determination of stress corrosion crack rates and service lives in a buried ERW pipeline. Int J Press Vessel Pip 84:739–748. https://doi.org/10.1016/j.ijpvp.2007.05.008
11. Agrawal AK, Narayanan RG, Kailas SV (2017) End forming behaviour of friction stir processed Al 6063–T6 tubes at different tool rotational speeds. J Strain Anal 52:434–449. https://doi.org/10.1177/0309324717724662
12. Brensing K, Sommer B (2008) Steel tube and pipe manufacturing processes. Salzgitter Mannersmann Rohrenwerke 1:1–63
13. Yano K, Oi K (2015) Method for producing a steel pipe using a high energy density beam 2:1–12
14. Löveborn D, Larsson JK, Persson KA (2017) Weldability of aluminium alloys for automotive applications. Phys Procedia 89:89–99. https://doi.org/10.1016/j.phpro.2017.08.011
15. Mishra RS, Mahoney MW (2007) Friction stir welding and processing. ASM International, Materials Park, Ohio
16. Kulkarni S, Ghosh PK, Ray S (2008) Improvement of weld characteristics by variation in welding processes and parameters in joining of thick wall 304LN stainless steel pipe. Iron Steel Inst Japan 48:1560–1569
17. Yang ZZ, Tian W, Ma QR et al (2008) Mechanical properties of longitudinal submerged arc welded steel pipes used for gas pipeline of offshore oil. Acta Metall Sin 21:85–93. https://doi.org/10.1016/S1006-7191(08)60024-1
18. Murugan N, Gunaraj V (2005) Prediction and control of weld bead geometry and shape relationships in submerged arc welding of pipes. J Mater Process Technol 168:478–487. https://doi.org/10.1016/j.jmatprotec.2005.03.001

19. Gunaraj V, Murugan N (1999) Application of response surface methodology for predicting weld bead quality in submerged arc welding of pipes. J Mater Process Technol 88:266–275. https://doi.org/10.1016/S0924-0136(98)00405-1
20. Chen X, Lu H, Chen G, Wang X (2015) A comparison between fracture toughness at different locations of longitudinal submerged arc welded and spiral submerged arc welded joints of API X80 pipeline steels. Eng Fract Mech 148:110–121. https://doi.org/10.1016/j.engfracmech.2015.09.003
21. Wen SW, Hilton P, Farrugia DCJ (2001) Finite element modelling of a submerged arc welding process. J Mater Process Technol 119:203–209. https://doi.org/10.1016/S0924-0136(01)00945-1
22. Lin M, Gao K, Wang C, Volinsky AA (2012) Failure analysis of the oil transport spiral welded pipe. Eng Fail Anal 25:169–174. https://doi.org/10.1016/j.engfailanal.2012.05.008
23. Gong K, Wu M, Xie F, Liu G (2020) Effect of dissolved oxygen concentration on stress corrosion cracking behavior of pre-corroded X100 steel base metal and the welded joint in wet—dry cycle conditions. J Nat Gas Sci Eng 77:103264. https://doi.org/10.1016/j.jngse.2020.103264
24. Simion P, Dia V, Istrate B, Munteanu C (2014) Controlling and monitoring of welding parameters for micro-alloyed steel pipes produced by high frequency electric welding. Adv Mater Res 1036:464–469. https://doi.org/10.4028/www.scientific.net/AMR.1036.464
25. Chung PC, Ham Y, Kim S et al (2012) Effects of post-weld heat treatment cycles on microstructure and mechanical properties of electric resistance welded pipe welds. Mater Des 34:685–690. https://doi.org/10.1016/j.matdes.2011.05.027
26. Park G, Kim B, Kang Y et al (2016) Characterization of bond line discontinuities in a high-Mn TWIP steel pipe welded by HF-ERW. Mater Charact 118:14–21. https://doi.org/10.1016/j.matchar.2016.05.005
27. Kato C, Otogtmo Y, Kado S, Hisamatsly Y (1978) Grooving corrosion in electric resistance welded steel pipe in sea water. Corros Sci 18:61–74. https://doi.org/10.1016/S0010-938X(78)80076-6
28. Bi Z, Wang R, Jing X (2012) Grooving corrosion of oil coiled tubes manufactured by electrical resistance welding. Corros Sci 57:67–73. https://doi.org/10.1016/j.corsci.2011.12.033
29. Miranda R, Costa A, Quintino L et al (2009) Characterization of fiber laser welds in X100 pipeline steel. Mater Des 30:2701–2707. https://doi.org/10.1016/j.matdes.2008.09.042
30. Donnell DO, Kettermann C (2018) Performance implications of high energy density welding of corrosion resistant alloy heat exchanger tubing. Press Vessel Pip Conf 49255:13–20. https://doi.org/10.1115/PVP2010-25165
31. Feng JC, Rathod DW, Roy MJ et al (2017) An evaluation of multipass narrow gap laser welding as a candidate process for the manufacture of nuclear pressure vessels. Int J Press Vessel Pip 157:43–50. https://doi.org/10.1016/j.ijpvp.2017.08.004
32. Hui-chi C, Guijun B, Chen-nan S (2014) High-energy beam welding processes in manufacturing. In: Handbook of manufacturing engineering and technology. Springer, London, pp 617–639
33. Guo J (2015) Solid state welding processes in manufacturing. In: Handbook of manufacturing engineering and technology. Springer, London, pp 569–592
34. Zamani E, Hossien G (2012) Explosive welding of stainless steel—carbon steel coaxial pipes. J Mater Sci 47:685–695. https://doi.org/10.1007/s10853-011-5841-9
35. Xuegang W, Fengjie YAN, Qian YAN, Xingeng LI (2007) Amorphous diffusion bonding of steel pipe and its impact toughness. Front Mater Sci China 1:225–227. https://doi.org/10.1007/s11706-007-0040-y
36. Chludzinski M, Eugenio R, Ramminger D et al (2019) Full-scale friction welding system for pipeline steels. Integr Med Res 8:1773–1780. https://doi.org/10.1016/j.jmrt.2018.12.007
37. Fratini L, Piacentini M (2006) Friction stir welding of 3D industrial parts: joint strength analysis. Eng Syst Des Anal 42517:763–770. https://doi.org/10.1115/EDSA2006-95382
38. Iqbal MP, Jain R, Pal SK (2019) Numerical and experimental study on friction stir welding of aluminum alloy pipe. J Mater Process Technol 274:116258. https://doi.org/10.1016/j.jmatprotec.2019.116258

39. Jain R, Kumari K, Kesharwani RK et al (2015) Friction stir welding : Scope and recent development. In: Davim J (ed) Modern manufacturing engineering, material forming, machining and tribology. Springer, Cham, pp 179–229
40. Mosavvar A, Azdast T, Moradian M, Hasanzadeh R (2019) Tensile properties of friction stir welding of thermoplastic pipes based on a novel designed mechanism. Weld world 63:691–699. https://doi.org/10.1007/s40194-018-00698-6
41. Chen B, Chen K, Hao W et al (2015) Friction stir welding of small-dimension Al3003 and pure Cu pipes. J Mater Process Technol 223:48–57. https://doi.org/10.1016/j.jmatprotec.2015.03.044
42. Packer SM, Matsunaga M (2004) Friction stir welding equipment and method for joining X65 pipe. In: The fourteenth international offshore and polar engineering conference—ISOPE 2004, pp 55–60
43. Threadgill PL, Leonard AJ, Shercliff HR et al (2013) Friction stir welding of aluminium alloys. Int Mater Rev 54:49–93. https://doi.org/10.1179/174328009X411136
44. G.D.Usro, Longo M, Giardini C (2013) Mechanical and metallurgical analyses of longitudinally friction stir welded tubes: The effect of process parameters. Int J Mater Prod Technol 46. https://doi.org/10.1504/IJMPT.2013.056301
45. Urso GD, Longo M, Giardini C (2011) Characterization of friction stir welded tubes by means of tube bulge test. AIP Conf Proc 1353:1259–1264. https://doi.org/10.1063/1.3589689
46. Pang Q, Hu ZL, Pan X, Zuo XQ (2014) Deformation characterization of friction-stir-welded tubes by hydraulic bulge testing. Miner Met Mater Soc 66:2137–2144. https://doi.org/10.1007/s11837-014-1135-4
47. Peterson J, Hall J, Steel RJ, Babb J, Collier M, Packer SM (2016) Out of position fraction stir welding of casing and small diameter tubing or pipe 2:1–23
48. Tavassolimanesh A, Alavi Nia A (2017) A new approach for manufacturing copper-clad aluminum bimetallic tubes by friction stir welding (FSW). J Manuf Process 30:374–384. https://doi.org/10.1016/j.jmapro.2017.10.010
49. Lammlein DH, Gibson BT, Delapp DR et al (2011) The friction stir welding of small-diameter pipe: an experimental and numerical proof of concept for automation and manufacturing. Proc Inst Mech Eng Part B J Eng Manuf 226:383–398. https://doi.org/10.1177/0954405411402767
50. Hattingh DG, Von WLG, Bernard D et al (2016) Semiautomatic friction stir welding of 38 mm OD 6082–T6 aluminium tubes. J Mater Process Tech 238:255–266. https://doi.org/10.1016/j.jmatprotec.2016.07.027
51. Akbari M, Asadi P (2019) Optimization of microstructural and mechanical properties of friction stir welded A356 pipes using taguchi method. Mater Res Express 6. https://doi.org/10.1088/2053-1591/ab0d72
52. Hirsch J (2014) Recent development in aluminium for automotive applications. Trans Nonferrous Met Soc China 24:1995–2002. https://doi.org/10.1016/S1003-6326(14)63305-7
53. Hu ZL, Wang XS, Yuan SJ (2012) Quantitative investigation of the tensile plastic deformation characteristic and microstructure for friction stir welded 2024 aluminum alloy. Mater Charact 73:114–123. https://doi.org/10.1016/j.matchar.2012.08.007
54. Gunaraj V, Murugan N (2002) Prediction of heat-affected zone characteristics in submerged arc welding of structural steel pipes. Weld J 81:45

Chapter 8
Industry 4.0 in Welding

Debasish Mishra, Surjya Kanta Pal, and Debashish Chakravarty

Abstract This chapter aims to highlight the importance of the fourth industrial revolution, popularly known as "Industry 4.0", and its application in welding. While considering all the manufacturing processes, welding has been preferred for the implementation of Industry 4.0 because of its vast applications in various industrial sectors. The importance of welding and its automation has been the foundations of this chapter. The journey of welding till date from the ancient manufacturing has been discussed chronologically, including the evolution of welding in all the industrial revolutions. The role of Industry 4.0 and the utility of its digital tools in welding have been outlined. Towards the end, a case study has been presented which demonstrates the application of Industry 4.0 for online monitoring of weld quality and control of defects in friction stir welding technique.

Keyword Industry 4.0 · Manufacturing process · Welding · Digital intervention · Online monitoring and control · Friction stir welding

8.1 Welding as a Manufacturing Process

"Fabrication" refers to the creation of structure, component, machine, etc., by working on to the raw material through several manufacturing techniques. The major operations include metal cutting, forming, and assembly. "Assembly" refers to the final fitting of the sub-components by techniques, viz. adhesive bonding, fastening

D. Mishra
Advanced Technology Development Centre, Indian Institute of Technology Kharagpur, Kharagpur, West Bengal 721302, India
e-mail: debsmishra02@gmail.com

S. K. Pal (✉)
Department of Mechanical Engineering, Indian Institute of Technology Kharagpur, Kharagpur, West Bengal 721302, India
e-mail: skpal@mech.iitkgp.ac.in

D. Chakravarty
Department of Mining Engineering, Indian Institute of Technology Kharagpur, Kharagpur, West Bengal 721302, India

J. P. Davim (ed.), *Welding Technology*, Materials Forming, Machining and Tribology, https://doi.org/10.1007/978-3-030-63986-0_8

using screws or bolts and welding. Adhesive bonding and mechanical fastening are temporary methods, whereas welding is a permanent method of joining. The joining of materials by welding can be accomplished with or without involving melting of the workpieces to be joined. In the case of occurrence of melting, suitable filler material may be required for welding. The fusion welding techniques, viz. arc welding, spot welding, laser beam welding, etc., involve heat; i.e. they heat up the metals to be joined up to their melting points, and subsequently fuse them together. Heating has been an integral part of welding, and this is how welding is characterized. Interestingly, other techniques do exist as well, which are popularly referred as "cold welding" or "solid-state welding", which do not melt the metals to be joined, rather, deform them plastically for the joining. This helps welding applicable in manufacturing to wide range of materials. Thus, welding has been a core part in the production process of every metal industry.

8.1.1 Need of Automation in Welding

Automation is derived from terms such as "automatic" and "mechanization", which indicate the use of several mechanisms for handling the systems automatically. The major use of automation is to reduce or eliminate the intervention of human in the manufacturing activities. The art of automation is fulfilled by the use of various custom-designed electronic and computer-controlled devices.

An automated production line consists of a set of production processes which are being performed automatically maintaining a sequence. For instance, an automobile production line has various tasks which are involved in the final production: manufacturing of the chassis, assembly of the auto-components, intermittent welding, painting, followed by enhancing the aesthetics. The production line has individual workstations connected by transfer system to move the automobile being manufactured from one station to the other. This is an example of "fixed automation". Similarly, a computerized numerically controlled (CNC) machine is a typical example of "programmable automation". Here, the machine tool is controlled by numbers. The numbers represent the coordinates for defining the position of tool and are provided to the machine with the help of a program script written with the help of a computer. In order to ensure the position of the tool during the operation, a suitable control system is implemented.

The examples mentioned above highlight the linkage between the mechanical and electronic systems. In the case of automated production line, the automobile is the product, and machineries/robots and transfer system refer to the mechanical components. The controllers, servo motors, sensors, and automation software comprise the electronic items in the automated production line. They represent a highly integrated system making use of microprocessors in the controllers to control the mechanical component, sensors and actuators to acquire information and provide output. In the present scenario, machines are being designed, developed, and manufactured by integrating various electronic devices with the mechanical systems, making it more flexible and reliable. Earlier, machine downtime used to affect the rate of production. However, the predictive maintenance which uses sensors attached to the mechanical components, monitors machine health constantly, thereby reducing the machine downtime.

In the context of welding, maintaining the product quality is crucial, or in simple words, the product should be free from defects. In the sequence of manufacturing operations, welding is the last process. Thus, automation along with monitoring and control of welding is inevitable. Automation in welding includes the use of robots integrated with other infrastructures. The prime objective of automating the welding technique is to reduce the process variation, which often occurs with manual handling of the jobs. For instance, in case of the manual arc welding technique, a lot of spatter may be generated because of improper setting of parameters, or variations in the parameters during welding. This will degrade the job and may require time-consuming cleaning of the surface. Automation of the process will help in avoiding the formation of spatter to a greater extent with elimination of manual intervention. Thus, the robotic system for welding is essential because of the following: (a) elimination of human errors, (b) less variation in the parameters which is ensured by the use of a feedback system, (c) superior weld quality with almost no need of any rework and (d) higher productivity.

8.1.2 Focus of This Chapter

"Industry 4.0" or in other words, the "fourth industrial revolution" has been a buzzword in the recent times [1–3]. Conceptualized by Germans, it aims at digitalizing the traditional manufacturing process. The objective of this chapter is to highlight the pertinence of Industry 4.0 in welding. Figure 8.1 shows a "*Google Trends*" analysis on the keyword, Industry 4.0, from 2014 to 2019 [4]. The interest can be seen to be increasing over the years.

The remainder of the chapter is organized in the following manner: "Sect. 8.2" presents the evolution of the manufacturing (industrial revolutions) chronologically which is inclusive of the history of welding. "Monitoring and control" of manufacturing is a bottom line in the success of Industry 4.0. This has been discussed in "Sect. 8.3" for different welding processes. This section also describes the role of Industry 4.0, and its "digital tools" for implementation in welding. Few important concepts such as (a) data mining, (b) machine learning (ML)/deep learning (DL) and (c) artificial intelligence (AI) have been addressed in this section. "Sect. 4" presents a case study to highlight the utility of data in online process monitoring and control of friction stir welding (FSW) technique. Finally, the concluding remarks have been presented in "Sect. 5".

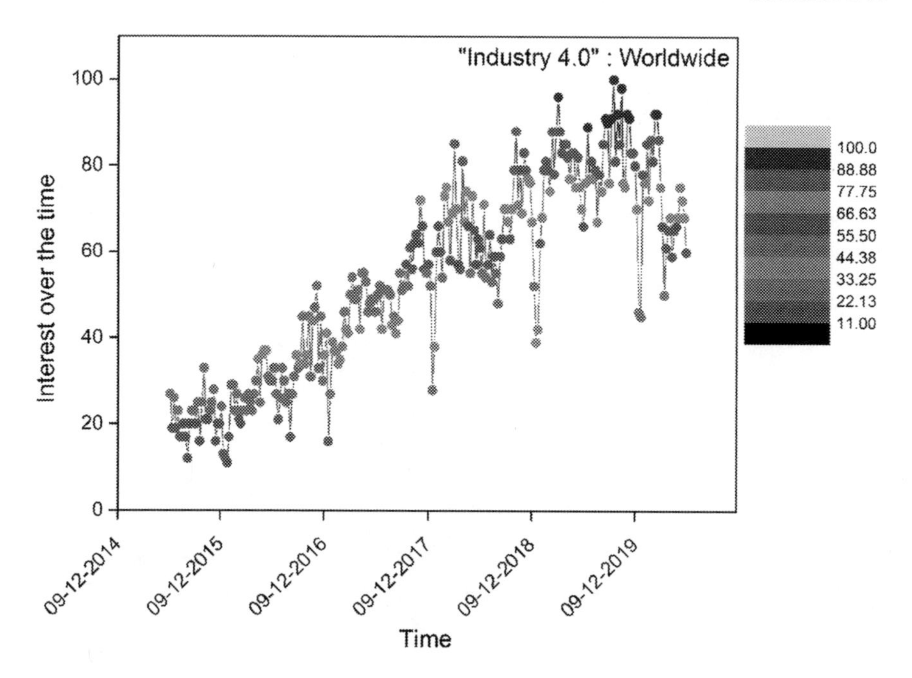

Fig. 8.1 Popularity of Industry 4.0 since inception

8.2 Industrial Transformation – Journey from Industry 1.0 to Industry 4.0

The manufacturing scenario has taken a paradigm shift in the recent times owing to globalization. Owing to the stakeholders' pressure, the industry at present must be agile and flexible [5]. Demand for high-precision products is on the rise due to which the traditional manufacturing style has become obsolete. Integration of advance manufacturing systems like sensory systems will help in achieving improved product quality at a lower cost. Sensory systems integrated to mechanical and electronic components have been able to create a revolution in the manufacturing sector. The following paragraphs discuss the evolution of manufacturing, i.e. the journey from the first to the current industrial revolution. The evolution of welding as a manufacturing process is also mentioned below.

8.2.1 Primitive Manufacturing and Industry 1.0 – the First Industrial Revolution

Human brain is one of the many wonders in the world. It has the capacity to create effective solutions to meet the daily needs. Dating back to the pre-industrialization period, manufacturing started when a skilled artisan along with a few disciples made goods and sold them eventually. These goods are modern-day handicrafts. These mostly occurred in households of rural areas. However, with the advent of the industrial age, manufacturing had its range from handicrafts to high technology products. The term "industrial revolution" indicates the various innovations in the process of production. It roots back to sixteenth century, where most of the jobs were performed manually with the help of labourers. This phase lasted for around 100 years until the first commercial and successful steam engine was designed and created [6]. The power that was obtained from driving the steam facilitated a change in the industrial context, which is known to be the first revolution, i.e. "Industry 1.0" [7]. With the advent of steam engine, there occurred a rapid increase in the rate of production. The factories built in this era had general purpose machines only run by mechanical prime movers, and the jobs were highly labour oriented.

Welding in this era resembled the forging technology, where two metal pieces were heated together to get welded. The idea was similar to the blacksmith shop. The process involved extreme heating of the metals, followed by forcing them to join together by applying pressure through a hammer. The process of joining metals in this era does not resemble to what is known today as the conventional method of welding. However, it just served as one of the ways of shaping and forming the metals.

8.2.2 Industry 2.0 – the Second Industrial Revolution

Humans have always tried to build machines that would perform their various works and also think of improving it. In the beginning of the nineteenth century, another revolution in the industry occurred with electricity being used as the primary source of power [8]. For instance, the manufacturing of an automobile in the previous era involved gathering all the components near the chassis. As such, the assembly process involved a huge amount of time in manufacturing a single automobile. The assembly line concept was introduced later, where the manpower was segregated based on the skill set, and each was assigned to a specific task from the entire process of production. The result of the moving assembly line was an increase in the overall productivity and decrease in the labour usage. This led to the mass production of goods that were now sold at low prices [8].

The conventional way of joining the metals is known to come into existence during this era. The first use of electric arc happened for two carbon electrodes with the help of a battery, and, the first use of fusion welding was documented where two lead plates were joined by using arc from the carbon electrodes. This was followed by the use of bare metal electrodes for arc welding that was consumed in the molten pool and becomes a part of the weld metal. However, the quality of the weldment was compromised because of the formation of oxides and nitrides. The importance of shielding the arc and molten metal from the atmosphere was realized. This led to the use of coated metal electrodes for arc welding, which became the precursor of the shielded metal arc welding technique (SMAW). However, the stability of the arc was still a problem. SMAW utilizes an electrode, which melts together with the metals to be joined forming a weld pool. It is popular in construction industries because of its applicability in any working condition, unless, it is protected from rain, sand and snow. The disadvantages of the process lie with the frequent change of electrodes during the process, which might induce several welding defects.

8.2.3 Industry 3.0 – the Third Industrial Revolution

The next revolution in the industrial scenario came into picture with the creation of a breakthrough machine called the computer. This computer was an analog decimal machine in its earliest form. This later got transformed into a fully-working computer with each passing generation, which ultimately led to the inception of concept of "computer integrated manufacturing" (CIM). The fusion of manufacturing system with electronic components and information technology led to the third revolution. This further reduced the labour intervention by automation of manufacturing processes [8]. The usage of computers for programming, planning and control of manufacturing operations increased the efficiency and effectiveness of the factories.

In this era, the problem of stabilizing the arc in the process of welding was achieved by the invention of metal electrode coated with a thin layer of lime [9]. One outstanding achievement of this era was the invention of the resistance welding method which led to development of other methods of joining like projection welding, spot welding, seam welding and flash butt welding. This was followed by the invention of the thermite welding technique. The gas tungsten arc welding (GTAW) was also introduced in this era, where the arc is delivered by tungsten, a metal which melts at very high temperature; thus, it is not consumed in the process. To fill the metals to be joined, a suitable filler material is utilized.

The gas metal arc welding technique (GMAW) came next which replaced the GTAW technique. The GMAW technique utilizes a spool of wire which acts as the electrode. This wire gets fused with the metals to be joined. The advantages of this technique include the constant spool of wire which eliminates the frequent change of electrodes as in SMAW, and also the feeding of a filler material as in GTAW. The technique became popular and was employed for welding of non-ferrous metals and steels owing to the higher deposition rate. However, the concern remained with the inert gases such as helium and argon, which were expensive.

Later, the same welding technique was successfully reported with an environment containing carbon dioxide gas and resulted in the short-circuit arc. Another addition was the use of oxygen along with the inert gas which introduced the spray-type arc transfer. The latter was the pulsed current, which allowed current variation. The plasma arc welding technique was also invented in this era, where the arc was made to pass via an orifice, making the arc concentrated and creating a plasma. It was utilized for metal spraying and cutting, as the temperature was found to be higher than that of the tungsten arc. The electron beam welding technique also got invented in this era, where a focused beam of electrons generated from a suitable source was utilized for joining the metals. The major application of this welding technique was in the automotive and aerospace.

Thus, it could be observed that, welding is one of the crucial manufacturing processes, which underwent a continuous development for the advent of an efficient and economic technique for joining metals. The recent developments in the history of welding have been friction welding, laser welding and FSW. While both friction welding and FSW belong to the category of solid-state welding method, laser welding is a fusion welding technique. The successful use of laser was first reported for cutting which was the inception of the idea for welding as well. The technique utilized a concentrated beam of light impinged on a tiny spot so that the area absorbs the light and becomes energetic resulting in the melting of the metals to be joined. The major advantages of this welding technique include consistent quality, high precision, high joining speed, etc. However, the cost incurred in the system poses a concern. Nevertheless, laser welding has been a preferred joining method for automobile manufacturing as it does not need any filler material, and thus, contributing to the lightweighting of the component.

The friction welding is advantageous as it does not require filler material or shielding gas; however, the geometry of the parts that can be joined limits its application. The FSW was introduced for welding aluminium alloys [10]. Lightweight metals having resistance to corrosion, high thermal and electrical conductivity like aluminium and magnesium became popular. Even, in the current scenario, these metals are being explored in a wide manner in automobile and aerospace sectors, where weight reduction is one of the prime objectives. Though GTAW and GMAW were capable in welding these metals, both require several tricks to be followed for achieving successful weldments. Alloys of aluminium and magnesium have low melting points, high specific heat and latent heat of fusion, high thermal conductivity and strong oxide layer [11]. It requires a great deal of exercise to select a proper filler material, current and voltage settings, and arc length. FSW, on the other hand, has been introduced as an easy and efficient method which utilizes the frictional heat generated by a rotating tool which plunges inside the metals to be joined [12]. It plastically deforms the materials and stirs them alongside the tool to create the joint. Since it does not involve melting, it has relatively lesser heat than that of GTAW and GMAW, which makes it easier for the alloys to get welded eliminating the difficulties of porosity, hot cracking, softening of the heat-affected zone (HAZ) [13].

8.2.4 Industry 4.0 – the Fourth Industrial Revolution

The world maintains its pace with the advent of innovative technologies and growing consumer needs for sustainable products. Industry 4.0 is the current industrial revolution which has brought several digital tools to be incorporated into manufacturing. The crux of Industry 4.0 lies in the data accumulation and analysis. It refers to the application of "Internet of Things" (IoT) and "Cyber-Physical Systems" (CPS) in the context of manufacturing. The IoT refers to various interconnected devices with an ability to transfer the data among them remotely, i.e. over a network by using some suitable communication protocol. The CPS refers to the system enabling the control of physical objects through computer and software. Thus, the production machines need to be equipped with several sensors which will aid in data acquisition of the manufacturing process. This data would help to derive inferences about the process in real time which will provide an idea on different activities and would enable a faster means of decision-making and control. The industrial transformations discussed above along with the evolutions in the welding have been shown in Fig. 8.2.

Taking a glance at the evolution of the welding methods and techniques, Fig. 8.3 depicts a qualitative description of these trends. In the very beginning, the aim was to achieve higher deposition rate, as it would have resulted in shorter cycle times. However, the product quality was low and the cost of production was high. Even this idea could be misleading, if the product requires repair, leading to loss of economy. The subsequent years have witnessed several advancements in the welding, which included single-shot technique, such as resistance spot welding (RSW), and continuous technique such as GMAW. This brought automation into picture which resulted in high-precision jobs, and reduced the rejection rate as well as the cost of production. Automation of welding has been in practice from quite a long time. The first use of robot for welding ushered in 1962, by *"General Motors"*, in their automobile industry for spot welding [14]. Then onwards, industrialists started recognizing the benefits of using robots in manufacturing operations, for sustainability of automobile

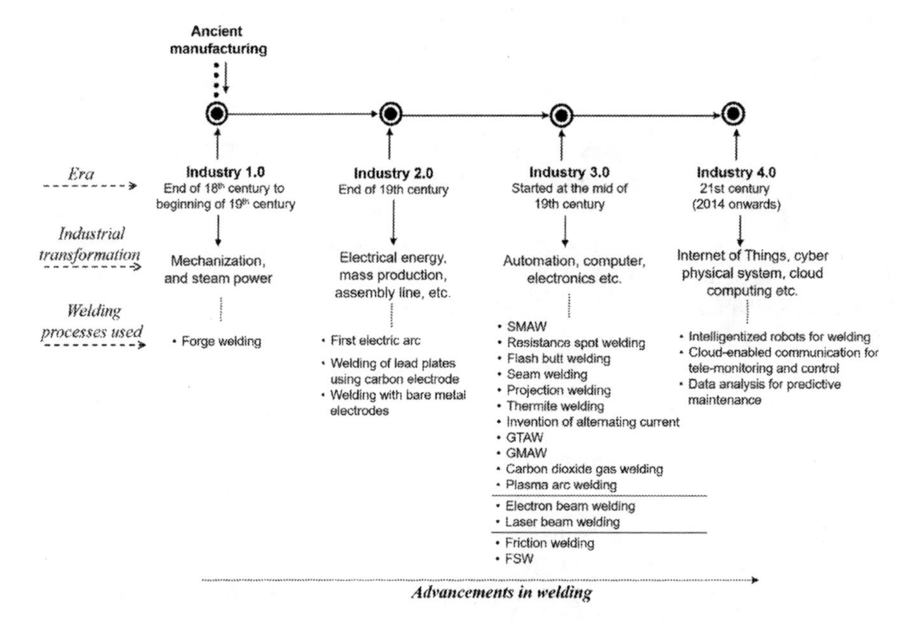

Fig. 8.2 Picture depicting transformations in industry and welding technologies

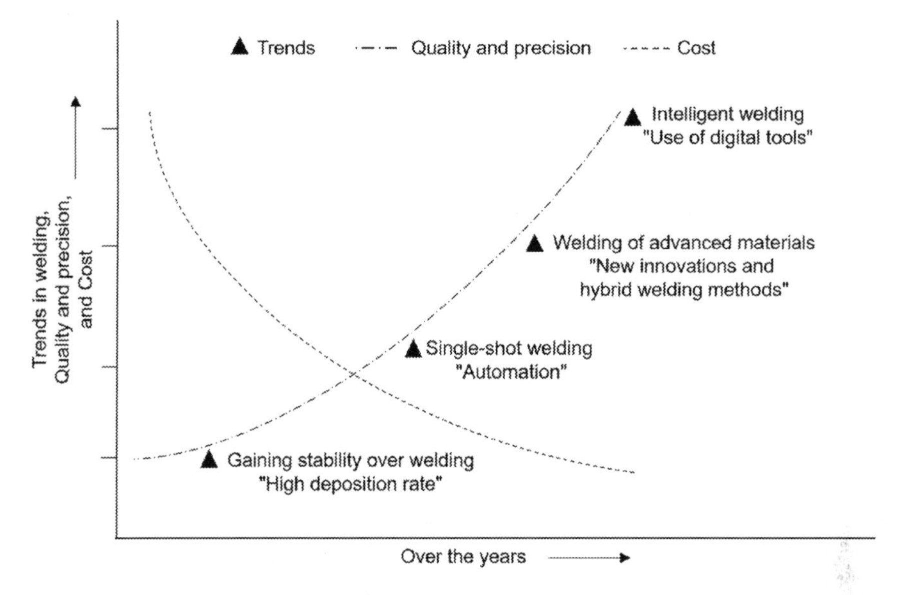

Fig. 8.3 Qualitative description of trends in welding

supply chain [15]. This becomes more important for welding, since the objective is to obtain a defect-free weld, which is highly possible via robots. Innovative welding techniques like laser beam welding and FSW were introduced, which are suitable for joining of several advanced materials [16–18]. They have gained popularity in aerospace, automobile and rail manufacturing industries. The latest trend has been the "intelligent welding" which aims to integrate the welding process with the digital tools for further reduction in the cost by the following: (a) reducing human intervention by employing intelligent robots, (b) real-time control of the weld quality with the help of sensors and embedded system, (c) real-time health monitoring of the welding equipment, (d) analysing the complete available data (process data, machine data and environmental data), etc.

Industry 4.0 can be referred as the "next-generation automation", which will have a lot of interventions from the digital tools towards automation. This is aimed at online prediction of the faults much before their occurrence, so that, the necessary prevention measures can be adopted [19]. Both, fault prediction and identification, can be met by harnessing the data from the process in real time. In a factory, the machines related to a common manufacturing process on the shop floor have to be connected to a common platform through the Internet. This platform would serve as a host system for collecting the sensory data from those connected machines. This data would be analysed through the various processing techniques to keep a close watch on the machines as well as the component being manufactured. The question now arises how this integrated platform is going to be established. The solution is the connecting devices such as sensors and transducers, data acquisition systems, actuators, micro-controllers, and computers. These devices will shape the future of

Industry 4.0. Thinking, framing and applying the principles of Industry 4.0 would be next to impossible without electronics.

Before utilizing the digital tools and concepts, it is necessary to know the different techniques being used for automation of welding processes. This understanding and information will help in appreciating the advantages while implementing Industry 4.0. The next section discusses this in detail.

8.3 Monitoring and Control of Various Welding Techniques

Monitoring refers to the art of "state estimation", i.e. determining few characteristics in a process to make useful decisions. So, it is imperative to understand how monitoring can be conducted in a manufacturing process.

Standardization of jobs in industry is crucial. Figure 8.4 depicts a flowchart for standardization of welding as a process. For instance, in case of arc welding, a suitable welding technique is first selected based on the materials to be welded. Second, the specification of the filler rod, type of shielding gas, etc., is identified. A lot of trial experiments follow next in order to determine the optimized parameters. These prerequisites form the "welding procedure specification" (WPS) [20]. Figure 8.5 shows a picture of WPS which can be observed to be a record consisting of a comprehensive information such as: welding technique, grade and thickness of the materials to be

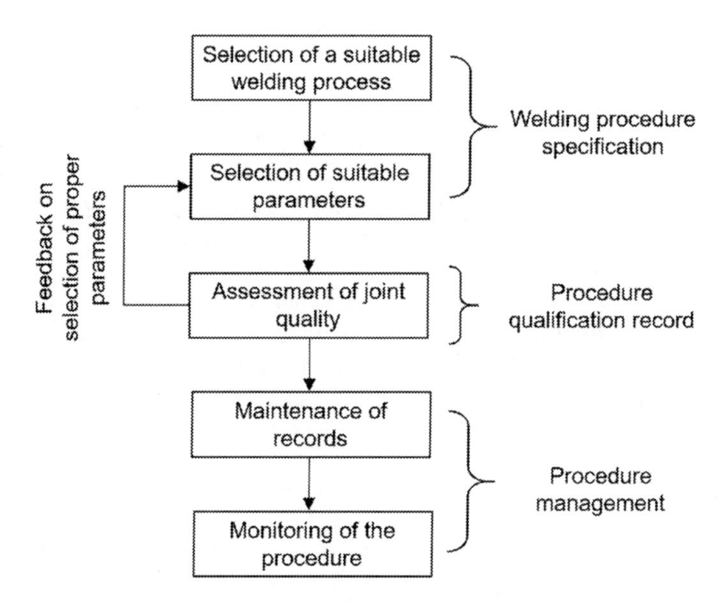

Fig. 8.4 Flowchart for welding procedure selection, qualification record and management of the procedure

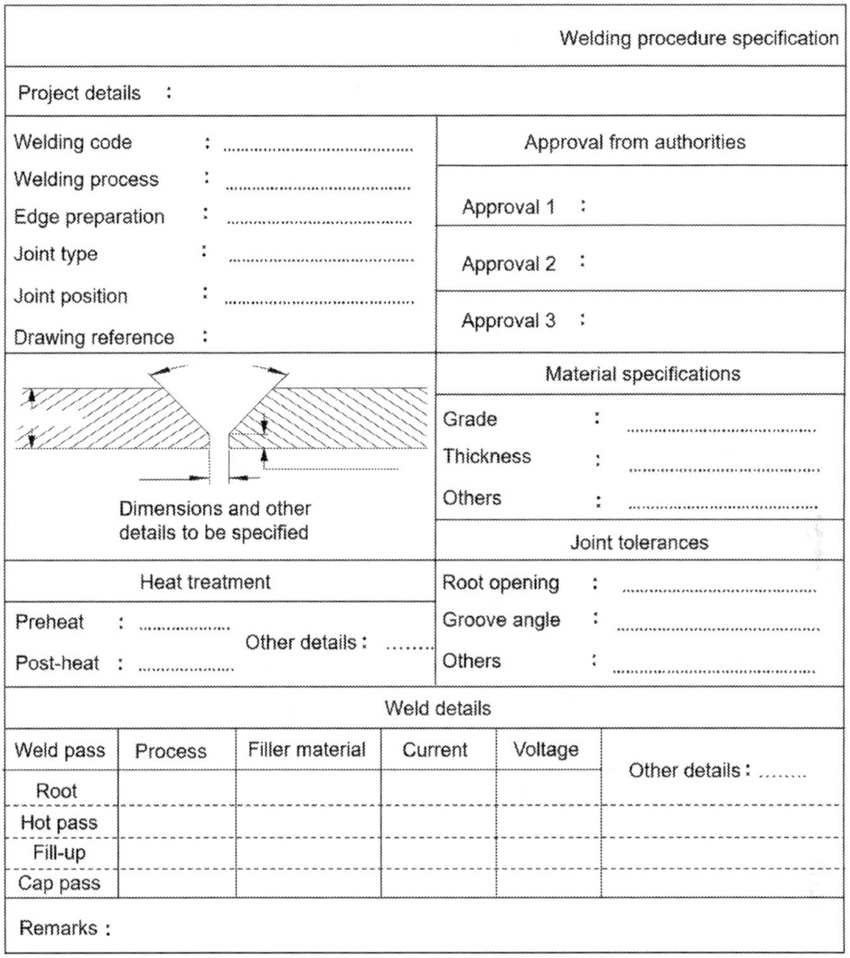

Welding procedure specification

Project details :

Welding code : ....
Welding process : ....
Edge preparation : ....
Joint type : ....
Joint position : ....
Drawing reference :

Approval from authorities

Approval 1 :

Approval 2 :

Approval 3 :

Dimensions and other details to be specified

Material specifications

Grade : ....
Thickness : ....
Others : ....

Joint tolerances

Root opening : ....
Groove angle : ....
Others : ....

Heat treatment

Preheat : ....
Post-heat : ....
Other details : ....

Weld details

Weld pass	Process	Filler material	Current	Voltage	Other details : ....
Root					
Hot pass					
Fill-up					
Cap pass					

Remarks :

Fig. 8.5 Picture depicting a document of welding procedure specification

welded, type of filler rod to be used, joint tolerances, heat treatment, values of welding parameters during each stage of welding and other details.

It is imperative to ensure the correctness of this document for which, the assessment of joint quality is performed next, as shown in Fig. 8.4. This document is referred as "procedure qualification record" (PQR), and it consists of results of several quality testing techniques for assessment of the joint performance. This includes destructive techniques, such as tensile testing, hardness, toughness, bend, and non-destructive techniques as well. A picture of this document has been shown in Fig. 8.6. On the basis of the obtained results, the optimized parameters are finalized and documented in the WPS. After successful implementation of these two documents, "procedure management" is performed on-site on a regular basis. The procedure management

Procedure qualification record	
Project details :	
Welding code : .... Welding process : .... Joint number : .... Joint type : .... Joint position :	Approval from authorities Approval 1 : Approval 2 : Approval 3 :
Details of base materials	Joint tensile test results
Yield strength : .... Ultimate tensile strength : .... % elongation : .... Hardness :	Yield strength : .... Ultimate tensile strength : .... % elongation : ....
	Other mechanical test results
Joint hardness results	Toughness : ....
Weld zone : .... HAZ : .... Other details	Bend test : .... Others : ....

NDT results		Macro examination		
Visual examination		Optical microscopy		
Magnetic particle testing		Grain size	Weld zone	HAZ
Dye penetrant testing				
Ultrasonic testing		Intermetallic compound		
Radiography testing				
Remarks :				

Fig. 8.6 Picture depicting a document of procedure qualification record

document contains the parameters and techniques used during welding. After acceptance of all the formal documentation, monitoring of the welding operations becomes essential. This will ensure that the optimized welding parameters are maintained each time, and the weldment retains the required properties without any welding defect. This marks the importance and need for monitoring of welding.

As mentioned in the beginning of this section, monitoring refers to determination of few characteristics in a process required to make useful decisions. Figure 8.7 presents a generic block diagram for monitoring and control of a manufacturing process. It resembles periodic tracking of an activity by gathering and analysing data and other available information. Monitoring is performed either directly or indirectly. They help in determining the quality of the job being manufactured. For

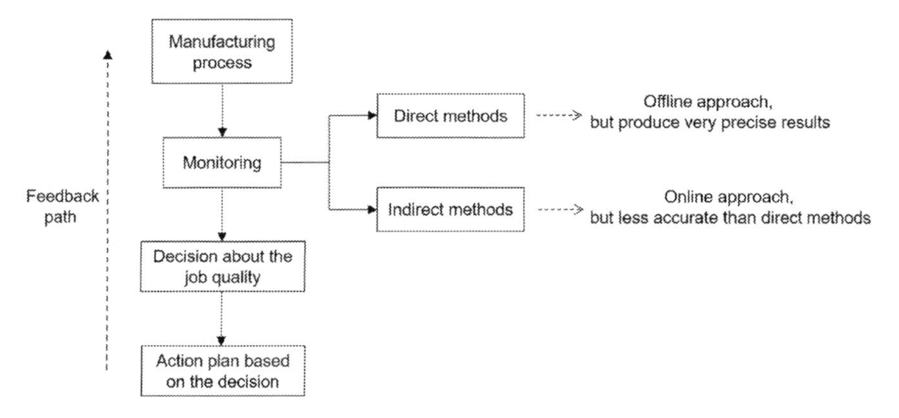

Fig. 8.7 Block diagram for monitoring

welding, the "characteristic" indicates the quality aspects, the values of which can help in judging the performance of the welding. These aspects may vary for different welding techniques and applications. Based on the analysis, decisions are taken to enhance the performance of the process. This decision forms the feedback channel.

Welding, as a manufacturing process, may be referred as a complex process, because of the dependency on several parameters. For instance, the voltage and current in arc welding, force and current in RSW, beam power and frequency in laser are thought to be the major dependencies. However, there are other factors as well which affect the weld geometry. These include the mismatch in the parts to be joined, variation of depth in the grooves, overlapping length in RSW, reflectivity index of the workpiece in laser welding. Thus, such parameters are to be taken into account for monitoring of welding. The major advantages of monitoring and control of welding include tracking of the weld seam, penetration depth, identification of welding defects, width of the weld zone, performance of the power source. The monitoring can either be performed offline or online. While the former refers to the estimation/determination of the weld quality after the weld is fabricated, the latter addresses the subject in real time. The direct and indirect forms of monitoring have been discussed next.

8.3.1 Direct Monitoring

The direct monitoring technique takes care of the actual measurements. This technique makes use of the quality variables as the monitoring variables. For instance, in case of welding, specimens are removed from the weld zone to determine the tensile strength, yield strength, hardness and grain size. The determination of these quality variables requires dedicated instruments such as universal tensile testing machine, hardness tester, optical instruments such as microscope, respectively. The results obtained from these instruments are accurate and usually depend upon the precision of the machine. However, they are usually destructive in nature, time-consuming and are carried out offline. As such, they do not help in getting inferences about the quality of the weld in real time.

8.3.1.1 Non-Destructive Techniques

An alternative to the above is the use of non-destructive testing (NDT) techniques, such as visual sensing, dye penetrant, magnetic particle testing, radiography, ultrasonic testing, which are being employed for quality monitoring of the welded samples [21]. The visual sensing makes use of instruments such as magnifying glasses, endoscopes, profilers to measure the surface conditions, such as weld dimensions, surface defect and alignment. Penetrant testing and magnetic particle testing are two other techniques used for identifying the surface anomalies in welded samples, which are invisible to bare eyes. The penetrant testing makes use of the capillary properties of liquid where the liquid with low surface tension infiltrates into clean and dry surface. A series of steps are applied in this technique which involves: pre-cleaning of the job, application of penetrant and developer, and post-cleaning of the job. The magnetic particle testing is applicable for ferromagnetic materials only. Here, the weld surface is magnetized by using a specialized powder, usually iron powder, and the observation is to identify the area where the powder gathers itself. This indicates possibility of defects being present in the sample. This is because of flux leakage at the spot of crack present in the welded sample, which makes the powder gather over that spot, in the presence of magnetic field. A major drawback of these techniques is their inability to detect the discontinuities inside the welded area. Though they are non-destructive in nature, they work offline, and thus, cannot fulfil the aim of Industry 4.0.

The challenge of identifying discontinuities within a material has been resolved by radiography technique. The X-ray radioactivity system has been one of the widely adopted techniques for the identification of the welding defects because of the ability of the X-rays or gamma rays in penetrating into the weldments. The entire weld surface is covered with a film, and the X-rays or gamma rays from a suitable radioactive source are allowed to pass through the welded sample. With the presence of cavity, hole or void in the welded sample, the radiation being received over the film varies, and this film is then processed and viewed under a special light emitting device to identify the degrees of variations. The major limitation of this method includes the execution of this activity in controlled conditions, i.e. because of the harmful radiations to humans, it is required to seal a major part of the area where the test would be carried out. Another limitation is that their inability to be applied in real-time application. Computed tomography is another such method which also utilizes the X-rays for detection of the welding defects.

An article reports the use of NDT techniques such as visual testing, penetrant testing and ultrasonic testing in the arc welding technique for monitoring of the welded samples, and ultrasonic testing was found to be more suitable because of its ability to detect most of the welding defects [22]. In order to overcome the limitation of being applicable as an offline inspection method, these NDT techniques have been clubbed with intelligent algorithms such as neural networks and fuzzy reasoning to implement them in real time. A review article [23] in this regard can be referred which discusses on the application of radiography, ultrasonic and eddy current inspection techniques. The use of NDT methods has also been explored for quality monitoring of FSW technique. While the X-ray technique has been found to be able to detect the voids, it failed to detect the other types of defect such as "kissing-bond" defect [24] and tiny root flaws. The tiny voids were captured by using the ultrasonic c-scan technique. This suggests that different NDT methods would be required for fulfilling the need of quality control. The ultrasonic phased array technique has also been utilized for inspection of the welds in FSW [25].

8.3.1.2 Vision-Based System

Other than these, the vision-based systems also come in the category of direct monitoring. These systems have been utilized to extract information about the molten pool in welding. The system makes use of light which is projected onto the workpiece to be welded, and the distance within the vision system and the workpiece forms one of the reference measurements. During welding, whenever a variation occurs, this distance varies leading to change in the position of the reflected beam which is captured by using a camera. The captured data is then processed digitally to infer useful information about the welding technique. The inferences include weld seam tracking, edge detection, groove, tracking of beginning and end positions of the weld, and information about the weld pool [26]. The vision-based systems can either be referred as active or passive, depending upon the nature of illumination.

The active vision-based system refers to the use of a source of light to illuminate the workpiece area, and the passive vision-based use the arc light for imaging [27]. A number of experiments have utilized the vision-based systems for tracking of the seam during arc and laser welding techniques [28–32]. These works have utilized active vision sensor, such as laser for detection. Beside the impressive results obtained in these works, they have been criticized for the high capital investment, inability in tracking complex seams and error in look-ahead detection [27]. This is so because the successful implementation requires proper illumination of the targeted areas, which limits them to be explored in laboratories rather than in industry. An alternative to this problem is the passive-based vision sensor which also has been explored for tracking the weld seam [27]. Using this sensor, the cost has been reported to be reduced by 15 times as compared to that of the active laser system [27]. However, this increases the pre-processing exercise, as the images suffer from the intensity of the arc light and have random noises.

Seam tracking is a major issue in the robotic welding [33]. As mentioned earlier, the robots have been identified as one of the core technologies for implementation of automation and control in the manufacturing sector. The robots which are utilized for arc welding have laser system attached to the torch. Robots track the seam and control the welding torch accordingly by means of suitable cameras, monitoring the laser light. These sensors help in identifying and locating the joint seam before the start of the welding. One of the major hurdles in this process is to ensure the safety of the sensors, which includes the laser and the camera, as they are fixed onto the torch. The sensor may also get affected by the harsh environment, which is another limitation. The torch embedded with camera and laser becomes inaccessible to all regions. Other than seam tracking, the optical detectors have been utilized for classification of weld into defective and defect-free in arc welding technique [34]. In this work, the utilized optical sensor consisted of two visible detectors, one for detecting the infrared rays, and the other one for the ultraviolet rays. Various descriptive statistics were observed from the signals for classifying the welds. For illumination, a diode-laser system was utilized.

Similarly, the use of vision-based system in FSW technique has spanned from classifying welds into defective or defect-free, and further classifying the welding defects. For the purpose of weld classification, a number of descriptive statistics such as *mean, standard deviation, entropy, energy, contrast* and *homogeneity* were derived from the weld images [35]. It was found that *standard deviation* was a better indicator for weld classification. Similarly, features such as *energy, variance* and *entropy* were derived from the wavelet coefficients of the weld images for weld classification [36]. The defect-free weld had high values of *energy* and *entropy*, and less *variance*, as compared to the defective weld. Another study proposed the feature extraction from the weld images by using "*maximally stable extremal region*" algorithm for weld classification [37]. This algorithm is based on the idea of identifying regions with almost no variation through a wide range of thresholds. The extracted features were fed as inputs to a ML model for weld classification. Another study proposed the use of contour and profile plots for weld classification where the grey level intensity was found to be distinctly varying for defect-free and defective welds [38]. For the classification of the welding defects, a study reports usage of image pyramid and reconstruction techniques [39, 40]. In the purview of ML, "support vector machine" (SVM) has been utilized for classification of the weld images because of their ability to generalize problems, even with fewer training samples [36, 37].

The vision-based system is advantageous as it does not directly interact with the weld pool and carry abundant information. However, because of several practical limitations, their use is also limited. The major drawback is the precise positioning of these instruments. The shop floor in an industry may consist of several machineries, all of which may be operating simultaneously. Thus, the presence of vibration is quite likely. The optical-based instruments are supposed to be placed in a vibration-free environment so as to ensure the accuracy of their measurements and repeatability as well. With a practical industry environment, ensuring both is quite difficult a task. Other practical limitations of these instruments include the delicate handling and loss of precision in dusty environment.

8.3.2 Indirect Monitoring

The indirect monitoring utilizes a state variable as a measure of the quality variable; i.e. it refers to the sensing of various physical quantities; the values of which are analysed to indirectly correlate with the weld quality. The necessity of the indirect approach of monitoring aims to address the difficulties related to direct monitoring techniques. These physical quantities are: current, power, force, torque, temperature, acoustic emission, sound, vibration, etc. Since attempts are made to correlate the weld quality with a physical quantity, the result is less accurate as compared to that of the direct monitoring approach. However, the indirect monitoring technique is more practical, economic and suitable for industrial applications. The biggest advantage of the indirect monitoring techniques is the in-line engagement with the manufacturing process. This provides the real-time capability. In order to build an efficient indirect monitoring system, it is crucial to identify the quality variable, and the suitable signal to identify/predict the quality. Accordingly, a sensor can be selected. The acquired signal is then processed online or offline using various digital signal processing techniques. In order to predict the quality, ML models are utilized. This prediction of the weld quality forms a decision. This decision is often compared with a reference value, to determine the deviation or error, and accordingly a suitable control system is designed for sending feedback to the process. The utility of each of these signals in welding has been discussed below.

8.3.2.1 Current/voltage and Power Signals

In arc welding, fusion occurs because of the heat derived from the conversion of electrical power. The arc is established between two conductors and sensing this arc helps in deriving useful information about the welding technique. The sensing of arc refers to capturing of the signature of the electrical parameters, such as current or voltage [41]. In case of robotic arc welding, the electrical parameters have been utilized for seam tracking [33]. In GMAW, the current signal has been found to be one of the most low-cost solutions for tracking the weld seam. This is because of the strong negative correlation existing between the arc current and arc length. Further, the lateral location of the weld seam has also been determined with the help of current signature [33]. The current signature has been found to have a peak whenever the torch is at the edge and valley while the torch passes over the seam. The voltage signature has also been utilized for seam tracking [33]. Studies have utilized these electrical parameters for predicting the weld quality. For instance, the joint strength of the weld has been predicted by using features extracted from the current and voltage signals [42, 43]. These studies also report the utility of studying the signal in the time domain [42], and also in the time–frequency domain [43]. The *root mean square* (RMS) value of current and voltage signals in the time domain, along with the welding parameters, was fed to a ML model for prediction of the weld strength. In the time–frequency domain, the RMS values of the coefficients of different wavelet packets extracted from the current signal were considered for predicting the weld strength. Similarly, the geometrical parameters of the weld, such as height, width of the bead and reinforcement, have been predicted using features extracted from the current and voltage signals [44]. The task of classifying welds into defective and defect-free has also been carried out by using these electrical signatures [45]. Various time-domain features have been noted from the signals for classification. Other findings by using these electrical parameters include quality of arc at the start of weld, variation in the distance between the electrode tip and job, problems in feeding of electrode, problems in joint fit-up, mode of metal transfer [46]. A recent study demonstrated the usage of these electrical parameters for process monitoring of arc welding by introducing certain anomalies such as cutting notches in workpiece for degrading the surface preparation, insufficient flow of shielding gas by reducing the flow rate and contaminating the job surface by applying grease [47]. Several time-domain features were found to be correlated with these disturbances. Further on the utility of the electrical parameters in monitoring of the arc welding technique and usage of various signal processing techniques and ML models for prediction of the weld quality, readers can refer to Ref. [41, 48].

The electrical parameters, namely electrode current and tip voltage, have also been utilized for online monitoring of the RSW technique [49, 50]. These parameters have been utilized to obtain information about the dynamic resistance and input impedance in the welding process [51]. While the dynamic resistance refers to the quantity with respect to change in the current or voltage during the welding, the input impedance is a measure of the opposition to the current in the circuit. The dynamic resistance has been found to be sensitive to the phases of the welding, and thus, forms a crucial quantity for monitoring the process.

Similarly, the electrical parameters have also been proven to be useful for monitoring FSW technique [52]. FSW is usually performed in dedicated machines which enable the variation of parameters through the use of motors [53]. As the welding technique involves distinct stages starting from tool rotation to plunging and movement of machine bed to retraction of the tool, the power consumption by the machine varies with these events. Any sort of anomaly occurring during the welding, hits the power signature [54]. The power signal has been utilized to identify the defects in a weldment by analysing it through the time–frequency signal processing technique [54, 55]. The current and voltage signals of the feed and spindle motors have also been utilized to predict the joint strength of the weldment by using a ML model, where the features were extracted from time-domain and time–frequency domain [56, 57]. Other than these, both power and force signals were utilized in FSW for monitoring the tool quality [1]. The research has been carried out by considering tools of varying health conditions, and the signals were studied in the time–frequency domain. Later, a ML model was utilized for classifying those tool conditions in real time.

An advantage of the electrical signatures is their inherent availability in the system, and non-interference with the manufacturing process; i.e. they do not create any disturbance. During sampling at large frequencies, very minute fluctuations occurring in the process can also be captured.

8.3.2.2 Arc Sound

The sound generated during arc welding has been found and reported to be containing useful information relating to the molten metal, arc column, and the metal transfer modes [58–61]. Conventionally, a welder recognizes the arc sound and typically gets used to with the different sounds which occur with varying modes of metal transfer. The primary source of sound is the arc itself, which is followed by the metal that gets transferred from the electrodes to workpiece or at times sticking of the electrode to the workpiece, and the spatter occurring during the welding. This marks importance of the sound signature. Studies have reported analysing the arc sound signal in time domain, where "*kurtosis*" has been found to be the feature having strong correlation with the deposition efficiency [62, 63]. In the frequency domain analysis, using *fast Fourier transform* (FFT), the arc sound was able to be differentiated between the continuous and pulsed-mode of metal inert gas welding processes [62]. This signal has also been utilized for monitoring of the defects in the welding [64]. Further study on the utilization of the arc sound signal for monitoring can be found from Ref. [58]. The sound signature has also been utilized in monitoring of the RSW technique for identifying the lifespan of the electrode and prediction of the joint strength [65].

A major drawback of this sensor is its utility in a machine shop floor where machines run simultaneously. The sound which will be arising from other sources will also get reflected over the sound signature of the welding process, and then, pre-processing of the signal would become difficult.

8.3.2.3 Acoustic Emission Signal

Following the arc sound, acoustic emission (AE) signal has also been successful in determining the weld quality. AE refers to the elastic transient waves which are generated within a material undergoing plastic deformation or fracture. The signal is known for its ability to detect the flaws and discontinuities within a material [66]. The intensity of the AE signal forms an indicator of these flaws. The signal carries a practical approach for implementation, as it does not get affected by the range of audible frequencies. However, the AE sensors are required to be mounted over the job to capture the emissions, which pose a limitation in the case of mass production. Further, as compared to the sonic signal, a number of AE sensors may be required to be mounted over the job for capturing the AE signature, which is only a microphone in case of sonic signal [65].

An earlier research reports the importance of AE signal for monitoring GTAW, submerged arc welding and RSW techniques [67]. The AE signal successfully captures the different phases of the spot welding technique, and thus, has been utilized to extract information about the nugget quality [68]. Features such as "*AE count*" and "*positive peak*" have been reported to have good correlation with the joint strength. AE signals have been successfully utilized in monitoring the FSW technique. This is because FSW involves plastic deformation of the workpieces to be joined caused by a rotating tool which plunges inside. Researchers have captured the AE signal for identification and localization of the void defect [69]. The analysis has been performed in the time–frequency domain, in which the "*energies*" of the wavelet coefficients were extracted to classify the welds. Further, the AE signal was utilized to analyse the effect of changing tool pin profiles and process parameters in FSW [70–72]. The signal processing techniques utilized in these studies included the analysis of the signal in time domain, and techniques such as *Fast Fourier Transform* (FFT), *Short-Time Fourier Transform* (STFT), and *Discrete Wavelet Transform* (DWT). The amplitude of the AE signal was found to have a good correlation with the varying parametric conditions [70].

The AE signal represents high-frequency phenomenon, and thus, has high signal-to-noise ratio. Because of the high-frequency nature, it carries a lot of information about the transient events occurring in manufacturing. However, the limitation is with the fixation of these sensors as they are of contact-type, and handling of the huge data.

8.3.2.4 Temperature Signal

The transfer of metal from electrode to workpiece in arc welding technique occurs in different modes. These modes have different characteristics, and the rate of metal transfer depends upon the value of electrical parameters as well. Thus, the change in the temperature during these modes is obvious. Instruments like the thermographic camera have been utilized for acquiring the temperature signal during arc welding technique [73]. In order to derive useful information about the welding process, certain disturbances were introduced, and the corresponding temperature signals were investigated. The penetration depth has been determined by studying the mean value of the temperature signal [74]. Likewise, the misalignment and irregularities in the surface have been detected by studying the variations in the signal [74]. A few welding defects have also been monitored by analysing this signal. The penetration depth has been controlled by analysing the radii of the surface isotherms [75], and size, area enclosed by the surface isotherm [76]. In spot welding technique, temperature signal has been acquired by using an infrared sensor [77]. A good correlation in between the temperature and current signatures has been found in this welding technique. Features from multiple signals, namely temperature, current, voltage, ultrasonic, force and dynamic resistance, were extracted and fed to a ML model for predicting the weld quality.

Infrared sensing is useful in addressing certain problems as it is non-invasive in nature, and the signal can be acquired with respect to point, line or area. However, the sensor requires a line of sight, which if interrupted, creates a disturbance in the measurements. The interruption could be caused because of introduction of other objects. The limited distance from the targeted area for fixing this sensor is another limitation.

The temperature signature has been useful for monitoring of FSW technique as well. This is because of the frictional interaction occurring between the tool and the workpieces. The stir zone has been found having the maximum temperature in the welded sample, followed by thermo-mechanically affected zone (TMAZ) and HAZ [78]. Thus, the mechanical and metallurgical properties of the welded sample are dependent upon the temperature [78]. The classification of the welded samples into defective and defect-free has been carried out by studying the temperature signature in time and time–frequency domains [79]. Significant variation in the signature was found which helped in classifying the defective ones from defect-free welded samples. FSW technique has also been controlled in real time by using the temperature signature [80]. A suitable controller was designed by collecting the temperature variations across a range of tool rotational speeds. This study has been further extended by considering force signature along with the temperature and tool rotational speed for better stability [81].

8.3.2.5 Force and Torque Signals

The force signal has been found to be a crucial parameter for monitoring of the spot welding and FSW techniques [50–52]. The force in spot welding technique is exerted by the electrodes, and in FSW, it is because of the tool that plunges into the workpieces. In spot welding technique, the force signal has been studied by using descriptive statistics which provided useful information about the expulsion of the material [50]. The force signal in FSW is responsible for the forging action, and, as and when this quantity becomes insufficient, welding defects may occur. Since FSW involves a rotating tool, the torque also carries information about the quality of the weld. These signals were utilized to detect the occurrence of welding defects. For instance, the gap between the mating edges of the two materials in FSW technique gives rise to "*gap defect*". The tool constantly remains in contact with the workpieces; thus, the force signature has been a suitable indicator. The study found a drop in the signal amplitude whenever the tool crossed the region with a gap. Further, the signal was also studied in the frequency domain by using FFT, and a significant difference was found in between the welds, with and without gaps [82]. The gap defect has also been detected by analysing the "*power spectral density*" (PSD) of the force signal computed by using *Discrete Fourier Transform* (DFT) [83]. A few studies have shown the capability of force signal for weld classification. This includes estimating the *fractal dimension* of the force signal by applying the *Higuchi's* algorithm [84]. The results obtained showed a higher value of fractal dimension for the defective weld, as compared to the defect-free weld. Secondly, analysis of the force signal for classifying the welds has been carried out by using "*Wavelet Packet Decomposition*" (WPD) and "*Hilbert Huang Transform*" (HHT) [85]. Features such as the "*instantaneous frequency*" and "*phase angle*" were extracted. While the *instantaneous frequency* was found to be higher in case of the defective weld with a negative-scale of *phase angle*, it was comparatively low for the defect-free weld with a positive-scale. Similarly, the torque signal has also been found to be effective in detection of the welding defects. Studies have used WPD to analyse the torque signal for weld classification by extracting features such as *dispersion, excess* and *asymmetry* [86]. Further, both force and torque signals have been studied by using

DWT and "*Continuous Wavelet Transform*" (CWT) for classification of welds [87, 88]. Feature extracted from the wavelet coefficients of the signals were the "*mean of square of errors*" and "*variance*". The frequency spectrum of the force signal has also been studied for classification of the welds [89]. A variation in the amplitude spectrums of the defective and defect-free welds was found. Further to assist in automating the FSW technique, the quality of the tool has been monitored in real time by using force and power signals [1]. The study considered tools with certain induced defects. The wavelet coefficients extracted from the said two signals were utilized to train a ML model in order to classify the tool conditions in real time.

In addition to monitoring, these signals have also been utilized to control the FSW technique. These studies have designed proportional integral derivative (PID) controllers [90, 91]. After acquiring the actual force signal, the process was controlled by calculating the error signal from the reference signal. The obtained result was used to vary a suitable input variable. A recent study has proposed the usage of multiple sensors, i.e. force torque and power, for monitoring and controlling the welding process [54]. The signals collected have been analysed in a cloud server which included: (a) predicting the ultimate tensile strength (UTS) of the weld joint in real time by using a ML model, (b) comparing the obtained UTS with a reference UTS value and (c) predicting new input parameters upon finding deviation between the predicted and fixed UTS values. The study has also shown the usage of multiple sensors resulting in higher accuracy over a single sensor.

8.3.3 Context Setting—A Summary

The previous sections have stated the use of direct and indirect methods for monitoring and control of the welding techniques. The direct monitoring methods produce accurate results, but work offline. Thus, they do not help in improving the process quality in real time. The indirect means of monitoring are more practical and can help in fulfilling the real-time needs. It can also be noted that a single sensor cannot address all the issues in a welding technique. However, among all the sensors that have been discussed, current/power is one such physical quantity which is common to all welding techniques, and is available inherently. For instance, this may refer to the measurement of arc current in arc welding, the power (intensity) of laser beam in laser welding, electrode current in spot welding or the current consumption by the motors in case of FSW. Thus, the variation in the electrical parameters may be considered as a reliable source for monitoring of the welding techniques.

As automation is a core part of Industry 4.0, the use of sensors in manufacturing becomes essential. The following paragraph discusses the principles of Industry 4.0, and their application in welding. Later, the digital tools required for implementation of these principles have been elaborated.

8.3.4 Role of Industry 4.0 in Welding

Industry 4.0 can be attributed with four major principles, as represented in Fig. 8.8. Each of these principles has been mentioned formally at first, followed by a discussion on their implementation in welding.

8.3.4.1 Interoperability

This is also termed as "Interconnection", and it refers to the mutual interaction among the CPS, human and machine over a network. Tele-monitoring is going to be the next generation technology as "digital transformation" permeates into manufacturing. In welding, this principle aims to have a connectivity between the welding power source and a central computer in which the variation of all the sensors' data can be captured over a network in real time. This has been termed as "remote monitoring" in general, and "tele-welding" in case of welding. This is useful because it eliminates the physical presence of human and local computers close to the machine area. The sensors' data will be stored remotely in the central computer over a network which can be accessed by multiple users simultaneously. This data can be utilized for online tracking of the welding process. Thus, several welding machines on a machine shop floor can be connected to one such central database which would help in assessing the entire shop floor's data in one platform. Hence, this principle bridges the machines, sensors, and humans in one tie over a network.

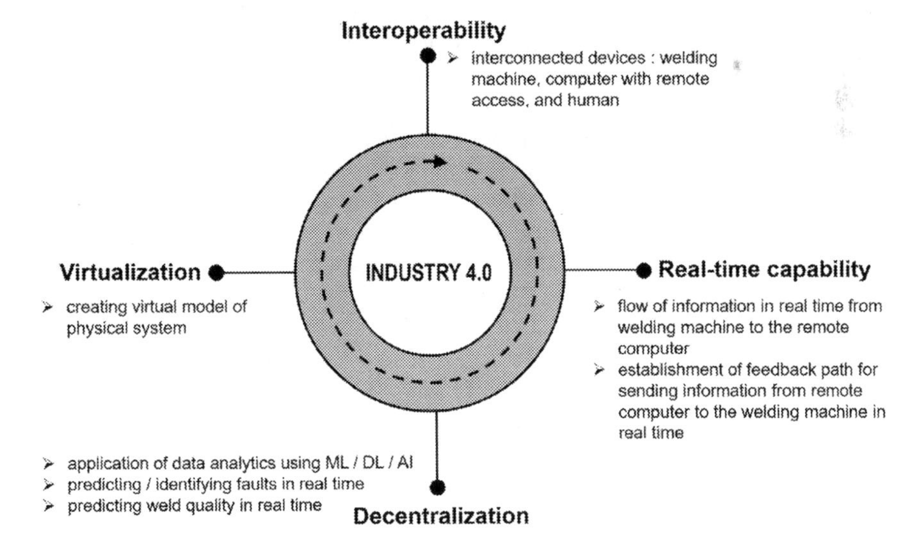

Fig. 8.8 Design and operating principles of Industry 4.0 and their utility in welding

8.3.4.2 Real-Time Capability

Once the communication is established between the machine and the computer/computing devices, the next thing to ensure is the real-time flow of the information. Here, information refers to the data which is acquired from the sensors engaged in the manufacturing process, both for process and machine health monitoring. "Delay" or lag in the data transmission is a concern in monitoring as it will assess the timeliness of the real-time system. This is followed by the real-time analysis of the data collected for finding useful inferences about the manufacturing process. This could be directly realized for the welding techniques. A recent article mentions the Industry 4.0 attributes in GTAW technique and refers to the utilization of a software termed, "Open Platform Communications Unified Architecture" [92]. This software is helpful for machine-to-machine communication and can be configured with a cloud server for data transmission. The software recognizes a machine from its *Internet protocol* (IP) address.

8.3.4.3 Decentralization

This principle refers to the decision-making in real time for manufacturing processes [93]. It aims at making the machines operate in an uninterrupted manner, also without any human intervention. In order to fulfil this, the machine needs to be embedded with several sensors. The reason of integrating several sensors has already been discussed. Once "interoperability" and "real-time capability" have been ensured, the exercise of extracting meaningful information comes into picture. This would require various open-source software for data analytics, followed by ML, DL and AI. The use of open-source platforms will provide: (a) flexibility in solving several problems, (b) agility for solving the problems in multiple ways, (c) speed by listing a range of services to choose from, (d) cost-effective solution and (e) easy use across an organization since it is available openly. This principle will help in developing an automated environment where the decisions are being taken by the system without any human involvement in real time. In reference to welding, the ML/DL can be utilized for predicting the weld quality, identifying the welding defects and faults in the power source/welding machine in real time. This will be followed by the corrective actions. For implementation of ML/DL, a "knowledge base" will be required. This knowledge base will consist of the data belonging to the production machine, acquired information from sensors, and environmental conditions prevailing during welding. This knowledge base needs to be enriched with each passing day so that the learning of the machine gets strengthened.

8.3.4.4 Virtualization

This principle aims at creation of virtual model for running real-time simulation of the physical processes. This virtual model exists and operates through sensors' information collected from the physical process in real time [94, 95]. One utility of this principle is to understand various phenomena that could occur during a manufacturing process by dealing with the virtual environment. For welding, a virtual model can help in understanding the response of the process to a wide range of parameters. For instance, in case of the spot welding technique, the life of the electrode can be simulated by considering influence of various parameters. Similarly, the life of the tool in FSW technique can be assessed for a wide range of materials. The other utility of this principle is the creation of "virtual machines" for different physical systems of a manufacturing shop floor in a single platform. This will help in integrating multiple operations on a single server and reduce the expense on hardware.

8.3.5 Digital Tools of Industry 4.0

In order to implement these principles, certain digital tools are required to be integrated for advanced automation in industry. These tools have been shown in Fig. 8.9 along with a short description and their underlying principles. The integration of these tools with the existing manufacturing system is aimed at: (a) real-time visualization of the production process, (b) automatic decision-making, (c) understanding the physics and determining the faults through simulation, and (d) improvement in the efficiency of the production [96]. The following section discusses these digital tools.

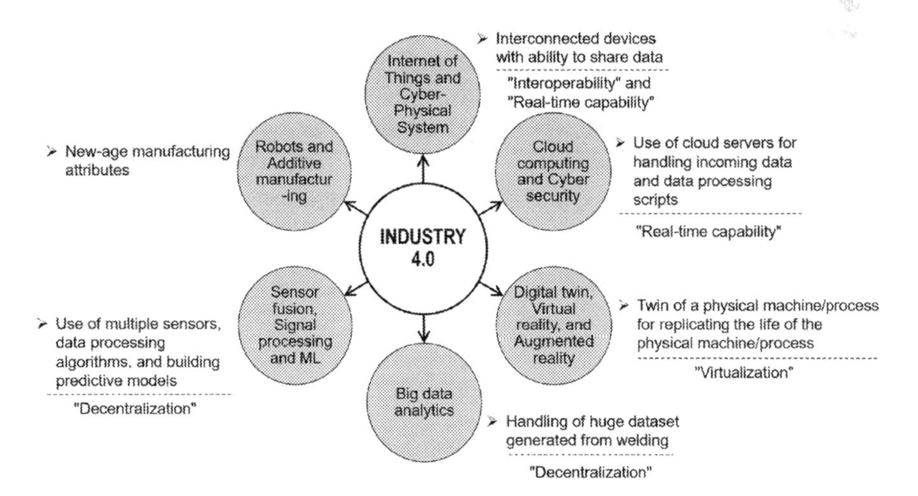

Fig. 8.9 Digital tools of Industry 4.0

8.3.5.1 Internet of Things and Cyber-Physical System

The IoT refers to various interconnected devices with an ability to transfer the data remotely by using some suitable communication protocol. It has been defined as the world of interconnected things with an ability of sensing, actuating and communicating among themselves and with the environment [97]. This resembles the machine (with sensors and data acquisition system) connected to a remote platform over the Internet, so that, they are able to send some information. The information refers to the data acquired from the sensors. This makes the machine “smart". The CPS refers to the system enabling the control of physical objects through computer and software. The term CPS has been referred as a unifying system, where the cyber parts are integrated with the physical parts, during both design and operation [97]. In manufacturing, the term “cyber” refers to the computer and software-based algorithms and “physical” refers to the machine. The interaction between these two systems implies enabling the control of the physical system via the cyber system. The feedback from the cyber world is sent to the embedded system of the physical machine in real time. This can be extended to welding.

8.3.5.2 Cloud Computing and Cyber Security

Cloud refers to a virtual server comprising of high-speed processing power, large memory, and other resources enabling computation [98]. The machines in a manufacturing shop floor can be connected to the cloud, and channels for transmitting the sensors’ data can be established [99]. This would help in realizing the remote monitoring or tele-monitoring [54]. Thus, the cloud server would act as the centralized platform. More often, “edge computing” or “fog computing” is also preferred which indicates to a server located closer to the machine [100]. The fog server is referred as a distributed and decentralized infrastructure, whereas the cloud server is a centralized system. The fog server does not resemble a separate entity, and it does not replace the cloud server, it complements the cloud instead. The fog server resembles multiple nodes which are located closer to the physical machine. They are used for short-term edge analysis due to their instant responsiveness. This enables the data to be received at the monitoring platform faster and also reduces the traffic in the cloud. Thus, the fog server acts as a mediator between the physical device and the cloud server. For an industry, there may be a need of numerous robots employed in series or parallel. These robots would be configured with several sensors. A cloud server connectivity for collecting the data from these sensors would make it easier for computations in a single platform. In addition, an edge device can be connected to each of the robot for initial screening, and subsequent transferring only the reliable data.

Since the Internet will be the most basic requirement to ensure the flow of data between devices, encrypted communication protocols will be required to secure and maintain an end-to-end communication. This is referred as cyber security.

The application of cloud computing in welding has been demonstrated in few articles. One of them utilizes a vision camera in arc welding technique for capturing the weld images in real time and transfers them to a cloud server for classification of those welds [101]. The acquired images were processed using the "OpenCV" platform which is another example of an open-source platform. The cloud server had the ML model trained for classifying the weld either into defective or defect-free. The potential of this work lies in the use of low-cost resources including computation in cloud for real-time monitoring of the welded samples. Another article shows the usage of cloud computing in FSW technique [54]. In this work, the cloud receives data from multiple sensors in real time and feeds them to ML models for online prediction and control of the weld quality. The usage of cloud eliminated the use of hardware for processing the signals.

8.3.5.3 Digital Twin, Virtual Reality, and Augmented Reality

These tools mark the importance of virtualization principle which aims at creating virtual model of physical system [102]. The digital twin is a replica of a physical counterpart existing in a digital world. The digital twin gathers information from the physical part in real time and applies analytics to make useful predictions. This tool will be useful for planning the welding process. For instance, the weldability of materials along with suitable parameters can be determined through welding simulations prior to actual welding. This might be useful for determining suitable conditions for joining dissimilar materials with significant differences in the physical properties. This may prove beneficial for automobile industry where welding is one of the major jobs. This sector is constantly exploring newer and advanced materials for manufacturing of the automobile. As such, it becomes crucial to determine weldability of these materials. The virtual reality (VR) allows users to experience a 360° enclosed view of the working environment, and the augmented reality (AR) enhances reality by adding digital information to it. Both VR and AR enable a cooperation between a user and a virtual world [103]. With these technologies, the design can be tested in a virtual environment, and the errors can be identified and rectified at the design level. These tools seem to be promising for imparting training to novice welders. This will help them understand the physics of the process, such as proper positioning and movement of the welding torch, importance of joint preparation and importance of shielding the molten pool in an interactive way. This aims to improve the quality of welding and productivity.

A digital twin has been proposed for FSW machine where the twin in real time gathers useful information from the physical FSW machine [1]. The twin evaluates the health of the machine by determining the error between the instructed and actual velocities of the spindle and feed motors in real time, classifies the health of the tool by checking the force and power variation and tracks the temperature of the hydraulic oil, oil flow rate, and oil turbidity. Similarly, a digital twin has also been proposed for spot welding which includes virtual simulations for weld designing, optimization, planning, production and evaluation [104].

8.3.5.4 Big Data Analytics

This tool refers to the large volume of dataset comprising of a combination of structured and unstructured data. Manufacturing industry generates huge amount of data each day which could be explored by implementation of big data. The handling and storage of this huge dataset would require cloud-based system. The constituents of big data are: volume, velocity, variety and veracity. The "volume" refers to the amount of data that could be gathered from a welding process. For instance, in case of the GMAW technique, the dataset may contain: (a) "basic data", which will include the physical data of the workpieces, environmental conditions, (b) "machine data", which will include the identification tag of the welding power source utilized, torch and consumables and (c) "sensors' data", which will include arc voltage, arc current, wire feed speed, flow rate of the shielding gas, pressure of the shielding gas, arc acoustics, acoustic emission, temperature and images of the welded sample. "Velocity" refers to the speed of data collection, and this will usually vary from one sensor to the other. Of course, this dataset will be heterogeneous, as it consists of data from different sources, which is referred as "variety". Finally, the term "veracity" aims at deriving the useful information from the dataset, and discarding the irrelevant ones.

A recent article mentions the importance of big data in optimization of welding process, documentation, production monitoring and quality management [105]. This could be realized for an industry like automobile manufacturing unit, where welding is one of the crucial operations in the production process, and it is highly imperative to identify the rate of rejection, possible reasons for the rejection, building predictive models using variety of data, and increasing the accuracy by using volume and velocity of the data. For similar welding machines performing similar jobs in a shop floor, this dataset can help in detecting, whether or not, a machine is degrading by considering into the quality of the welds. Similarly, the warehouse for welding consumables can be optimized by tracking the rate of consumption of the shielding gas and filler rods. A limit value can be set for the tank containing the shielding gas, which can alarm in real time, as the limit value approaches. This can help in ensuring the quality of the weld.

8.3.5.5 Sensor Fusion, Signal Processing and Machine Learning

Sensor fusion is a process of combining signals from various sources. Here, the information is extracted from multiple sources in order to integrate them into a single indicator. The information extracted is analysed by the help of sensor fusion, also known as data fusion algorithms, and have been found to be beneficial for manufacturing process [5]. The automation of welding cannot be fulfilled by using a single sensor, instead multiple sensors would be required. This is because a number of characteristics of a welding process are required to be monitored, which may not be fulfilled by using a single sensor. Further, benefits in terms of higher accuracy have been reported with the use of sensor fusion. In order to fuse the data, one of the widely utilized techniques is the "intelligent fusion" which uses neural networks. Earlier research shows the application of multi-sensor fusion in welding processes [54, 63, 77, 106]. A challenge in the sensor fusion task is to deal with the heterogeneous data. The detection of different types of phenomena and producing measurements for several attributes makes it cumbersome to fuse the information gathered. In addition, the sampling rate is not same for different sensors which make it difficult for real-time application. Even if the data are sampled at the same frequency, a delay is introduced in the process, owing to the volume of data being generated from these sensors.

There are several signal processing techniques which have been discussed in the preceding sections. The selection of a proper technique for extraction of meaningful information is vital [52]. An article shows the importance of multiple sensors such as optical, acoustic signal and arc voltage, for monitoring of GTAW technique [107]. A feature-level data fusion was applied, and a SVM model was built for weld classification. Likewise, in case of the spot welding technique, features from multiple signals, namely temperature, current, voltage, ultrasonic, force and dynamic resistance, were extracted and fed to a ML model for predicting the weld quality [77]. Another study demonstrated the importance of using multiple sensors for process monitoring over a single sensor, in FSW technique [54]. The sensors utilized in this study were force, torque, and power, which were fused in a ML model for predicting the weld quality. The accuracy of prediction was high in case of multiple sensors as compared to the information gathered from a single sensor.

AI and ML have been contributing significantly towards the automation of production processes. The most promising factors for the implementation of AI are the transformation of human knowledge and skill into computer software. AI will provide engineering assistance for the automation of low-level manufacturing issues. Over the years, the integration of machine tools, monitoring algorithms and machines with computer network have led to enhanced product quality. The control mechanisms that govern the manufacturing processes in every step are: sensory systems, data mining, ML models, AI and expert systems. One of the essential requirements in manufacturing is the decision on the most suitable or the optimal combination of evaluation criterion. Sufficient information does not exist to determine the best possible combination of design parameters. In case of non-automated system, the experienced personnel in the organization is the most sought after during decision-making process. AI involves expert systems, which once fed with adequate information are capable of taking care of decisions at the shop floor. The intelligent and automated machines working on AI can be utilized by integrating these machines with the intelligent controlling systems that can monitor and diagnose the ongoing operations. This will lead to better product quality along with competitive advantage. The ML models have been utilized for classification of welds, pattern recognition, prediction of weld quality and health monitoring. A review on the use of various ML models in monitoring of the arc welding process can be found in Ref. [108]. This review suggests the artificial neural network (ANN) to be the best suited for dealing with the noisy and nonlinear data. A recent article reviewed the application of ML models in monitoring of the laser welding technique [109]. This review covered models such as feedforward NN, backpropagation NN, and SVM. Apart from these three, there are many other ML models available for data engineering. However, the task is to identify the best suitable model applicable to a specific problem in real time.

8.3.5.6 Robots and Additive Manufacturing

Robots are critical to welding automation. In the present context, robots are required not only for large business, but also for "*micro, small, and medium enterprises*" (MSMEs), considering the advent of Industry 4.0. Among a handful of advantages, the following are crucial: (a) robot-assisted welding increases productivity, (b) the application of robot will bring in consistency and repeatability, (c) superior weld quality by enhancing the accuracy in case of robot-assisted welding, (d) minimizing the cost overhead, and (e) higher level of customization is possible with the use of robots because of flexibility [110]. Transforming the factory floor operations from analog to digital is a challenging process. The major challenges are: (a) handling huge volume of data with the existing plant capacity, (b) insufficient resources and experience to analyse this volume of data and (c) the inherent complexity. However, to build a smart factory, industry will seek assistance from robots, who will be the best-suited physical and cognitive assistant. Robots have the ability to perform uninterruptedly, collect and analyse data simultaneously and also help in attaining a superior weld quality. This makes them indispensable.

Additive manufacturing (AM) is the process of making components by deposition of materials in layers. It differs from the conventional way of component manufacturing which includes removal of material from a stock to eventually achieve the desired form. The AM technology allows users to manufacture components without using any tool, and thus, any complicated geometry can be fabricated. Owing to this, the AM process substantially reduces the lead time and increases the production efficiency by reducing the loss of material. Since the arc welding technique utilizes a filler material to deposit the molten material in the groove, research is being carried out in this direction and is being referred as "arc welding-based AM" [111].

8.3.6 A Concept Block Diagram for Implementing Industry 4.0 in Welding

Figure 8.10 depicts a possible block diagram for implementing Industry 4.0 in a welding process. The cloud/edge connectivity would drive tele-welding. While the edge platform can be utilized for small computations such as initial screening of the data, the cloud platform will contain the database of the welding operations being performed. The cloud platform will also have real-time data processing scripts, which will be continuously operating on the data for extracting useful information such as predicting the quality of the weld. The reference quality can be plugged into the script, which will be utilized for real-time evaluation, forming a closed-loop control

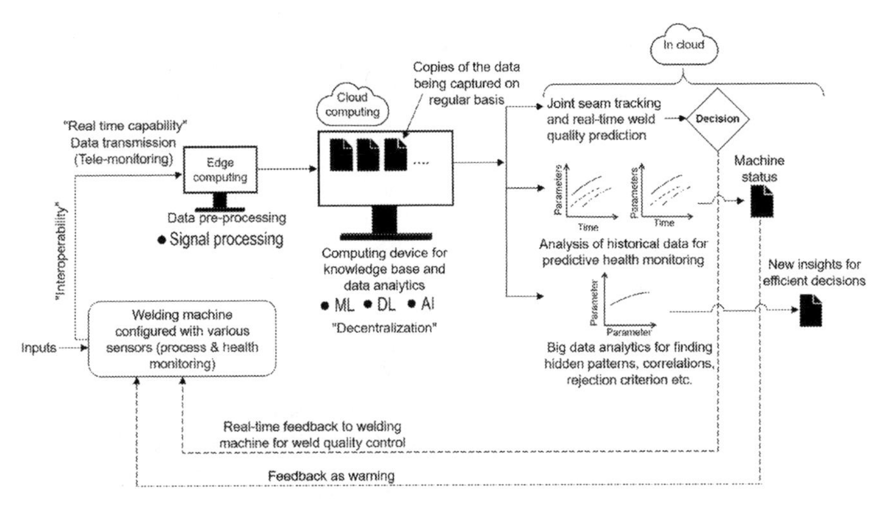

Fig. 8.10 Industry 4.0 application in welding

system. The error can be utilized to trigger a feedback to the welding machine for correction in the defect in real time. This will help in ensuring the weld quality at all times. In addition, sensors would also be engaged for real-time evaluation of the machine health, which can send alarms for any possible faults. Further, the enormous data available from the welding machines (data from process sensors and health monitoring sensors, and environmental data) can be analysed to find hidden patterns, correlations and job rejection reasons. With these, the concept aims at increasing the productivity and decreasing the machine downtime, through the use of automation and AI. As there would be uninterrupted collection of data from the equipment on the shop floor, the traceability would be faster. Further, the connected machines and factories would drive collaboration among producers and suppliers along the supply chain.

8.4 Case Study—Application of Industry 4.0 in Welding

In this section, a case study presents the application of few concepts of Industry 4.0 in FSW technique. Specifically, the study includes utilization of the data acquired in the welding process for online prediction of the weld quality, and control of the same. FSW has been already introduced in the beginning of this chapter. For further insights on FSW technique, readers may refer to Ref. [112–114]. From the view point of automation of FSW, the different sensors and signal processing techniques employed by researchers have already been mentioned in this chapter.

8.4.1 Selection of Manufacturing Attribute—Problem Formulation

As mentioned earlier, welding is the last process in the sequence of manufacturing. Thus, it is highly essential to ensure that the weld joint is error-free. This indicates that the weldment must have the required mechanical and metallurgical properties and is free from welding defects.

For FSW, the WPS can be considered to be consisting of three categories of parameters: (a) "material parameters", representing the material to be joined along with their physical and chemical properties, (b) "design parameters", mentioning about the dimensions of the tool to be selected for the specified materials along with the profile of the shoulder and pin and (c) "joining parameters", specifying the range of welding speed, rotational speed of tool, tilt angle and plunge depth. Procedure monitoring of FSW process resembles tracking of the joining parameters [52]. This can be achieved by using various sensors, as discussed in the preceding sections.

Figure 8.11 represents the schematic diagram of the FSW machine considered for this case study. It is a computerized numerically controlled machine with individual motors for rotational axis and linear axis. The machine is instrumented with a load cell which acquires the variation in force during the welding. The importance of force in FSW has already been highlighted in the previous discussions. A computer, as shown in the figure, is connected to the machine which allows user to provide the input parameters (joining parameters) and stores the data acquired from the load cell.

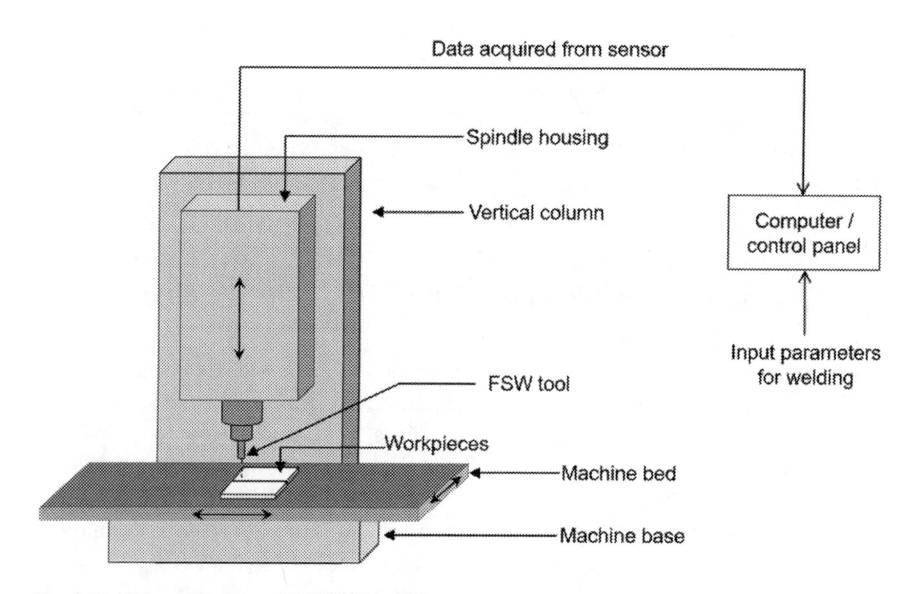

Fig. 8.11 Schematic diagram of FSW machine

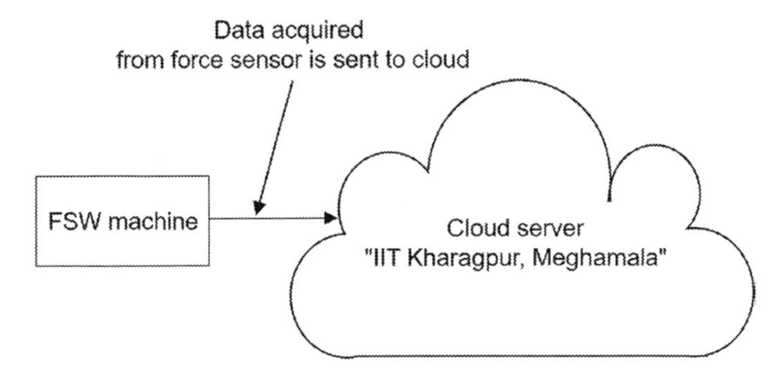

Fig. 8.12 Application of CPS in welding

The objective is to automate and control this FSW process. Among the pre-discussed joining parameters, welding speed and rotational speed of tool can be controlled in the present machine, as they have individual motors connected to their axes. The below-mentioned subsection elaborates the approaches taken for developing the online architecture for weld quality prediction and control. This architecture will be a step towards automation of this welding technique. This case study is indicative of how the principles of Industry 4.0 will help in monitoring and control of welding and can be extended to other manufacturing processes.

8.4.2 Application of Internet of Things, Cyber-Physical System and Tele-Welding

Figure 8.12 depicts a picture which displays the connection between the FSW machine and a cloud server. The remote server (cloud service) has been provided by "Indian Institute of Technology Kharagpur" and is named as "Meghamala". The remote connection was established by using "transmission control protocol", by referring to the IPs of the machine and the cloud server. The specific point for sending and retrieval of messages was ensured by providing a specific port number. Thus, the FSW machine can now communicate to another device over the Internet. The communication here refers to the transmission of the data acquired from the sensor in real time. However, at this point, this communication is only one-way; i.e. there is no reception of data from the cloud. The real-time transmission of data over a network enables tele-welding. "Meghamala" consists of the data processing scripts and related software for deriving insights from the data. Thus, the "Meghamala" becomes the "cyber" part and the FSW machine resembles the "physical" part. All together, they can be referred as CPS, provided the feedback or sharing of data from the cyber part is also enabled. This necessarily indicates control of the FSW machine based on the obtained weld quality by the cyber part in real time. The involved analytics and feedback have been discussed in the subsequent sections

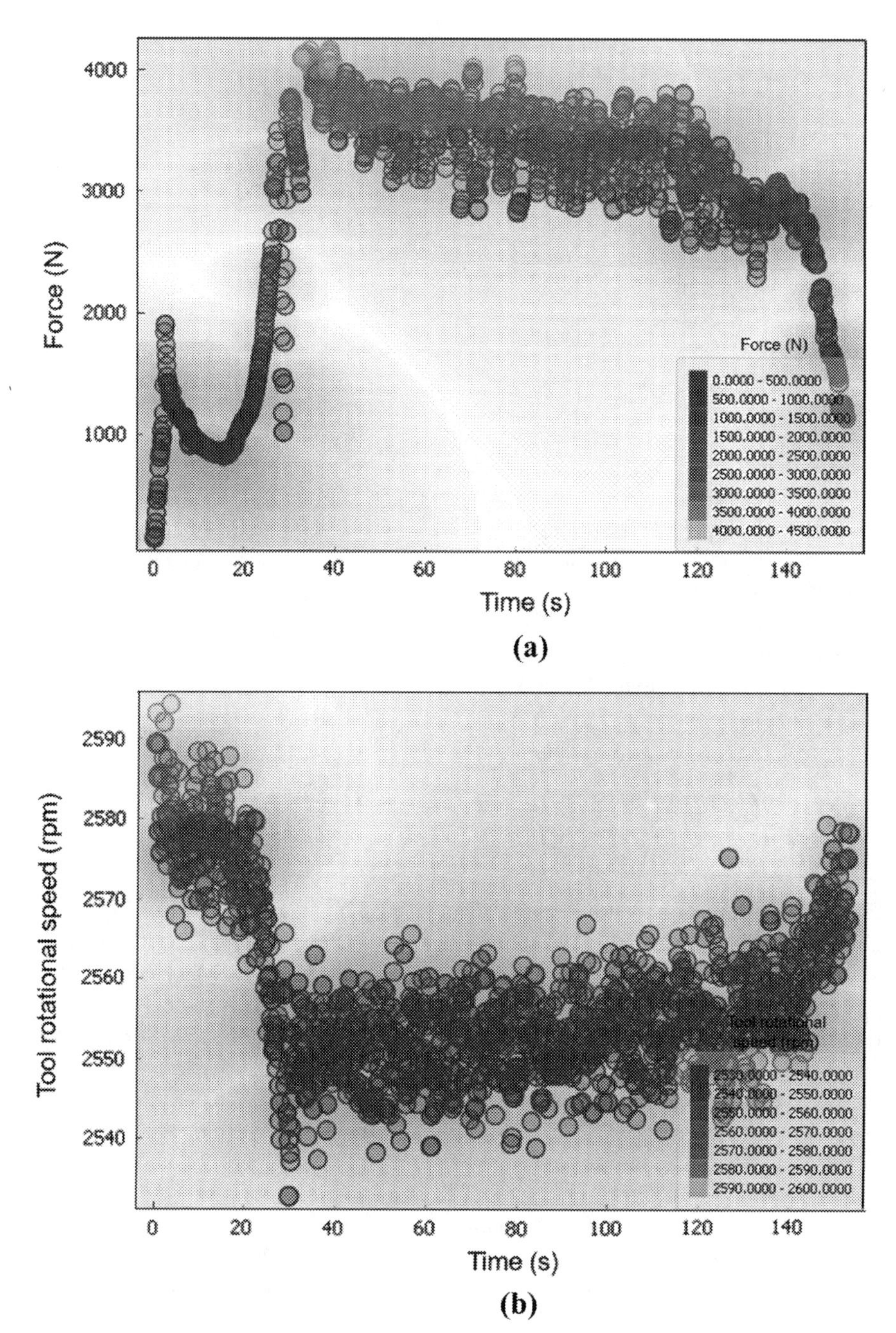

Fig. 8.13 Plots of: **a** variation of raw force signal with time, **b** variation of rotational speed signal corresponding to **a** and **c** variation of welding speed signal corresponding to **a**

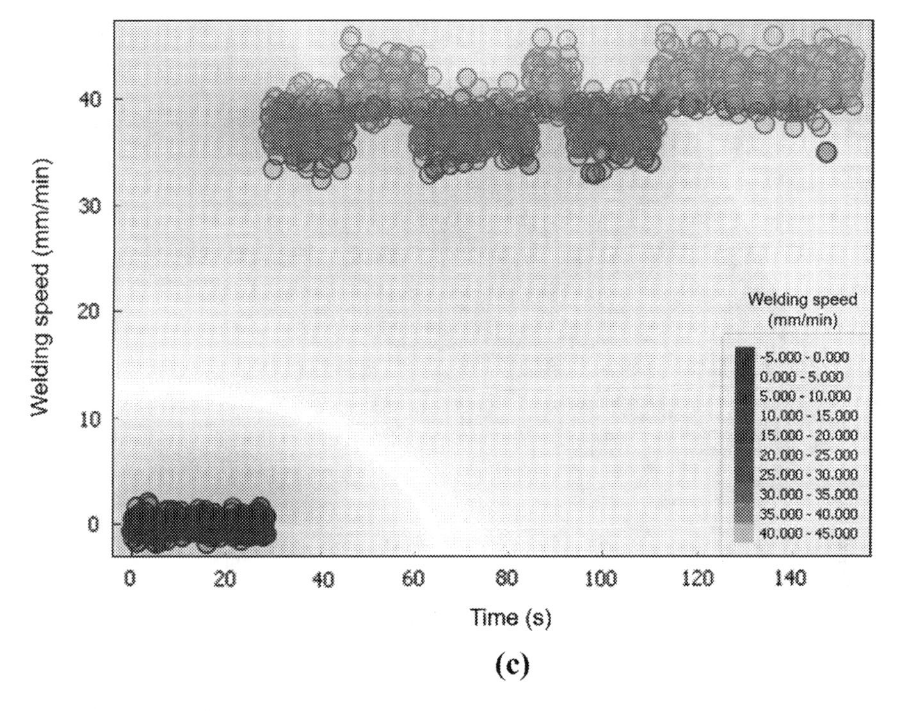

Fig. 8.13 (continued)

8.4.3 Data Collection, and Domain Knowledge

Figure 8.13a shows the raw data acquired from a force sensor, for a sample welded with a 2600 rpm (rotational speed of tool) and 40 mm/min (welding speed). The coloured plot depicts the stages involved in FSW operation which begins from plunging of the tool into the workpieces to the retraction of the same. The first peak denotes the plunging of the pin, followed by a second peak referring to the plunging of the shoulder. The rise in the peak is because of the cold material which experienced the action of the rotating tool. Dwelling of the tool reflects the dip in the signal amplitude owing to the thermal softening of the workpieces. The welding stage follows next, where the rotating tool is accompanied with the linear motion of the machine bed. Both these actions stir the deformed material alongside the tool, and finally, the tool retracts out. The corresponding variation in the signatures of tool rotational speed and welding speed has been shown in Fig. 8.13b and c, respectively. The initial rotational speed can be observed to be around 2600 rpm during the plunging stage, which later kept fluctuating in the range from 2540 to 2570 rpm throughout the welding. The welding speed refers to the linear motion of the machine bed, which can be seen fluctuating in the range from 32 to 45 mm/min, during the welding stage. The initial welding speed is zero as there is no movement of the machine bed during the plunging stage.

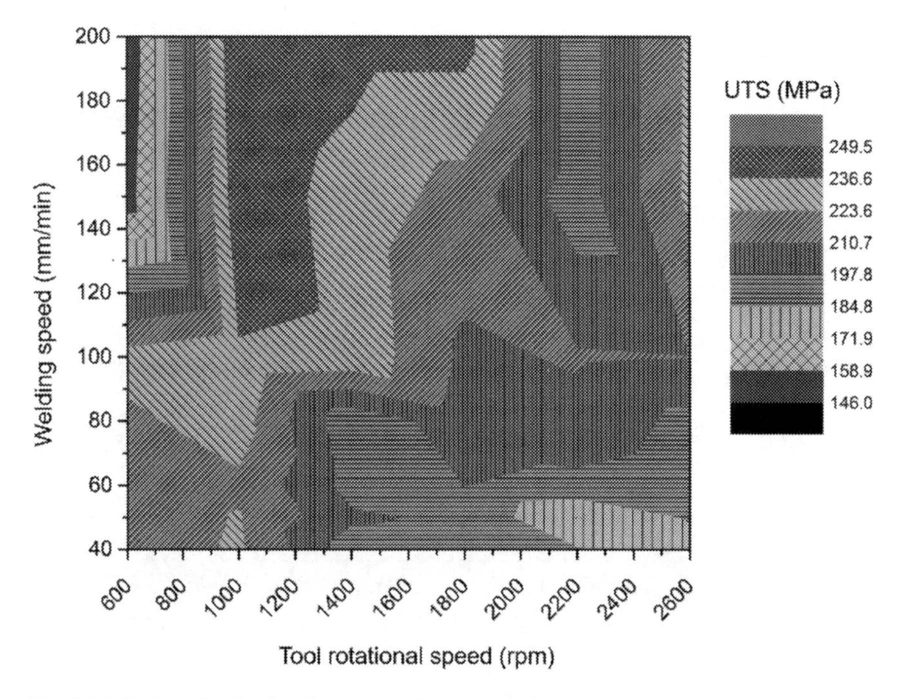

Fig. 8.14 Surface plot for the obtained tensile strength values

The discussions above provide an understanding of the behaviour of the process signals in FSW. In order to predict the weld quality, a predictive model was essential to build. This would lead to the fulfilment of the decentralization principle. It refers to the decisions taken by analysing certain data acquired from a manufacturing process. The decision here also refers to individual decisions at multiple devices. In this case study, two devices are involved: one is the physical system (FSW machine), and the other is the cyber component (cloud server). The tasks performed at the physical part were: (a) welding, (b) acquisition of the signal and (c) transmission of the acquired signal. The cyber part is held responsible for: (a) analysis of the data, (b) extraction of useful information and (c) feedback to the physical part.

For the predictive model, several experiments were needed to be performed. These experiments were performed in the machine, depicted in Fig. 8.11. Several combinations of joining parameters were selected as elaborated in the WPS for FSW. Workpieces of AA6061 (an alloy of aluminium) were butt welded. The reasons for selecting this alloy are its versatile applications in various sectors of manufacturing [115, 116]. This is because of its properties such as high strength, low density, high electrical and thermal conductivity, and resistance to corrosion. [117].

In order to assess the welded joints (i.e. PQR of FSW joints) obtained with the selected combinations, ultimate tensile strength of the welds was determined. The tensile strength is one of the vital mechanical properties, and it refers to the maximum load a welded joint can bear. The tensile strength was determined in a universal tensile

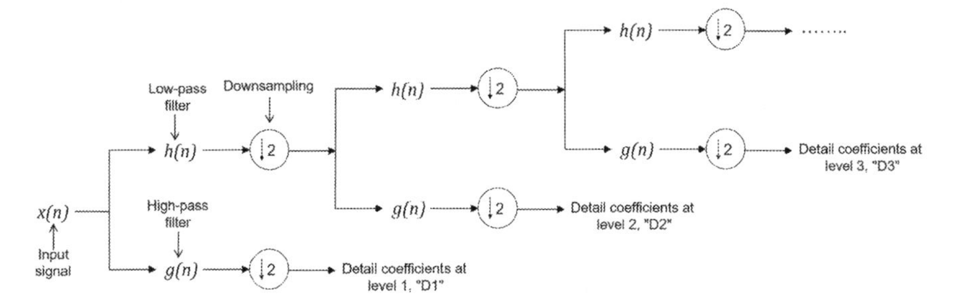

Fig. 8.15 Signal decomposition by using DWT

testing machine, a direct monitoring technique. Figure 8.14 shows a surface plot of the obtained tensile strength values with respect to the selected parametric combinations. While the base AA6061 had a tensile strength of 270 MPa, a wide range of strength values, i.e. from 146 MPa to 249.5 MPa, were recorded. The parametric combinations yielding the best and worst tensile strength values can be referred from the colour map. The tensile strength can be observed to be lower with high and low values of welding speed and rotational speed, respectively. With these values, sufficient amount of heat is not available to plastically deform the workpieces. This in turn leads to formation of poor welds because of improper flow of the material, and also welding defects such as voids are found in the weld. Thus, it is always essential to identify the optimum combination of parameters for welding. For this case, the weld fabricated with 1000 rpm and 200 mm/min have the highest tensile strength value, i.e. 249.5 MPa. This tensile strength map, obtained with the wider range of parametric combinations, was utilized to come up with an efficient predictive model having enough knowledge of the parametric combinations.

Data of force variation, rotational speed and welding speed, for all the parametric combinations were also recorded during the welding. An initial database was created consisting of these parameters, force and the obtained tensile strength values.

8.4.4 Signal Processing, Feature Extraction and Building of a Database

The acquired force signature was studied in the time–frequency domain by applying DWT. *Wavelets* can be defined as a wave which exists for a limited duration of time, have an irregular shape and a zero mean value. Usually, in order to transform a signal from one domain to another domain, the signal is multiplied with the basis vectors to determine the similarity index in the transform domain. As the sinusoids are the basis vectors in *Fourier transform*, mother wavelets are the basis vectors in case of *wavelet transform* [118].

Figure 8.15 shows the process of signal decomposition in DWT, in which the original signal is being divided into various frequency bands. These frequency bands

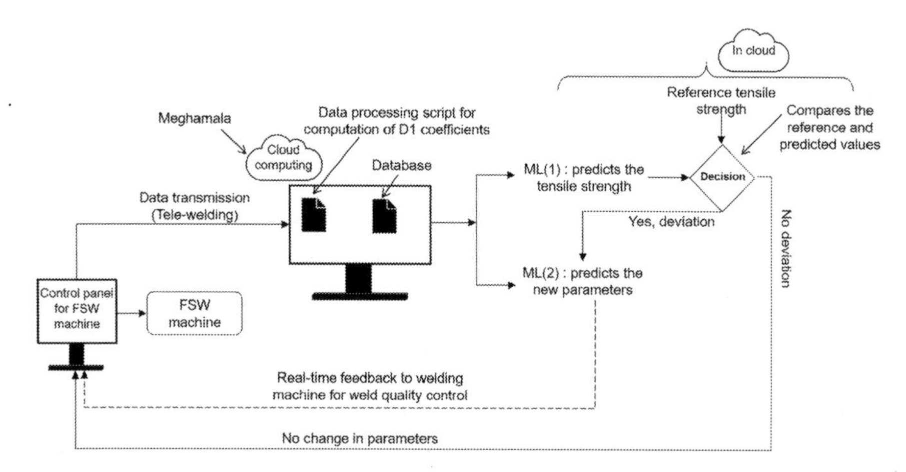

Fig. 8.16 Schematic diagram for real-time weld quality prediction and control

differ in resolution, which is the outcome of passing the original signal though a series of filters, referred as high-pass and low-pass. The result of high-pass filter is the "detail coefficients" which resembles the detail information. The result of low-pass filter is the "approximation coefficients" which resembles a coarse approximation. It refers to the convolution of the original signal with these filters. Subsequent analyses produce the coefficients at various levels, named as level 1, level 2, level 3, and so on. The low-frequency components which slowly change over the time are represented by approximation coefficients, and the ones which rapidly change over the time are represented by the detail coefficients.

DWT is advantageous because of its capability to yield information about the time as well as frequency. Thus, they are apt for analysing the transient signals. Thus, the acquired force signals were also studied by applying DWT. The detail coefficients obtained at the first level (D1) were considered only, so as to minimize the computation time. Thus, the database for ML (predictive model) consisted of the D1 coefficients of force signal as the features, along with the values of the tensile strength as the target variable. A ML model was then created for predicting the tensile strength.

8.4.5 Real-Time Weld Quality Prediction and Control

Figure 8.16 depicts the schematic diagram of the architecture developed for real-time weld quality prediction and control. In the preceding sections, the inputs to the FSW machine and tele-welding were elaborated. The data transmission to cloud occurred in batches. After receiving a batch of data in the cloud, it is processed for computation of the D1 coefficients which forms the input for ML(1). This model predicts the tensile strength of the weld in real time. This covers a part of the objective of this case study.

For the second part of the objective, i.e. online control of the weld quality, the data processing script is fed with a reference tensile strength value. This reference value is based upon the application for which the weld is being fabricated. A decision box compares the outcome of the ML(1) with this reference value. The reference value would come from the PQR of the FSW joint, the corresponding joining and other parameters can be found from the WPS. Thus, during the welding, if any defect is produced, then the reference tensile strength value cannot be retained. Hence, the decision box comparison is justified. When the predicted value is more than the reference value, this indicates that the weld joint is proper and the procedure is maintained. In this scenario, the parameters remain unchanged (indicated as "No, deviation" in Fig. 8.16). If the predicted value is lower than the reference value (indicated as "Yes, deviation" in Fig. 8.16), the corresponding batch of data is sent to another ML model, ML(2). The inputs to the ML(2) were also the D1 coefficients of force, but the outputs were rotational speed and welding speed. These predicted parameters are sent as feedback to the FSW machine.

Thus, the cloud server consists of the models, ML(1) and ML(2), data processing script, database, and this combination represents the cyber component. They analyse the data acquired from the physical FSW machine, and control the machine based on certain decisions. This accomplishes the definition of CPS. The trained models were plugged in the developed architecture for real-time control.

Figure 8.17 shows the picture of a sample welded to test the architecture. The initial joining parameters' combination was a lower value of rotational speed and a higher value of welding speed. As observed from Fig. 8.14, such a combination exhibits lower tensile strength. This is because of the insufficient heat resulting into defective welds. The same can be seen in the sample indicated as "defective region" (Fig. 8.17). A much higher value of tensile strength was provided, intentionally, to the data processing script as the reference. The architecture detected the induced situation and predicted new parameters for welding. These parameters were fed back to the machine, leading to change of the parameters in real time. A defect-free weld is then obtained, indicated as "defect-free region" in Fig. 8.17.

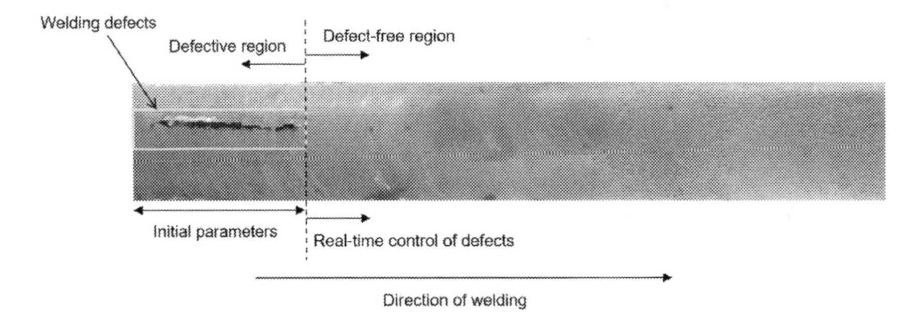

Fig. 8.17 Real-time weld quality prediction and control of defects

8.4.6 *Discussion*

The case study presented above is a step towards automation of the FSW process. The developed architecture predicts the weld quality, i.e. tensile strength of the weld in real time, and controls the quality, if it is found deviating from a reference standard. The developed architecture will help industry by eliminating the rejection of material. More quality measuring parameters such as grain size and hardness for assessing the quality of the weldment can be introduced in the architecture, depending upon the application of the welded joint. Different parameters may require different signals as a typical signal may capture a characteristic more proficiently than other. As such, different signal processing techniques may be required to derive the useful information from the signals. To implement Industry 4.0, the foundation lies in the utilization of sensor and proper signal processing techniques at the pinnacle for deriving the useful information.

8.5 Conclusion

This chapter attempted to review the evolution of all the four industrial eras, right from Industry 1.0 to Industry 4.0. This provided critical insights on the future potential of Industry 4.0 to attain manufacturing excellence. It is important to identify the optimal process parameters for every manufacturing process for the elimination of the production bottlenecks. Further, it is also crucial to constantly monitor the machines operating at the shop floor to gain a sense of control. This chapter provides a holistic view of how the digital tools of Industry 4.0 can aid in monitoring and control of a welding technique. The case study in this chapter can be extended by introducing other disturbances in the process, and finding ways to eliminate them in real time. Over time, the underlying concepts highlighted in this chapter can be generalized with subsequent research on various manufacturing processes.

Acknowledgements and funding This chapter is an outcome of the project funded by the Department of Heavy Industry under the Ministry of Heavy Industries and Public Enterprises, Government of India and TATA Consultancy Services (Grant No: 12/4/2014—HE&MT).

References

1. Roy RB, Mishra D, Pal SK et al (2020) Digital twin: current scenario and a case study on a manufacturing process. Int J Adv Manuf Technol. https://doi.org/10.1007/s00170-020-05306-w
2. Lee J, Davari H, Singh J, Pandhare V (2018) Industrial Artificial Intelligence for industry 4.0-based manufacturing systems. Manuf Lett 18:20–23. https://doi.org/10.1016/j.mfglet.2018.09.002

3. Lee J, Bagheri B, Kao HA (2015) A Cyber-Physical Systems architecture for Industry 4.0-based manufacturing systems. Manuf Lett 3:18–23. https://doi.org/10.1016/j.mfglet.2014.12.001
4. Google Trends. https://trends.google.com/trends/explore?date=today 5-y&q=industry 4.0. Accessed 12 Jun 2020
5. Kumile CM, Bright G (2012) SENSOR FUSION CONTROL SYSTEM FOR COMPUTER INTEGRATED MANUFACTURING. South African J Ind Eng 19:179–194. https://doi.org/10.7166/19-1-114
6. Industrial Revolution. In: Wikipedia, Free Encycl. https://en.wikipedia.org/wiki/Industrial_Revolution. Accessed 7 Jun 2020
7. Ghobakhloo M (2018) The future of manufacturing industry: a strategic roadmap toward Industry 4.0. J Manuf Technol Manag 29:910–936. https://doi.org/10.1108/JMTM-02-2018-0057
8. Yin Y, Stecke KE, Li D (2018) The evolution of production systems from Industry 2.0 through Industry 4.0. Int J Prod Res 56:848–861. https://doi.org/10.1080/00207543.2017.1403664
9. Strohmenger AP (1914) Electric fusion of metals
10. Iqbal MP, Tripathi A, Jain R et al (2020) Numerical modelling of microstructure in friction stir welding of aluminium alloys. Int J Mech Sci 185. https://doi.org/10.1016/j.ijmecsci.2020.105882
11. Sahu S, Thorat O, Mahto RP, et al (2019) A Review and Case Study on Mechanical Properties and Microstructure Evolution in Magnesium–Steel Friction Stir Welding. pp 101–109
12. Thomas WMW, Norris I, Nicholas ED, et al (1991) Friction stir welding—process developments and variant techniques. SME Summit 1–21
13. Mypati O, Sadhu A, Sahu S et al (2020) Enhancement of joint strength in friction stir lap welding between AA6061 and AISI 304 by adding diffusive coating agents. Proc Inst Mech Eng Part B J Eng Manuf 234:204–217. https://doi.org/10.1177/0954405419838379
14. (2017) The Evolution of Robotic Welding. https://www.robotics.org/blog-article.cfm/The-Evolution-of-Robotic-Welding/33. Accessed 22 May 2020
15. Panigrahi SS, Bahinipati B, Jain V (2019) Sustainable supply chain management: A review of literature and implications for future research. Manag Environ Qual an Int J 30:1001–1049. https://doi.org/10.1108/MEQ-01-2018-0003
16. Mypati O, Mishra D, Sahu S et al (2020) A Study on Electrical and Electrochemical Characteristics of Friction Stir Welded Lithium-Ion Battery Tabs for Electric Vehicles. J Electron Mater 49:72–87. https://doi.org/10.1007/s11664-019-07711-8
17. Sahu SK, Mishra D, Mahto RP et al (2018) Friction stir welding of polypropylene sheet. Eng Sci Technol an Int J 21:245–254. https://doi.org/10.1016/j.jestch.2018.03.002
18. Sahu SK, Pal K, Mahto RP, Dash P (2019) Monitoring of friction stir welding for dissimilar Al 6063 alloy to polypropylene using sensor signals. Int J Adv Manuf Technol 104:159–177. https://doi.org/10.1007/s00170-019-03855-3
19. Farkas A (2018) Impact of Industry 4.0 on robotic welding. IOP Conf Ser Mater Sci Eng 448:7. https://doi.org/10.1088/1757-899X/448/1/012034
20. Norrish J (2006) Advanced Welding Processes. Woodhead Publishing Limited
21. Gibson BT, Lammlein DH, Prater TJ et al (2014) Friction stir welding: Process, automation, and control. J Manuf Process 16:56–73. https://doi.org/10.1016/j.jmapro.2013.04.002
22. Terán HC, Arteaga O, Alcocer FS et al (2019) Application of multiple methods of ndt for the evaluation of welded joints in a steel bridge astm-a-588. Solid State Phenom 287:8–12. https://doi.org/10.4028/www.scientific.net/SSP.287.8
23. Kah P, Mvola B, Martikainen J, Suoranta R (2014) Real time non-destructive testing methods of welding. Adv Mater Res 933:109–116. https://doi.org/10.4028/www.scientific.net/AMR.933.109
24. Mahto RP, Kumar R, Pal SK (2020) Characterizations of weld defects, intermetallic compounds and mechanical properties of friction stir lap welded dissimilar alloys. Mater Charact 160:110115. https://doi.org/10.1016/j.matchar.2019.110115

25. Kleiner D, Bird CR (2004) Signal Processing for Quality Assurance in Friction Stir Welds. 46:2003–2005
26. Kah P, Shrestha M, Hiltunen E, Martikainen J (2015) Robotic arc welding sensors and programming in industrial applications. Int J Mech Mater Eng 10. https://doi.org/10.1186/s40712-015-0042-y
27. Xu Y, Yu H, Zhong J et al (2012) Real-time seam tracking control technology during welding robot GTAW process based on passive vision sensor. J Mater Process Technol 212:1654–1662. https://doi.org/10.1016/j.jmatprotec.2012.03.007
28. Yu J-Y, Na S-J (1998) A study on vision sensors for seam tracking of height-varying weldment. Part 2: Applications. Mechatronics 8:21–36. https://doi.org/10.1016/s0957-4158(97)00024-x
29. Zou Y, Wang Y, Zhou W, Chen X (2018) Real-time seam tracking control system based on line laser visions. Opt Laser Technol 103:182–192. https://doi.org/10.1016/j.optlastec.2018.01.010
30. Zou Y, Chen X, Gong G, Li J (2018) A seam tracking system based on a laser vision sensor. Meas J Int Meas Confed 127:489–500. https://doi.org/10.1016/j.measurement.2018.06.020
31. Kovacevic R, Zhang Y, Li L (1996) Monitoring of Weld Joint Penetrations Based on Weld Pool Geometrical Appearance. Weld. Journal-Including Weld. Res. Suppl. 75:317–329
32. Kovacevic R, Zhang YM (1997) Real-Time Image Processing for Monitoring of Free Weld Pool Surface. J Manuf Sci Eng Trans ASME 119:161–169. https://doi.org/10.1115/1.2831091
33. Scherler M Survey of Robotic Seam Tracking Systems for Arc Welding. https://www.robotics.org/content-detail.cfm/Industrial-Robotics-Tech-Papers/Survey-of-Robotic-Seam-Tracking-Systems-for-Arc-Welding/content_id/956. Accessed 15 May 2020
34. Sforza P, De Blasiis D (2002) On-line optical monitoring system for arc welding. NDT E Int 35:37–43. https://doi.org/10.1016/S0963-8695(01)00021-4
35. Murthy V, Ullegaddi K, Mahesh B, Rajaprakash BM (2017) Application of Image Processing and Acoustic Emission Technique in Monitoring of Friction Stir Welding Process. Mater Today Proc 4:9186–9195. https://doi.org/10.1016/j.matpr.2017.07.276
36. Bhat NN, Kumari K, Dutta S et al (2015) Friction stir weld classification by applying wavelet analysis and support vector machine on weld surface images. J Manuf Process 20:274–281. https://doi.org/10.1016/j.jmapro.2015.07.002
37. Sudhagar S, Sakthivel M, Ganeshkumar P (2019) Monitoring of friction stir welding based on vision system coupled with Machine learning algorithm. Meas J Int Meas Confed 144:135–143. https://doi.org/10.1016/j.measurement.2019.05.018
38. Rajashekar R, Rajaprakash BM (2013) Analysis of Banded Texture of Friction Stir Weld Bead Surface by Image Processing Technique. 5:30–39
39. Ranjan R, Khan AR, Parikh C et al (2016) Classification and identification of surface defects in friction stir welding: An image processing approach. J Manuf Process 22:237–253. https://doi.org/10.1016/j.jmapro.2016.03.009
40. Parikh C, Ranjan R, Khan AR et al (2017) Volumetric defect analysis in friction stir welding based on three dimensional reconstructed images. J Manuf Process 29:96–112. https://doi.org/10.1016/j.jmapro.2017.07.006
41. Uhrlandt D (2016) Diagnostics of metal inert gas and metal active gas welding processes. J Phys D Appl Phys 49. https://doi.org/10.1088/0022-3727/49/31/313001
42. Pal S, Pal SK, Samantaray AK (2008a) Artificial neural network modeling of weld joint strength prediction of a pulsed metal inert gas welding process using arc signals. J Mater Process Technol 202:464–474. https://doi.org/10.1016/j.jmatprotec.2007.09.039
43. Pal S, Pal SK, Samantaray AK (2008b) Neurowavelet packet analysis based on current signature for weld joint strength prediction in pulsed metal inert gas welding process. Sci Technol Weld Join 13:638–645. https://doi.org/10.1179/174329308X299986
44. Pal S, Pal SK, Samantaray AK (2008c) Sensor based weld bead geometry prediction in pulsed metal inert gas welding process through artificial neural networks. Int J Knowledge-Based Intell Eng Syst 12:101–114. https://doi.org/10.3233/KES-2008-12202
45. Quinn TP, Smith C, McCowan CN et al (1999) Arc Sensing for Defects in Constant-Voltage Gas Metal Arc Welding. Weld J (Miami, Fla) 78:322–328

46. Barborak D, Conrardy C, Madigan B, Paskell T (1999) "Through-arc" process monitoring techniques for control of automated gas metal arc welding. Proc - IEEE Int Conf Robot Autom 4:3053–3058. https://doi.org/10.1109/robot.1999.774062
47. Benakis M, Du C, Patran A, French R (2019) Welding Process Monitoring Applications and Industry 4.0. In: 2019 IEEE 15th International Conference on Automation Science and Engineering (CASE). IEEE, pp 1755–1760
48. Pal SK, Kharagpur T (2008) DEVELOPMENT AND VALIDATION OF VARIOUS SOFT COMPUTING BASED MODELS FOR PULSED by Sukhomay Pal
49. Ru-xiong LI (2012) Quality monitoring of resistance spot welding based on process parameters. Energy Procedia 14:925–930. https://doi.org/10.1016/j.egypro.2011.12.1034
50. Xu G, Wen J, Wang C (2009) Zhang X (2009) Quality monitoring for resistance spot welding using dynamic signals. IEEE Int Conf Mechatronics Autom ICMA 2009:2495–2499. https://doi.org/10.1109/ICMA.2009.5246513
51. Ma Y, Wu P, Xuan C et al (2013) Review on techniques for on-line monitoring of resistance spot welding process. Adv Mater Sci Eng 2013. https://doi.org/10.1155/2013/630984
52. Mishra D, Roy RB, Dutta S et al (2018) A review on sensor based monitoring and control of friction stir welding process and a roadmap to Industry 4.0. J Manuf Process 36:373–397. https://doi.org/10.1016/j.jmapro.2018.10.016
53. Kesharwani RK, Panda SK, Pal SK (2015) Experimental Investigations on Formability of Aluminum Tailor Friction Stir Welded Blanks in Deep Drawing Process. J Mater Eng Perform 24:1038–1049. https://doi.org/10.1007/s11665-014-1361-5
54. Mishra D, Gupta A, Raj P et al (2020) Real time monitoring and control of friction stir welding process using multiple sensors. CIRP J Manuf Sci Technol. https://doi.org/10.1016/j.cirpj.2020.03.004
55. Roy RB, Ghosh A, Bhattacharyya S et al (2018) Weld defect identification in friction stir welding through optimized wavelet transformation of signals and validation through X-ray micro-CT scan. Int J Adv Manuf Technol 99:623–633. https://doi.org/10.1007/s00170-018-2519-3
56. Das B, Sukhomay Pal SB (2014) Monitoring of friction stir welding process through signals acquired during the welding. 5th International & 26th All India Manufacturing Technology, Design and Research Conference (AIMTDR 2014) December 12th–14th, 2014. IIT Guwahati, Assam, India, pp 1–7
57. Das B, Pal S, Bag S (2017a) Weld quality prediction in friction stir welding using wavelet analysis. Int J Adv Manuf Technol 89:711–725. https://doi.org/10.1007/s00170-016-9140-0
58. Tam J (2005) Methods of Characterizing Gas-Metal Arc Welding Acoustics for Process Automation
59. Saini D, Floyd S (1998) An Investigation of Gas Metal Arc Welding Sound Signature for On-Line Quality Control High-speed signal data acquisition and computer-aided analysis of sound signature may reveal conditions that generate weld defects. 172–179
60. Pal K, Pal SK (2011a) Monitoring of weld penetration using arc acoustics. Mater Manuf Process 26:684–693. https://doi.org/10.1080/10426910903496813
61. Tarn T-J, Chen S-B, Fang G (2011) Robotic Welding. Intelligence and Automation, Springer, Berlin Heidelberg, Berlin, Heidelberg
62. Pal K, Bhattacharya S, Pal SK (2009) Prediction of metal deposition from arc sound and weld temperature signatures in pulsed MIG welding. Int J Adv Manuf Technol 45:1113–1130. https://doi.org/10.1007/s00170-009-2052-5
63. Pal K, Bhattacharya S, Pal SK (2010a) Multisensor-based monitoring of weld deposition and plate distortion for various torch angles in pulsed MIG welding. Int J Adv Manuf Technol 50:543–556. https://doi.org/10.1007/s00170-010-2523-8
64. Pal K, Bhattacharya S, Pal SK (2010b) Investigation on arc sound and metal transfer modes for on-line monitoring in pulsed gas metal arc welding. J Mater Process Technol 210:1397–1410. https://doi.org/10.1016/j.jmatprotec.2010.03.029
65. Podržaj P, Polajnar I, Diaci J, Kariž Z (2005) Estimating the strength of resistance spot welds based on sonic emission. Sci Technol Weld Join 10:399–405. https://doi.org/10.1179/174329305X44107

66. Sala A, Tonolini F, Villa G, Nardoni G (1987) Monitoring of acoustic emission during the welding process. Weld Int 1:655–658. https://doi.org/10.1080/09507118709453014
67. Jolly WD (1969) Acoustic emission exposes cracks during welding. Weld J 48:21–27
68. Luo Y, Li JL, Wu W (2013) Nugget quality prediction of resistance spot welding on aluminium alloy based on structureborne acoustic emission signals. Sci Technol Weld Join 18:301–306. https://doi.org/10.1179/1362171812Y.0000000102
69. Chen C, Kovacevic R, Jandgric D (2003) Wavelet transform analysis of acoustic emission in monitoring friction stir welding of 6061 aluminum. Int J Mach Tools Manuf 43:1383–1390. https://doi.org/10.1016/S0890-6955(03)00130-5
70. Soundararajan V, Atharifar H, Kovacevic R (2006) Monitoring and processing the acoustic emission signals from the friction-stir-welding process. Proc Inst Mech Eng Part B J Eng Manuf 220:1673–1685. https://doi.org/10.1243/09544054JEM586
71. Jiménez-Macías E, Sánchez-Roca A, Carvajal-Fals H et al (2014) Wavelets application in prediction of friction stir welding parameters of alloy joints from vibroacoustic ANN-based model. Abstr Appl Anal 2014. https://doi.org/10.1155/2014/728564
72. Subramaniam S, Narayanan S, Ashok SD (2013) Acoustic emission-based monitoring approach for friction stir welding of aluminum alloy AA6063-T6 with different tool pin profiles. Proc Inst Mech Eng Part B-Journal Eng Manuf 227:407–416. https://doi.org/10.1177/0954405412472673
73. Chin BA, Madsen NH, Goodling JS (1983) Infrared Thermography for Sensing the Weld Process. 84–86
74. Alfaro SCA, Franco FD (2010) Exploring infrared sensoring for real time welding defects monitoring in GTAW. Sensors 10:5962–5974. https://doi.org/10.3390/s100605962
75. Khan MA, Madsen NH, Chin BA (1984) Infrared Thermography As A Control For The Welding Process. In: Burrer GJ (ed) Thermosense VI: Thermal Infrared Sensing for Diagnostics and Control. pp 154–163
76. Nagarajan S, Banerjee P, Chen W, Chin BA (316AD) Control of the Welding Process using Infrared Sensors. IEEE Trans Robot Autom 8:400
77. Cullen JD, Athi N, Al-Jader M et al (2008) Multisensor fusion for on line monitoring of the quality of spot welding in automotive industry. Meas J Int Meas Confed 41:412–423. https://doi.org/10.1016/j.measurement.2007.01.006
78. Mahto RP, Kumar R, Pal SK, Panda SK (2018) A comprehensive study on force, temperature, mechanical properties and micro-structural characterizations in friction stir lap welding of dissimilar materials (AA6061-T6 & AISI304). J Manuf Process 31:624–639. https://doi.org/10.1016/j.jmapro.2017.12.017
79. Das B, Pal S, Bag S (2019) Probing defects in friction stir welding process using temperature profile. Sadhana - Acad Proc Eng Sci 44:1–9. https://doi.org/10.1007/s12046-019-1068-2
80. De Backer J, Bolmsjö G, Christiansson A-K (2014) Temperature control of robotic friction stir welding using the thermoelectric effect. Int J Adv Manuf Technol 70:375–383. https://doi.org/10.1007/s00170-013-5279-0
81. Backer J (2014) Feedback Control of Robotic Friction Stir Welding
82. Fleming P, Lammlein D, Wilkes D et al (2008) In-Process Gap Detection in Friction Stir Welding. 1:62–67. https://doi.org/10.1108/02602280810850044
83. Yang Y, Kalya P, Landers RG, Krishnamurthy K (2008) Automatic gap detection in friction stir butt welding operations. Int J Mach Tools Manuf 48:1161–1169. https://doi.org/10.1016/j.ijmachtools.2008.01.007
84. Das B, Pal S, Bag S (2016a) Defect Detection in Friction Stir Welding Process Using Signal Information and Fractal Theory. Procedia Eng 144:172–178. https://doi.org/10.1016/j.proeng.2016.05.021
85. Das B, Pal S, Bag S (2016b) A combined wavelet packet and Hilbert-Huang transform for defect detection and modelling of weld strength in friction stir welding process. J Manuf Process 22:260–268. https://doi.org/10.1016/j.jmapro.2016.04.002
86. Das B, Pal S, Bag S (2017b) Torque based defect detection and weld quality modelling in friction stir welding process. J Manuf Process 27:8–17. https://doi.org/10.1016/j.jmapro.2017.03.012

87. Kumar U, Yadav I, Kumari S et al (2015) Defect identification in friction stir welding using discrete wavelet analysis. Adv Eng Softw 85:43–50. https://doi.org/10.1016/j.advengsoft.2015.02.001
88. Kumari S, Jain R, Kumar U et al (2016) Defect identification in friction stir welding using continuous wavelet transform. J Intell Manuf 1–12. https://doi.org/10.1007/s10845-016-1259-1
89. Boldsaikhan E, Corwin EM, Logar AM, Arbegast WJ (2011) The use of neural network and discrete Fourier transform for real-time evaluation of friction stir welding. Appl Soft Comput J 11:4839–4846. https://doi.org/10.1016/j.asoc.2011.06.017
90. Cook GE, Crawford R, Clark DE, Strauss AM (2004) Robotic friction stir welding. Ind Robot an Int J 31:55–63. https://doi.org/10.1108/01439910410512000
91. Longhurst WR, Strauss AM, Cook GE, Fleming PA (2010) Torque control of friction stir welding for manufacturing and automation. Int J Adv Manuf Technol 51:905–913. https://doi.org/10.1007/s00170-010-2678-3
92. Reisgen U, Mann S, Middeldorf K et al (2019) Connected, digitalized welding production—Industrie 4.0 in gas metal arc welding. Weld World 63:1121–1131. https://doi.org/10.1007/s40194-019-00723-2
93. Gräler I, Pöhler A (2018) Intelligent Devices in a Decentralized Production System Concept. Procedia CIRP 67:116–121. https://doi.org/10.1016/j.procir.2017.12.186
94. Keller M, Rosenberg M, Brettel M, Friederichsen N (2014) How Virtualization, Decentrazliation and Network Building Change the Manufacturing Landscape: An Industry 4.0 Perspective. Int J Mech Aerospace, Ind Mechatron Manuf Eng 8:37–44. https://doi.org/10.1016/j.procir.2015.02.213
95. Carvalho N, Chaim O, Cazarini E, Gerolamo M (2018) Manufacturing in the fourth industrial revolution: A positive prospect in Sustainable Manufacturing. Procedia Manuf 21:671–678. https://doi.org/10.1016/j.promfg.2018.02.170
96. Erboz G (2017) How to Define Industry 40: The Main Pillars of Industry 4.0. Manag Trends Dev Enterp Glob Era 761767
97. Press I, Perez R (2018) Internet of Things A to Z. Internet Things a to Z. https://doi.org/10.1002/9781119456735
98. Durao F, Carvalho JFS, Fonseka A, Garcia VC (2014) A systematic review on cloud computing. J Supercomput 68:1321–1346. https://doi.org/10.1007/s11227-014-1089-x
99. Rossi B (2016) Why cloud technology is central to Industry 4.0. https://www.information-age.com/cloud-technology-central-industry-4-0-123462532/. Accessed 20 May 2020
100. Sakovich N Fog computing vs. cloud computing for IoT projects. https://www.sam-solutions.com/blog/fog-computing-vs-cloud-computing-for-iot-projects/. Accessed 20 May 2020
101. Haffner O, Kučera E, Kozák Š, Stark E (2017) Application of Pattern Recognition for a Welding Process. Communiation Pap 2017 Fed Conf Comput Sci Inf Syst 13:3–8. https://doi.org/https://doi.org/10.15439/2017f115
102. Miskinis C What separates digital twin based simulations VS a reality that is augmented. https://www.challenge.org/insights/digital-twin-vs-augmented-reality/. Accessed 20 May 2020
103. Berg LP, Vance JM (2017) Industry use of virtual reality in product design and manufacturing: a survey. Virtual Real 21:1–17. https://doi.org/10.1007/s10055-016-0293-9
104. Zhang W Welding Simulation & Digital Twins. https://www.swantec.com/welding-simulation-digital-twins/. Accessed 20 May 2020
105. International F (2018) Big Data in Welding Technology. 1–10
106. Chen B, Wang J, Chen S (2009) Modeling of pulsed GTAW based on multi-sensor fusion. Sens Rev 29:223–232. https://doi.org/10.1108/02602280910967639
107. Zhang Z, Wen G, Chen S (2016) Multisensory data fusion technique and its application to welding process monitoring. Proc IEEE Work Adv Robot its Soc Impacts, ARSO 2016-Novem:294–298. https://doi.org/https://doi.org/10.1109/ARSO.2016.7736298
108. Pal K, Pal SK (2011b) Soft computing methods used for the modelling and optimisation of Gas Metal Arc Welding: A review. Int J Manuf Res 6:15–29. https://doi.org/10.1504/IJMR.2011.037911

109. Cai W, Wang J, Zhou Q et al (2019) Equipment and machine learning in welding monitoring: A short review. ACM Int Conf Proceeding Ser Part F1476:9–15. https://doi.org/10.1145/3314493.3314508
110. Goel R, Gupta P (2020) Robotics and Industry 4.0. pp 157–169
111. Han Q, Li Y, Zhang G (2018) Transactions on Intelligent Welding Manufacturing
112. Jain R, Kumari K, Kesharwani RK, et al (2015) Friction Stir Welding: Scope and Recent Development. pp 179–229
113. Mishra D, Sahu SK, Mahto RP, et al (2019) Friction Stir Welding for Joining of Polymers. Springer Singapore, pp 123–162
114. Mishra RS, Ma ZY (2005) Friction Stir Welding and Processing
115. Fridlyander IN, Sister VG, Grushko OE, et al (2002) Aluminum alloys: Promising materials in the automotive industry. Met Sci Heat Treat 44:365–370. https://doi.org/https://doi.org/10.1023/A:1021901715578
116. Davis JR (2001) Light Metals and Alloys. Alloy Underst Basics 351–416. https://doi.org/https://doi.org/10.1361/autb2001p351
117. Campbell FC (2012) Introduction and uses of lightweight materials. Light Mater Underst basics 1–31
118. Gao RX, Yan R (2011) Wavelets. Springer, US, Boston, MA

Chapter 9
Comparative Study of Laser Weldability of Titanium Alloys

J. M. Sánchez-Amaya and C. Churiaque

Abstract The weldability of titanium alloys depends on their chemical composition and microstructure. Thus, pure titanium, α alloys and α + β alloys are claimed to have excellent weldability, although metastable β alloys have bad weldability. These general weldability considerations are applied regardless of the joining technology, but specific novel welding techniques, as laser beam welding, can improve this property. The main objective of this paper is to determine experimentally and precisely the range of laser weldability of the most common three families of titanium alloys. To carry out this study, butt welds were prepared using LBW under conduction regime, in specimens of different titanium alloys (α, α + β, and β) with the same thickness and size. The analyses of required input laser energy to generate full penetration welds, metallographic examinations of welds and mechanical evaluation of the joints were performed on welded samples. Results showed that much higher input laser energy density was required to achieve full penetration welds in the β alloy than in the α and α + β alloys. In addition, in both α and α + β samples, microhardness values of the fusion zone of welds were similar to the base metal. However, the microhardness values at fusion zone of β alloy were slightly lower than those measured at base metal. Tensile strength tests of these welds generated good results for both α and α + β samples (the specimens did not break at the welded area, presenting UTS and YS values similar to the base metal). Tensile specimens of β welds, however, presented worse results, as they broke at the weld, the load values being lower than those obtained for its base metal. Nevertheless, LBW induced some improvements in welds of β alloy, in comparison with other welding techniques. All these results have allowed us to state an order of laser weldability in conduction mode, according to which, β alloy would have a worse laser weldability than α and α + β alloys.

Keywords Laser beam welding · Titanium alloys · Laser weldability

J. M. Sánchez-Amaya (✉) · C. Churiaque
Department of Materials Science and Metallurgical Engineering and Inorganic Chemistry, School of Engineering, University of Cádiz, Cádiz, Spain
e-mail: josemaria.sanchez@uca.es

J. P. Davim (ed.), *Welding Technology*, Materials Forming, Machining and Tribology, https://doi.org/10.1007/978-3-030-63986-0_9

9.1 Introduction

As in other metal alloys, the weldability of titanium alloys depends, among other properties, on their chemical composition [1]. Titanium alloys are conventionally classified into three categories, according to their microstructure: α alloys (consisting of various grades of commercially pure titanium alloys), α + β alloys (presenting α phase and a small volume fraction of β phase in equilibrium, although martensite transformation can also occur upon fast cooling) and β alloys (which do not transform martensitically upon fast cooling from the β phase field) [2]. The weldability of titanium alloys is known to be strongly influenced by the category previously described. Thus, it is reported in [3] that unalloyed titanium and all α titanium alloys have good weldability, toughness and strength. α + β alloys show good formability and weldability, their properties being highly influenced by heat treatments. β alloys are claimed to present worse weldability due to the degradation of strength after welding [4]. Ahmed et al. also reported in [1] that pure titanium, α alloys and α + β alloys have excellent weldability, although metastable β alloys have limited weldability due to the high content of β stabilizing elements. These general weldability considerations are applied regardless of the joining technology, but specific welding techniques can modulate this ability. Therefore, it is possible to improve the weldability employing novel joining processes.

Among the different novel welding technologies, laser beam welding (LBW) presents different advantages and is specially suitable to join titanium alloys, as it allows a high localization and low size of the melting pool, the required energy being considerably reduced. Many research papers have demonstrated the applicability of LBW to join different titanium alloys, especially for automotive and aerospace industries [5, 6]. Some examples of these studies, developed by the authors of the present contribution, focusses on different aspect of the process when LBW is applied under conduction regime, as the analyses of LBW heat source shape in Ti_6Al_4V alloy by FEM simulation [7], the influence of surface pre-treatments in LBW of Ti_6Al_4V alloy [8], the analysis of microstructure and properties of laser welded Ti5553 (β alloy) samples [9], or the corrosion [10] and tribocorrosion [11] behaviour of laser treated CPTi and Ti_6Al_4V samples.

As stated before, the weldability of some titanium alloys is rather bad, although innovative welding technologies, as LBW, can be employed to improve it. The objective of this communication is to determine experimentally and precisely the range of laser weldability of the most common three families of titanium alloys. To carry out this study, full penetration butt welds were prepared using LBW under conduction regime, in specimens of different titanium alloys with the same thickness and size. Specifically, samples of an α (CpTi) alloy, an α + β alloy (Ti_6Al_4V) and a β alloy ($Ti_5Al_5V5Mo_3Cr$) have been studied. This selection allowed the authors study the laser weldability of representative titanium alloys families existing in the market.

9.2 Experimental Procedure

Butt welding treatments were performed on samples of three titanium alloys, whose composition is detailed in Table 9.1. The size of titanium samples were 50 mm × 50 mm × 1.5 mm, generating butt welded samples of 100 mm × 50 mm × 1.5 mm (Fig. 9.1a), size allowing one to obtain small size standard tensile specimens (Fig. 9.1b), procedure already used by the authors in [8]. LBW of samples were performed under conduction regime using a HPDL (ROFIN-SINAR DL028S, maximum laser power of 2800 W) and a laboratory-made conditioning chamber in which a continuous argon flow of 25 LN/min was employed (Fig. 9.2). The upper part of the chamber has a protective lens alloying the laser beam reach the samples placed inside. In this experimental setup, the laser beam worked at the focal distance (42 mm from the focusing lens), providing a spot size on surface of 1.19 × 1.5 mm^2. Laser treatments consisted of a linear laser scan at a constant travel speed (*S*) and laser power (*P*) for each alloy. The specific values of both parameters were experimentally fitted for each titanium alloy to obtain full penetration welds. Thus, CpTi required values of $P = 2.2$ kW and $S = 10$ mm/s, Ti_6Al_4V values of $P = 2.5$ kW and $S = 10$ mm/s, and $Ti_5Al_5V_5Mo_3Cr$ values of $P = 2.75$ kW and $S = 5$ mm/s.

The beads size and shape (depth and width), the microstructure and the microhardness of the different zones were analysed. For this study, mounted cross sections of samples were evaluated, after polishing and etching for 10 s with Kroll's reagent (6 mL HNO_3, 2 mL HF, 92 mL H_2O). The morphology and size of the beads were analysed with a Leica microscope (Model. MST53) controlled by LAS V4.2 software. As this software can correlate measured pixels and real distances of images,

Table 9.1 Chemical composition of the three titanium alloys used, in wt%. Element (wt%)

Element (wt%)									
Alloy	Al	V	Fe	C	Mo	Cr	O	N	Ti
CpTi (α titanium alloy)	–	–	0.03	0.01	–	–	–	–	99.95
Ti_6Al_4V (α + β titanium alloy)	5.67	4.50	0.18	0.01	–	–	–	–	89.59
Ti–5Al–5V–5Mo_3Cr (β titanium alloy)	5.03	5.10	0.38	–	5.06	2.64	0.14	<0.01	81.65

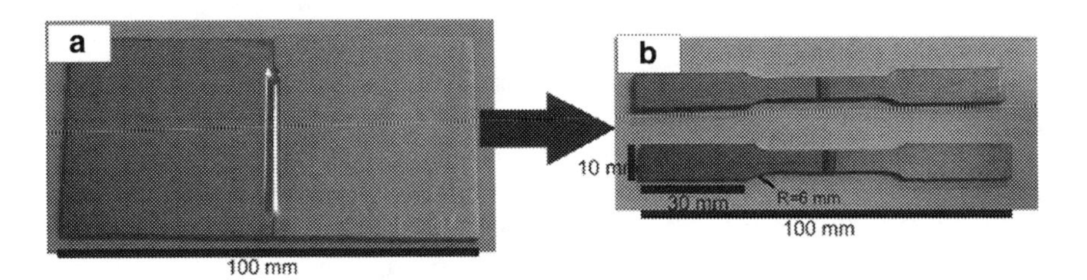

Fig. 9.1 Laser butt welds of titanium alloys (**a**) and extraction of "small size" standard tensile specimens (**b**)

Fig. 9.2 Experimental arrangement system used to perform LBW tests

it allows the measurements of depth and width of the welds (calibrated equipments allowed one to verify this correlation).

Microhardness measurements were accomplished with a Duramin microhardness tester of Struers, employing a charge of 2.945 N (0.3 HV). Finally, selected welds were subjected to tensile tests in a Shimadzu universal testing machine (100 kN) to evaluate the weld strength, fixing a deformation speed of 0.005 mm/min at the elastic deformation regime and 1.6 mm/min at the plastic deformation regime. These tensile tests were performed on standard samples extracted from the welds (Fig. 9.1b), to fulfil the requirements of the "small size" specifications of ASTM E8/E8M-11.

9.3 Results and Discussion

Butt joints of the above-mentioned titanium alloys were laser welded. The minimum laser energy required to generate laser welds with total penetration in 1.5 mm thick specimens were determined experimentally. Figure 9.3 includes the macrographs of the butt welds obtained with the different alloys studied.

Table 9.2 includes the values of the parameters used to obtain the welds. The depth and width of the molten area of welds are also indicated in the table. It was verified that, in all cases, the welding morphology corresponds to conduction mode. A macrographic characterization of the welds has been performed. The silver colour of the face side welds (Fig. 9.3) fulfil the AWS D.17 standard. It can be confirmed in Table 9.2 that similar energy density (E) values are needed to obtain full penetration

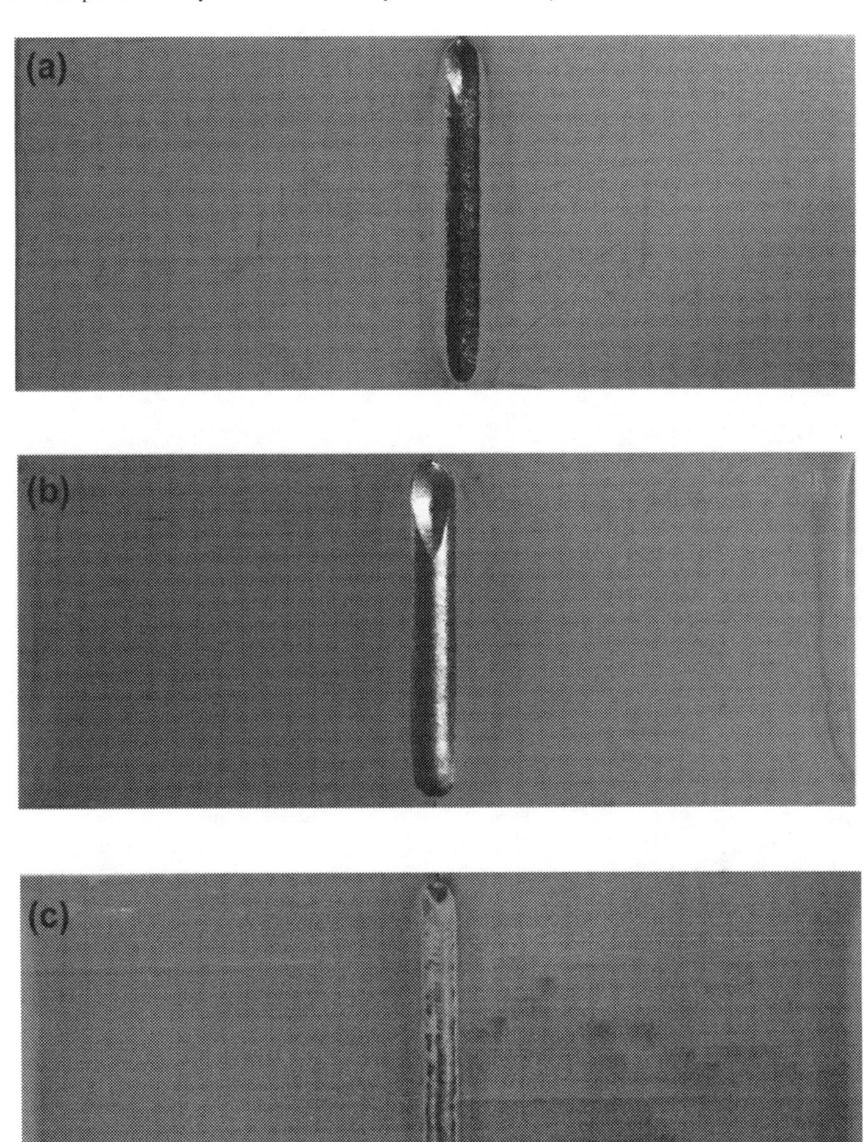

Fig. 9.3 Macrographic images of butt welds of **a** CpTi, **b** Ti_6Al_4V y **c** Ti–5Al–5V-5Mo–3Cr

Table 9.2 HPDL equipment laser parameters used to obtain full penetration butt weld on 1.5 mm thick titanium alloys

Sample identification	Laser power (W)	Welding speed (mm/s)	Energy density (kJ/cm^2)	Depth (FZ) (mm)	Width (ZF) (mm)
CpTi	2200	10	18.3	1.5	4.8
Ti_6Al_4V	2500	10	20.8	1.5	4.7
$Ti_5Al_5V_5Mo_3Cr$	2750	5	45.8	1.5	5

The depth (*d*) and width (*w*) values of fusion zones (FZ) measured in cross-sectional images are also included in the table

welds under conduction regime in CpTi (18.3 kJ/cm^2) and Ti_6Al_4V (20.8 kJ/cm^2). However, to reach full penetration welds in $Ti_5Al_5V_5Mo_3Cr$, much higher energy density is required (45.8 kJ/cm^2). Taking into account this criteria, it can be claimed that both α (CpTi) and $\alpha + \beta$ (Ti_6Al_4V) alloys have better laser weldability than β ($Ti_5Al_5V_5Mo_3Cr$) alloys, since a much greater input energy density is required to obtain a butt weld with total penetration. These results are in good agreement with those reported in [1], in which the same general weldability order (regardless of the welding technology used) is reported for these alloys.

Subsequently, cross sections of the three welds have been micrographically characterized, as shown in Fig. 9.4. No geometric defects have been found in the welds of any of the three alloys, complying with AWSD17.1 regarding the existence of geometric imperfections.

The results obtained by analysing the microstructure of BM, HAZ and FZ of welds of the three alloys are described below. Figure 9.5 includes the metallographic images of the CpTi butt weld. It is observed that the BM is formed by equiaxial α grains with an average size of about 10 μm. The FZ presented elongated and serrated α grains, with an average size of 200 μm. In addition, it has areas with very fine acicular α grains. The HAZ has a microstructure similar to the MB, although with larger grains. It is appreciated in the HAZ that the grain size decreases as approaching to BM.

Figure 9.6 includes metallographic images of the Ti_6Al_4V butt weld. In this alloy, the BM consists of α equiaxial grains with intergranular β phase. The FZ is composed of a microlaminar $\alpha + \beta$ microstructure in α matrix. The grains of the FZ are larger and longer than those formed at the HAZ. The HAZ is formed by areas with equiaxial α grains with intergranular β phase and zones with $\alpha + \beta$ microlaminar microstructure in an α matrix.

Micrographs of the different areas of the $Ti_5Al_5V_5Mo_3Cr$ butt weld are shown in Fig. 9.7. The BM zone of this alloy has an α/β microstructure in which the α particles, with an average size of 5 μm, are distributed within the β matrix. The grains in the FZ have columnar dendritic morphology, which indicates that the alloy has a high concentration of β stabilizing elements. The average grain size in the FZ is approximately 100 μm. The HAZ is constituted by equiaxial β grains, showing an epitaxial growth from the HAZ to the FZ. The grains of the HAZ near the FZ have an average size of about 350 μm. It is appreciated that in this area the grain size increases as approaching to the BM.

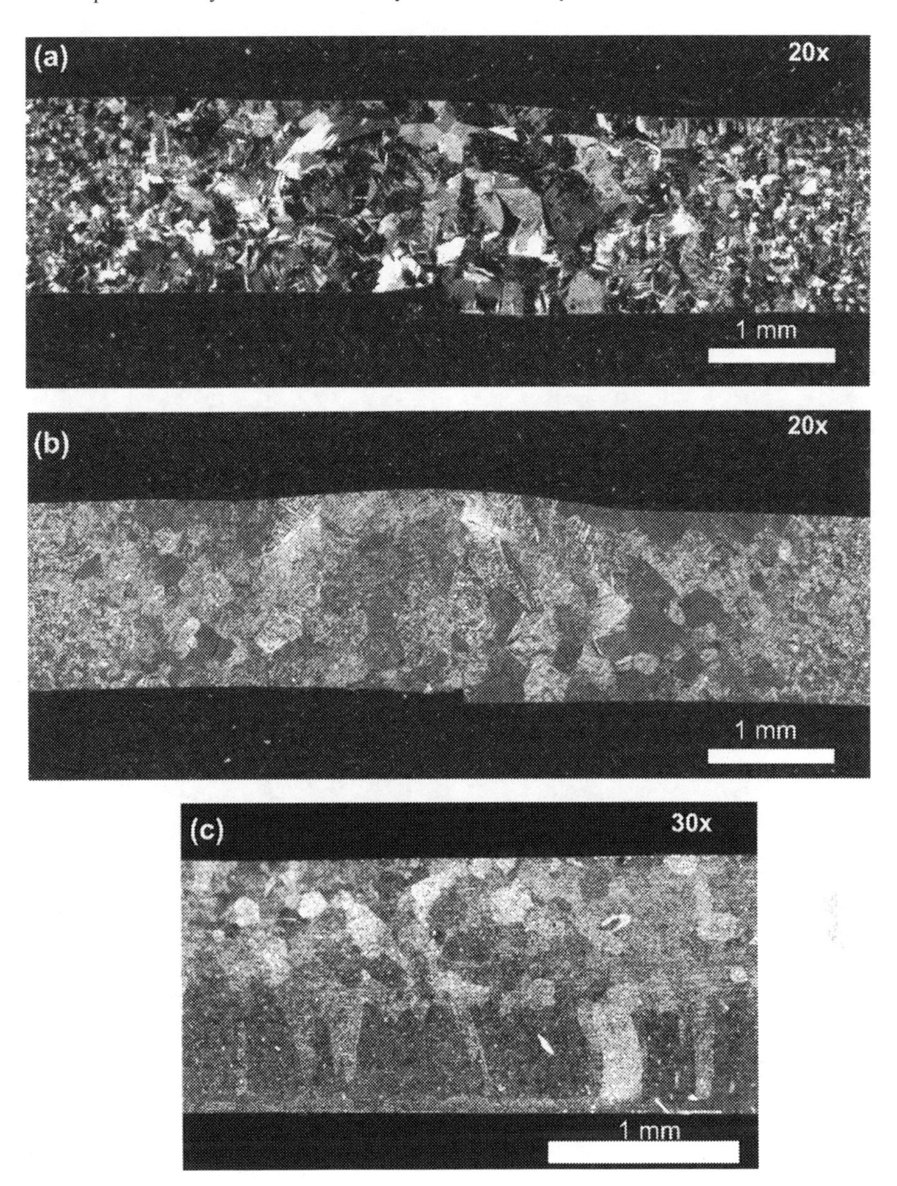

Fig. 9.4 Micrographic images of cross section of butt welds of **a** CpTi, **b** Ti_6Al_4V y **c** Ti–5Al–5V–5Mo–3Cr

Microhardness profiles have been measured for the characterization of the different weld zones, as shown in Fig. 9.8. In this figure, it can be seen that in the CpTi alloy, the values in the FZ range between 290–300 HV, in the HAZ between 280–300 HV and 260–280 HV for the BM. In the Ti_6Al_4V alloy, the FZ has values between 380 and 400 HV, the HAZ between 360–400 HV and the BM between 360–380 HV.

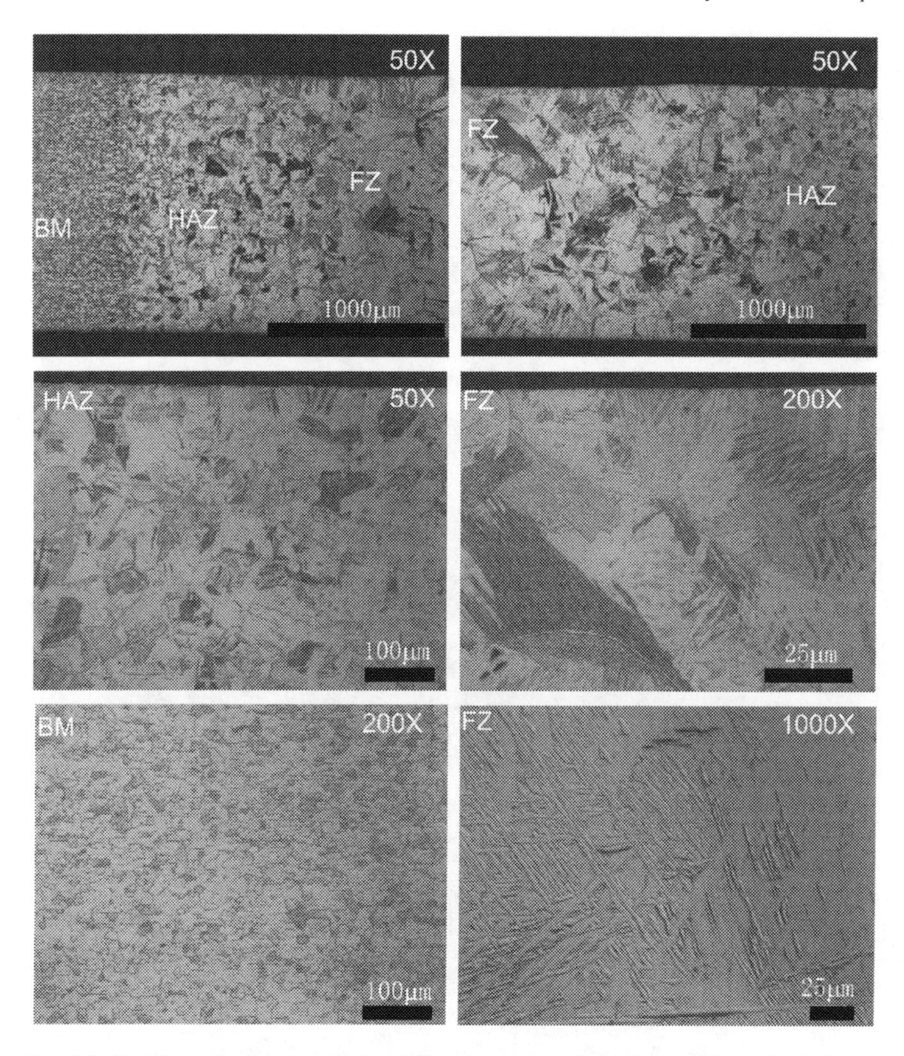

Fig. 9.5 Metallographic images of the different zones of CpTi butt weld, at the indicated magnifications

Finally, in the $Ti_5Al_5V_5Mo_3Cr$ alloy, the microhardness ranges from 290–320 HV in the molten zone; between 300–360 HV in the HAZ and between 370–390 HV in the BM. These results show that the microhardness values are similar for the three zones (FZ, HAZ and BM) in the CpTi and Ti_6Al_4V welds. However, in $Ti_5Al_5V_5Mo_3Cr$ welds, the values in the FZ are slightly lower than in the base metal. This same trend in microhardness values has been observed at electron beam welds (EBW) of these three alloys [12–14]. Thus, for this metastable β titanium alloy, FZ has lower microhardness than the BM. This is due to the fact that precipitation hardening is

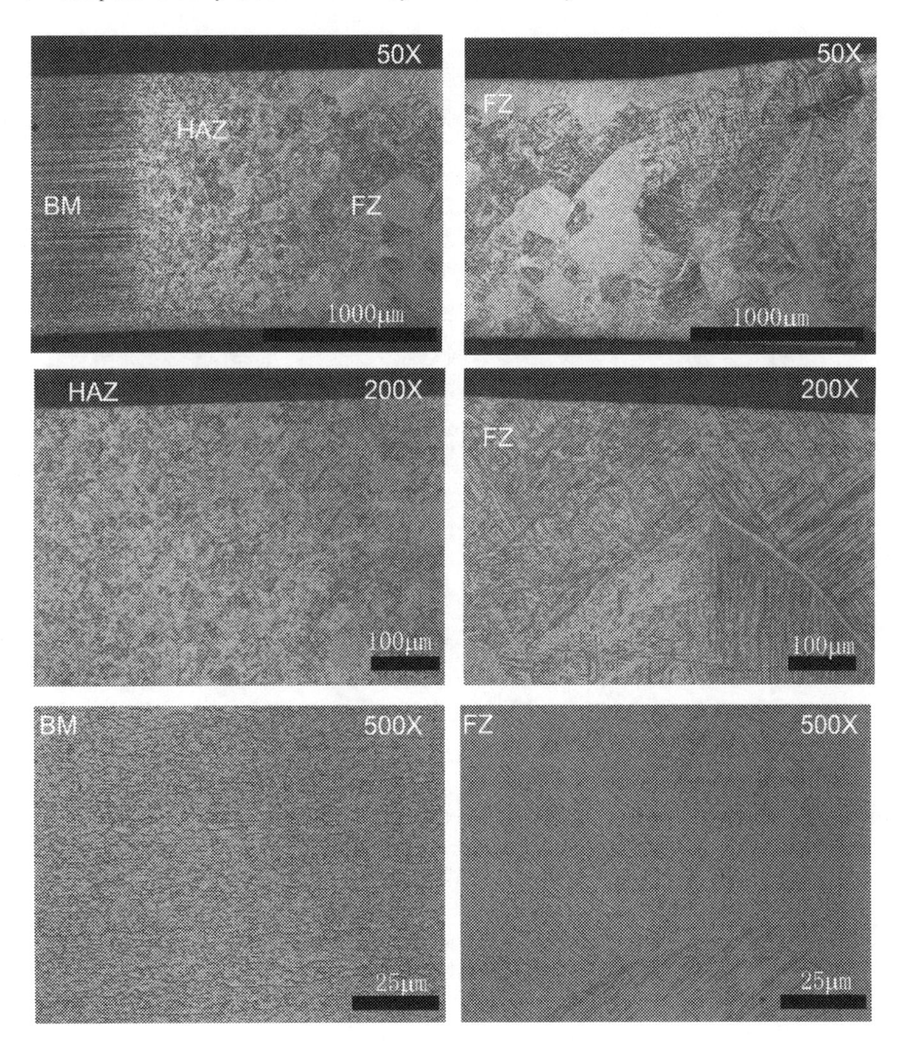

Fig. 9.6 Metallographic images of the different zones of Ti_6Al_4V butt weld, at the indicated magnifications

suppressed by the excess of stabilizers of phase β, since the concentration of molybdenum equivalent ([Mo]eq) is approximately 12. Bania [15] affirm that martensite α´ does not precipitate if [Mo]eq is greater than 10. In addition, the low hardness observed in both the FZ and the HAZ may be due to the formation of the α phase in the HAZ and the presence of metastable/retained β phase in the FZ [16]. In any case, it seems that the heat treatment related by the laser welding process provokes the loss of the previous precipitation hardening process at the $Ti_5Al_5V_5Mo_3Cr$ alloy.

Additionally, the tensile strength of these laser welds has been evaluated. Tensile tests were performed to unwelded base metal standard specimens for the three alloys

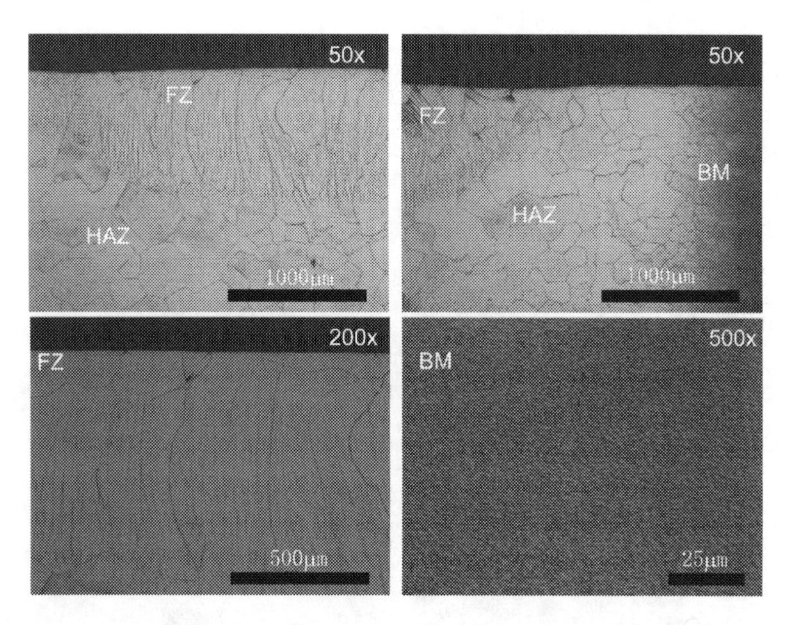

Fig. 9.7 Metallographic images of the different zones of $Ti_5Al_5V_5Mo_3Cr$ butt weld, at the indicated magnifications

(tests taken as references) and to specimens extracted from the welds (being the own weld at the middle of the standard specimen), in which the joint is in the middle of the standard specimen (as shown in Fig. 9.1b). Table 9.3 includes the mechanical properties obtained from the results of the tensile tests. Images of the welded specimens after the tests have been included in Fig. 9.9. The results obtained indicate that the elastic limit and tensile strength are reasonably high in all welds, which indicates that the welding procedure used is reliable. It can also be seen in Fig. 9.9 that the welds of CpTi and Ti_6Al_4V break through the base metal, while the $Ti_5Al_5V_5Mo_3Cr$ specimen breaks through the molten zone. In accordance with common mechanical requirements regulations, it can be concluded that CpTi and Ti_6Al_4V welds fulfil these requirements. On the contrary, the results obtained for the $Ti_5Al_5V_5Mo_3Cr$ weld do not accomplish the requirements, since the specimen fractured at the FZ. However, it is appreciated that the ultimate tensile strength (UTS) and yield strength (YS) values are relatively high and comparable to those described in the literature. In fact, these results are in good agreement with the results of tensile tests performed at $Ti_5Al_5V_5Mo_3Cr$ welds obtained by Shariff et al. [16], which indicate that $Ti_5Al_5V_5Mo_3Cr$ welds have UTS values around the 75% of the reference value for the $Ti_5Al_5V_5Mo_3Cr$ base metal, and a variable ductility reduction. On the other hand, it should be mentioned that LBW generates better welds in this β alloy than other welding techniques, as EBW or GTAW [12], confirming that LBW is a convenient technology to overcome the limited weldability of $Ti_5Al_5V_5Mo_3Cr$ alloy.

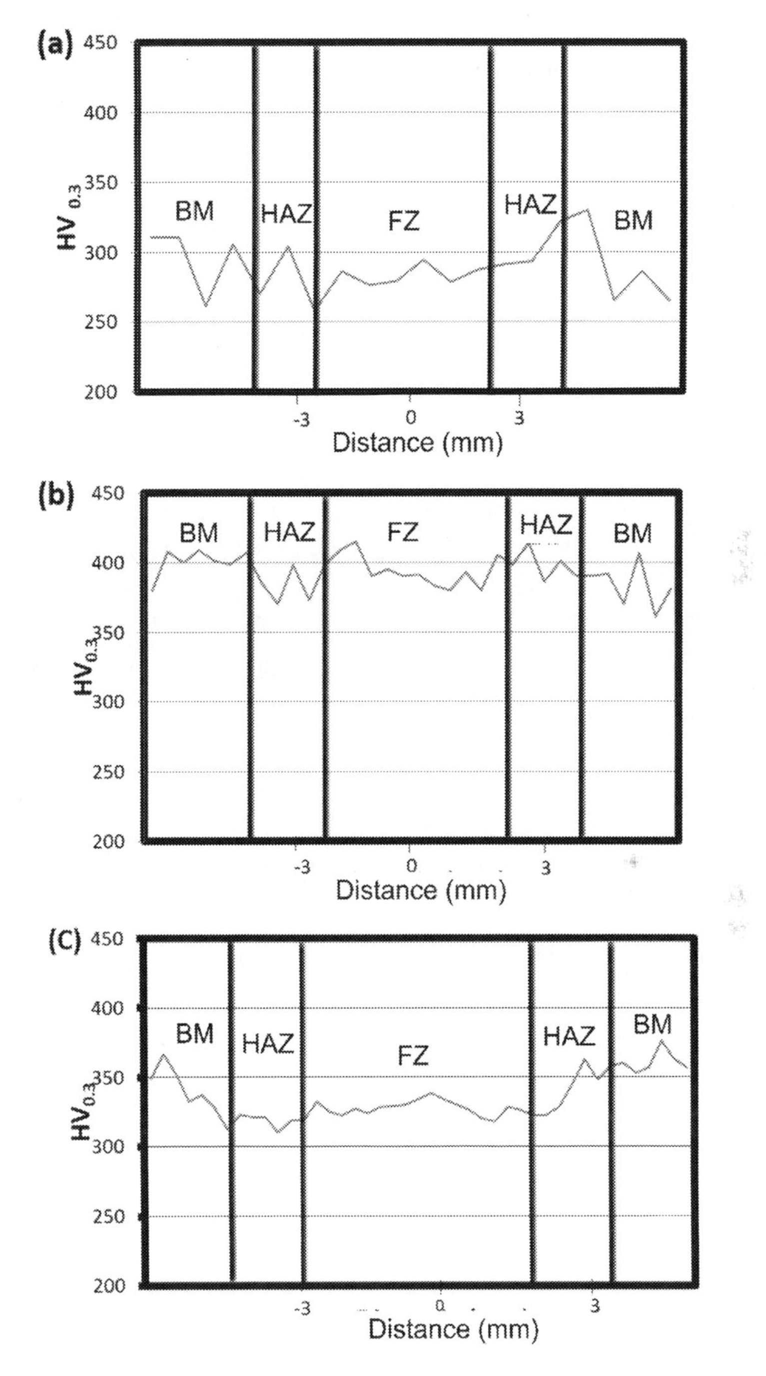

Fig. 9.8 Microhardness profiles measured at butt welds of titanium alloys: **a** CpTi, **b** Ti_6Al_4V and **c** $Ti_5Al_5V_5Mo_3Cr$

Table 9.3 Results obtained from tensile tests performed on base metal (BM) and laser welds (WELD) of the three titanium alloys indicated

Sample identification	YS (MPa)	UTS (MPa)	ε (%)	Fracture Zone	Fulfil regulations
CpTi-BM (reference)	588	683	27	Base Metal	Yes
CpTi-WELD	630	693	23	Base Metal	Yes
Ti_6Al_4V-BM (reference)	941	983	13	Base Metal	Yes
Ti_6Al_4V-WELD	979	1029	8	Weld	Yes
$Ti_5Al_5V_5Mo_3Cr$-BM (reference)	1028	1053	12	Base Metal	Yes
$Ti_5Al_5V5Mo_3Cr$-WELD	816	818	4	Weld	No

Table reports the Ultimate Tensile Strength (UTS), Yield Strength (YS), Ductility (ε), fracture zone, and indication of regulations accomplishment

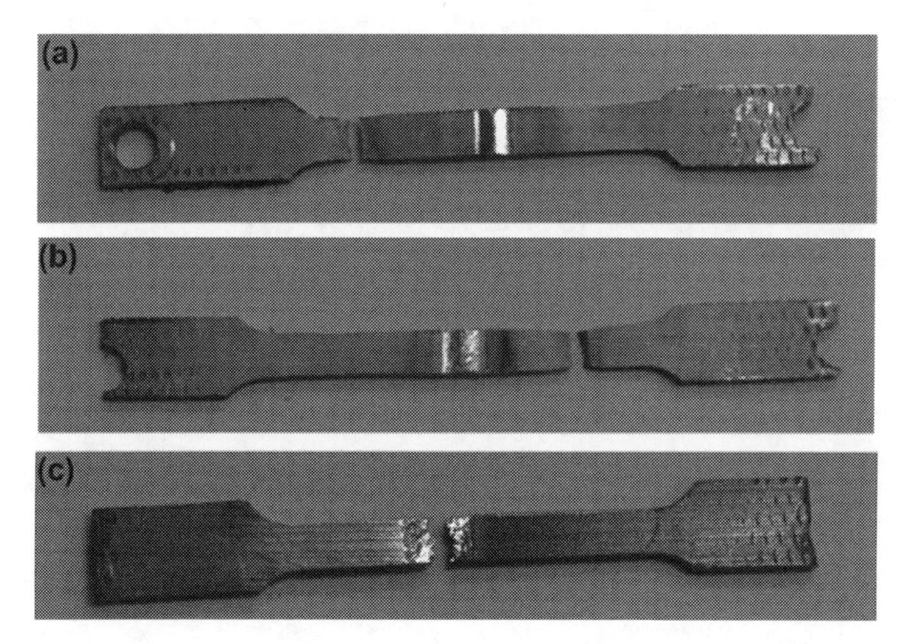

Fig. 9.9 Titanium weld specimens subjected to tensile tests: **a** CpTi, **b** Ti_6Al_4V and **c** $Ti_5Al_5V_5Mo_3Cr$

9.4 Conclusions

The laser weldability of different titanium alloys has been compared in the present contribution. For this study, full penetration butt welds, obtained by LBW under conduction regime, have been achieved for 1.5 mm thickness samples of three different titanium alloys (CpTi, Ti_6Al_4V and $Ti_5Al_5V_5Mo_3Cr$). It has been found that it is necessary to use much higher input laser energy density (more than double) to achieve welds with full penetration in the $Ti_5Al_5V_5Mo_3Cr$ alloy than in the Ti_6Al_4V and CpTi alloys.

The welded samples show an acceptable colour appearance according the current regulations, not presenting any of the welds geometric defects. The metallographic analysis carried out have allowed us to identify the different micro-constituents of the welds. The analysis of the microhardness profiles indicates that the three welding zones (FZ, HAZ and BM) have similar microhardness values in the Ti_6Al_4V and CpTi samples. However, in $Ti_5Al_5V_5Mo_3Cr$ samples, the microhardness values at FZ are slightly lower than those measured at base metal. This is related to the loss of the precipitation hardening mechanism present at the $Ti_5Al_5V_5Mo_3Cr$ base metal.

Finally, the tensile strength of the different welded samples has been evaluated. In the welds of CpTi and Ti_6Al_4V, the specimens do not break at the welded area, presenting UTS and YS values similar to the base metal. On the other hand, the $Ti_5Al_5V_5Mo_3Cr$ specimens break at the weld, the load values being lower than those obtained in the base metal.

All these results have allowed us to state an order of laser weldability in conduction mode, according to which the $Ti_5Al_5V_5Mo_3Cr$ alloy would have a worse weldability than the CpTi and Ti_6Al_4V alloys.

Acknowledgements Authors would like to thank to Prof. J. Botana for providing the facilities and equipments to perform the experimental tests, and also to T. Pasang for providing the $Ti_5Al_5V_5Mo_3Cr$ alloy samples.

References

1. Ahmed T, Rack HJ (1998) Phase transformations during cooling in $\alpha + \beta$ titanium alloys. Mater Sci Eng A 243(1–2):206–211
2. Lütjering JC, Williams G (2007) Titanium, 2nd edn. Springer, New York
3. Destefani JD (1992) Introduction to titanium and titanium alloys. In: Properties and selection: nonferrous alloys and special purpose materials, vol 2. ASM International, Materials Park, OH
4. Richter K, Behr W, Reisgen U (2007) Low heat welding of titanium materials with a pulsed Nd:YAG laser. Mater Wiss Werkst 38(1):51–56
5. Sánchez-Amaya JM, Amaya-Vazquez MR, Botana FJ (2013) In: Katayama S (ed) Handbook of laser welding technologies, Chapter 8: laser welding of light metal alloys: aluminium and titanium alloys. Amazon, UK, pp 215–254
6. Auwal ST, Ramesh S, Yusof F, Manladan SM (2018) A review on laser beam welding of titanium alloys. Int J Adv Manuf Technol 97:1071–1097
7. Churiaque C, Amaya-Vazquez MR, Botana FJ, Sánchez-Amaya JM (2016) FEM simulation and experimental validation of LBW under conduction regime of Ti_6Al_4V alloy. J Mater Eng Perform 25(8)
8. Sánchez-Amaya JM, Amaya-Vázquez MR, González-Rovira L, Botana-Galvin M, Botana FJ (2014) Influence of surface pre-treatments on laser welding of Ti_6Al_4V alloy. J Mater Eng Perform. 23(5):1568–1575
9. Sánchez-Amaya JM, Pasang T, Amaya-Vazquez MR, Lopez-Castro J, Churiaque C, Tao Y, Botana FJ (2017) Microstructure and mechanical properties of Ti5553 butt welds performed by LBW under conduction regime. Metals (Basel) 7(7):269

10. Amaya-Vazquez MR, Sánchez-Amaya JM, Boukha Z, Botana FJ (2012) Microstructure, microhardness and corrosion resistance of remelted TiG_2 and Ti_6Al_4V by a high power diode laser. Corros Sci 66:36–48
11. Silva DP, Churiaque C, Bastos IN, Sánchez-Amaya JM (2016) Tribocorrosion study of ordinary and laser-melted Ti_6Al_4V Alloy. Metals (Basel) 6(10):253
12. Pasang T, Sánchez Amaya JM, Tao Y, Amaya-Vazquez MR, Botana FJ, Sabol JC, Misiolek WZ, Comparison of Ti–5Al–5V–5Mo–3Cr welds performed by laser beam, electron beam and gas tungsten arc welding (conference paper). Procedia Eng 63:397–404
13. Pasang T, Sabol JC, Misiolek WZ, Mitchell R, Short AB (2012) Metallurgy and deformation of electron beam welded similar titanium alloys. In: AIP conference proceedings 2012, vol 1431, no September 2011, pp 982–990
14. Mitchell R, Short A, Pasang T, Littlefair G (2010) Characteristics of electron beam welded Ti & Ti alloys. Adv Mater Res 275:81–84. 9(1):76–99
15. Bania PJ (1993) Beta titanium alloys in the 1990s. In: Eylon D, Boyer RR, Koss DA (eds) TMS, Warrendale, PA, pp 3–14
16. Shariff T, Cao X, Chromik RR, Baradari JG, Wanjara P, Cuddy J (2011) Laser welding of Ti–5Al–5V–5Mo–3Cr. Can Metall Q 50(3):263–272

Chapter 10
A Novel High-Efficiency Keyhole Tungsten Inert Gas (K-TIG) Welding: Principles and Practices

Yonghua Shi, Yanxin Cui, Shuwan Cui, and Baori Zhang

Abstract As an innovation of the conventional Tungsten Inert Gas (TIG) welding, Keyhole Tungsten Inert Gas (K-TIG) welding is highly efficient, well-known for its penetration ability and welding quality. In this chapter, the K-TIG welding is introduced, including its equipment, operation methods, the influence of welding parameters and application extensions. Then, the keyhole stability rules are discussed, so do the arc forces and heat input. To obtain a deeper insight into the welding process, experiments on Ti alloy are carried out, from which the arc current signal, arc voltage signal and arc acoustic signal are collected synchronously. Some characteristics are extracted from the collected signals, by which the welding process is analyzed quantitatively. After the careful analysis of K-TIG welding process, a welding penetration recognition model is established, which shows a good performance. Finally, to evaluate K-TIG welding, experiments are performed on duplex stainless steels and Ti alloys. The effect of the welding parameters on the weld geometry profile and the misorientation angle distribution of grain boundary are discussed. The microstructure and mechanical properties of weldments are evaluated. All of the testing results verify the perfect performance and welding quality of K-TIG welding.

Keywords K-TIG · Welding parameters · Keyhole stability · Arc forces · Heat input · Signal processing · Penetration recognition · MADGB · Microstructure · Mechanical properties

10.1 Overview

10.1.1 High Current Tungsten Inert Gas (TIG) Welding

As a highly efficient way of welding, GTAW has been widely used to produce high-quality joints with various materials. By using low fume generation and requiring

Y. Shi (✉) · Y. Cui · S. Cui · B. Zhang
School of Mechanical and Automotive Engineering, South China University of Technology, Guangzhou 510640, China
e-mail: yhuashi@scut.edu.cn

J. P. Davim (ed.), *Welding Technology*, Materials Forming, Machining and Tribology, https://doi.org/10.1007/978-3-030-63986-0_10

no fluxes, GTAW is considered as a clean welding process. Usually, low welding currents is needed so the arc pressure performed on the surface of the liquid metal is not enough. In this case, the bottom part of the metal weldment absorbs the heat from the top in a way of heat conduction. For thicker weldment, edges preparation is necessary to form V- or X-type edges. Otherwise, filler material and multiple passes have to be added to complete the joint which leads to the time and economy cost [1].

GTAW can be operated under high current (>300 A) which results in the greater penetration of the joint. As the current goes up, the arc pressure increases and forcing the melting material going down which is called the displacement of the weld pool. Joins between square-edged sections can be produced with this method. During the process of this variant, a larger amount of heat can be transferred to the weldment when the electrode works as the cathode. Besides, these advantages enable it to become a popular manually welding method.

Goryachev seems to be the first researcher to carry out high-current GTAW tests to form displacement of weld pool under the arc [2]. Liptak successfully welded the aluminium plate with the thickness of 32 mm [3]. After that, more materials including stainless steel, low-alloy steel and magnesium alloy were tested to be welded with this method [4].

However, the control of the weld penetration is hard, especially if the weld is not fully penetrating. Besides, some defects like hollow bead may appear when the current is over 350 A, and the process may become unstable when it exceeds 500 A. These situations can be improved by using hollow electrode [5].

Deep penetration welding with keyhole is usually seen as another kind of GTAW with a cooling system. One of its advantage is to overcome these problems and has gained significant industrial credibility. The displacement of the molten material under high current results in a hole under the welding arc. Thus, the arc is thought to be buried under the weld pool for a full penetration of the weldment. Therefore, K-TIG welding has been seen as a suitable way to realize some welding with special needs and has potential application in industrial production.

10.1.2 Keyhole Tungsten Inert Gas (K-TIG) Welding

The keyhole TIG welding connects metal weldment by forming small holes in the weld bead. It happens if adequate heat is input melting the metal, and the increased arc pressure widens the holes and push it to the bottom of the weldment. Then, the keyhole will be formed. The formation of the keyhole can form a narrow channel between the surface and the inside of the plate, which can directly transfer the energy to the joint and improve the energy transfer efficiency compared. Besides, a little hole at the weldment's bottom side becomes a new way for the gas leaking away.

Its physical characteristic makes it unique and suitable for those material whose thermal conductivity is low. Stainless steel and titanium alloy both belong to this range of the materials, and some other expensive materials require a high quality and good appearance. As a result, forming keyhole enables the fabricator to produce good weld with a high productivity compared with submerged arc, MAG welding or some other conventional welding method [6] performed single-pass welding of AISI 304, including 12 mm plate at 300 mm/min, 8 mm plate at 500 mm/min and 3 mm plate at 1000 mm/min.

The formation of the keyhole is caused by the high pressure under the arc centre, which is similar to the recoil pressure discovered in the LBW and EBW [7–9], but for a great portion that this pressure owes to the mechanical impact of the PAW jet. Some trails of keyhole TIG welding were conducted by Norrish using improved TIG device to realize TIG welding with dual gas [10].

K-TIG welding has a lot of advantages. First, joints can be completed in just one pass without filling wires of metal as well as edge preparations. Second, it can be applied to many kinds of materials including steel and titanium alloy. Compared with other keyhole mode welding methods, the cost of the equipment is small because it is a variant of GMAW, and the equipment is similar. Moreover, it can achieve full penetration of the thick plates welding, and the welding quality is better than GTAW.

10.1.3 Equipment of K-TIG Welding

The commercial devices for high-current GTAW can be used for K-TIG welding. The whole welding system includes a DC constant current power source and a water-cooling welding torch. The welding process is usually performed automatically because the torch electrode is set to at a negative potential with respect to the weldment.

The continuous using of the K-TIG torches over 1000 amps is guaranteed by the design of the water cooling including protective cover of the torch and the shielding mask. The entire process of getting cool within the torch is essential for keeping a stable arc with different welding current, because of the continuous temperature change over different places on the torch electrodes. Fan designed a new water-cooling system, which not only includes water-cooling channel inside the torch but also under the weldment [11]. This strategy is proved to reduce the volume of the welding pool, as shown in Fig. 10.1.

10.1.4 HDR Imaging of K-TIG Welding

There are many parameters that can reflect the welding condition. The monitoring of these parameters does help to the adjustment during the process as well as ensuring the welding quality. Novel technology can be introduced in the observation of the keyhole behaviour. High-dynamic range imaging was used in the measurement of the entrance of the keyhole geometry [12]. First, the relationship between the radiance of the K-TIG welding scene in reality and the pixel value is measured and described as a response function. The distribution of the radiance in a high dynamic was then

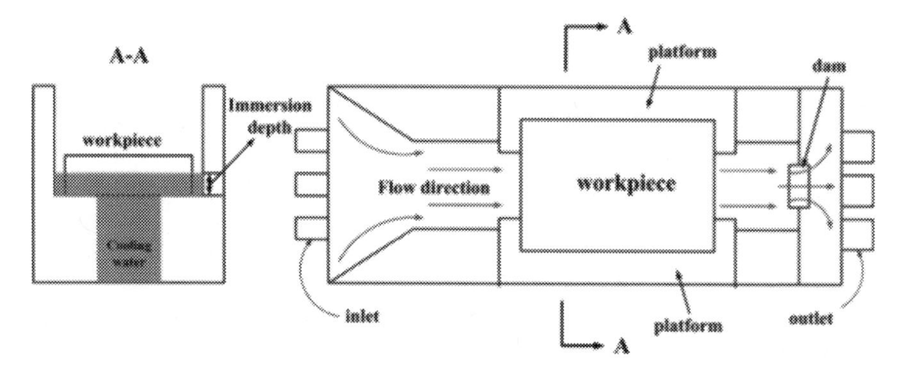

Fig. 10.1 Fan's new water-cooling system [11]

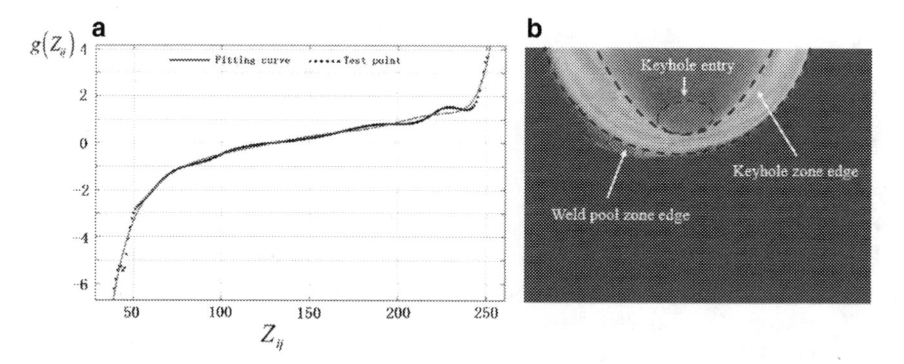

Fig. 10.2 HDR imaging in K-TIG welding. **a** Response function of the camera **b** Reconstruction of the welding scene radiance [12]

reconstruction with images sequences. As a result, much as possible features from the entrance of the keyhole can be recorded, and the identification of the keyhole entrance was achieved, as shown in Fig. 10.2. This study overcomes the obstacles brought by the strong arc light in visual monitoring from the front of the keyhole.

10.2 Operation of K-TIG Welding

10.2.1 Operating Windows

The practice of the K-TIG welding involves a specific set of welding parameters that optimize the welding process, which is known as operating windows. Without it, the K-TIG welding technology was in capable of being used in industrial welding

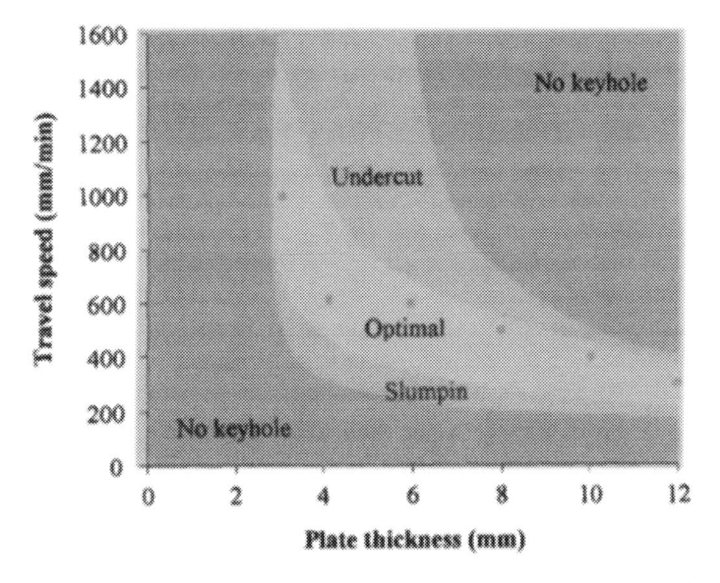

Fig. 10.3 Speed thickness operating window of austenitic stainless steel welding [6]

application. Because of this, a large variety of material with different parameters was tested for the K-TIG welding.

The jobs done in Ref. [6] is shown in Fig. 10.3. This group of tests was all done using the same electrode type, electrode geometry and shielding gas. The travel speed and the thickness were varied to examine the results. He also concluded that a opposite change should be seen in welding speed when the thickness of the samples increases. And the lack of the support from the pool causes the failure of the keyhole. Moreover, a high speed may be acceptable for the weldment which has a thickness of less than 6 mm. But the appearance of the weld seem becomes worse when the parameters are getting close to the boundaries of the operational window.

10.2.2 K-TIG Welding Process Parameters

Some factors have influence on achieving the keyhole TIG welding; for example, if there are no backing plates or the arc gases could not escape from the bottom of the weldment, then the process will become a failure. Besides, a combination of gas composition, material properties, thickness of the plate and some other parameters is important in this process. The increase of the thickness disables the keyhole formation. The operating window is found by experiment to describe the sensitivity of the process for welding variables.

10.2.2.1 Welding Current

Research was done to figure out the effect of the welding current on the quality of the weld seam. Tests on the SAF 2205 showed that the appearance of the weld changed as the currents increased. There is a minimum welding current for starting the keyhole welding process, which is called 'threshold current'. In other words, when the current is increased above the threshold will bring lower welding quality because of the undercut.

This principle is also proved by the research done by Fei [13]. The 9 mm armour grade quenched, and tempered steel was joined with keyhole TIG welding at same speed of 28 cm per min. A threshold current is about 545 A. When it was higher or lower than this current, over-penetration or under-penetration would be found after the weld jointing, as shown in Fig. 10.4. However, current is just one of the many quality considerations for ensuring good welding quality, so when welding current is very close to the threshold current, other welding parameters should be considered.

The frequency of the welding current had influence in the improvement of weld joint microstructure and mechanical properties. By applying an alternated current with 38.6 kHz, Fang carried out experiment joining Q345 steel with the thickness of 5.5 mm [13]. The threshold current to achieve stable keyhole TIG welding drops from 430 to 340 A compared to the ordinary K-TIG welding. Moreover, the current window also expanded.

Cui proposed a welding method of K-TIG called 'One pulse one open keyhole' to solve the welding difficulty caused by narrow parameters window in K-TIG welding. The pulse waveform is periodic, and each unit of periodic contains a base current of 350 A and a peak current of 350 A lasting for 500 ms [15].

10.2.2.2 Travel Speed

The changes of the ravel speed and the thickness of the weldment result in different consequences. It is common that once the travelling speed is not high enough, the weld pool will collapse. The minimum travelling speed is called the threshold of the travel speed. The threshold of the travel speed decreases with the increasing of the weldment thickness and the decreasing of the thermal conductivity. As a basic rule, the maximum travelling speed is always higher in the welding of thin plates compared with the thick ones. That is, because the threshold of the travel speed depends on the characteristic of the arc as well as the performance of the weldment. For example, the range of the speed was 300–350 mm/min in the welding of SAF2205 if the keyhole was stably formed. But for AISI 304 plate with thickness of 12 mm, the speed needed to be 175 mm/min [6]. These shows that a higher travelling speed can be performed during the thin plate welding, but the limitation depends on the material and the arc.

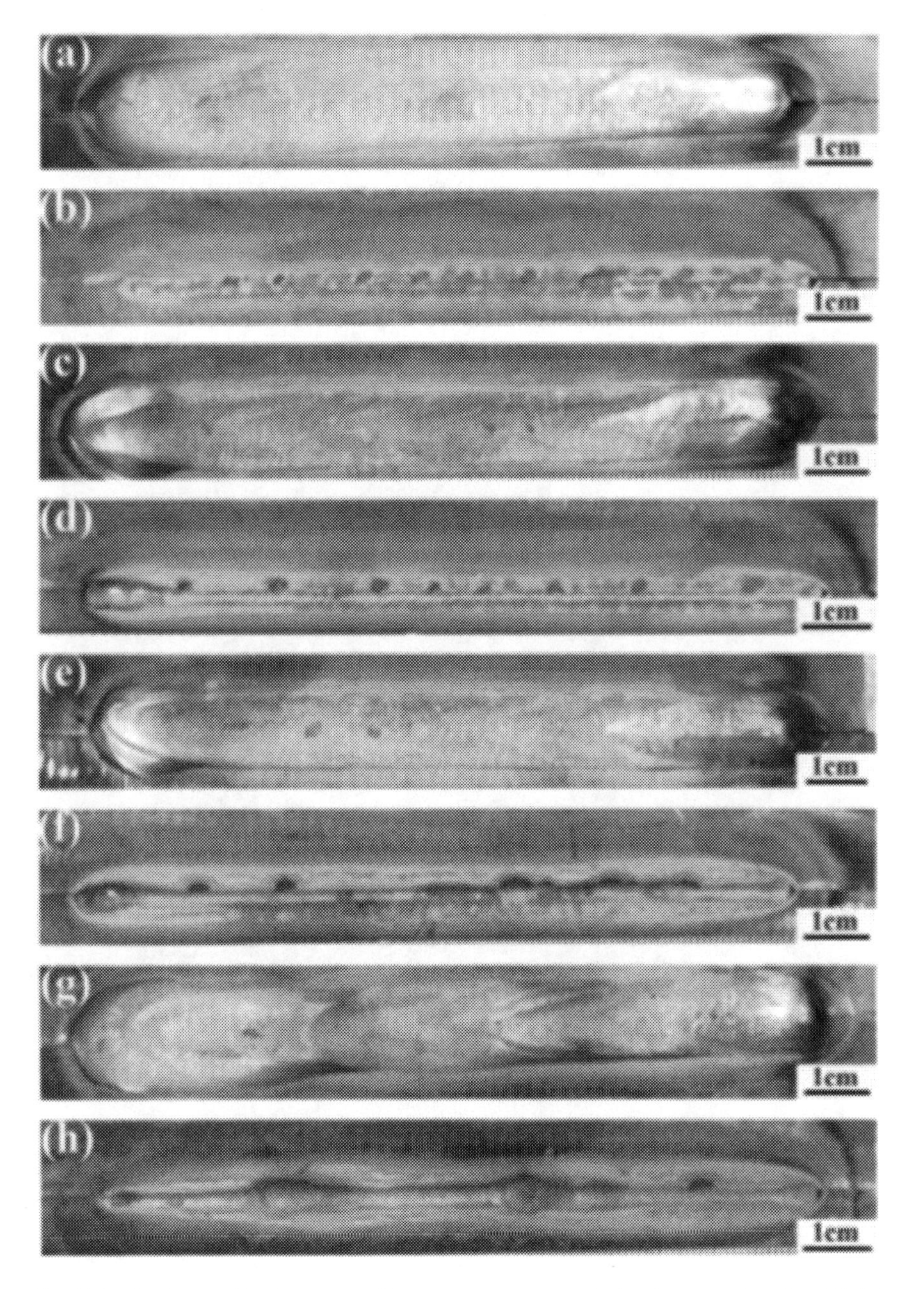

Fig. 10.4 The front and back pictures of the armour grade quenched and tempered steel **a**–**b** 525 A **c**–**d** 545 A **e**–**f** 565 A **g**–**h** 575 A [14]

10.2.2.3 Voltage

Voltage is also a key factor in the welding. According to the tests about on 5.6 mm SAF 2205 with K-TIG welding, the voltage must be set above 11.3 V to form a keyhole. The threshold current was over 620 A at 11.3 V, but it dropped to 510 A when the voltage was over 14 V. The decrease of the threshold current was obvious when the voltage was changing within a range between 12 and 13 V. That means that the parameters are sensitive in this range, as shown in Fig. 10.5.

Moreover, the role of the voltage during keyhole TIG welding was shown in the experiment of 12 mm AISI 304 plates. The travelling speed is 300 mm/min, and the argon is used as shielding gas with a mix of 10% added nitrogen. The tip of the torch electrode has an angle of 45°, and the current was controlled accurately, making the range of the change within ±5 A. The degree of difficulty containing a stable keyhole was recorded, as shown in Table 10.1. The voltage and the current suitable for the K-TIG welding were changing synchronously but an optimal set of the parameters of

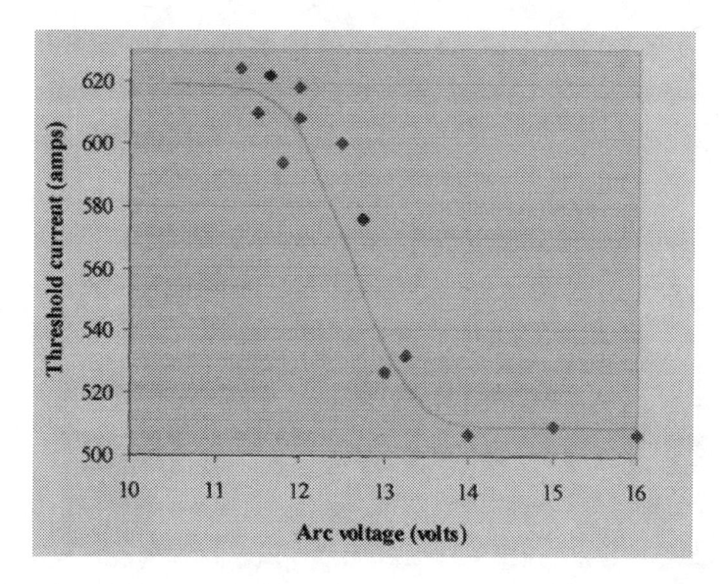

Fig. 10.5 Relationship between the threshold current and the voltage when a keyhole welding is performed in a 5.1 mm plate of SAF 2205 [6]

Table. 10.1 Effect of arc voltage on keyhole formation in 12 mm austenitic stainless steel [6]

Arc voltage (V)	Current (A)	Degree of difficult
15	750	Very difficult
16	730	Difficult
17	700	Comfortable
18	680	Becoming harder
19	650	Difficult

the voltage and current could be got through serious of experiments under a certain welding condition.

10.2.2.4 Shielding Gas

One of the influence of the shielding gas flow is that it will change the transformation of the heat and momentum, so the changing of its composition has influence on the threshold current. For pure argon, adding helium into the pure argon increases the threshold current while adding nitrogen or hydrogen reduces it, as shown in Fig. 10.6. When other parameters were fixed, the threshold current in pure argon was 40% higher than in mixture of Ar-10 N.

Different strategies of the gas backing were compared in the K-TIG welding in C–Mn steel carried out by Liu. The weldments were jointed with gas shielding. In addition, more tests are carried out with argon as covering and jet flow on the back, respectively. It showed that unstable keyhole existed without gas backing, and a narrow range of current from 410 to 450 A was suitable for keeping a stable keyhole. When the gas jets of argon flowed towards the back of the weldment, the current range could be expanded to 420–560 A. This finding gives a possible method to get a stable weld pool during full penetration welding [16].

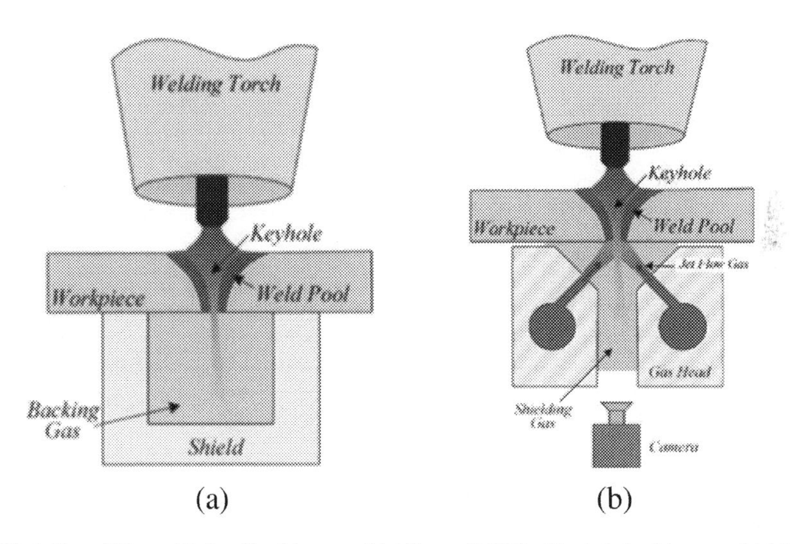

Fig. 10.6 Two different kinds of backing gas shielding. **a** K-TIG with static backing gas shielding. **b** K-TIG with jet flow gas [16]

10.2.2.5 Electrode Geometry

The geometry of the electrode is also the variable which influents the welding process. Based on the experience gained from GTAW welding, conical tips were only used in K-TIG welding, but two parameters of the electrode were investigated, including the diameter of the electrode and tip angle. Tests were done on the 5.6 mm SAF 2205. The threshold current when using an electrode tip angel of 90° was high. It is about 1.4 times of the threshold current of the 45° tip electrode. Between the angle of 45° and 60°, almost half of this changing occurred. While diameter of the torch electrode is decreased from 6.4 to 3.2 mm, threshold current was increased over 10%. In conclusion, both the increasing of the torch electrode diameters and the decreasing of the angle of the electrode tip led to a result that the threshold current decreased. Moreover, the decreasing of the tip angle also reduced the current but with a greater effect.

The cooling rate of the electrode was found to be an important factor influencing the weld quality. A GTAW torch with a water cooling of a high rate = n was used to achieve higher cooling rate, as shown in Fig. 10.7. The comprising tests were made by changing the currents during welding from 100 to 500 A in order to find out the difference of the welding performance of the ordinary cooling system and the highly cooling system. The results showed that the cooling tungsten had a more focusing arc which causing a smaller wide of heat region and deeper penetration of the weld seam.

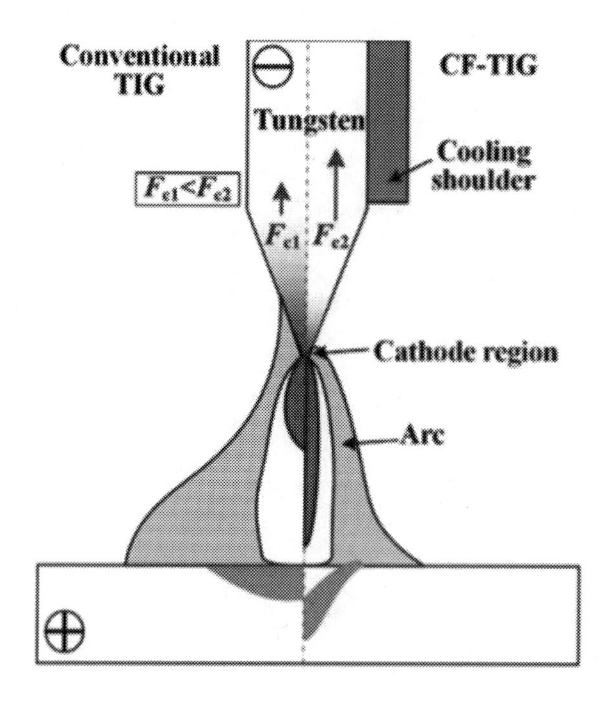

Fig. 10.7 Comparison of the design of conventional TIG electrode and the K-TIG electrode with cooling shoulder [17]

Wang proposed a new method of TIG welding by using double-pulsed variable polarity with ultrasonic frequency on 7 mm AA2219 aluminium alloy [18]. In the pulse peak stage of low frequency, the formation of the stable keyhole was achieved. Besides, the double pulses show a function of stirring the weld pool which resulted in grain structures with fully equiaxed within the weld zone. A consistent distribution can be seen in both the microstructure within the welding joint and its microhardness.

10.2.2.6 Material

More extensions of the trials on other materials were studied by various researchers. 8 mm AISI 430 steel plates were successfully welded with different modified of K-TIG welding, including using high-frequency pulse arc, adding austenite interlayer and imply post-welding heat treatment [19]. It was found that the heat input decreased, the primary columnar austenite generated.

Keyhole TIG welding was tested in amour grade quenched and tempered (Q&T) steel, and full penetrated seam was got on 9 mm thick plates, with speed of 28 cm/min [14]. Huang used K-TIG welding to joint different materials such as AISI 304 stainless steel and Q345 low-alloy steel of 8 mm thickness with just one pass of welding [20]. The microstructure and welding properties of dissimilar joints were analyzed, and it is concluded that a good weld seam jointed with different steels can be achieved with a relatively high welding speed within an operating window of 50 A.

Mid-thickness AISI316L stainless steel was welded by Feng [21]. It was found that the weld seam consists of different phases including austenite, dendritic structures and δ-ferrite. The seam and the base metal have a similar strength but the impact property within the seam was lower.

10.2.3 Application Extensions of K-TIG

10.2.3.1 Pipes Welding

Flat position weldment is the original application of the keyhole TIG welding, but actually the range of application can be extended including pipes welding. The keyhole TIG welding has be also used in girth welding of rotated pipes, which can effectively produce joints of high quality, as shown in Fig. 10.8. It is the preferred method over other welding technology because it eases the control of the welding parameter and produces good quality joints. However, some additional constraints are brought into the pipes welding compared with the flat plate welding.

One problem is the start of weld which will be overlapped at the end of the process, and defects may be left. In addition, the sudden termination of the K-TIG welding leads to porosity, underfill and burn-through. Especially, when the current drops down below the threshold and the pipes keep rotating, porosity is usually seen in the joints. That is, because the current reduction results in the close of the keyhole,

Fig. 10.8 K-TIG welding of pipe [6]

preventing the gas to escape. Therefore, proper adjustments of the current and the rotating motion are the solutions to this. By decreasing current slowly and keep the rotation of the pipes.

Another problem is the preheating of the pipe due to the high heat input for keyhole formation. The pipe diameter is limited due to this process, which is assumed to depend on the characteristics of the metal as well as the thickness of the weldment.

10.2.3.2 Underwater K-TIG Welding

Local-dry keyhole TIG welding underwater with DSS plates was studied by Cui in 2019 [22]. The keyhole welding system was added with a local-dry cavity that can exclude water to create high-quality welded joints during the welding procedure, as shown in Fig. 10.9. The impact toughness of the welded joints has 78% of the onshore weld metal, and some improvement was found in the weld joint than the base metal.

10.2.3.3 Wire Feeding K-TIG Welding

Filler material was used in the K-TIG welding of armour steel joint [23]. Fei was the first one to introduce this to improvement the properties of the weld seam as well as the microstructure. The results showed that the wire feeding speed can be set to 3 m/min, and the 6.2 mm armour steel plates can be penetrated fully at the

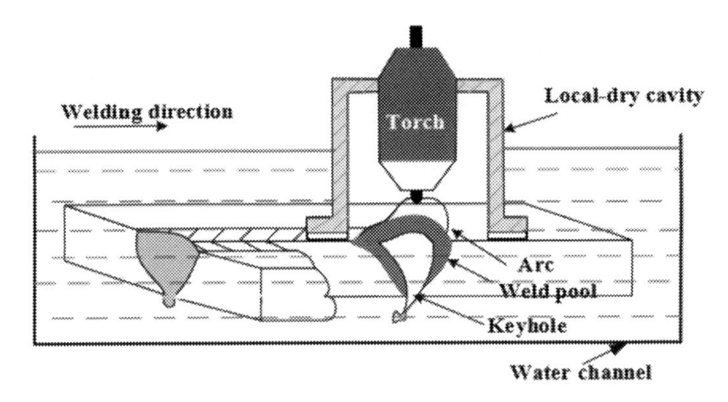

Fig. 10.9 Underwater local-dry K-TIG welding equipment [22]

same time. It explodes the application of K-TIG process and showed potential to the production of high-quality armour steel joint [24]. Other researchers using dynamic wire feeding system, which used a robotic manipulator in two steps of 180° in vertical down progression. The system helps to reach good process stability and robustness for joints.

10.2.4 Keyhole Stability

The modelling of the keyhole GTAW is important when analyzing the balance in the welding keyhole. Setting the liquid density to be ρ, the acceleration of gravity to be g. r represents the keyhole radius. h represents the liquid head height, and γ stands for the surface tension coefficient. The factors that cause the keyhole to collapse are ρgh (Hydrostatic) and surface tension pressure. And the keyhole is hold in balance with the recoil pressure coming from the arc. A saddle shape model of the keyhole is used to describe the profile of the keyhole; thus, the surface tension can be approximately calculated. There are two rules that keep the keyhole stable instead of collapse or under-penetration.

10.2.4.1 Fist Keyhole Stability Rule

In order to get the full penetration of the weldment, the maximum height of the keyhole h_m must be greater than the plate thickness h_w so that the force caused by the surface tension is enough to support the liquid metal in the full penetrated keyhole. The maximum depth supported by surface tension is given by [25]

$$h_w \leq h_m = \frac{\gamma}{\rho g}\left(\frac{1}{r_a} + \frac{1}{r_b}\right) \quad (10.1)$$

Table. 10.2 Different range of the value of $h_w w$ corresponding to various states of the keyhole

Case	State of the keyhole	Product of h_w and w
$\frac{2\gamma}{\rho g} \geq h_w w$	A stable weld pool is obtained	<28
$\frac{2\gamma}{\rho g} < h_w w \leq \frac{4\gamma}{\rho g}$	The stability of the keyhole depends on r_b and will experience a fail when the travel speed reach its limits	28–56
$h_w w > \frac{4\gamma}{\rho g}$	The weld pool will collapse	>56

where r_a and r_b are the principle radii of the curve in keyhole model. Supposed that r_a is very large which causes the weld pool collapsing, then the pool will be supported with a high welding speed:

$$\rho g h_w \leq \frac{2\gamma}{r_a} \tag{10.2}$$

Based on the above analysis and given that $w = 2r_a$, the formation and stability of the keyhole are listed in Table 10.2.

10.2.4.2 Second Keyhole Stability Rule

The optimal thickness of the weldment can be described by the geometry of the weld pool. w_f is the width of the front face while w_r represents the width of the beam on the root. The critical value of h_w is given by

$$h_w = \sqrt{w_r w_f} \tag{10.3}$$

If $\sqrt{w_r w_f} > h_w$, then the molten pool will unzip, and the surface energy will experience an overall decrease. Otherwise, when $\sqrt{w_f w_r} < h_w$, the unzipping will not proceed.

10.2.4.3 Keyhole Failure

When the thickness of the weldment is too high, the keyhole will become unstable and crash because of the leaking of liquid metal via the exit of the keyhole. The leaking liquid metal freezes with the shape of 'stalactites', and this phenomenon appears repeatedly. The metal of weld pool drained out when the depth of pool increases to a critical level, and the keyhole is formed again after that. The metal becomes bigger when the oxygen is excluded from the bottom of the weldment or replacing the weldment of AISI 304 to 3CR12 or C–Mn steel.

The 'first stability rule' for keyhole GTAW fails in the modelling of the thin plate (3 mm) K-TIG welding. The liquid metal becomes a serious of bead when flows along the side of the weld seam. At the same time, the keyhole is not close. The failure of the keyhole can be seen as a process of splitting the liquid into two part and pulling them towards to different sides of the weld seam.

10.2.5 Arc Forces in K-TIG Welding Keyhole

The interface between the liquid and solid metal can be found within the cavity of the keyhole. The material that filled in this cavity is called the weld pool whose surface is exposed to the air or shielding gas atmosphere. In K-TIG welding, a crater is generally found at the end of the joint, which has similar shape as keyhole. The terminal crater is caused by the absence of the metal at the end of the joint. The volume of the cavity could be measured by suddenly stop the welding process, and the deficit can also be determined. During the K-TIG welding, arc force is important in causing the cavity, and it can also be described in formula.

10.2.5.1 Radial Force Due to Magnetic Field

When the current is flowing through the arc plasma, the radial incremental compressive force f_{jbr} and internal pressure are balanced with each other. r_{arc} is the radius of the arc flowing, which is presented as r for convenient. The current $I(r_{\text{arc}})$ can be expressed in a different form which is a function of the r_{arc}, and the total radial force F_{jbr} is given by [26]

$$F_{jbr} = \int_{0}^{I_0} \frac{\mu I(r_{\text{arc}})}{4\pi} \mathrm{d}I(r_{\text{arc}}) = \frac{\mu I_0^2}{8\pi} \tag{10.4}$$

10.2.5.2 Axial Force Due to Magnetic Field

The geometrically divergent within the arc causes electromagnetic force parallel to the arc axis. The model of the arc starts from $z = e$ and ends at $z = a$. Letting R_e and R_a be the arc radius at these two planes, the axial magnetic force of the is given by integration:

$$F_{jbz} = \frac{\mu I_0^2}{4\pi} \ln\left(\frac{R_a}{R_e}\right) + \int_{0}^{R_e} \frac{\mu I^2(r, e)}{4\pi r} \mathrm{d}r - \int_{0}^{R_a} \frac{\mu I(r, a)}{4\pi r} \mathrm{d}r \tag{10.5}$$

The combination of $J_z \times B_\theta$ and $J_r \times B_\theta$ is equal to the net force on the molten pool. Therefore, the arc force caused by the arc which is axial symmetry and has a arbitrary distribution of current which can be descried as:

$$F_{jb} = F_{jbr} + F_{jbz} = \frac{\mu I_0^2}{8\pi}\left(1 + 2\ln\left(\frac{R_a}{R_e}\right)\right) + \int_0^{R_e} \frac{\mu I^2(r,e)}{4\pi r}\mathrm{d}r - \int_0^{R_a} \frac{\mu I^2(r,a)}{4\pi r}\mathrm{d}r \tag{10.6}$$

10.2.5.3 Force Changes with Current

Actually, the metal material along the arc axis is moved away completely during the K-TIG welding. As a result, the formation of the keyhole means that the current density is zero, where the liquid metal is pushed away by the arc. As a result, the model of the distribution of the current can be modified as

$$I = I_0\left(\frac{r}{R_e}\right)^n \tag{10.7}$$

where $r \leq R_a$. When the n is set to be very large, the force is given by

$$F_{jb} = \frac{\mu I_0^2}{8\pi}\left(\frac{3}{2} + 2\ln\frac{R_a}{R_e}\right) \tag{10.8}$$

10.2.6 The Input and Conductivity of Heat

The keyhole stability of K-TIG welding is largely which depends on the weld pool size, which has great relationship with the heat input and cooling rates on the weldment. Therefore, the modelling of the temperature fields becomes important in K-TIG welding studies. Rosenthal firstly proposed the mathematical description of the temperature in cutting and welding process. Then, the moving point source was also studied.

10.2.6.1 Temperature Fields

With several assumptions in the quasi-stationary temperature fields model provided by Rosenthal's [27], the temperature fields two dimension can be described as:

$$T = \frac{Q}{2\pi kG}\exp\left(-\frac{vE}{2\alpha}\right)K_0\left(\frac{vR}{2\alpha}\right) \tag{10.9}$$

While in three dimension, it is modified as:

$$T = \frac{Q}{2\pi k R} \exp\left(-\frac{vE}{2\alpha}\right) \exp\left(\frac{vR}{2\alpha}\right) \tag{10.10}$$

where Q is the input energy; k is the thermal conductivity; α is diffusivity; G is the thickness of the weldment; v represents the moving speed of the heat source. R is the Euclidean distance which has different value in various dimension. The value of E is $X - vt$.

10.2.6.2 Temperature Distribution with High Welding Speeds

In some cases, the travel speed so high that vt is thought to be very large. As a result, the sum of R and E becomes:

$$R + E = \sqrt{(vt)^2 + Y^2 + Z^2} - vt \approx \frac{R^{*2}}{2vt} \tag{10.11}$$

Letting $\lambda = \frac{k}{\rho c}$, where density is represented as ρ; the thermal capacity is represented as c and $R^{*2} = Y^2 + Z^2$. Then, the distribution of the temperature is modified as

$$T = \frac{Q}{2\pi \lambda \rho c R} \exp\left(-\frac{R^{*2}}{4\lambda t}\right) \tag{10.12}$$

10.2.6.3 Melting Efficiency

According to Well's study [28], for the thin plate, the K-TIG welding could be seen as 2D fusion process. The melting point isotherm T_c was equated with the weld pool boundary. When its tangent is parallel to the travelling direction, the isotherm is at its maximum width. In this case, the losses inherent can be estimated during the fusion process. First, the ratio of total heat input Q to the heat required to melt the material Q_m was defined as 'melting ratio' or M_r.

For thin plate, the situation can be considered as two dimension:

$$M_r = 2\left(1 + \frac{2\alpha}{5vD}\right) \tag{10.13}$$

where D is the width of the beam. On the other hand, for thickness plate, it can be seen as three dimension:

$$M_r^{-1} = \frac{Q_m}{Q} = \frac{r_m + 2{r_m}^2)}{2(2r_m + 1)^2} \exp\left(\frac{-r_m}{r_m + 1}\right) \tag{10.14}$$

where r_m is the maximum of the value of $\frac{\upsilon R}{2\alpha}$.

10.2.6.4 Recent Research on Heat Input of K-TIG Welding

Fei's study showed that the heat input was another factor except for travel speed that influenced the undercut defect forming, because it changed the plasma jet trajectory inside the keyhole [29]. Large heat input led to the formation of hump in the weld centre and exacerbation of undercut formation, as shown in Fig. 10.10.

Liu applied a classic heat source model that widely used in PAW keyhole welding simulation to the analysis of K-TIG welding [30]. Then, the distribution of the heat input was given below.

For internal constrained arc heat source:

$$q_i(x, y) = \exp\left(-\frac{3r^2}{\sigma_i^2}\right)\frac{3\eta U I \varphi_1}{\pi\sigma_i^2\left(1 - \exp\left(-\frac{3r_1^2}{\sigma_i^2}\right)\right)} \quad (r_1 > r) \tag{10.15}$$

For outer free arc heat source

$$q_o(x, y) = \exp\left(-\frac{3r^2}{\sigma_k^2}\right)K_q\frac{3\eta U I \varphi_2}{\pi\sigma_k^2 exp\left(-\frac{3r_1^2}{\sigma_k^2}\right)} \quad (r_1 \le r) \tag{10.16}$$

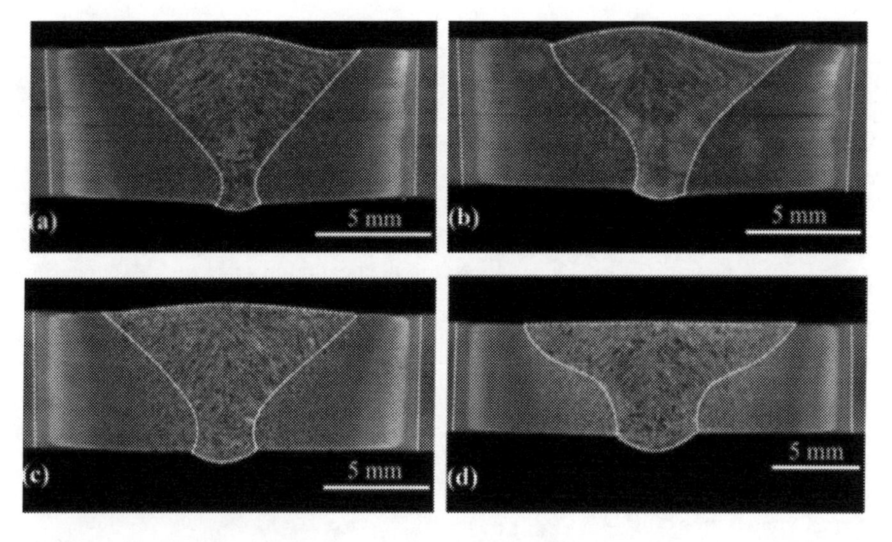

Fig. 10.10 Macrographs of four typical welds with different heat input. **a** 7.78 kJ/cm **b** 8.1 kJ/cm **c** 9.18 kJ/cm **d** 11.16 kJ/cm [29]

where $K_q = \frac{\frac{h}{L}+2}{3}$, $\varphi_1 + \varphi_2 = 1$. They are energy distribution coefficients; arc voltage is represented as U; thermal efficiency is represented as ηwelding current is represented as I. σ_i and σ_k are heat distribution parameters; the changing height of the keyhole is represented as h, and r_1 stands for the arc radius.

10.3 Signal Characteristics in K-TIG Welding

Keyhole Tungsten Inert Gas welding is a much complex physical process, during which the thermal field, optical field, acoustic field, force filed, magnetic field and electrical field are coupling with each other. There is a certain relation between different physical phenomena, indicating the real-time welding quality somehow.

In this chapter, experiments on Ti alloy are carried out (Table 10.3). The arc voltage signal, arc acoustic signal and arc current signal in K-TIG welding are collected and analyzed. Figure 10.11 shows the normalized signals, which are collected in the 500

Table. 10.3 Experimental parameters for Ti alloy

Materials	Welding current	Workpiece dimensions	Welding speed	Flow rate	Shielding gas
TC4	500–560 A	300 × 100 × 12 mm	300 mm/min	15 L/min	Ar

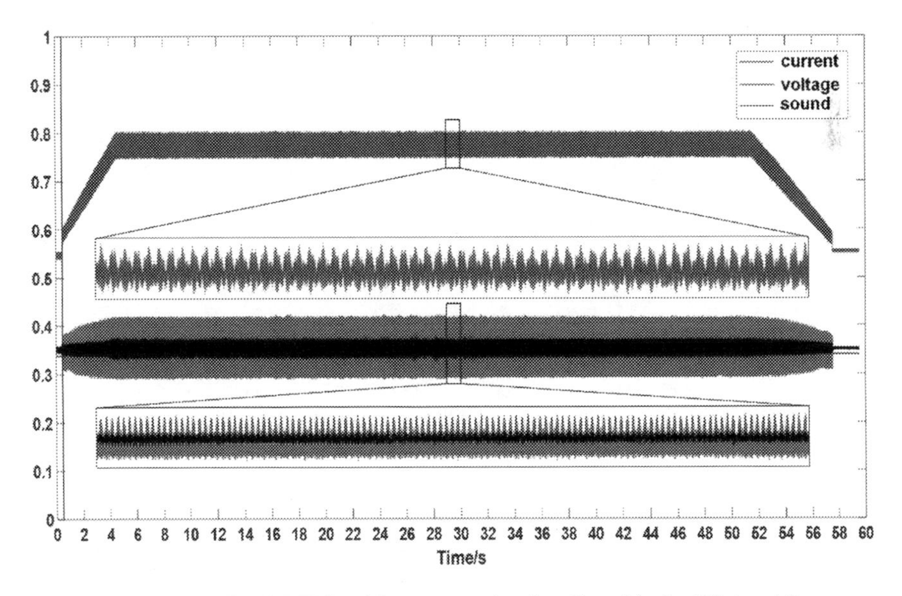

Fig. 10.11 The normalized K-TIG welding process signals collected in the 500 A welding current experiment of Ti alloy [31]

A welding current experiment of Ti. Characteristics extracted from these signals are discussed. Penetration recognition method is then introduced.

10.3.1 Electrical Signal Characteristics

The phenomenon of 'DC bias' caused by zero drift of the signal conditioner and the DC coupling collection pattern is illustrated in the signals originally collected. In order to remove the DC bias, a 50 Hz high-pass filter is adopted. To avoid the interference of arc ignition and ending process, a piece of signal(around 12 s) is selected during the stable welding process.

10.3.1.1 Statistical Characteristics in Time Domain

Generally speaking, K-TIG welding works in a constant current. However, due to the variation of the height of the arc column or many other factors, the collected electrical signal presents the volatility. Some statistical features can be helpful to analyze the electrical signal in time domain.

The degree of scatter of electrical signal can be expressed as the variance D. The bigger the variance, the more dispersed the signal.

$$D = \frac{\sum_{i=0}^{n} (s_i - \bar{s})^2}{n - 1} \tag{10.17}$$

where n is the window size 3000; $\bar{s}$ is the mean value; and s_i is the ith point in the signal.

The signal energy can be expressed as the root-mean-square R.

$$R = \sqrt{\frac{\sum_{i=0}^{n} s_i^2}{n}} \tag{10.18}$$

The morphological characteristics of the signal can be expressed as the kurtosis K.

$$K = \frac{E(s - \bar{s})^4}{\sigma_t^4} \tag{10.19}$$

where $E(s - \bar{s})^4$ is the fourth order central moment of the signal; σ_t is the standard deviation of the signal.

According to Eq. (10.17), the variance of the electrical signal is drawn in Fig. 10.12. The variance of arc current is bigger than that of arc voltage, as the

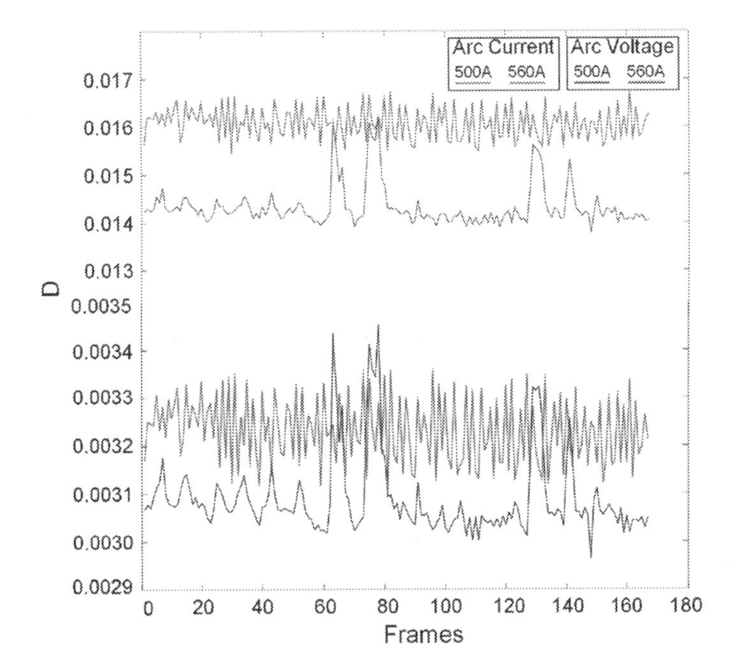

Fig. 10.12 Variance of the electrical signals

sensitivity of hall current sensor and the high current welding status. When the backside keyhole not formed, the variance of the electrical signal is smaller than that in the status when the backside keyhole formed. The 'peak' in arc voltage appears simultaneously with that in arc current.

According to Eq. (10.18), the root-mean-square of the electrical signal is drawn in Fig. 10.13. The root-mean-square R represents the signal energy, which results in the bigger R in arc current than that in arc voltage. The curve in Fig. 10.13 shows the short-time energy variation of the electrical signal.

According to Eq. (10.19), the kurtosis of the electrical signal is drawn in Fig. 10.14. The numerical range of kurtosis is similar no matter in arc current signal or arc voltage signal. The kurtosis characteristic is supposed to reveal the variation of the weld pool surface, i.e. the weld pool oscillation [32]. Therefore, the similar numerical range of kurtosis shows the fact that the electrical signal is somehow in relation with the weld pool oscillation. However, the kurtosis in arc voltage signal is much more separable than that in arc current signal.

10.3.1.2 Characteristics in Frequency Domain [31]

Figure 10.15 illustrates the signals collected in the experiments shown in Table 10.3 in the frequency domain. Figure 10.15 demonstrates the correlation among arc current, voltage and arc sound. Four remarkable frequency bands exist in frequency domain:

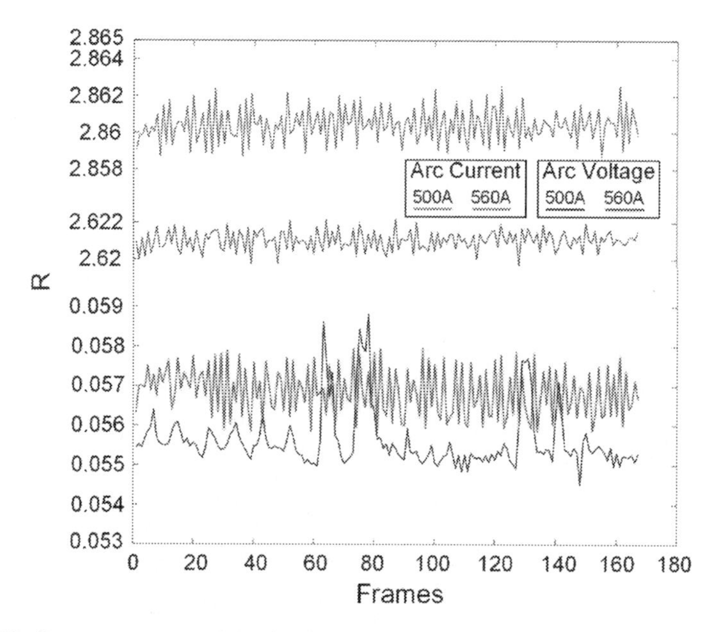

Fig. 10.13 Root-mean-square of the electrical signal

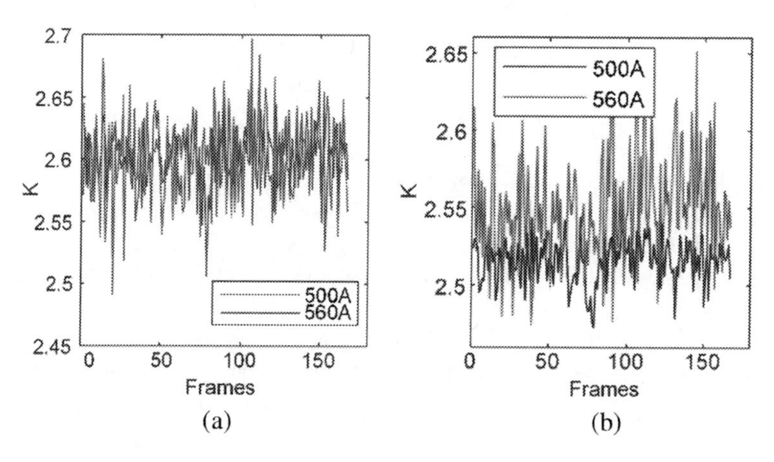

Fig. 10.14 Kurtosis of the electrical signal. **a** Kurtosis of arc current. **b** Kurtosis of arc voltage

2–4, 6–7.5, 15.5–16.5 and 19–20 kHz, among which the first frequency band at 2–4 kHz has the most power.

As Fig. 10.15 shows the largest power, among the four frequency bands is concentrated within the 2–4 kHz band. Figure 10.16 shows the power spectrum in the range of 2–4 kHz of current signals in the tests by Ti. According to Fig. 10.16, not only similar are the power spectra in Band 1 under different weld states, but a frequency delay exists. By changing the welding current from 500 to 560 A, different states of

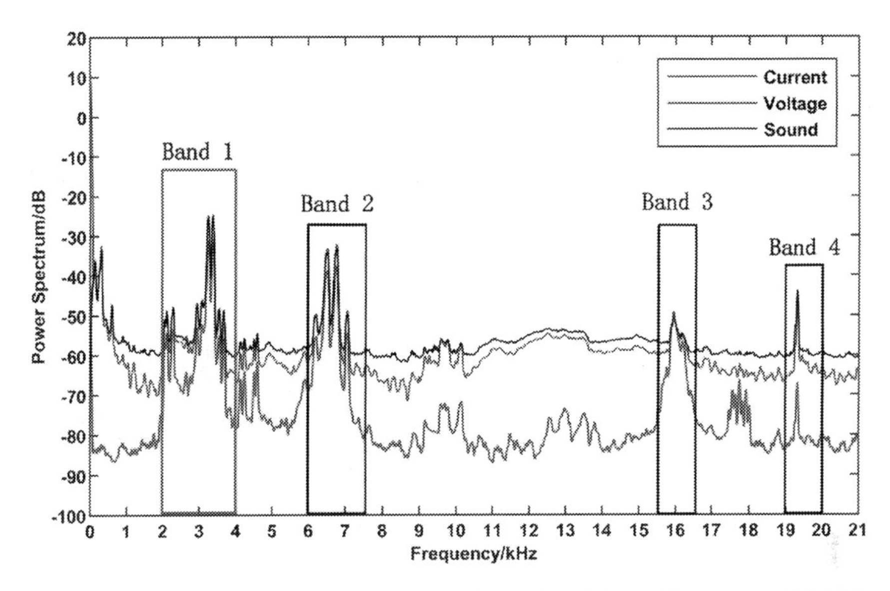

Fig. 10.15 Power spectrum of signals collected during testing with a welding current of 500 A by Ti [31]

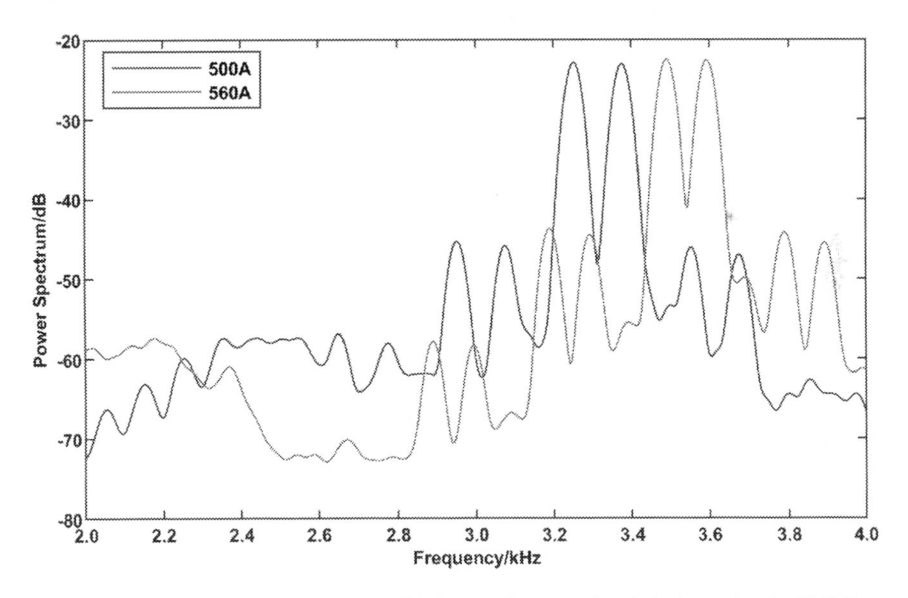

Fig. 10.16 Power spectrum in the range of 2–4 kHz of current signals in the testing by Ti [31]

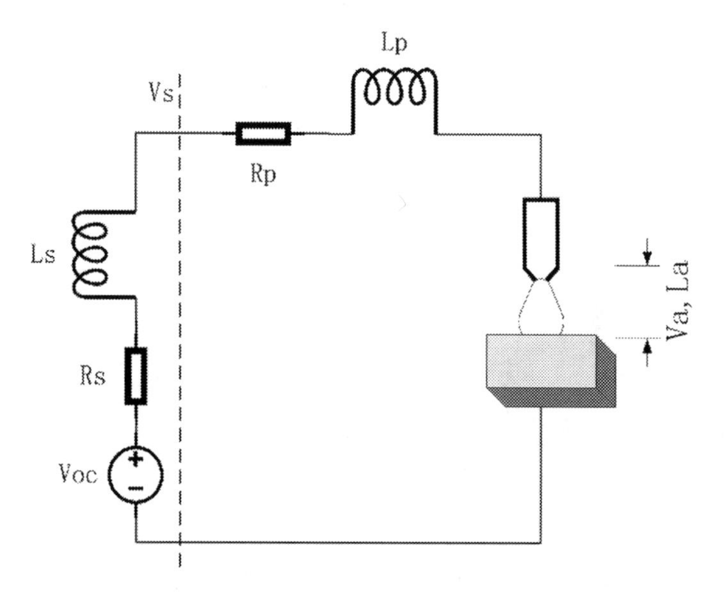

Fig. 10.17 Welding circuit [31]

the weld seam, i.e. partial and complete penetration, can be artificially maintained. For K-TIG welding, a rear keyhole is present in the full penetration state while it is not present in the partial penetration state.

To give an exposition to the above phenomenon, 'frequency delay', circuit of the welding system (shown in Fig. 10.17) must be analyzed.

Jarvis [33] pointed out the arc resistance Ω as Eq. (10.20).

$$\Omega = \frac{\omega_g L_a}{A_e \sin\theta} \cong \frac{\omega_g L_a}{I \sin\theta} \tag{10.20}$$

where the resistivity of Ar is represented as ω_g; the tip angle is represented as θ; the arc column length is represented as L_a; and the emission area of the cathode is A_e.

Defining $k = \frac{\omega}{\sin\theta}$, V_a, the arc voltage can be derived as Eq. (10.21):

$$V_a = V_{ao} + kL_a \tag{10.21}$$

Due to the DC welding property, the inductance in the welding circuit is convincing enough to be ignored. Hence, V_s, the output voltage of the welding power supply, is able to be derived as the following equation:

$$V_s = V_p + V_a = V_{ao} + IR_p + kL_a \tag{10.22}$$

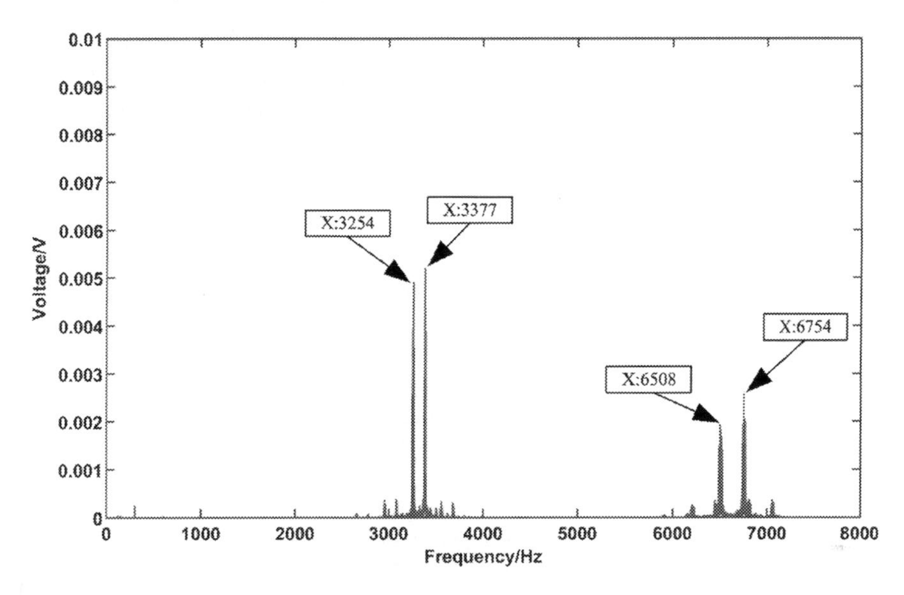

Fig. 10.18 Frequency spectrum of the arc column length [31]

where V_{ao} is a constant, and R_p is the resistance of the welding cable. In addition, the monitored voltage signal is V_s as the direct connection between the voltage hall sensor and the output of the welding power supply.

The result can be simplified and derived as the following equation through performing a Fourier transform in Eq. (10.22):

$$\mathcal{F}(kL_a) = \mathcal{F}(V_s) - \mathcal{F}(V_{ao}) - \mathcal{F}\big(IR_p\big) \tag{10.23}$$

On the basis of Eq. (10.23), there is relationship between the frequency information of the arc column length L_a and V_s. The frequency spectrum of the arc column length L_a is shown in Fig. 10.18. Through analyzing arc voltage signal, four frequency features can be extracted. It can be deduced that the frequencies of 3377 and 3254 Hz are predominant, while the frequencies of 6754 and 6508 Hz are harmonic.

In order to analyze the frequency features, the voltage signal is filtered by a bandpass filter (3254–3377 Hz). The filtered arc voltage signal is shown in Fig. 10.19. It is obvious that an evident period characteristic concealed in the arc voltage signal after filtering. This evident period characteristic can be pointed out through pondering the filtered arc voltage signal, and Fig. 10.19 shows the upper envelope of the filtered arc voltage signal. Frequency characteristic of arc column length L_a is presented in this upper envelope.

Figure 10.20 illustrates the frequency spectrum of the upper envelope of the filtered arc voltage signal. A remarkable frequency feature has been dug out, which means the variation of the arc length. Regardless of the surface irregularity of weld plates

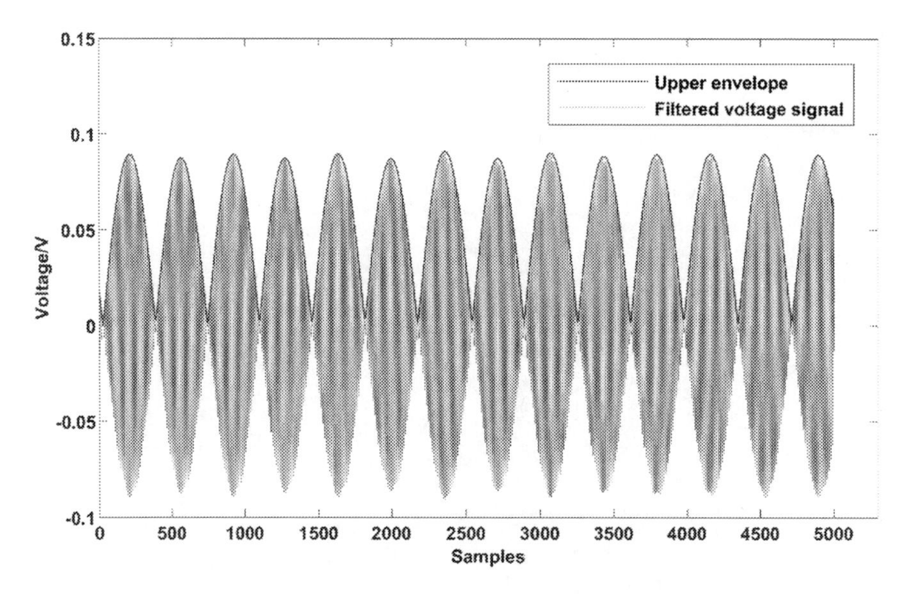

Fig. 10.19 Upper envelope of the filtered arc voltage signal [31]

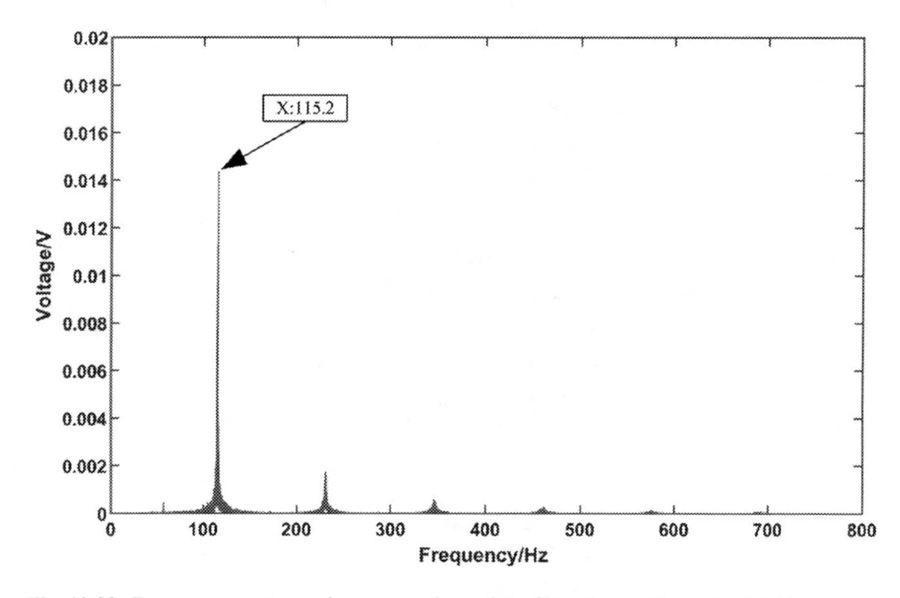

Fig. 10.20 Frequency spectrum of upper envelope of the filtered arc voltage signal [31]

and the displacement of the fixed weld gun, the variation of L_a is the effect of weld pool oscillation.

Seven images of weld pool within 0.5 s are shown sequentially in Fig. 10.21. Due to the reflection of light from an intense arc, the light area in the weld pool changes

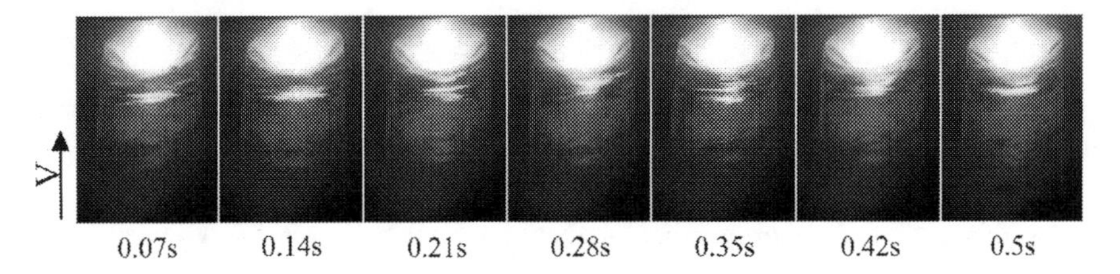

Fig. 10.21 Weld pool oscillation process during K-TIG welding [31]

with the conversion of the molten metal from a convex to a concave surface. The oscillation process is shown in Fig. 10.21.

10.3.2 Acoustic Signal Characteristics

Figure 10.11 shows the collected arc sound signal. Two dominating noises exist in the monitored arc sound signals. First is the DC bias voltage due to the zero drift of signal conditioner and the DC coupling circuit. The second is the ambient noise from KUKA controller, the cooling water source and cooling fans of the welding power source, etc.

The power spectrum of the noise and arc sound signal is shown in Fig. 10.22. The energy of noise mainly focuses on 0–8 kHz, and it encounters with another peak in 16 kHz. It accounts for the ambient noise of arc sound signal during most of the frequency bands. A conclusion is able to be drawn that the dominating noise in arc sound signal is the environmental noise illustrated in Fig. 10.22. A simple passband filter could not work well and denoise the arc acoustic signal.

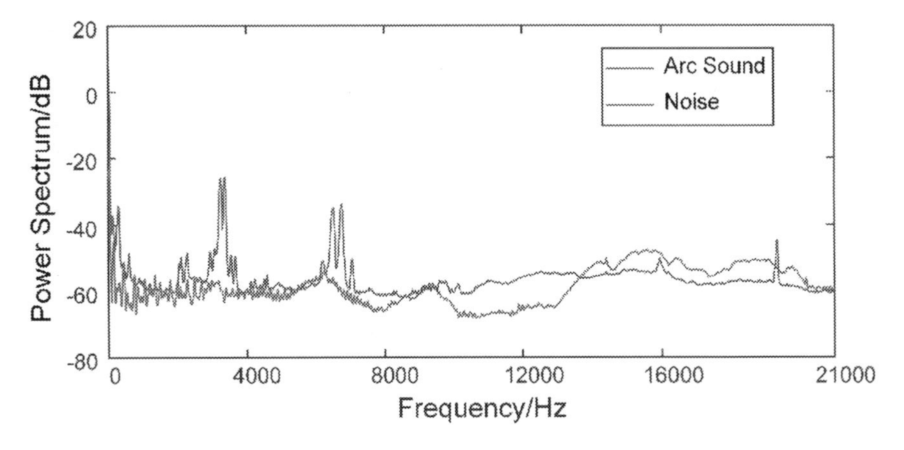

Fig. 10.22 Power spectrum of arc acoustic signal and noise [34]

In the light of the aforementioned analysis, the monitored arc acoustic signal could be derived as:

$$S(t) = y(t) + D = s(t) + n(t) + D \tag{10.24}$$

where the DC bias is expressed as D; the pure arc sound signal is $s(t)$; the DC-removal arc acoustic signal is represented by $y(t)$; $S(t)$ is the monitored signal; and the ambient noise is expressed as $n(t)$. The 50 Hz high-pass filter is used to eliminate DC offset and power frequency interference.

Since the noise throughout the entire arc acoustic signal is stable, a pure arc acoustic signal $s(t)$ can be obtained by doing subtraction between the additive noise $n(t)$ and the original monitored signal $S(t)$. Therefore, to denoise the arc sound signal, the Spectral Noise Subtraction (SNS) method is brought in. Firstly, transferring both sides of Eq. (10.24) through FFT:

$$Y(\omega) = S(\omega) + N(\omega) \tag{10.25}$$

Then, the power spectrum of Eq. (10.25) can be derived as:

$$|Y(\omega)|^2 = |S(\omega)|^2 + |N(\omega)|^2 + 2Re\left[S(\omega)\overline{N(\omega)}\right] \tag{10.26}$$

where because of the existence of the additive and stable noise $s(t)$, $Re\left[S(\omega)\overline{N(\omega)}\right]$ is zero in one frame. Therefore, the amplitude of $S(\omega)$ can be derived as the following equation:

$$|S(\omega)| = \sqrt{|Y(\omega)|^2 - |N(\omega)|^2} \tag{10.27}$$

Nevertheless, from the point of mathematic, the consequence of $|Y(\omega)|^2 - |N(\omega)|^2$ is probably negative, by which a new noise called 'musical noise' may be brought in. The flow chart of the reformative spectral noise subtraction algorithm is shown in Fig. 10.23 [35]. Figures 10.24 and 10.25 show the denoised arc sound signal. The noise is suppressed, and the pure arc acoustic signal is maintained.

10.3.2.1 Statistical Characteristics in Time Domain

To avoid the interference of arc ignition and arc ending process, a piece of signal (10 s) is selected during the stable welding process. The statistical characteristics are detailed in Sect. 10.3.1.1. However, considering that an experienced welder can weld perfectly by acoustic sensing, the logarithmic response of the ear should be taken into account. The logarithmic energy (Le) is introduced to express the signal energy.

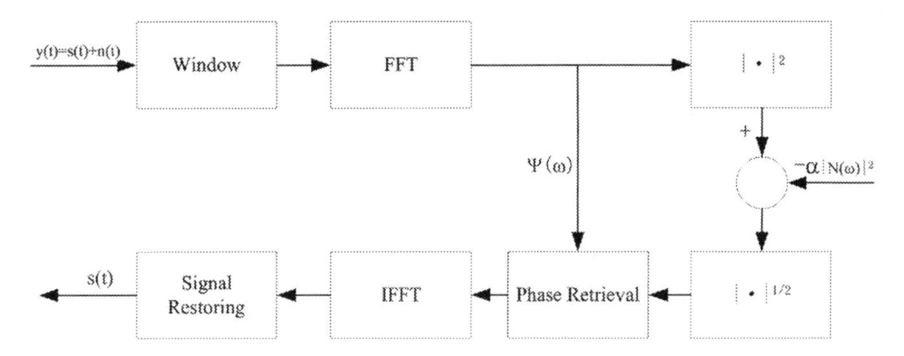

Fig. 10.23 Flow chart of the reformative spectral noise subtraction algorithm [34]

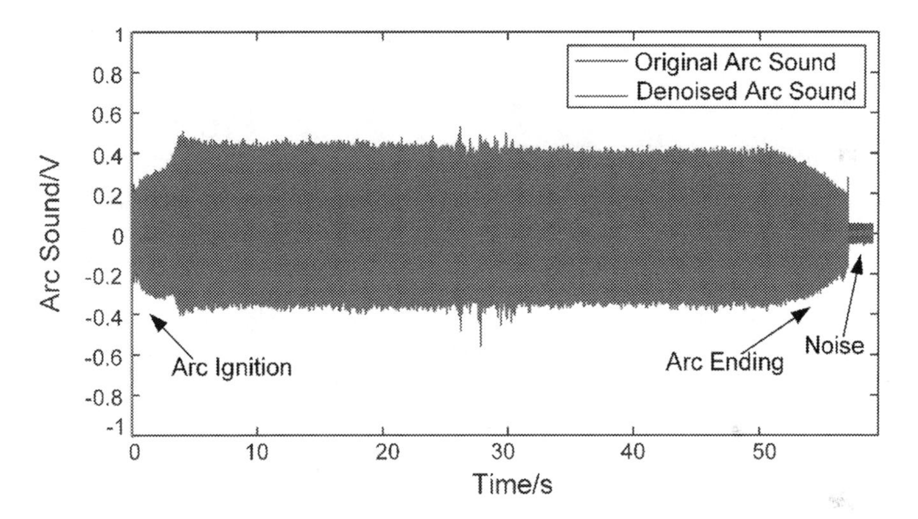

Fig. 10.24 Contrast of original and denoised arc acoustic signal in time domain [34]

$$Le = \log E = \log \sum_{i=1}^{n} s^2(i) \tag{10.28}$$

The variance of the acoustic signals is drawn in Fig. 10.26. The numerical range of variance in 500 A is smaller than that in 560 A, which means that the arc sound in full penetration state is more furious than that in partial penetration state.

The kurtosis of the acoustic signals is drawn in Fig. 10.27. The logarithmic energy of the acoustic signals is drawn in Fig. 10.28. We can draw a similar conclusion with that in Fig. 10.26 that the arc sound in full penetration state is more furious.

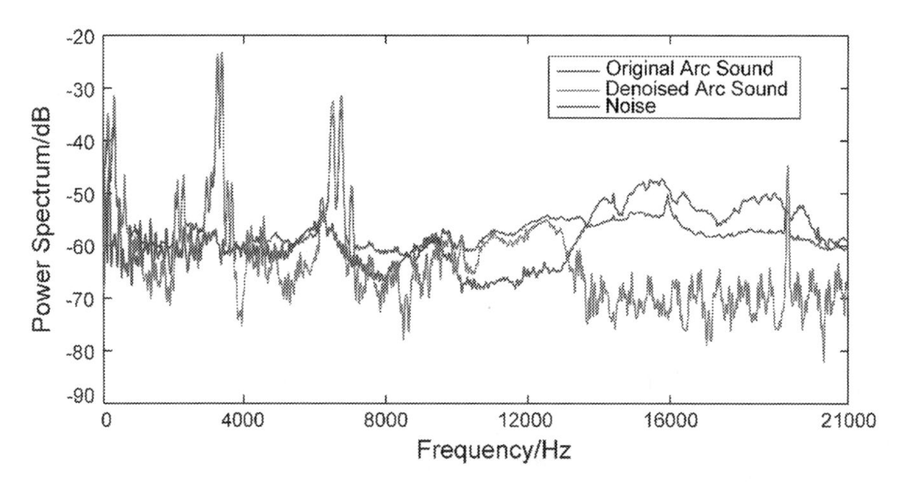

Fig. 10.25 Contrast of original and denoised arc acoustic signal in frequency domain [34]

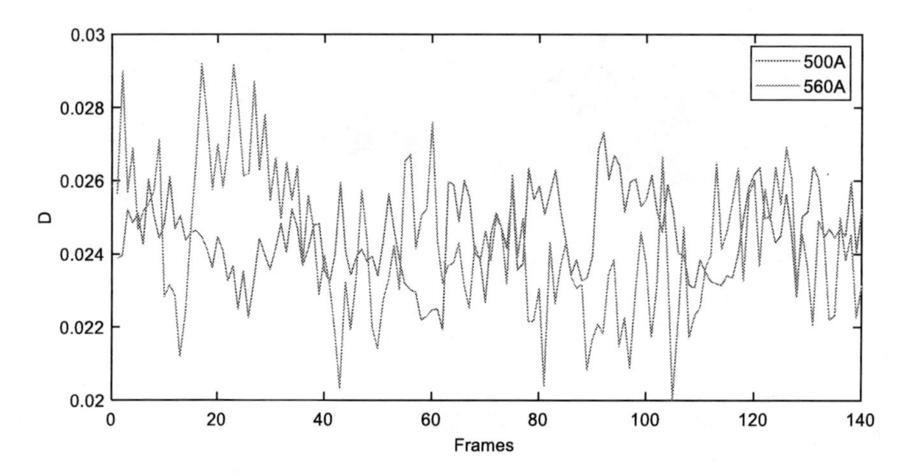

Fig. 10.26 Variance of the acoustic signals

10.3.2.2 Mel-Frequency Cepstral Coefficients

Considering the character of logarithmic response of the human ear, the MFCCs can reveal more details of the target signal in the frequency domain.

Figure 10.29 shows the 24 order MFCCs of the arc sound signal. The first of the coefficients is the logarithmic energy of the acoustic signal in each frequency part. The others are the 24 order MFCCs.

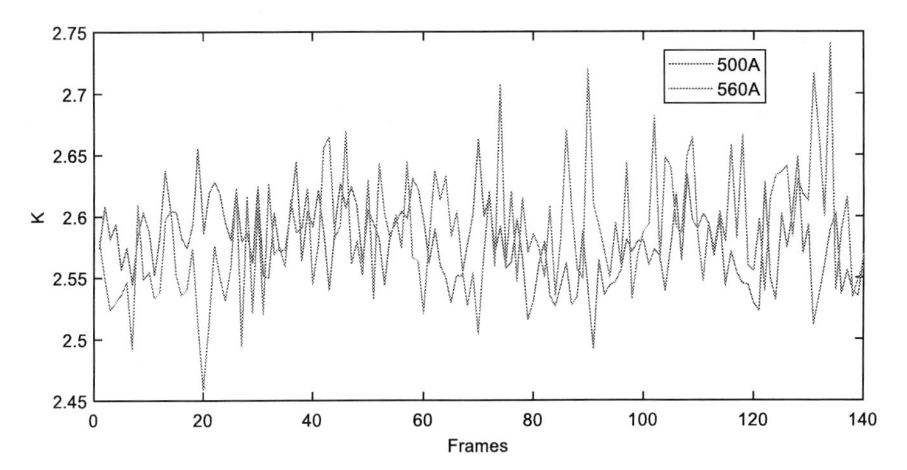

Fig. 10.27 Kurtosis of the acoustic signals

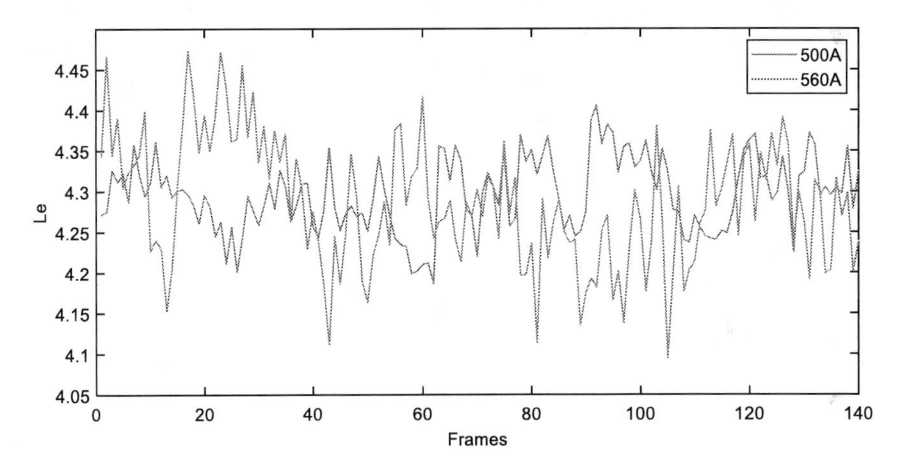

Fig. 10.28 Logarithmic energy of the acoustic signals

10.3.3 Penetration Recognition

After plenty of welding tests, we have found three penetration types. They are shown in Fig. 10.30: excessive penetration, full penetration and partial penetration. Three levels of welding current are utilized, 420 A, 500 A and 580 A, which contributes to the three penetration types, respectively. We have done some experiments on 304 stainless steel. The arc current, arc voltage and arc acoustic signals are collected simultaneously. Details about the experiments are shown in Table 10.4.

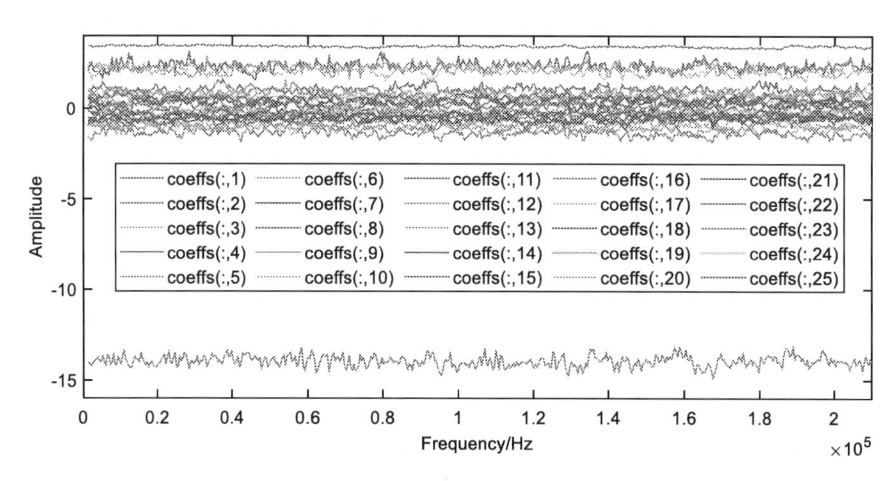

Fig. 10.29 MFCCs of the arc sound signal [34]

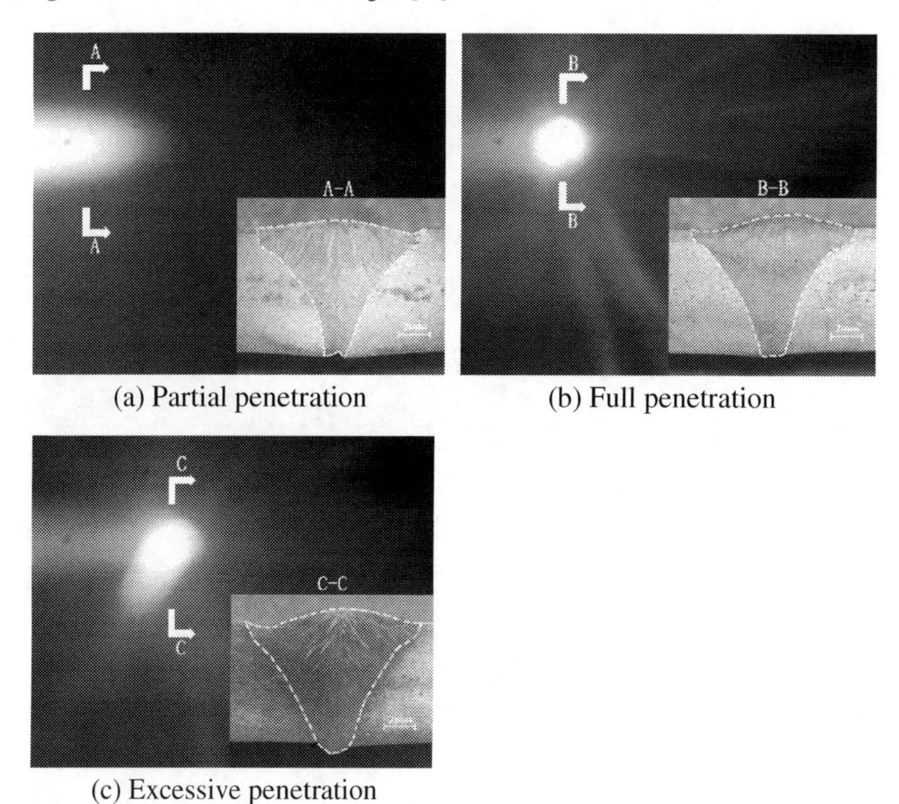

Fig. 10.30 Classification of three penetration type [34]

Table. 10.4 Experimental parameters for 304 stainless steel [34]

Materials	Workpiece dimensions	Welding current	Welding speed	Shielding gas	Flow rate
304	300 × 200 × 12 mm	420–500–580 A	210 mm/min	Ar	15 L/min

10.3.3.1 Feature Extraction and Dimension Reduction

So many machine learning algorithms there are to identify the welding penetration. To predict the welding penetration, one of the eager learning algorithms, SVM, is adopted because of its high generalization ability. Features from different signals, which are representing the physical process from different point of view, would have a better performance than those from a single signal. Therefore, in order to promote the performance of the proposed SVM model, some features from arc current signal, arc voltage signal and arc acoustic signal are put forward to train the SVM model.

To represent the vibration period of excitation source, pitch period, which is notable enough among plenty of acoustic signal features, has been selected. With little calculating times, the pitch period feature can be calculated by the short-time correlated function. Meanwhile, the aforementioned features, MFCCs, are also utilized in this section. Three frequency bands have been found in the whole welding process in Figs. 10.22 and 10.25, i.e. 3–5 kHz, 6.5–8.5 kHz and 18–20 kHz. Equation (10.29) shows the energy features of the three frequency bands, which are utilized to recognize different penetration states. For electrical signals, two frequency bands 3–4.5 kHz and 7–8.5 kHz are selected, and the kurtosis of them is extracted.

$$E = \int_{f_1}^{f_2} S(\omega)\mathrm{d}\omega \tag{10.29}$$

In the end, we can extract 18 features from arc acoustic, voltage and current signals. In order to lessen the data dimension, the Principal Components Analysis (PCA) method is adopted. The accumulated contribution of the components can explain the variance. We reserved the components that explain 95% of the variance. These are shown in Fig. 10.31. The first eight PCA components are retained.

10.3.3.2 Classification Model Based on ECOC-SVM-GSCV

A welding penetration recognition model based on ECOC-SVM-GSCV is put forward and illustrated in Fig. 10.32. The Gaussian kernel type is used in the proposed model. Grid search method is utilized to optimize the two vital parameters, γ and c, which decide the performance of the proposed model. However, there is a multi-class problem so that a limitation exists because the traditional SVM is a binary classifier. To solve this multi-class problem, the error-correcting output codes method is used

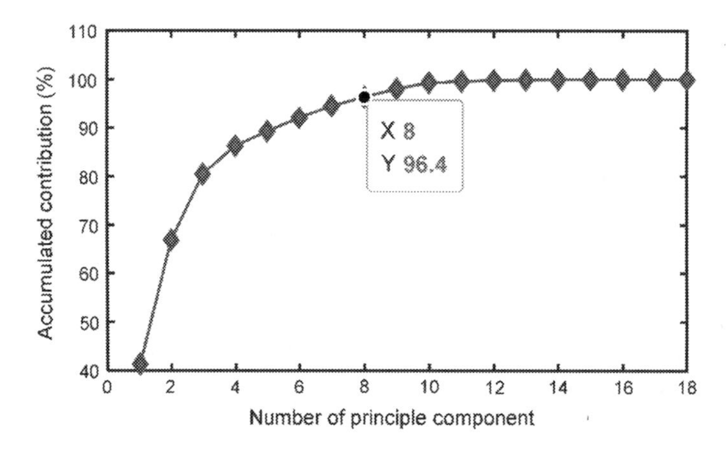

Fig. 10.31 Accumulated contribution of PCA components [34]

to establish a multi-class SVM-based recognition model [36]. Table 10.5 shows the coding design. When a new data is input, it will be classified by three classifiers sequentially. According to the coding design shown in Table 10.5, then the proposed model will output a code, including the outputs of three classifiers. After calculating the Hamming distance between the output and the coding design, the class of this new data is able to be determined on the principle of selecting the smallest distance.

As Fig. 10.32 shows, 18 features are put forward from arc current, arc voltage and arc sound signals. One sample consists of these 18 features. The number of samples is 1080, which are divided into three equal parts, including 360 excessive penetration samples, 360 full penetration samples and 360 partial penetration samples. To collect these samples, experiments have done, and three kinds of weld beads are obtained. Each class of samples is obtained from a corresponding weld bead. Experiments parameters are detailed in Table 10.4. To fully train the classification model, ten-fold cross-validation is used. The final accuracy is 98.7% as we set $c = 10$ and $\gamma = 10$.

10.4 K-TIG Welding of Duplex Stainless Steels

S32101 duplex stainless steel has good mechanical properties and outstanding corrosion resistance. It is widely used in bridges, building structures, large ships, petrochemical equipment, nuclear power equipment and other fields [37]. Because the environment of some application areas is radioactive and has a long service life, the performance requirements of welded joints are also higher.

In this chapter, K-TIG welding was used to weld S32101 duplex stainless steel plates without filling welding wire or opening a groove. The thickness of the duplex stainless steel is 10.8 mm, and its chemical composition is presented in Table 10.6. The influence of welding input on weld geometry profile, microstructure and MP

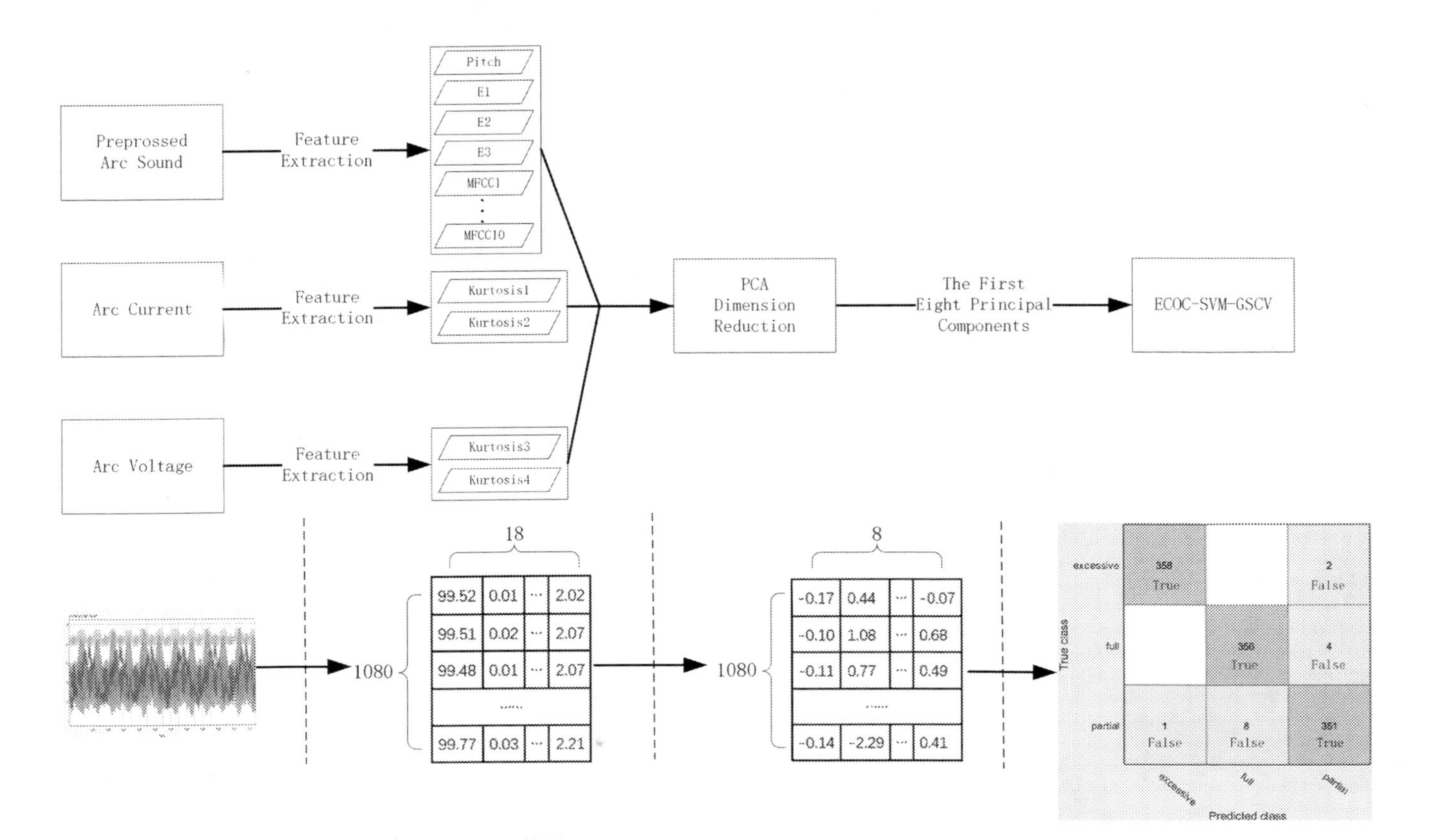

Fig. 10.32 Recognition model of welding penetration based on SVM [34]

Table. 10.5 Coding design [34]

	Classifier 1	Classifier 2	Classifier 3
Partial penetration	1	1	0
Full penetration	−1	0	1
Excessive penetration	0	−1	−1

Table. 10.6 Chemical composition of duplex stainless steel (wt.%)

Element	Cr	Ni	Mn	Si	C	N	Fe
wt.%	21.52	1.56	4.98	0.49	0.017	0.21	Bal.

of welded joints is introduced. Before welding, the surfaces of the workpieces to be welded were mechanically cleaned to remove oil and impurities to ensure the quality of the welded joints. To ensure the quality of the weld, the pure argon gas is passed on the surface of the workpiece during welding.

10.4.1 Effect of the Welding Parameters on the Weld Geometry Profile

Bead depth (D) and bead width (W) are the main parameters that characterize the geometry profile (GP) of the weld. Weld GP can reflect the quality of the weld. In the K-TIG welding process, D and W can be change with the change of welding parameters. Therefore, it can optimize the welding process through the weld GP under different welding parameters. Find the best welding parameters for welding duplex stainless steel. In this chapter, the K-TIG welding parameters are presented in Table 10.7.

Table. 10.7 Process parameters of K-TIG welding of S32101 duplex stainless steel [38]

Test no.	Welding current (A)	Welding speed (mm/min)	Arc voltage (V)
1	470	210	16.5
2	490	210	17.0
3	510	210	17.5
4	530	210	18.0
5	550	210	18.5
6	530	240	18.0
7	530	270	18.0
8	530	300	18.0

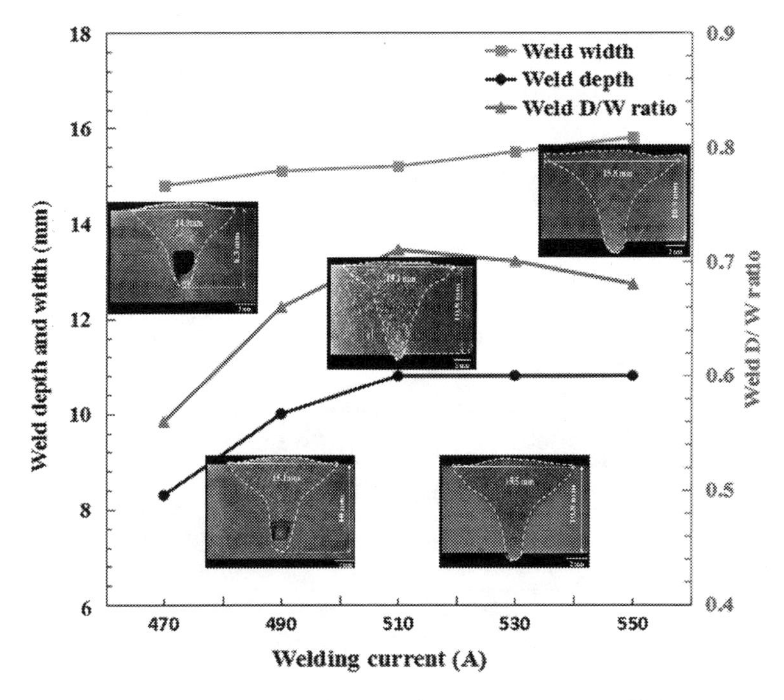

Fig. 10.33 Geometry profile of the weld under different welding currents [38]

10.4.1.1 Welding Current

Under different welding currents, the GP of the weld is displayed in Fig. 10.33. When other parameters remain unchanged and only the welding current is changed, the *W* and *D* increase as the welding current increases. When the welding current is low, the penetration ability of the arc is weak. Therefore, welding pores appear in the weld when the welding current is 470 A. As the welding current increases, the penetration ability of the arc gradually increases. A keyhole mode was produced during the welding process, and the pores disappeared. However, when the welding current is large enough, due to excessive heat input, the balance between the static pressure of the liquid metal in the welding pool and the surface tension is destroyed [38]. The weld pool collapsed, and *D* of the weldment on the back was too high. Therefore, suitable welding current is a key factor affecting the welding quality.

10.4.1.2 Welding Speed

Figure 10.34 displays the GP of weld under different welding speeds. When the other parameters remain unchanged, the W and D increase as the welding speed decreases. When the welding speed is varied between 210 mm/min and 240 mm/min, the welds

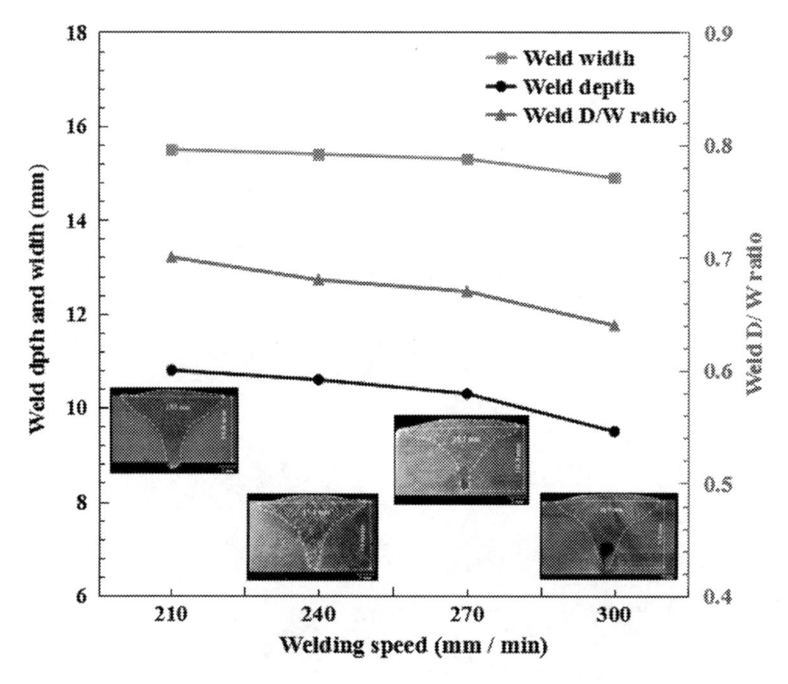

Fig. 10.34 GP of the weld under different welding speeds [34]

are completely penetrated. Because the lower welding speed corresponds to the higher welding heat input, the penetration ability of the arc is also stronger [39].

10.4.2 Microstructure

The microstructure of each area in the welded joint is displayed in Fig. 10.35a. The base material (BM) is composed of austenite and ferrite, as displayed in Fig. 10.35b. There is a clear difference between the two-phase morphology. The austenite is elongated along the rolling direction and distributed in the ferrite. In the transition zone from BM to WM (Fig. 10.35c), the area of Heat Affected Zone (HAZ) is too narrow. This is mainly because compared with the traditional TIG welding method, the energy of the arc is more concentrated in the K-TIG welding process.

Under the influence of the welding thermal cycle, the microstructure of the HAZ has changed significantly, as shown in Fig. 10.35d. The ferrite in the HAZ becomes coarser, and some of the austenite precipitate around the ferrite to form the Grain Boundary Austenite (GBA). The austenite in the WM mainly exists in the form of GBA, Intergranular Austenite (IGA) and Widmanstätten Austenite (WA). There are no intermetallic phases in the WM, which was listed in Fig. 10.35e

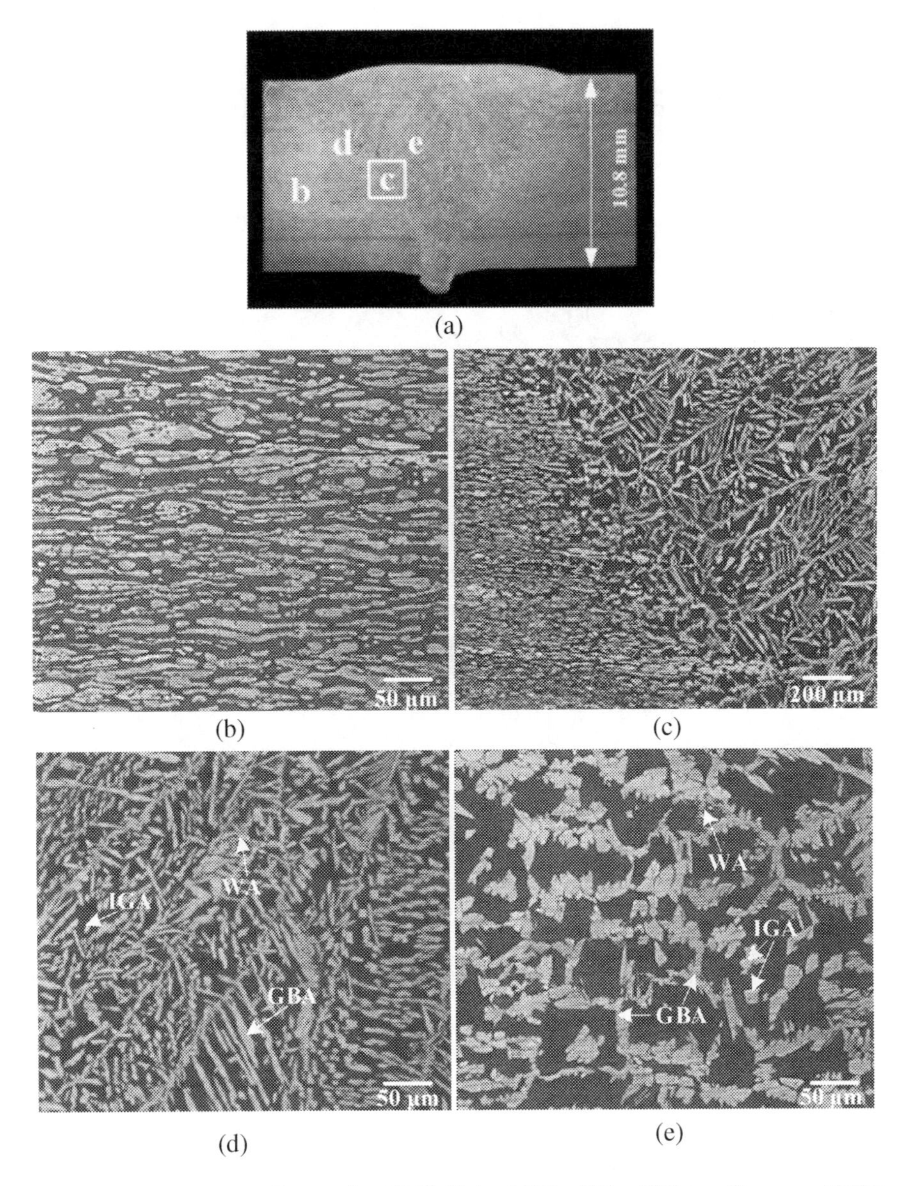

Fig. 10.35 Microstructure (Test no. 4). **a** Welded joint, **b** BM, **c** BM to WM transition zone, **d** HAZ, **e** WM

Figure 10.36 shows the Euler diagram of the welded joint. Table 10.8 lists the average grain sizes and the proportion of the phase in welded joint. The grain size and the shape in the HAZ and WM are obviously compared with the comparative example and the BM. In the BM, the grain size is small, and the proportion of the two phase is close. In the HAZ, the grain size is not uniform. The proportion of two phases

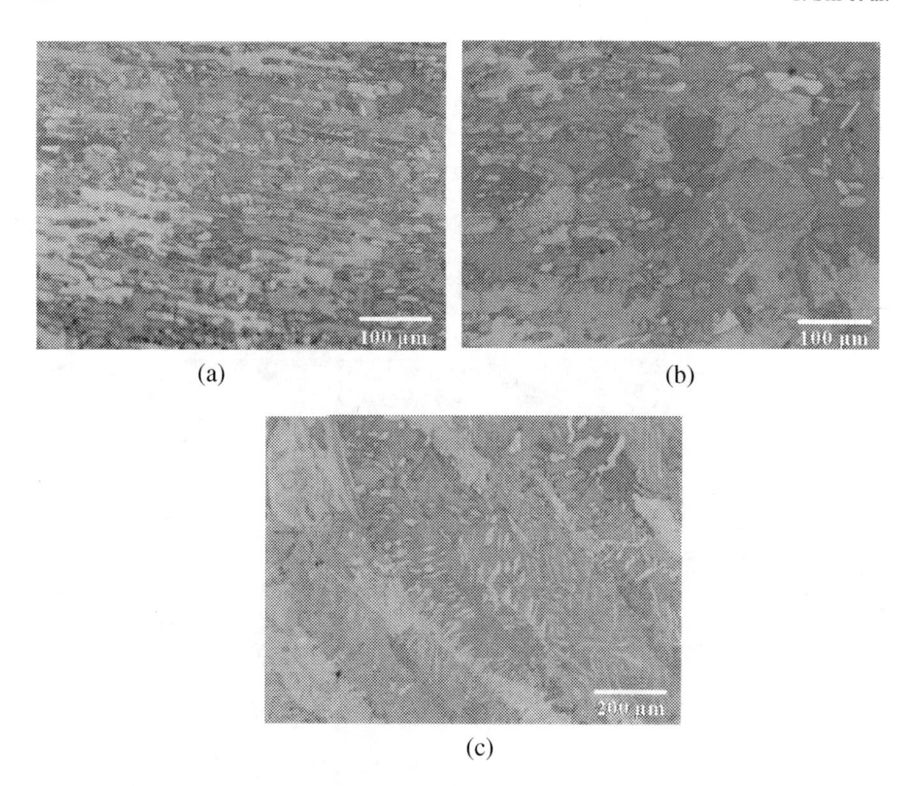

Fig. 10.36 Euler diagram of each area in S32101 duplex stainless steel K-TIG welded joint (Test 1–4). **a** BM, **b** HAZ, **c** WM

Table. 10.8 Average grain sizes and proportion of the phase in the welded joint

Phase		BM	HAZ	WM
Ferrite	Grain size (μm)	10.01	16.21	26.15
	Fraction (%)	51.9	63.8	59.8
Austenite	Grain size (μm)	8.52	13.41	22.94
	Fraction (%)	48.1	36.8	40.2

has changed significantly, in which the proportion of austenite is reduced to 36.8%. A large number of large-sized grains appeared in the WM, and columnar crystal grew along the direction of the maximum temperature gradient. The proportion of austenite becomes 40.2%.

10.4.3 Misorientation Angle Distribution of Grain Boundary (MADGB)

A large number of studies has shown that the MADGB has a significant effect on the mechanical properties of polycrystalline materials. Base on the orientation angles (θ), Grain Boundaries (GB) are generally divided into two types. One is a low-angle grain boundary (LAGB, $2° \leq \theta \leq 5°$), and the other is a high-angle grain boundary (HAGB, $\theta \geq 15°$) [40]. When the θ is 60°, it is called $\Sigma 3$ coincidence site lattice (CSL) boundary. It has lower GB energy and less impurity segregation. The $\Sigma 3$ CSL boundary can hinder the migration of HAGB, and HAGB can obstruct the expansion of brittle cracks [41].

MADGB maps of the ferrite and austenite in S32101 duplex stainless steel K-TIG welded joint were shown in Figs. 10.37 and 10.38. The MADGB of ferrite in the BM is relatively uniform and the MADGB of ferrite in the HAZ, the number of HAGB decreased slightly. In the WM, the MADGB is mainly in LAGB. In Fig. 10.38a, the MADGB of austenite in the BM is mainly in LAGB, and the frequency of $\Sigma 3$

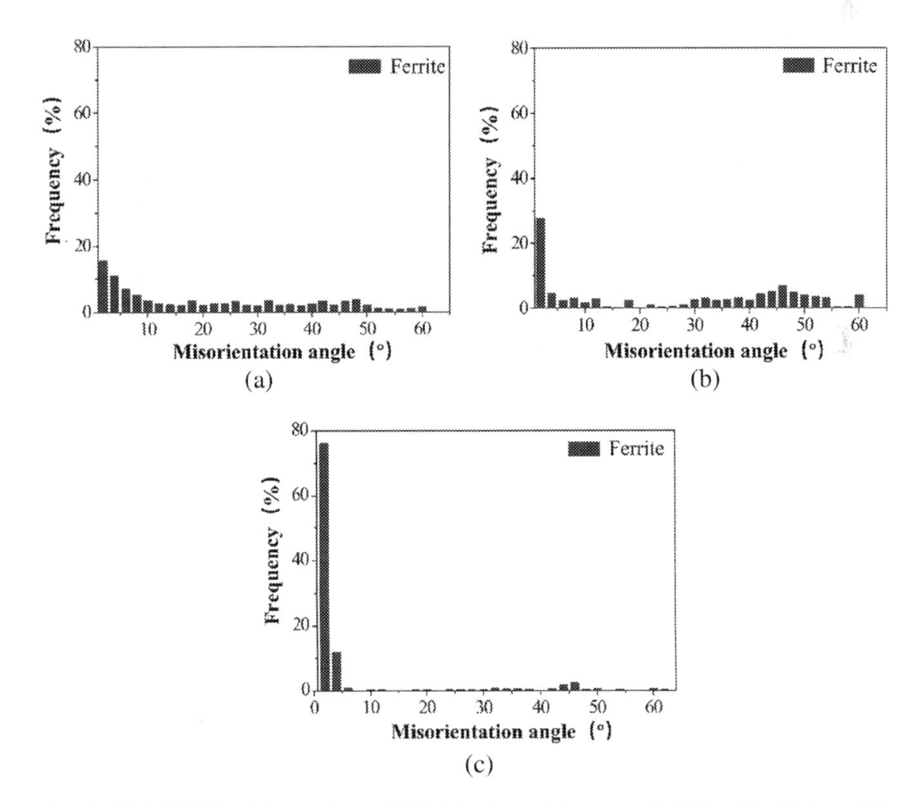

Fig. 10.37 MADGB of the ferrite in S32101 duplex stainless steel K-TIG welded joint. **a** BM, **b** HAZ, **c** WM

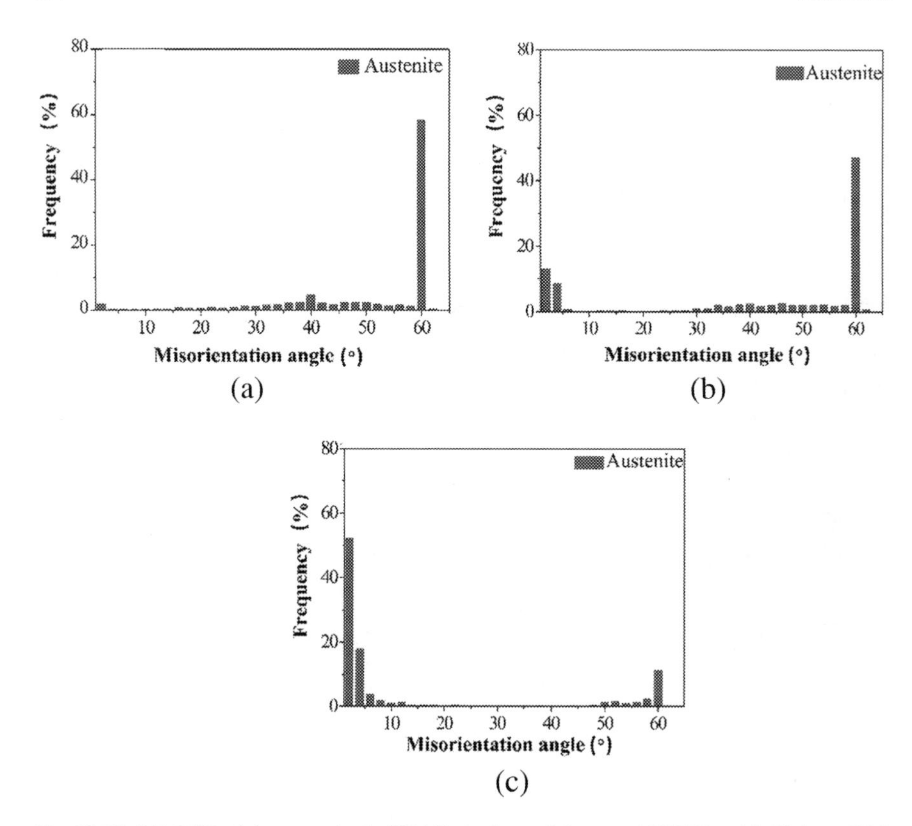

Fig. 10.38 MADGB of the austenite in S32101 duplex stainless steel K-TIG welded joint. **a** BM, **b** HAZ, **c** WM

CSL boundary is higher, reaching 56.3%. In the HAZ, the MADGB of the austenite has a bimodal distribution characteristic. The frequency of the Σ3 CSL boundary and HAGB decreased, and the frequency of the LAGB increased. Affected by the welding thermal cycle, the frequency of Σ3 CSL boundary of the austenite in the WM decreases to 11.3%. The ability to hinder HAGB migration is reduced.

10.4.4 Mechanical Properties

The sample dimensions for MP tests are shown in Fig. 10.39. The tensile test, bend test and Charpy impact test were conducted according to ASTM E8 [42], ASTM E290 [43] and ASTM A370 [44], respectively.

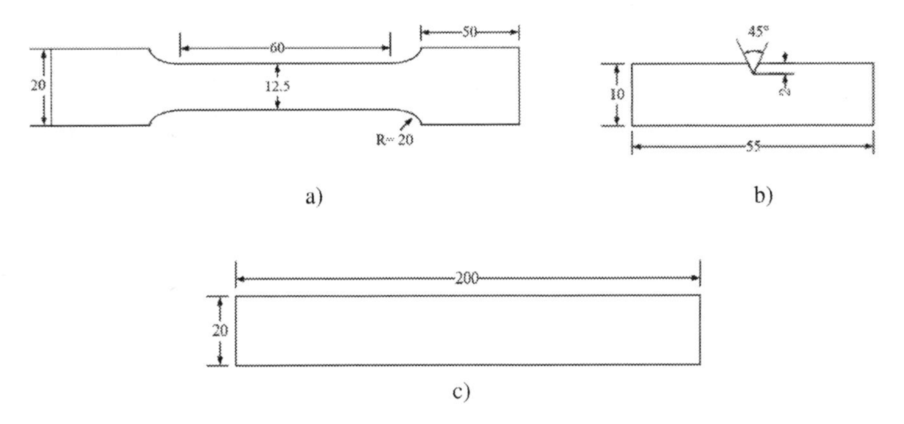

Fig. 10.39 Specimen dimensions. **a** Tensile test, **b** Charpy impact test, **c** Bend test [38]

10.4.4.1 Microhardness

Figure 10.40 displays the Vickers microhardness of the welded joint. The microhardness curves of different areas of the welded joint are in the shape of '*M*', and the centre of the weld is taken as the axis of symmetry, and the values on both sides are almost symmetrical. The microhardness of the HAZ is the highest, and the microhardness of the BM is the lowest. The microhardness of WM is between BM and HAZ. This is mainly because the microhardness of ferrite is greater than that of austenite, and the proportion of ferrite in the HAZ is the largest.

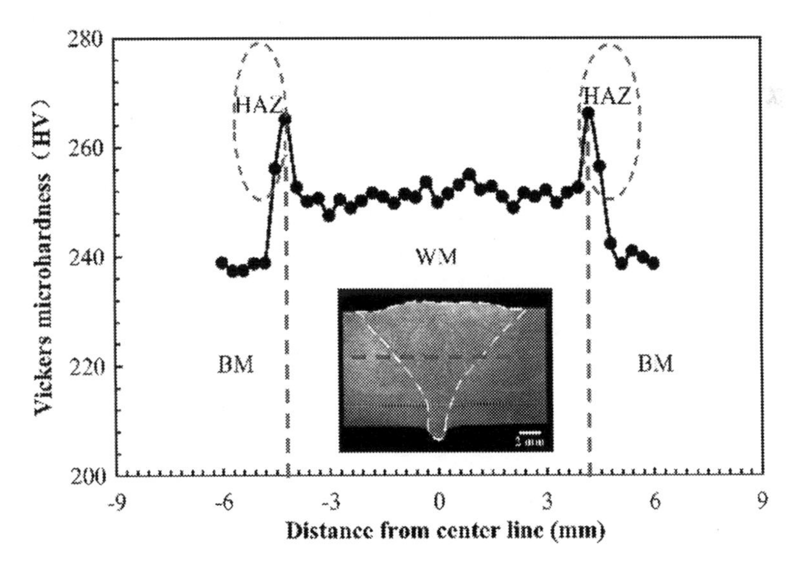

Fig. 10.40 Microhardness distribution in the welded joint

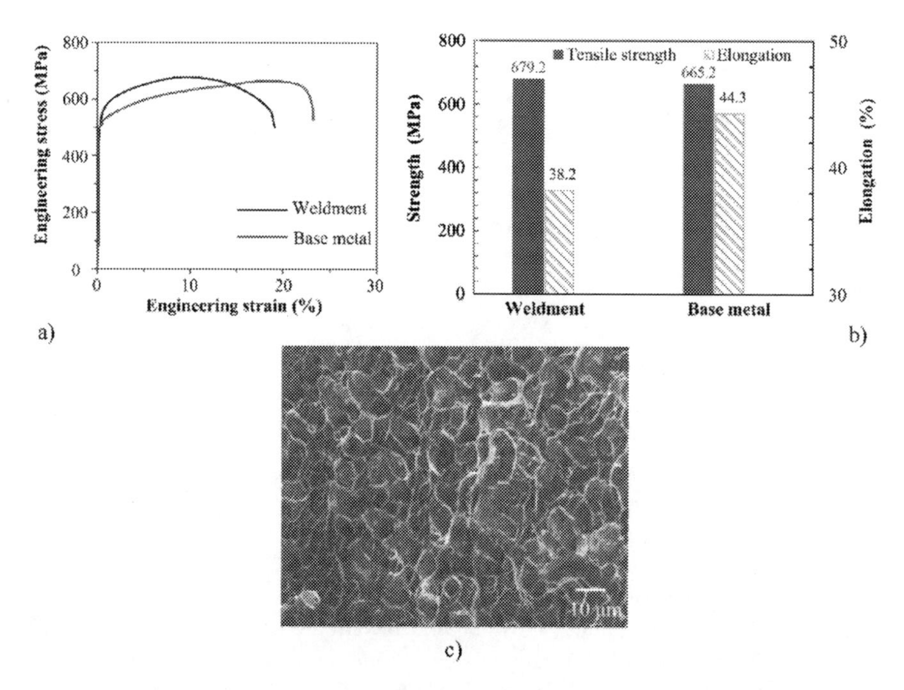

Fig. 10.41 Tensile property test. **a** Tensile curves, **b** test results, **c** SEM figure of tensile fracture

10.4.4.2 Tensile Property

The tensile property of the welded joint no. 4 was tested, and the test results are displayed in Fig. 10.41. The tensile strength of the weldment and BM is 679.2 and 665.2 MPa, respectively. The elongation of the weldment and BM is 38.2% and 44.3%, respectively. Compared with BM, the strength of the weldment is higher, and the elongation of the weldment is lesser. Scanning Electron Microscope (SEM) figure of tensile fracture is shown in Fig. 10.41c. There are a large number of dimples and tearing edges in the tensile fracture surface. It is a mode of the ductile fracture.

10.4.4.3 Bend Property

Figure 10.42 shows the weldment specimen after bending test. No obvious cracks were found on the surface of the weldment. It indicates that the S32101 duplex stainless steel K-TIG welded joint has good ductility.

Fig. 10.42 Bend test specimen (Test no. 4) [38]

10.4.4.4 Impact Property

By observing the microstructure of the K-TIG welded joint, it can be seen that the HAZ is very narrow. Therefore, the WM and BM with test no. 4 were tested for Charpy impact property. The value of absorbed energy (AV) is listed in Table 10.9. The AE of the WM is obviously lower than AE of the BM, but it reaches 83.4% of the BM, which meets the requirements. The impact property of WM is mainly affected by two factors. One is that the proportion of ferrite in the WM is more than that of the BM. On the other hand, the frequency of Σ3 CSL boundary in the WM is lower than that of the BM.

10.5 K-TIG Welding of Titanium Alloy

TC4 is an $\alpha + \beta$ two-phase titanium alloy, which has the advantages of low density, high strength and high corrosion resistance, so it is widely used in aerospace, automotive, medical biology, energy and other fields [45]. Welding is an indispensable process technology in the processing and manufacturing of titanium alloys. However, the titanium alloy has a strong activity in the molten state, and it is very easy to react with a variety of substances during the welding process to deteriorate the performance of the welded joint. In addition, during the thermal cycle of welding, the

Table. 10.9 AE of the WM and BM [38]

	Absorbed energy, Akv (J)			
	1	2	3	Average value
Weldment	169	175	180	174.7
Base metal	204.9	215.3	207.9	209.4

Table. 10.10 Chemical composition of the BM (wt.%)

Element	Al	V	Fe	C	N	O	H	Ti
wt.%	6.11	4.06	0.12	0.012	0.012	0.156	0.0015	Bal.

microstructure of titanium alloy welded joints has changed significantly, which may affect the performance of the welded joints.

In this chapter, we mainly introduce the mechanical properties of the 12 mm thickness TC4 titanium alloy K-TIG Weld Metal (WM) under different welding current conditions and the evolution characteristics of the welded joint microstructure. The chemical composition of the TC4 titanium alloy is presented in Table 10.10.

10.5.1 Weld Geometry Profile

Before welding, the surfaces of the workpieces were cleaned by using acetone. The workpieces were mechanically fixed and maintained a gap of 0.5 mm. Pure argon gas is introduced into the surfaces of the workpiece in K-TIG welding. The specific welding parameters are listed in Table 10.11.

Figure 10.43 shows the TC4 titanium alloy K-TIG welding joint under different welding currents. It proved that the welding current affect the quality of the TC4 titanium alloy K-TIG weld. When the welding current is 490 A, the penetration ability of the arc is poor, which can not be fully penetrated, as shown in Fig. 10.43a. When the welding current is 510, 530 and 550 A, the welds are fully penetrated. When the welding current is 570 A, the corresponding welding heat input is too large. At this time, the dynamic balance in the molten pool is destroyed. The welding seam collapses, and the excess height on the back of the welding seam is too large, as shown in Fig. 10.43e.

Figure 10.44 displays the profiles of the K-TIG welds under different welding currents (510–550 A). Both were well formed, and the weld surfaces were bright and clear. When the welding currents were 510–530 A, X-ray nondestructive testing (NDT) photos of the welds were taken and are shown in Fig. 10.44. The welds in the picture were uniform, had a light black broadband, did not have large white spots

Table. 10.11 Process parameters of TC4 titanium alloy K-TIG welding

No	Welding current (A)	Welding speed (mm/s)	Arc voltage (V)
1	490	3.5	17.0
2	510	3.5	17.5
3	530	3.5	18.0
4	550	3.5	18.5
5	570	3.5	19.0

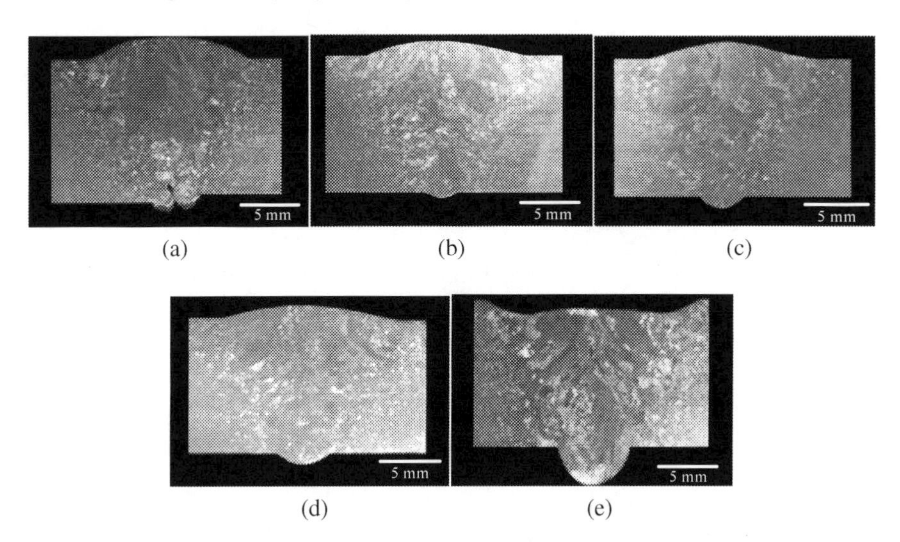

Fig. 10.43 Transverse cross section of TC4 titanium alloy K-TIG welded joint under different welding currents. **a** 490 A, **b** 510 A, **c** 530 A, **d** 550 A, **e** 570 A

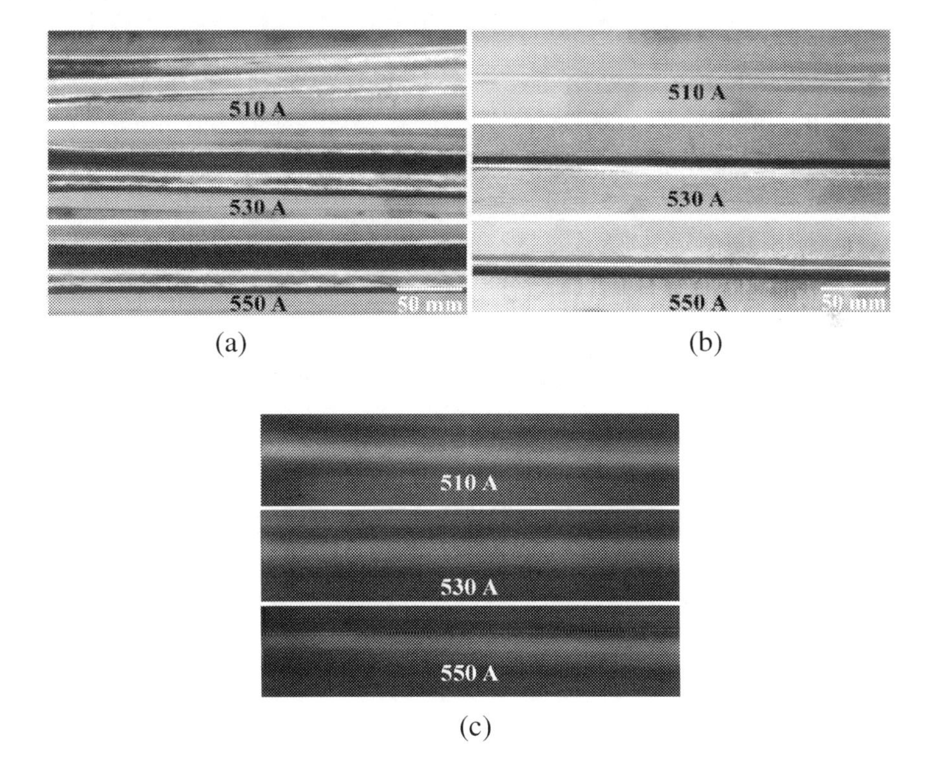

Fig. 10.44 Appearance of the TC4 Titanium alloy K-TIG weldment. **a** Upper surface, **b** Lower surface, **c** Image of X-ray NDT

or lines, all of which indicates good welds. There were no defects, such as pores or cracks.

10.5.2 Mechanical Properties

To analyze the influence of welding current on the MP of TC4 titanium alloy K-TIG WM, the tensile property and impact property of titanium alloy K-TIG WMs with different welding currents are tested in this section. The tensile test and Charpy impact test of BM and WM were conducted base on ASTM E8 [46] and ASTM A370 [44], respectively. The schematic diagram of the specimens is displayed in Fig. 10.45.

10.5.2.1 Tensile Property

After the K-TIG welding, the tensile property of the BM and the TC4 titanium alloy K-TIG WM with welding currents of 510, 530 and 550 A was tested. The tensile strength and elongation of the BM and the WM are displayed in Fig. 10.46. The tensile strength and elongation of the BM are 1054.5 MPa and 18.3%, respectively. The test results show that when the welding current is increased from 510 to 550 A, the tensile strength of the TC4 titanium alloy K-TIG WMs gradually decreases. When the welding current is 550 A, the tensile strength of the WM is the lowest, but it already meets the tensile strength requirements of the TC4 titanium alloy plate. When the welding current increases, the elongation of the TC4 titanium alloy K-TIG WM gradually increases.

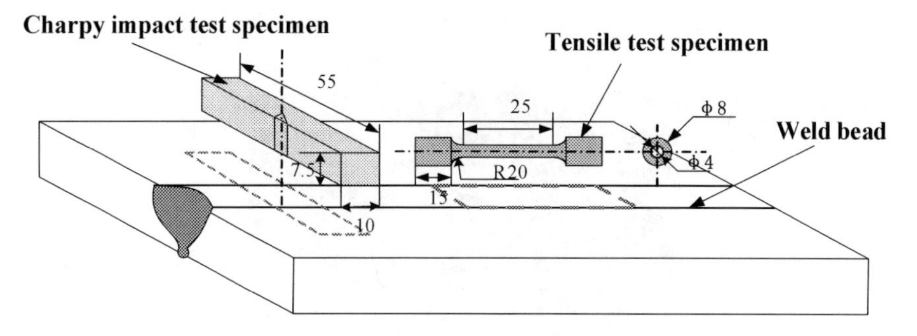

Fig. 10.45 Dimensions of the test specimens

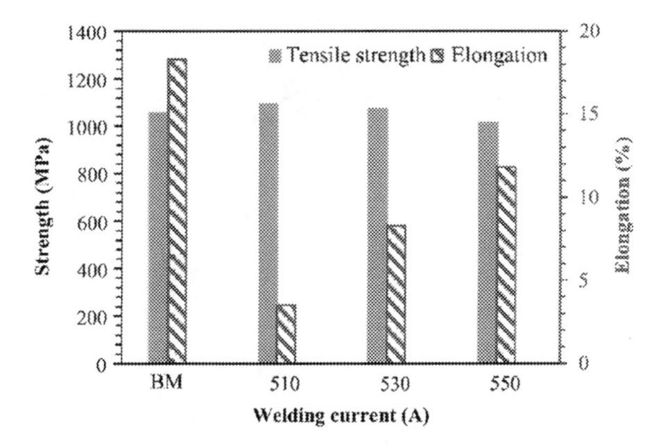

Fig. 10.46 Tensile property of the BM and WM

Table. 10.12 AE of the BM and WM [45]

Welding current (A)	Absorbed energy, Akv (J)
510	20.4
530	32.1
550	33.6
BM	24.3

10.5.2.2 Impact Property

In this section, the notch samples are mainly made at the BM and WM, and the Charpy impact tests are used to examined the AE of the BM and WM. The AE of the BM and the WM obtained under different welding currents is shown in Table 10.12. When the welding current are 510 A, 530 A and 550 A, the mean value of the AE of the WM are 20.4 J, 32.1 J and 33.6 J, respectively. The impact test results show that when the welding current is 510–550 A, with the increase of the welding current, the AE of the WM is gradually increased. When the welding current is 530 A or 550 A, the mean value of the AE of the WM is higher than that of the BM. When the welding current is 510 A, the mean value of the AE of the WM is slightly lower than the BM, but reached 84% of the BM.

10.5.3 Microstructural Characterization

The microstructure of TC4 titanium alloy can directly affect the tensile and impact properties of the material. During the welding process, affected by the temperature, the structure of the HAZ and the WM will change significantly. The HAZ and the WM are the main areas that affect the performance of the welded joint. In this section, we take a K-TIG welded joint with a welding current of 550 A as an example to introduce the evolution characteristics of the microstructure of each area of the TC4 titanium

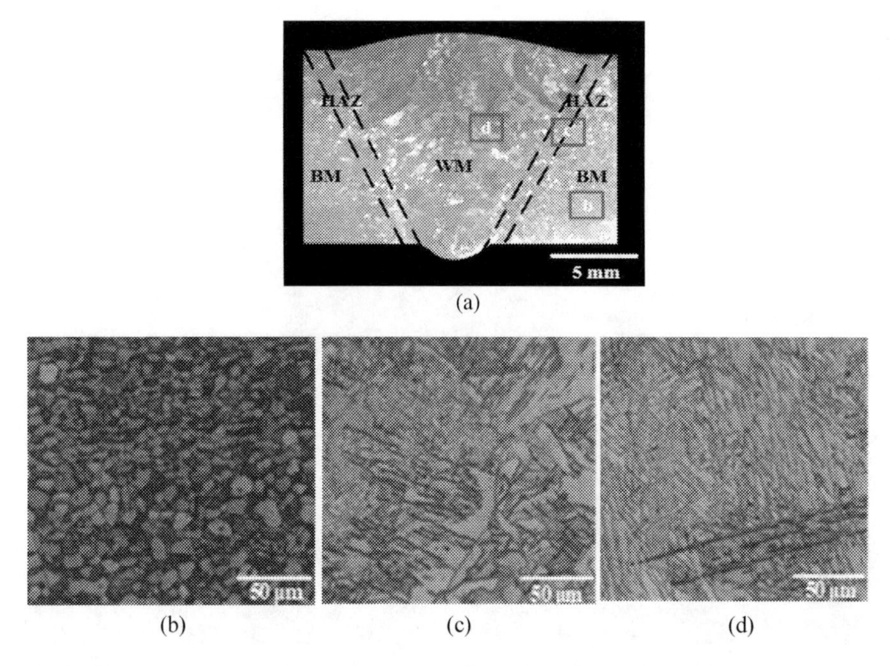

Fig. 10.47 Microstructure. **a** Transverse cross section **b** BM, **c** HAZ, **d** WM [45]

alloy K-TIG welded joint. Figure 10.47b–d displays the band contrast images of the TC4 titanium alloy K-TIG welded joint. The microstructure of the BM is consists of equiaxed primary α grains, α laths and few retained β grains, as shown in Fig. 10.47b. In Fig. 10.47c, the morphology of α phase in the HAZ has obviously changed. Under high temperature conditions, the α grains grow significantly. Because the temperature in the HAZ is higher than the β transus temperature (995 °C), there are a large number of α laths in the HAZ. The α laths are distributed in parallel and unidirectional form. The residence time of the WM at high temperature is relatively long, and the grains of WM grow rapidly. A large number of α laths are formed in the WM. The α laths in the WM mainly exist in multi-directional form, with scattered distribution, as shown in Fig. 10.47d.

The EDS spectra show the contents of Al, V and Ti in the BM and WM, as shown in Fig. 10.48. Compared with the BM, it was found that the elements such as Al, V and Ti in the WM did not change significantly. The test results proved that the chemical compositions of the BM and the WM are almost the same.

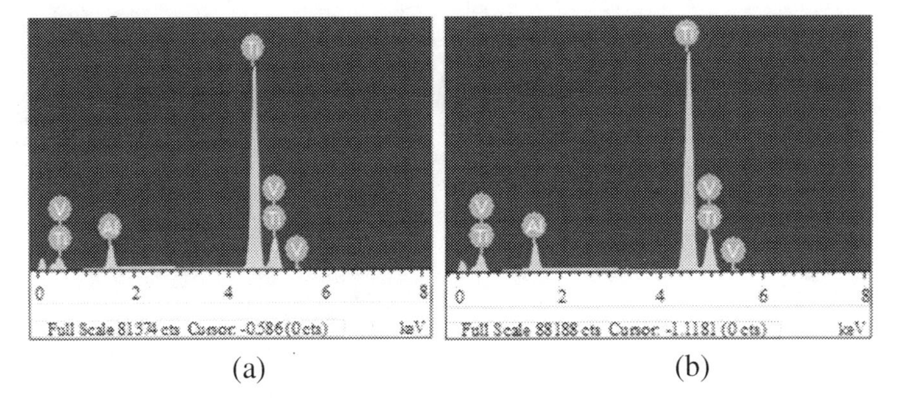

Fig. 10.48 EDS spectra. **a** BM, **b** WM [45]

10.5.4 MADGB of the TC4 Titanium Alloy K-TIG Welded Joint

Figure 10.49 shows the MADGB maps of the α phase in the each area of the TC4 titanium alloy K-TIG welded joint. Each area was scanned with a size of 0.23 mm $\times$ 0.2 mm and 1 mm $\times$ 0.8 mm, respectively. In Fig. 10.49, HAGB and LAGB are represented by green and black lines, respectively. In the BM, the HAGB proportions in two test areas are 60.8% and 53.8%, respectively. The HAGB proportions of the HAZ in two test areas are higher than that of the BM. The HAGB proportions of the WM in two test areas are 96.5% and 93.9%, respectively. The test results indicate that the HAGB proportions of the WM is the highest.

10.6 Conclusions

In this chapter, the K-TIG welding is introduced, and experiments are performed on duplex stainless steels and Ti alloys. The following conclusions can be drawn.

1. A newly designed equipment is needed for operating K-TIG, and the HDR monitoring technology can improve the monitoring performance of the welding process.
2. Several factors are found to have effects on the K-TIG welding process, including the current, travel speed, voltage, shielding gas, electrode geometry and the material category.
3. A wide field of K-TIG welding is showed by introducing several typical industrial applications.
4. The fundamental theory of the K-TIG welding is introduced by analyzing the arc force and thermal field within the weld pool, revealing the mechanism of keyhole formation and molten pool change.

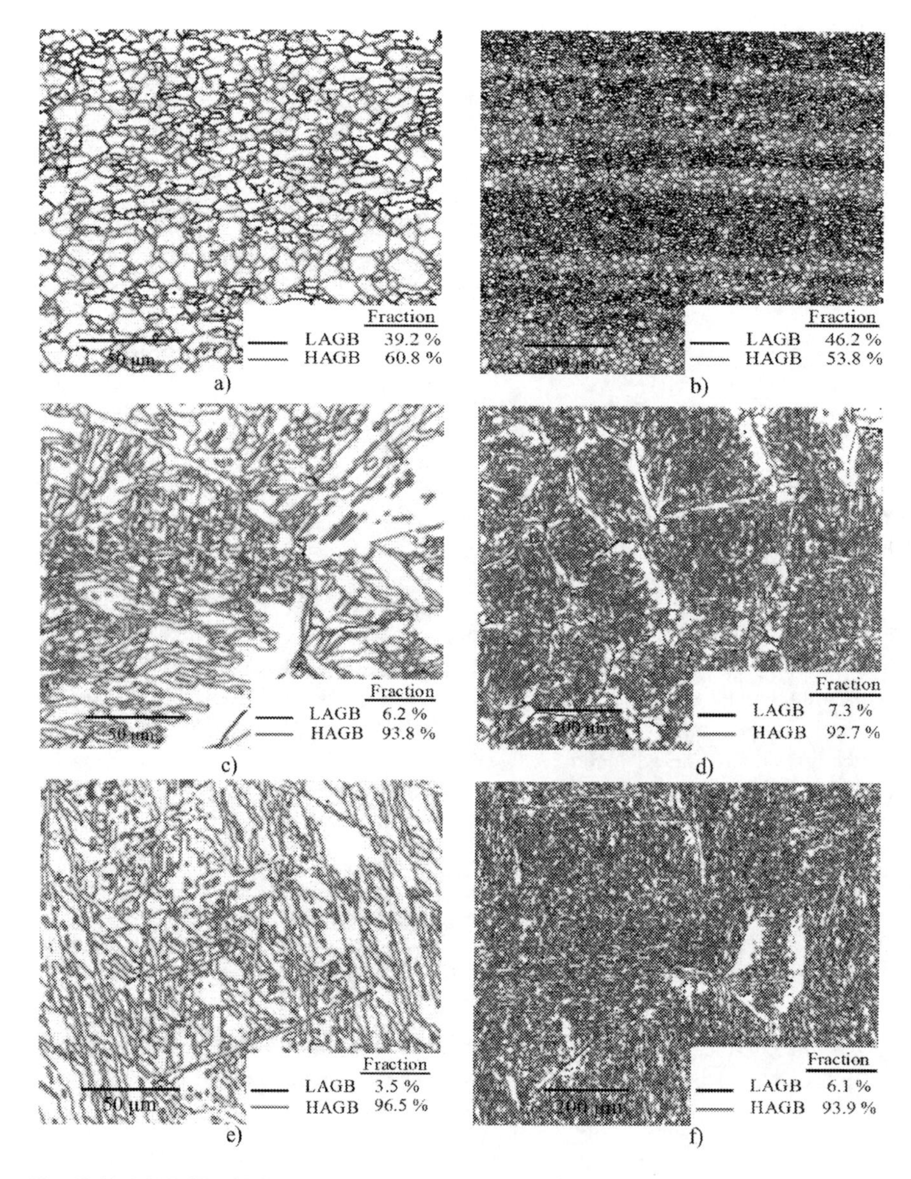

Fig. 10.49 MADGB of TC4 titanium alloy welded joint. **a**, **b** BM, **c**, **d** HAZ, **e**, **f** WM [45]

5. A unique frequency band 2–4 kHz is found, which can represent different welding states including partial penetration and full penetration.
6. The physical significance of frequency features of arc length is put forward by analyzing welding circuit and arc voltage signal. The variation of the arc voltage signal shows the weld pool oscillation. A signal processing algorithm is proposed to extract the oscillation frequency.

7. An SNS method is introduced to denoise the collected arc sound signal and extract the pure arc sound.
8. A welding penetration recognition model based on ECOC-SVM-GSCV is established. 18-dimension features including pitch, MFCC and kurtosis are extracted, and the data dimension is reduced by PCA. After ten-fold cross-validation and grid search optimization, a ECOC-SVM-GSCV model with an accuracy of 98.7% can be obtained.
9. The 10.8 mm thickness S32101 duplex stainless steel and 12 mm thickness TC4 titanium alloys can be welded by K-TIG welding in a single pass without opening a groove and filling metals.
10. In S32101 duplex stainless steel welded joint, the tensile strength of the WM is higher than that of the BM. The elongation of the WM is smaller than that of the BM. The impact toughness of the WM is lesser than that of the BM, but it meets the mechanical property requirements of the weldment.
11. In TC4 titanium alloys K-TIG welded joint, the strength of the WM is 93.82% of that of the BM. The HAGB proportion of the WM is higher than that of the BM, and the impact toughness of the WM is superior to that of the BM.

References

1. Lathabai S, Jarvis B, Barton K (2001) Comparison of keyhole and conventional gas tungsten arc welds in commercially pure titanium. Mater Sci Eng A 299(1–2):81–93
2. Goryachev AP, Zelenin VA (1964) Autom Weld 17(12):21–26
3. Liptak JA (1965) Gas tungsten arc welding heavy aluminium plate. Weld J 44(6):276s–281s
4. Nesterov A, Bulgachev E, Boitsev N (1987) Increasing the stability of the process of welding structures of aluminium and magnesium alloys using an immersed arc. Weld Int 1(7):659–660
5. Yamauchi N, Taka T, Ohi M (1981) Development and application of high current TIG process (SHOLTA) welding process. Sumitomo Search 25:87–100
6. Jarvis B (2020) Keyhole gas tungsten arc welding: a new process variant [Online]. Research Online
7. Lancaster J (2014) The physics of welding. Elsevier Science, Kent
8. Andrews JG, Atthey DR (1976) Hydrodynamic limit to penetration of a material by a high-power beam. J Phys D Appl Phys 9:2181–2194
9. Matsunawa A, Kim J, Seto N, Mizutani M, Katayama S (1998) Dynamics of keyhole and molten pool in laser welding. J Laser Appl 10(6):247–254
10. Norrish J (2006) Advanced welding processes. Woodhead Publishing, Cambridge
11. Fan W, Ao S, Huang Y, Liu W, Li Y, Feng Y, Luo Z, Wu B (2017) Water cooling keyhole gas tungsten arc welding of HSLA steel. Int J Adv Manufact Technol 92(5–8):2207–2216
12. Zhang B, Shi Y, Cui Y, Wang Z, Hong X (2020) Prediction of keyhole TIG weld penetration based on high-dynamic range imaging. J Manufact Process Adv Line Publication. https://doi.org/10.1016/j.jmapro.2020.03.053
13. Fang Y, Liu Z, Cui S, Zhang Y, Qiu J, Luo Z (2017) Improving Q345 weld microstructure and mechanical properties with high frequency current arc in keyhole mode TIG welding. J Mater Process Technol 250:280–288
14. Fei Z, Pan Z, Cuiuri D, Li H, Wu B, Ding D, Su L, Gazder A (2018) Investigation into the viability of K-TIG for joining armour grade quenched and tempered steel. J Manufact Process 32:482–493

15. Cui S, Liu Z, Fang Y, Luo Z, Manladan S, Yi S (2017) Keyhole process in K-TIG welding on 4 mm thick 304 stainless steel. J Mater Process Technol 243:217–228
16. Liu Z, Fang Y, Qiu J, Feng M, Luo Z, Yuan J (2017) Stabilization of weld pool through jet flow argon gas backing in C-Mn steel keyhole TIG welding. J Mater Process Technol 250:132–143
17. Liu Z, Chen S, Cui S, Lv Z, Zhang T, Luo Z (2019) Experimental investigation of focusing cathode region by cooling tungsten. Int J Therm Sci 138:24–34
18. Wang Y, Qi B, Cong B, Zhu M, Lin S (2018) Keyhole welding of AA2219 aluminum alloy with double-pulsed variable polarity gas tungsten arc welding. J Manufact Process 34:179–186
19. Xie Y, Cai Y, Zhang X, Luo Z (2018) Characterization of keyhole gas tungsten arc welded AISI 430 steel and joint performance optimization. Int J Adv Manufact Technol 99(1–4):347–361
20. Huang Y, Luo Z, Lei Y, Ao S, Shan H, Zhang Y (2018) Dissimilar joining of AISI 304/Q345 steels in keyhole tungsten inert gas welding process. Int J Adv Manufact Technol 96(9–12):4041–4049
21. Feng Y, Luo Z, Liu Z, Li Y, Luo Y, Huang Y (2015) Keyhole gas tungsten arc welding of AISI 316L stainless steel. Mater Des 85:24–31
22. Cui S, Xian Z, Shi Y, Liao B, Zhu T (2019) Microstructure and impact toughness of local-dry Keyhole Tungsten inert gas welded joints. Materials 12(10):1638
23. Fei Z, Pan Z, Cuiuri D, Li H, Wu B, Su L (2018) Improving the weld microstructure and material properties of K-TIG welded armour steel joint using filler material. Int J Adv Manufact Technol 100(5–8):1931–1944
24. Riffel KC, Silva RHG, Dalpiaz G, Marques C, Schwedersky MB (2019) Keyhole GTAW with dynamic wire feeding applied to orbital welding of 304L SS pipes. Soldagem Inspeção. 24:e2418
25. Champion FC, Davy N (1947) Properties of matter. Blackie & Son Ltd., London
26. Converti J (1981) Plasma jets in welding arcs. Ph.D. thesis. Mechanical Engineering, M.I.T., Cambridge, MA
27. Rosenthal D (1941) Mathematical theory of heat distribution during welding and cutting. Weld J 1941:220–234
28. Wells A (1952) Heatflow in welding. Weld J 31:263–267
29. Fei Z, Pan Z, Cuiuri D, Li H, Wu B, Ding D, Su L (2019) Effect of heat input on weld formation and tensile properties in Keyhole Mode TIG welding process. Metals 9(12):1327
30. Chen S, Liu Z, Zhao X, Lv Z, Fan X (2019) Cathode-focused high-current arc: heat source development with stable keyhole in stationary welding. Int J Heat Mass Transf 143:118475
31. Cui Y, Shi Y, Hong X (2019) Analysis of the frequency features of arc voltage and its application to the recognition of welding penetration in K-TIG welding. J Manufact Process 46:225–233
32. Na L (2014) Recognition and real time control of penetration characteristics of Al alloy pulsed GTAW process based on the arc audio information. Ph.D. thesis. Shanghai Jiao Tong University
33. Jarvis BL (2001) Keyhole gas tungsten arc welding: a new process variant. Ph.D. thesis. University of Wollongong
34. Cui Y, Shi Y, Zhu T et al (2020) Welding penetration recognition based on arc sound and electrical signals in K-TIG welding. Measurement 163:107966
35. Berouti M, Schwartz R, Makhoul J (2003) Enhancement of speech corrupted by acoustic noise. In: IEEE international conference on acoustics, speech, and signal processing (ICASSP), pp 208–211
36. Allwein E, Schapire R, Singer Y (2001) Reducing multiclass to binary: a unifying approach for margin classifiers. J Mach Learn Res 1(2):113–141
37. Alvarez-Armas I (2008) Duplex stainless steels: brief history and some recent alloys. Recent Patents Mech Eng 1:51–57
38. Cui SW, Shi YH, Sun K et al (2018) Microstructure evolution and mechanical properties of keyhole deep penetration TIG welds of S32101 duplex stainless steel. Mater Sci Eng A 709:214–222
39. Zhang M, Wang XN, Zhu GJ et al (2014) Effect of laser welding process parameters on microstructure and mechanical properties on butt joint of new hot-rolled nano-scale precipitation-strengthened steel. Acta Metall Sin (Engl Lett) 27(3):521–529

40. Cui SW, Shi YH, Cui YX et al (2018) The impact toughness of novel keyhole TIG welded duplex stainless steel joints. Eng Fail Anal 94:226–231
41. Wronski S, Tarasiuk J, Bacroix B et al (2012) Investigation of plastic deformation heterogeneities in duplex steel by EBSD. Mater Charact 73:52–60
42. ASTM International (2016) Designation: E8/E8M-16a standard test methods for tension testing of metallic materials. ASTM International, West Conshohocken
43. ASTM International (2014) Designation: E290-14 Standard test methods for bend testing of material for ductility. ASTM International, West Conshohocken
44. ASTM International (2017) Designation: A370-17 Standard test methods and definitions for mechanical testing of steel products. ASTM International, West Conshohocken
45. Cui SW, Shi YH, Zhu T et al (2019) Microstructure, texture, and mechanical properties of Ti-6Al-4V joints by K-TIG welding. J Manufact Process 37:418–424
46. ASTM International (2016) Designation: E4-2016 standard practices for force verification of testing machines. ASTM International, West Conshohocken

Chapter 11
Fatigue Analysis of Dissimilar Metal Welded Joints of 316L Stainless Steel/Monel 400 Alloy Using GTAW

Cherish Mani, B. Sozharajan, R. Karthikeyan, and J. Paulo Davim

Abstract Fatigue performance analysis of weld joints of Monel 400 and stainless steel 316L (SS 316L) using Gas Tungsten Ac Welding (GTAW) process is presented. Dissimilar metal welding is considered as the most perplexing task when compared to the similar metals joining by welding. Dissimilar weld joints based on austenitic stainless steels are extensively utilized in many applications of high temperature systems of energy conversion. Monel alloys are widely used in many fields especially in oil and gas and has excellent resistance capacity to many corrosive environments. ERNiCrFe-5 is used as filler for welding of the dissimilar metal combination chosen. The welded joints of steel structures are easily affected by fatigue/alternative loading once it is subjected to service. Experimental results were used to fit the S-N curve, and corrected endurance strength and fatigue life were predicted based on Eurocode3. Fatigue damage theory has been used to quantify the accumulated damage. Fracture morphology has been analysed using SEM to assess the crack initiation and type of failure.

Keywords Welding of dissimilar metals · GTAW · SS 316L · Monel 400 · Microhardness · Fatigue analysis

11.1 Introduction

The welding of dissimilar metals, when compared with the conventional welding, is the joining of two identical metals that would generally not weld together as they have diverse mechanical and chemical characteristics. Certain characteristics of the metals to be welded are their point of evaporation, melting point, material strength, compatibility and chemical composition. Regardless of these impediments, there is a requirement to weld such dissimilar materials for different purposes inclusive

C. Mani · B. Sozharajan · R. Karthikeyan (✉)
Department of Mechanical Engineering, BITS, Pilani, Dubai Campus, Dubai, UAE
e-mail: rkarthikeyan@dubai.bits-pilani.ac.in

J. Paulo Davim
Department of Mechanical Engineering, University of Aveiro, Aveiro, Portugal

J. P. Davim (ed.), *Welding Technology*, Materials Forming, Machining and Tribology, https://doi.org/10.1007/978-3-030-63986-0_11

of saving the cost, reducing the weight, increase in device efficiency and so on. Welding of dissimilar metals maximizes the advantages that each metal produces while limiting their disadvantages [1]. Dissimilar metal welding has been broadly utilized in the industries of oil and gas, power generation, nuclear energy, military defence, shipbuilding, medical, electronics, automotive and aerospace because of phenomenal qualities, for example, flexibility, shorter cycle time and low distortion. The practical scenario by cyclic loading, material will result in fatigue failure at load below yield strength in low cycle failure and ultimate strength for high cycle failure. Crack initiation is caused by imperfections in materials, and if the crack reaches a critical size, it will propagate into failure [2].

Study by Giovanni Meneghettia et al. (2019) has experimentally studied on the fatigue properties of dissimilar welds in low carbon steel/ductile iron which recommends that the arc-welded joints are having homogenous joint and partake better fatigue performances [3]. Four-point bending loading and axial loading are used for testing fully penetrated and partially penetrated butt-welded joints. 10–30 Hz frequency and 0.05–0.5 R ratio have been used for the study. 97.7% probability of survival at 2 million cycles has been used for comparison of experimental results according to Eurocode3. The fatigue performance was found to be better than the international standards suggested for steel welded joints. Hafiz Waqar Ahmada et al. (2020) investigated the Nickel 617/12 Cr steel welded joints with and without buttering technique on 12 Cr steel side. 10 Hz frequency and 0.1 R ratio were used for ambient condition as well as corrosive environment in 3.5% Na cl. Buttering has helped in reducing crack growth rate in ambient condition. However, in corrosive environment, it has not caused any improvement [4]. Wen-Ke Wang et al. (2020) examined high cycle fatigue behaviours on dissimilar metal welded of CrMoV and NiCrMoV using buttering technique. Buttering using TIG and SAW welding are used for the welding which is followed by PWHT. The HCF tests were conducted at 120 Hz with stress ratio of -1. Specimens were made in base metal, weld zone and HAZ. Basquin's equation has been used for fitting the experimental results. Both HAZ revealed finer microstructures having but lower fatigue curve than base metals. The fusion interface area of CrMoV base metal showed the weakest zone in dissimilar joint, Also the behavioural mechanism of carbides in fatigue failure model was also studied and effect of microcracks specially appear around the coarse carbides in the crack tip plastic zone [5].

Yi-Bo Shang has analysed the fatigue crack propagation behaviour of different regions weld A508/316 L weld joints and inferred fatigue failure is originated in the weld region of material. Columnar grain direction affects the trans granular failure in the weld region and in the interface the resistance to crack propagation is observed due to the presence of mixed martensite region. The presence of coarse martensite promotes trans granular fracture [6]. K. Kuwabara analysed dissimilar metals welding of austenitic stainless steel 304 with Inconel 82, SS 308L, 309L and 308MoL using TIG and MIG welding at low cycle loading. Non uniformity in strain hardening caused by plastic deformation reduces the fatigue strength and crack propagation resistance of welded joints is on par with the base metals considering the effect of residual stresses on closing and opening of cracks [7].

Eurocode3 has been used for comparison of S-N curve obtained using experimental data related to welded joints by many researchers. Pijpers et al. [8] studied butt welding of S690 and S1100 and compared experimental S-N data with Eurocode3 based curve. The experimental curve is above the code values in both cases. Costa et al. [9] experimented with high strength steel and common steel and compared the S-N curve with Eurocode3 to evaluate safety margin. Zong et al. [10] investigated fatigue performance of fillet weld joints of construction steel and compared the performance using Eurocode3 and finite element analysis.

SS 316L grades have excellent toughness even at cryogenic temperatures. It also has higher creep, stress to rupture and tensile strength at elevated temperatures. They are weldable and possess reasonable corrosion resistance, and hence, they are being used in high temperature and corrosive environment in combination with nickel-based alloys [11]. Monel 400 alloy has the excellent mechanical properties over a wide temperature ranges including sub-zero temperatures as well as corrosion resistance [12].

The present study is aimed at fatigue behaviour analysis of GTAW welded joints of SS 316L/Monel 40 alloy using Eurocode3 and 95% survival probability with supported findings from microhardness survey, optical microscopy and scanning electron microscopy.

11.2 Materials and Methods

316L stainless steel and Monel 400 alloy were chosen as the materials for the current study, and they were welded using gas tungsten arc welding process and a Ni–Cr based filler material. The chemical composition of the materials used for the study are listed in Table 11.1.

11.3 Welding

The welding plates used were having thickness of 3.2 mm and were welded using TIG 275 Lincoln welding machine with argon as shielding gas. 30 V groove type butt welding with a root gap of 2 mm has been used for the study (Fig. 11.1). Table 11.2 shows the weld groove details and process parameters.

11.4 Microhardness

The hardness values are measured with the help of Matsuzawa MMT-X digital microhardness tester as per ASTM E-384. The Vickers microhardness for the heat-affecting zone (HAZ), base metal (BM) and weld zone (WZ), for dissimilar weldments was

Table 11.1 Chemical composition

Composition (Elements)	SS 316L (Base metal-1)	Monel 400 (Base metal-2)	ENiCrFe-5 (Filer wire)
C	0.03	0.3	0.04
Cr	17	–	16
Fe	65.64	2.5	06–10
Mn	2	2	1
P	0.045	–	0.03
S	0.03	0.024	0.015
Si	0.75	0.5	0.35
Mo	2.5	–	–
Cu	0	31.67	0.5
Ti	–	–	
Co	–	–	0.12
Ni	12	63	70(min)

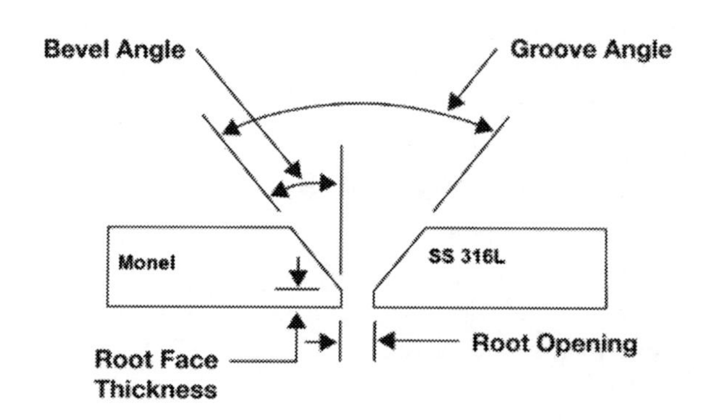

Fig. 11.1 Weld Groove

Table 11.2 Weld parameters and characteristics

Bevel angle	30°
Filler wire	ENiCrFe-5
Welding current	80 A
Root face thickness	1 mm
Root opening	2 mm
Polarity	DCEN
Welding speed	2.5 mm/s
Shielding (Argon)	8 LPM
Backing gas (Argon)	5–7 LPM
Tungsten size and type	1/8″, 2% throated tungsten

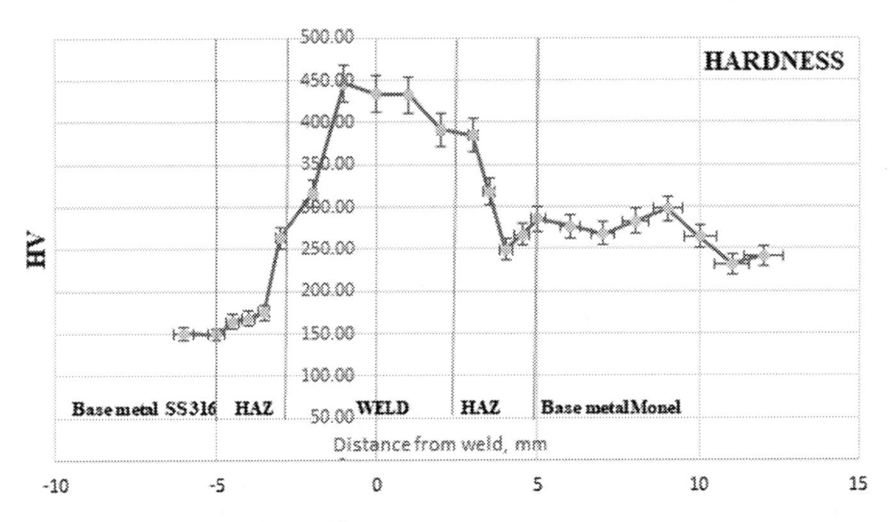

Fig. 11.2 Microhardness along weld

evaluated, and from Fig. 11.2, hardness values are varying from base metal 1 to base metal 2. Weld region records the maximum hardness (440–460 HV) due to the dendritic structure formed during re-solidification. HAZ has lesser hardness in Monel side due to grain coarsening, and Monel base metal has higher hardness than HAZ due to the presence of refined grains. However, in stainless steel side, although the HAZ hardness is lesser than weld metal, it is higher than the base metal. This may be attributed to the precipitation in the weld fusion zone. The similar trends of hardness were reported by Mishra et al. [13], and a linear relation between fatigue limit and hardness has been reported by Casagrande et al. [14].

11.5 Optical Microscopy

The examination was carried out on five different regions including base metal and heat-affected zone of SS 316L and Monel 400, and weld zone at 400X magnification. The microstructural features generally change based on the heat input, electrode materials and weld speed. The microstructural changes are caused due to grain coarsening, grain refinement and precipitation which in turn alter the mechanical properties [15]. In Fig. 11.3, SS 316L base metal illustrates austenitic polyhedral grain structure with random twinning, and Fig. 11.4 reveals austenitic dendrites structure with a lighter colour of the etched ditches. SS 316 L HAZ have undergone precipitation due to the weld heat input, and these are highly susceptible for intergranular corrosion. Figure 11.5, microstructure shows dendrites structure of the weld material, whereas Fig. 11.6 illustrates coarse dendrites structure with partially melted zone with lower dilution of filler material. In Fig. 11.7, polyhedral grain structure of

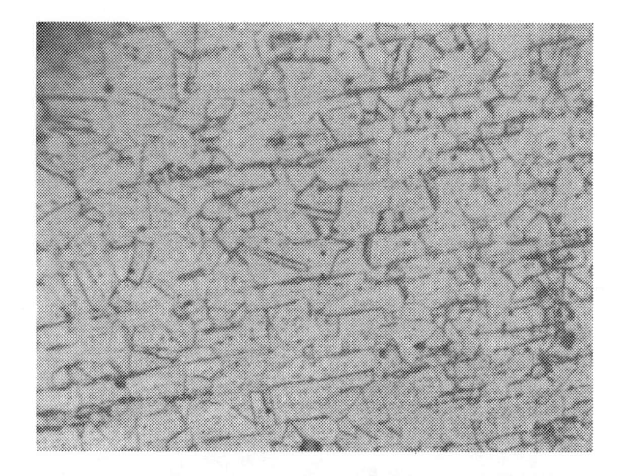

Fig. 11.3 SS 316L base metal OM

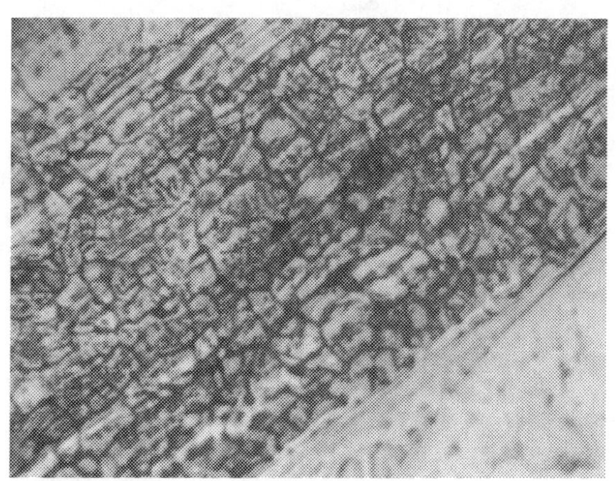

Fig. 11.4 HAZ of SS 316 L

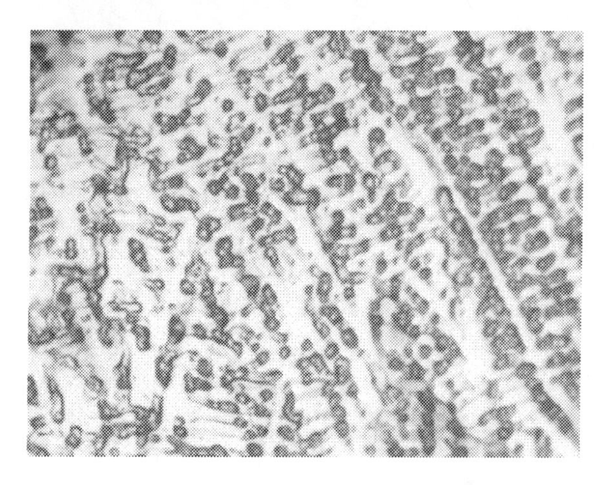

Fig. 11.5 Weld zone

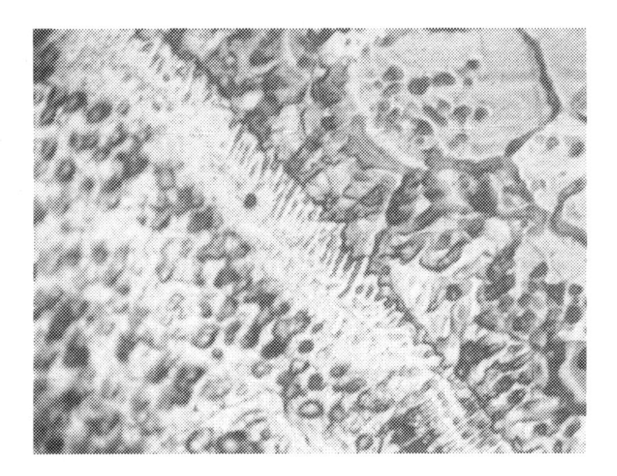

Fig. 11.6 HAZ/fusion zone Monel

Fig. 11.7 Monel base metal

Monel alloy with little amount of twinning is randomly observed. Etching has also revealed rolling direction in Monel base metal region.

11.6 Fatigue Testing

Shimadzu Servopulser (100 kN) dynamic testing machine has been used for fatigue testing with constant amplitude stress-controlled cycle loading with R ratio = 0.1. R ratio is the ratio of minimum to maximum stresses with a frequency range of 20–45 Hz. The maximum stress level was fixed based ($K \times \sigma_y$) on the yield stress of the material (284.1 MPa). Initial loading factor K varies from 0.6 to 0.8. The loading coefficient is adjusted to obtain the fatigue limit for the tests [16]. The experiments were conducted up to 2 million cycles depending on the loading condition.

11.6.1 Quantitative Fatigue Analysis

Quantitative analysis is used to determine the slope and intercept for S-N data and also the 95% survival probability as well as the Eurocode3 design curve parameters [16]. Basic formula used to describe the fatigue performance of material is represented by

$$S^m N = C \tag{11.1}$$

S—Nominal stress amplitude, and N is the number of cycles to failure. The slope and intercept m and C can be determined by using regression analysis. Logarithm can be taken for Eq. (11.1) to determine the unknown values.

$$\log N = \log C - m \log S \tag{11.2}$$

Further, 95% survival probability has been evaluated using Eq. (11.3) where σ is the standard error for the experimental data.

$$\log N = \log C - 1.645\sigma - m \log S \tag{11.3}$$

Fatigue life of steel structural component especially welded steel components is affected by major influencing factors such as geometry discontinuity, material property and fabrication, environment, residual stress and stress range. Eurocode3 takes into consideration of the above parameters and provides the endurance limit with the require safety margin. It uses limit state method which mainly considers constant amplitude stress range and number of cycles to failure for designing the curve [17]. If the number of cycles $N \leq 5 \times 10^6$, fatigue strength design curve in terms of stress range, and stress amplitude is given as follows:

$$\Delta\sigma_{\mathrm{R}}^{m} N = \Delta\sigma_{\mathrm{C}}^{m} 2 \times 10^6 \tag{11.4}$$

where $m = 3$ and $\Delta\sigma_{\mathrm{C}}$ is the stress amplitude if $N = 2 \times 10^6$

Similarly if the number of cycles is $5 \times 10^6 \leq N \leq 10^8$, then

$$\Delta\sigma_{\mathrm{R}}^{m} N = \Delta\sigma_{\mathrm{C}}^{m} 5 \times 10^6 \tag{11.5}$$

$m = 5$, and $\Delta\sigma_{\mathrm{C}}$ is the stress amplitude when $N = 10^8$.

Experimental conditions and results and quantitative analysis results are summarized in Table 11.3. In Table 11.3, $\Delta\sigma_{\mathrm{C}} = 90\,\mathrm{MPa}$ when $N = 2 \times 10^6$. Table 11.4 shows the comparison of results using different analyses.

Table 11.3 Experimental and theoretical fatigue results of welded specimen

S. no	σ_{max} (MPa)	σ_{min} (MPa)	N (Cycles)	$\Delta\sigma_R$ (Experimental) (MPa)	$\Delta\sigma_R$ (95% survival probability) (MPa)	$\Delta\sigma_C$ (MPa)	$\Delta\sigma_R$ (Eurocode3) (MPa)
1	198.87	19.887	189,296	178.983	162.499	90	197.487
2	188.24	18.824	325,836	169.416	134.411	90	161.035
3	180.12	18.012	349,140	162.108	137.321	90	164.786
4	170.46	17.046	525,326	153.414	118.422	90	140.532
5	153.21	15.321	616,595	137.889	112.685	90	133.225
6	146.46	14.646	797,994	131.814	104.027	90	122.251
7	142.05	14.205	1,112,355	127.845	93.849	90	109.439
8	131.64	13.164	1,210,598	118.476	91.419	90	106.394
9	124.18	12.418	1,455,459	111.762	86.345	90	100.058
10	113.64	11.364	1,825,385	102.276	80.491	90	92.782
11	101.89	10.189	2,108,628	91.701	76.971	90	88.427
12	85.23	8.523	2,431,914	76.707	73.642	90	84.321

Table 11.4 Comparison of endurance strength/fatiguelimit values

Fatigue design curve	Endurance strength/Fatigue limit (S_e) (MPa)
Experimentally fitted	97.723
95% survival probability	78.523
Eurocode3	89.125

11.6.2 S-N Curve

Test on 15 welded specimens were carried out, and 12 effective test yielding data are presented in Table 11.3. All tested specimens are failed at weld region due to high hardness and geometrical discontinuity. Due to increase in hardness, ductility decreases, and it becomes more brittle. Due to geometrical discontinuity, more stress concentration occurs at weld region. These two factors are the primary reasons to for crack initiation in weld material.

Experimental S-N values were plotted in Fig. 11.8 with 95% survival probability curve using Eq. (11.3). Eurocode3 design curve determined using equations were also plotted alongside.

The endurance strength at 2×10^6 cycles was determined. Experimental fatigue limit was 8.78% higher than Eurocode3 design values and 19.647% higher than corresponding 95% survival probability curve value. From this, it can be inferred that Eurocode3 design curve was generally safe and applicable. But it does not provide

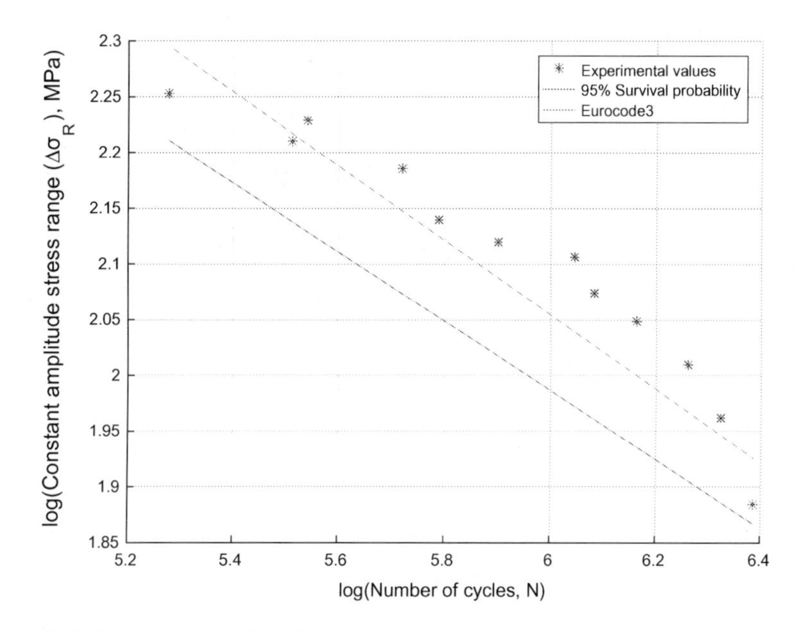

Fig. 11.8 Fatigue test results of welded specimen

enough safety margin for batch of specimens compared to 95% survival probability [10]. Hence, it is better to choose 95% survival probability fatigue design curve with endurance strength of 78.523 MPa which is comparatively conservative to assess the fatigue performance of TIG welded SS316 L/Monel 400 specimens. Endurance strength (78.523 MPa) to ultimate tensile strength (548.06 MPa) ratio is found to be 0.143. It shows that ultimate tensile strength is 6.98 times higher than endurance strength of the weld material.

11.6.3 Fatigue Damage

Based on the service conditions, the service life can be determined using fatigue damage theory. It calculates the remaining service life of a structure if the fatigue life (N_f) is known at a particular stress level. The prediction model for damage accumulation (D) is obtained by [16]

$$N_f = \frac{(\beta + 1)\left(\sigma_{\max}^{\beta+1} - \sigma_{\min}^{\beta+1}\right)^{-1}}{2B(\beta + 2)} \tag{11.6}$$

$$D = 1 - \left(1 - \frac{n}{N_f}\right)^{1/(\beta+2)} \tag{11.7}$$

where $\sigma_{\max}$ and $\sigma_{\min}$ = maximum and minimum stresses and n = number of fatigue cycles.

Using regression analysis and experimental data, the material constants B and β can be determined. The fatigue damage curve with $\beta = 4.014$ could be drawn according to Eq. (11.7) as shown in Fig. 11.8. Initially, slope of the curve gradually increased due to increasing of fatigue cycles. So, damage curve increased linearly when cyclic ratio ranges between 0 and 0.5. It shows crack initiation in the specimen with maximum damage value of 0.12. After cyclic ratio 0.5, damage curve becomes nonlinear up to cyclic ratio 0.9. It shows crack propagation in the specimen with maximum damage value of 0.33. From a cyclic ratio of 0.9, the cumulative damage drastically increases; hence, the effective bearing area of specimen decreases which results in transient fracture. Crack propagation stage is not fully developed which accounts for the reduced portion of the fatigue life during the cyclic ratio from 0.9 to 1.0. [16] (Fig. 11.9).

11.7 SEM for Fractures Surfaces

The fracture surface look normally undistinguished to visual examination, whereas it is also challenging to categorize the mode of failure from macroscopic features.

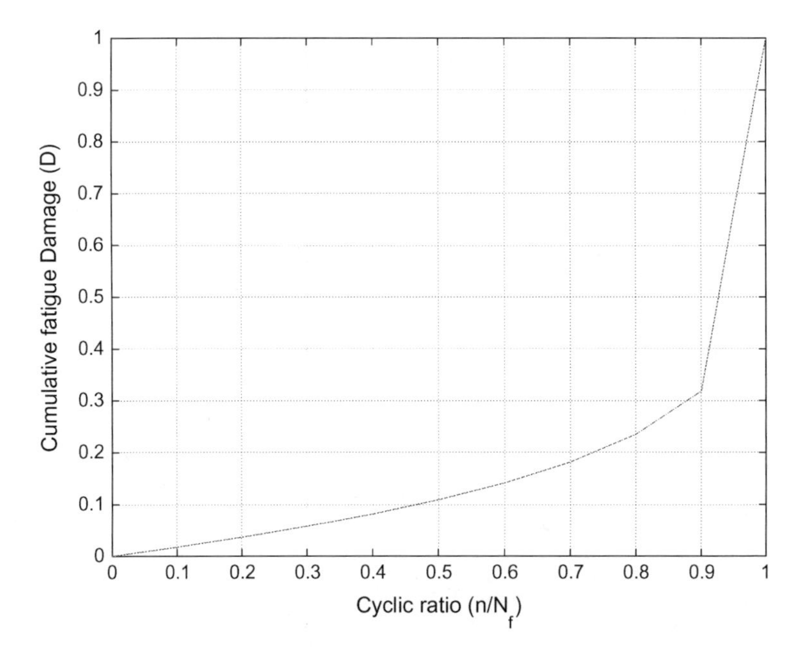

Fig. 11.9 Fatigue damage test result of welded specimen

Hence, in this scenario, it is recommended to examine the fracture surface using SEM to identify fracture surface. VEGA 3 series PC-controlled scanning electron microscopes with a tungsten heated filament was deployed to examine the fracture surfaces. The fractured fatigue test specimens are displayed on Fig. 11.10, wherein the facture was identified at weld fusion zone.

The specimen has undergone significant plastic deformation prior to fracture. The crack was initiated at the weld region near to the interface regions and had gradually traversed to the region of higher hardness as reported earlier by other researchers [18, 19]. The diverse microcleavage fracture morphology could be observed on fracture fractographs in Figs. 11.11, 11.12, 11.13, 11.14, 11.15 and 11.16.

Figure 11.11 shows the elongation or the plastic deformation of the dissimilar weld specimen. Figure 11.12 illustrates fatigue striations within fibrous regions along

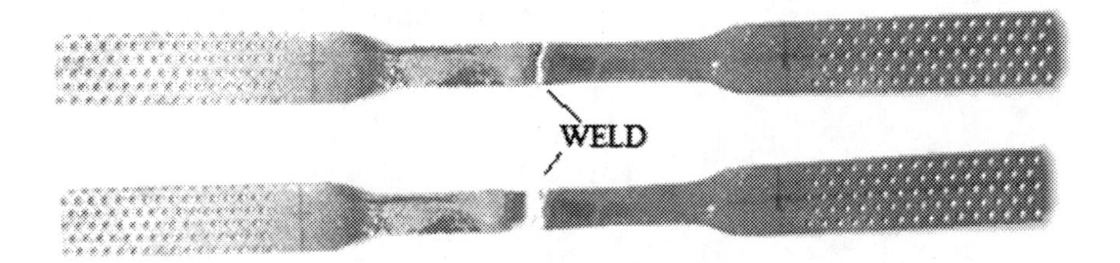

Fig. 11.10 Fractured weld test specimens

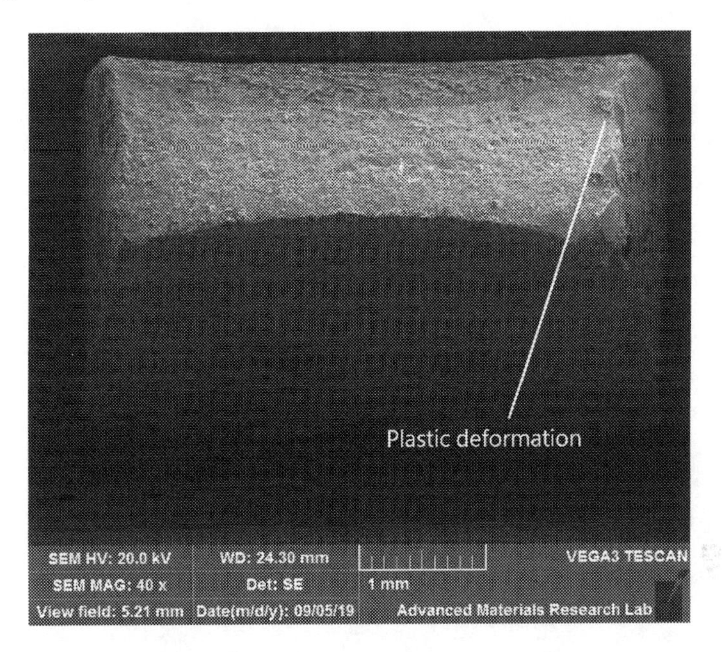

Fig. 11.11 Fracture surface 1 at 40×

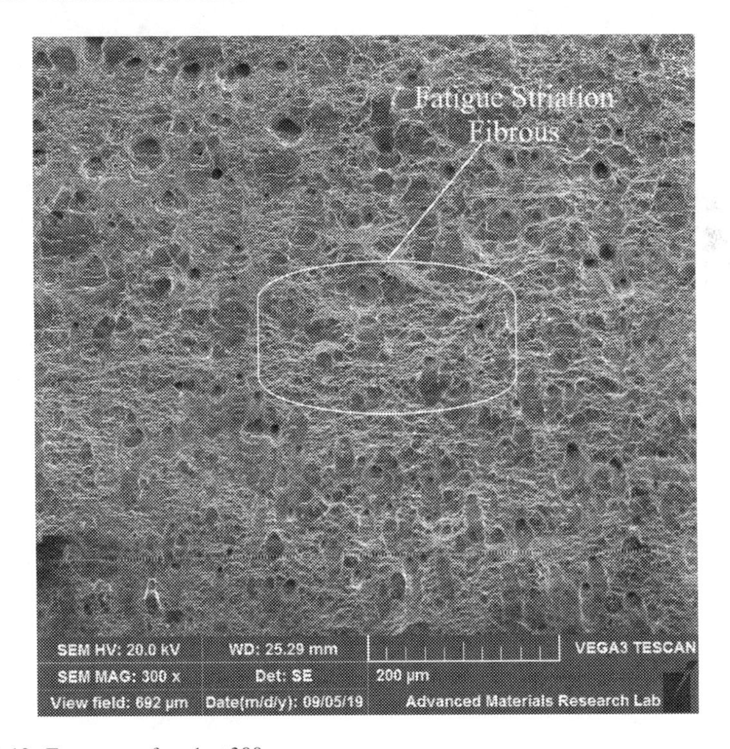

Fig. 11.12 Fracture surface 1 at 300×

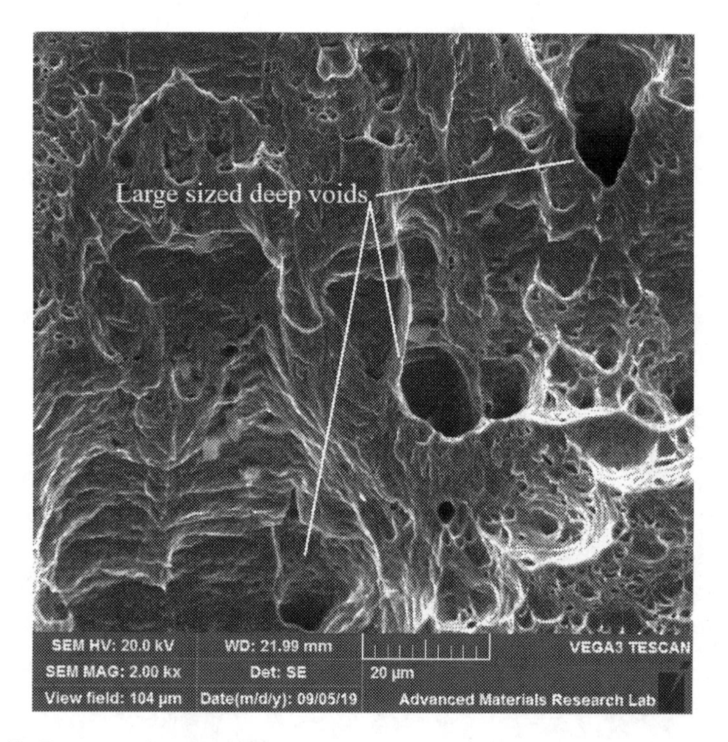

Fig. 11.13 Fracture surface 1 at 2000×

with isolated voids formed near the surface of their respective crack initiation regions. Some signs of delamination were also noticed. Figure 11.13 shows large-sized and deep voids near the crack initiation region at relatively higher magnification. Similar observations were reported by Malhotra and Shahi [20]. The examination of tear ridge and facets in Fig. 11.14 illustrates cleavage features. The tear ridge might result from the boundary of two neighbouring grains with distinct disorientation. Moreover voids which could be considered as forerunners of dimples can be found within the facets in layers similar to the report by Dang [21].

Figure 11.15 shows the crack initiation after elongation on fatigue loading. The beach marks on Fig. 11.16 are clear indication of fatigue when they can be seen on fractograph with a rough and flat surface. Fatigue cracks on butt weld dissimilar joint were close to the coarse grain region in the fusion zone. Fatigue strips on the fracture and the presence of opaque-trans granular singularities are observed on Fig. 11.17 associated with dimples (microvoids) which are characteristic of a ductile type of fracture that emerge to slippage in the materials which address the primary mode of failure. A distribution of relatively larger microvoids are observed in high magnification in Fig. 11.17. This ductile trans granular mode of fracture is related with relatively high ductility metals which were also reported by Bahgat et al. [22].

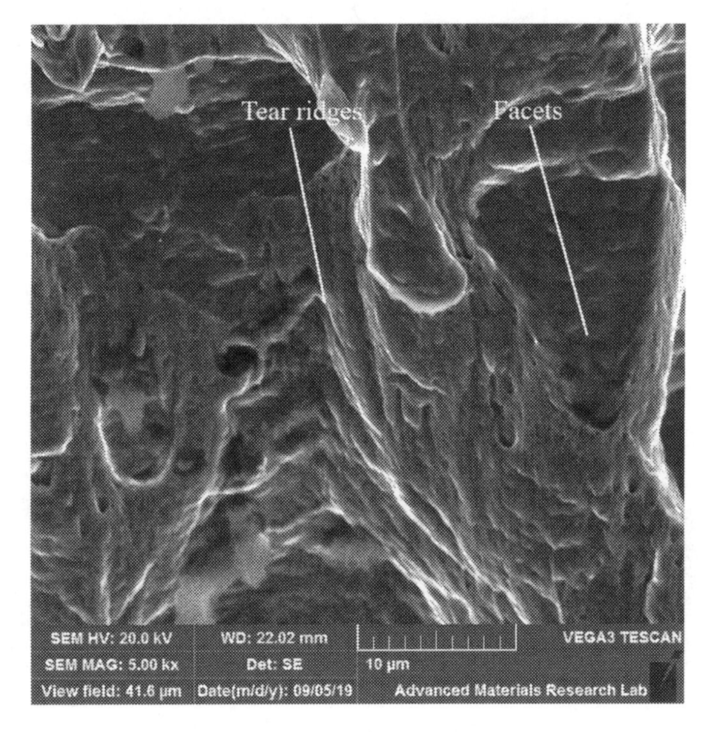

Fig. 11.14 Fracture surface 1 at 5000×

11.8 Conclusions

Dissimilar metal welds using gas tungsten arc welding, and the welded joints were subjected to fatigue testing. The endurance limit for the butt joint specimens was found to be 14.3% of ultimate tensile strength at 95% survival probability level. The maximum fatigue damage value was reported to be 0.33 from quantitative fatigue damage analysis. The fatigue failure mainly occurred in the weld region due to increased hardness and geometrical discontinuity. However, the joints have undergone significant plastic deformation before failure. The fractography revealed striations as well as dimples.

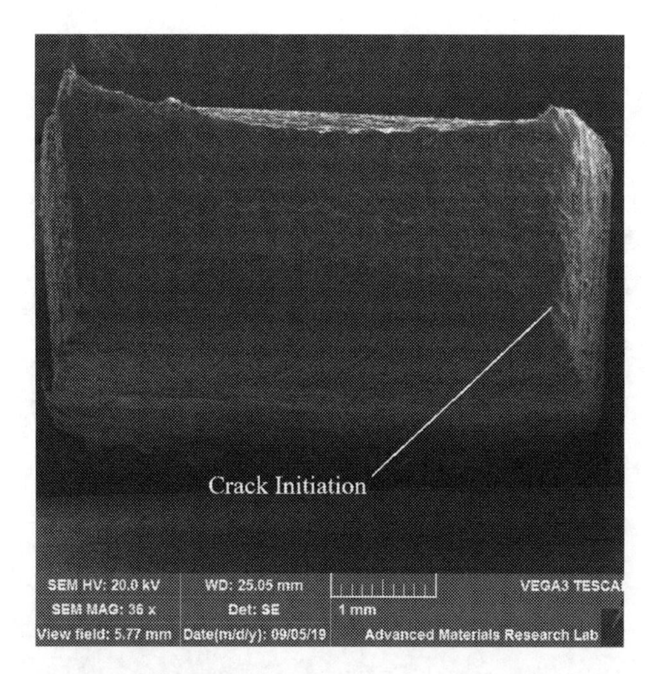

Fig. 11.15 Fracture surface 2 at 40X

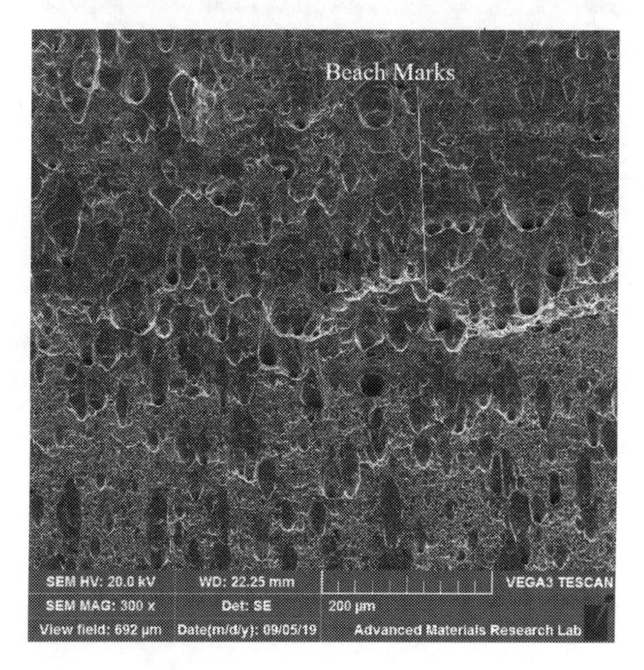

Fig. 11.16 Fracture surface 2 at 300×

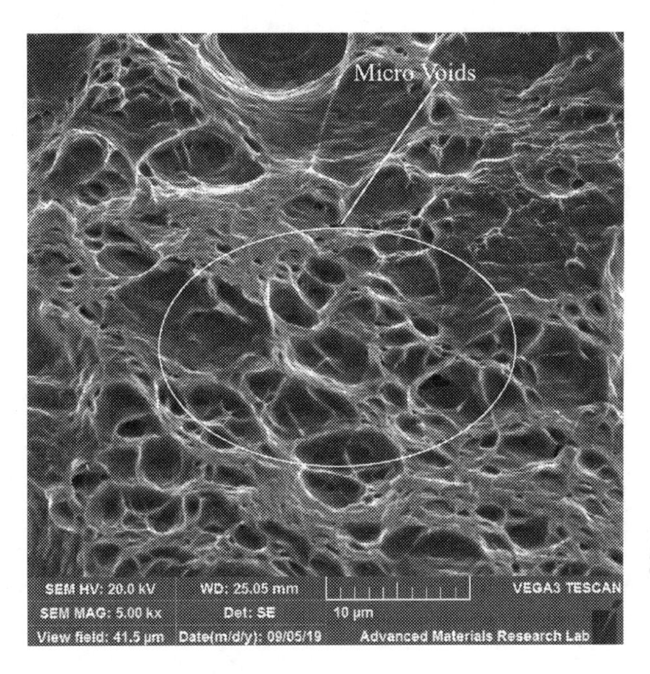

Fig. 11.17 Fracture surface 2 at 5000×

References

1. Enz J (2017) Laser beam welding of high-alloyed aluminum-zinc alloys. Dissertation
2. Makhlouf ASH, Aliofkhazraei M (2015) Materials failure analysis with case studies from the aerospace and automotive industries. Elsevier Science & Technology, Erscheinungsort nicht ermittelbar
3. Meneghetti G, Campagnolo A, Berto D et al (2019) Fatigue properties of austempered ductile iron-to-steel dissimilar arc-welded joints. Procedia Struct Integr 24:190–203. https://doi.org/10.1016/j.prostr.2020.02.016
4. Ahmad HW, Chaudry UM, Tariq MR et al (2020) Assessment of fatigue and electrochemical corrosion characteristics of dissimilar materials weld between alloy 617 and 12 Cr steel. J Manuf Processes 53:275–282. https://doi.org/10.1016/j.jmapro.2020.02.038
5. Wang W-K, Liu Y, Guo Y et al (2020) High cycle fatigue and fracture behaviors of CrMoV/NiCrMoV dissimilar rotor welded joint at 280 & #xB0;C. Mater Sci Eng A 786:139473. https://doi.org/10.1016/j.msea.2020.139473
6. Shang Y-B, Shi H-J, Wang Z-X, Zhang G-D (2015) In-situ SEM study of short fatigue crack propagation behavior in a dissimilar metal welded joint of nuclear power plant. Mater Des 88:598–609. https://doi.org/10.1016/j.matdes.2015.08.090
7. Kuwabara K, Takahashi Y, Kawaguchi S et al (1992) Fatigue strength of dissimilar welded joints for the main vessel of an LMFBR. Nucl Eng Des 133:335–344. https://doi.org/10.1016/0029-5493(92)90160-w
8. Pijpers RJM, Romeijn FSK, Kolstein MH, Bijlaard A (2007) The fatigue strength of butt welds made of S960 and S1100. In: 3rd international conference on steel and composite structures. Steel and composite structures proceedings. Taylor & Francis, London
9. Costa J, Ferreira J, Abreu L (2010) Fatigue behaviour of butt welded joints in a high strength steel. Proc Eng 2:697–705. https://doi.org/10.1016/j.proeng.2010.03.075

10. Zong L, Shi G, Wang Y-Q et al (2017) Experimental and numerical investigation on fatigue performance of non-load-carrying fillet welded joints. J Constr Steel Res 130:193–201. https://doi.org/10.1016/j.jcsr.2016.12.010
11. ASM International, Stainless Steels for Design Engineers (#05231G)
12. Kalaimathi M, Venkatachalam G, Sivakumar M (2014) Experimental investigations on the electrochemical machining characteristics of Monel 400 alloys and optimization of process parameters. Jordan J Mech Indust Eng 8:143–151
13. Mishra D, Vignesh M, Raj BG et al (2014) Mechanical characterization of Monel 400 and 316 stainless steel weldments. Proc Eng 75:24–28. https://doi.org/10.1016/j.proeng.2013.11.005
14. Casagrande A, Cammarota G, Micele L (2011) Relationship between fatigue limit and Vickers hardness in steels. Mater Sci Eng A 528:3468–3473. https://doi.org/10.1016/j.msea.2011.01.040
15. Kianersi D, Mostafaei A, Amadeh AA (2014) Resistance spot welding joints of AISI 316L austenitic stainless steel sheets: phase transformations, mechanical properties and microstructure characterizations. Mater Des 61:251–263. https://doi.org/10.1016/j.matdes.2014.04.075
16. Guo H, Wan J, Liu Y, Hao J (2018) Experimental study on fatigue performance of high strength steel welded joints. Thin-Walled Struct 131:45–54. https://doi.org/10.1016/j.tws.2018.06.023
17. European Committee for Standardization (2015) EN 1993-1-9 Eurocode3, design of steel structures—part 1-9: fatigue. CEN, Brussels
18. Wang H, Wang G, Xuan F, Tu S (2013) Fracture mechanism of a dissimilar metal welded joint in nuclear power plant. Eng Fail Anal 28:134–148. https://doi.org/10.1016/j.engfailanal.2012.10.005
19. Santos JLD, Monteiro SN, Cândido VS et al (2017) Fracture modes of AISI type 302 stainless steel under metastable plastic deformation. Mater Res 20:596–602. https://doi.org/10.1590/1980-5373-mr-2017-0051
20. Malhotra D, Shahi AS (2020) Metallurgical, fatigue and pitting corrosion behavior of AISI 316 joints welded with Nb-based stabilized steel filler. Metall Mater Trans A 51:1647–1664. https://doi.org/10.1007/s11661-019-05623-0
21. Dang N, Chen S, Liu L et al (2019) Analysis of hybrid fracture in α/β titanium alloy with lamellar microstructure. Mater Sci Eng A 744:54–63. https://doi.org/10.1016/j.msea.2018.12.007
22. Bahgat Radwan A, Abdullah AM, Roven HJ et al (2015) Failure analysis of 316L air cooler stainless steel tube in a natural gas production field. Int J Electrochem Sci 10:7606–7621

Chapter 12
Industrial Pipeline Welding

Spyros Papaefthymiou

Abstract The growth path of advanced and developing countries demands huge amounts of raw materials and energy sources. In applications such as transportation of flammable oil/gas(es), which frequently involve high pressures, render steel the optimum pipe material. Modern steel grades ensure safe and reliable long service life of the pipeline. The most sensitive part is the weld in which defects may be introduced. Pipelines are exposed to harsh conditions (e.g., arctic environment, sour service, high pressures, etc.). Pipes are typically welded using high frequency induction welding (HFIW) and submerged arc welding (SAW), both high productivity techniques. This chapter overviews requirements, production methods, and metallurgical phenomena that occur during pipe manufacturing and may affect weld quality. Scholars, students, and engineers are introduced to the challenges of industrial welding and to the significance of steel in matching application requirements of harsh environments.

12.1 Introduction to Energy Sector Pipeline Projects

Modern pipelines face new challenges due to the characteristics of both environment and transporting medium. In order for pipelines to safely operate and to transport the maximum possible amounts of oil and gas, pipeline steels needed to excel in their properties and durability. Pipeline manufacturers had to overcome constructional, welding, and metallurgical limitations to meet the demands. Pipelines currently operate under sour gas (H_2S) aqueous environments due to the increased amount of H_2S in today's oil and gas reserves. Thus, the sour gas resistance embraces the typical corrosion or erosive material losses and makes imperative a focused pipeline design that addresses that kind of corrosion attacks. Without proper care, the presence of sour environment can lead to stress corrosion or sulfide stress cracking (SSC)

S. Papaefthymiou (✉)
School of Mining and Metallurgical Engineering, Division of Metallurgy and Materials Science, Laboratory of Physical Metallurgy, National Technical University of Athens, 9 Her. Polytechniou str, GR-15780 Zografos, Athens, Greece
e-mail: spapaef@metal.ntua.gr

J. P. Davim (ed.), *Welding Technology*, Materials Forming, Machining and Tribology, https://doi.org/10.1007/978-3-030-63986-0_12

and/or hydrogen-induced cracking (HIC) in the pipeline steel leading to its premature failure accompanied by great environmental and financial risks.

Microstructure and metallurgical characteristics of the applied pipeline steel and the weld are of paramount importance for the structural integrity of the pipeline under extreme conditions. Furthermore, pipeline steels nowadays operate in deep/ultra-deep seas and arctic environments. These operational conditions dictate the limits for modern pipelines and for the utilized pipeline steel grades. Thus, in addition to the seismic, the geological, and the all kind of strength and strain operational conditions, including all kind of operations for laying out the pipeline in the seabed (from shallow to deep-sea waters), the extreme service conditions need to be addressed in the pipeline design. Low temperature operations on deep/ultra-deep-sea and arctic environments that represent extreme environments and bring the current steel compositions to their limits have urged the material development to higher toughness steel grades. Apart from the necessity for excellent weldability, the high cleanliness and the high strength-elongation combinations with targeted microstructures are now in focus. Today, pipeline manufacturers need to take also the unknown into consideration. Pipeline design, steel and manufacturing requirements, the possible transport of more corrosive mediums, and the service in even extremer environments urge steel and pipeline manufacturers to continuous improvements through research with the aim to match these technological and service aspects.

According to U.S. Energy Information Administration [1], the main source of energy will still be fossil fuels. It is expected that the consumption of petroleum will be decreased from 32% (estimated at 2018) to 27% up until 2050 while gas consumption is expected to double until 2035. The energy security which relies on reliable and affordable access to energy sources was greatly supported by the construction of thousands of kilometers long pipelines. In 2006, the total length of high pressure pipelines was estimated to be 3,500,000 km worldwide from which 64% is used for gas transportation while 19 and 17% for the transportation of petroleum and crude oil.

12.2 Matching Tomorrow Challenges Today

12.2.1 Extreme Conditions

Modern pipelines operate under specific environments and transport mediums that are considered extreme conditions. These relate operational and welding conditions and the properties of the transported medium.

- Operational conditions:
  Deep and ultra-deep-sea that demand higher strength steel grades to be used **reeling operation** for laying out the pipeline into the seabed that demand high strain **corrosion and erosion environment** that limits the phases present into the steel's microstructure.

- Transported medium:
  Oil and gas with presence of more than 1.5% $\mathbf{H_2S}$ dictates the application of super clean steel grades free from inclusions. Additionally, high dislocation density and a very refined microstructure can create hydrogen-induced cracking (HIC) related failure and is usually tricky to avoid when high strength micro-alloyed very refined thermo-mechanically controlled rolled steel grades are used for the pipeline construction.
- Welding conditions:
  Melting and instant solidification under unprotected atmosphere redefine the fusion zone and its surrounding microstructure. Oxidation phenomena related to the exposure of the melt to the oxygen of the atmosphere lead to oxide penetrators within the fusion line (weld zone) and limit the pipelines toughness especially in low temperatures (arctic environment), reeling applications, strain- and stress-related constructions, and pipeline operation.

In order to meet these challenges, pipeline manufacturers rely on two main factors, the steel grades to be used and the forming and welding conditions during pipe construction.

Pipeline steel grades for the next decade (2021–2030) conform to the following main standards:

- API 5L—Specification of line pipe,
- EN ISO 3183 (PSL 1–3) for steel pipe for pipeline transportation systems and
- DNV-OS-F101 for offshore standard—Submarine pipeline systems.

Steel and pipe need to withstand high stress, strain, and corrosive environments. Sour service, deep-sea applications with off shore construction methods including reeling demands in combination with low temperature, and/or service under arctic conditions set the prerequisites for both steel and pipeline operations in the years to come. These extreme conditions and the fact that manufacturers need to face the unknown challenge both industrial pipeline welding and steel manufacturers urge them to work closely together to match these conditions and to overcome material and construction limitations.

The next paragraphs elaborate briefly on these conditions.

12.2.2 Sour Service

Process conditions related with oil and gas that contains high levels of hydrogen sulfide (H_2S) or sulfuric acid (H_2SO_4) are called sour service. Oil and gas contain various concentrations of H_2S [2]. In combination with moisture, atomic hydrogen (H) is generated, which diffuses easily in the steel matrix. Figure 12.1 illustrates the adsorption, diffusion, and desorption phenomena resulting in hydrogen-induced cracking (HIC) initiation and propagation [3]. Interstitial sites, voids, and dislocations represent trapping sites for atomic hydrogen [3].

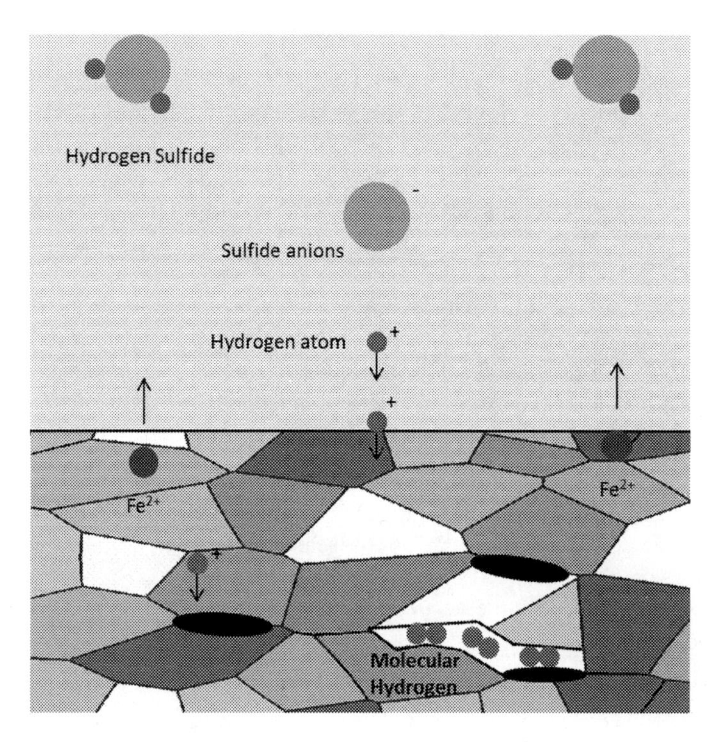

Fig. 12.1 Schematic illustration of HIC process

Inclusions as well as precipitates have high affinity to the H atoms, due to their high binding energy for them. It is nearly impossible for the H atoms to escape; thus, they are regarded as strong and irreversible traps [4]. On the contrary, the H atoms can either be trapped in grain boundaries and dislocations or their movement can be facilitated thereof [5]. By decreasing the grain size, additional boundaries exist, assisting H atom mobility. The lower grain size has higher density of junction points that obstruct the H atom mobility. Thus, the minimum diffusion of H atoms can be achieved by the optimal control of grain size [4]. Usually, in coarser structures, less H is trapped in nodes and junction points. This fact facilitates the H flux and consequently its accumulation at the crack tip, leading to crack propagation [4, 5].

12.2.3 Arctic Environment

Laying out pipelines in areas where subzero temperatures prevail is accompanied with great challenges with respect the steel properties. It is known that as temperature drops below zero, the lattice atoms' kinetic energy is forced to nil (zero or nothing). As a consequence, the atoms move closer together, and the thermal expansion coefficient

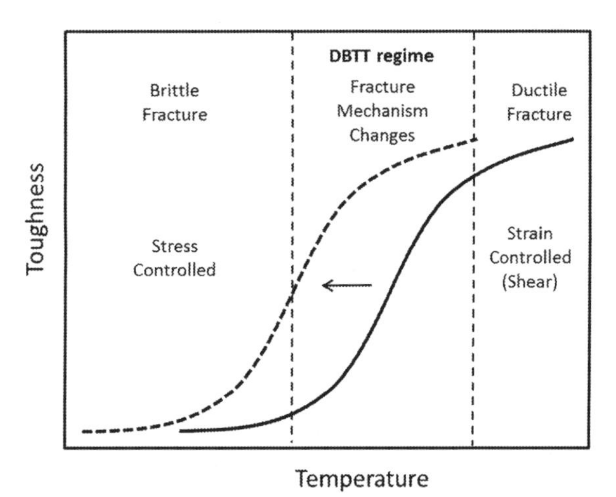

Fig. 12.2 Fracture toughness versus temperature

and the lattice parameters tend to become smaller, which finally determine the steel's mechanical properties.

When reaching the nil ductility temperature, both tensile strength and yield strength in metallic alloys increase. Metals that crystallize in the body-centered cubic, BCC, show great dependence of their yield and tensile strengths on temperature as a consequence of narrow dislocation width and high Peierls forces. Additionally, they display a toughness and ductility loss in a narrow temperature region below room temperature. Thus, the temperature at which the material behavior losses its ductility is of great significance. The ductile-to-brittle transition temperature (DBTT), see a typical graph in Fig. 12.2, is crucial for pipeline steels operating at arctic conditions. The DBTT should be kept as low as possible. For controlling the DBTT, steel manufacturers use proper (micro-) alloying additions, control the average steel grain size through sophisticated thermo-mechanical processing, select low-carbon contents, and in general keep the steel super clean and with a controlled shape and size of its inclusion content.

Empirical equations [6] are describing the strength and the DBTT of ferritic-pearlitic steels:

$$\sigma\left(\frac{klbf}{in^2}\right) = 7.8 + 4.7\%\text{Mn} + 12\%\text{Si} + 51\sqrt{\%N_{\text{free}}} + 2.5d^{-\frac{1}{2}} + ps \qquad (12.1)$$

$$\text{TT}(°\text{C}) = -19 + 44\%\text{Si} + 700\sqrt{\%N_f} - 11.5d^{-\frac{1}{2}} + 2.2\%P \qquad (12.2)$$

where Mn, Si: manganese and silicon, respectively; P is pearlite, N_{free} is free nitrogen, d is the average grain size, ps is the precipitation strengthening factor, σ is the proof strength, and TT: the DBTT.

As previously described, the size and morphology of inclusions is significant as such penetrators are considered notches and lower the steel toughness, e.g., the steel's capacity to absorb energy. Steels with high number of penetrators show reduced ability to resist failure at points of local stress concentration [7].

Another factor to be considered in pipelines is the fatigue limit of the used steel, especially at low temperatures and in systems subjected to dynamic loads and/or cyclic stresses. For instance, pressure changes or extreme stresses imposed by external pressures and other challenges imposed by packed snow or high winds need to be considered. Pipelines can easily be exposed to either frost heave or thaw settlement and could experience large bending forces that typically create large longitudinal strains [7].

For all these reasons, the steels used in such applications should have the necessary high strength and high ductility/toughness potential to suit such conditions. Yield strength increase results in loss of fracture toughness leading to low formability and potentially to cracking. However, modern high strength low alloyed (HSLA) steels (X80–X100 according to the API standards) tend to overcome this issue. Micro-alloyed thermo-mechanical rolled and/or controlled processed steels show microstructure-properties combinations suitable for such applications. High strength–toughness combination increases formability in favor of the pipeline [7].

12.2.4 Deep-Sea Projects

The design of the submarine pipeline includes restrictions on the strength of the pipe both externally and internally. During installation, the pipe is empty internally to reduce as much as possible the tensile stress received by the pipeline at the bottom due to the weight of the hanging part of the pipeline to be installed. As the pipe moves deeper, the external hydrostatic pressure becomes higher and higher, and it is possible that the pipe will collapse. In addition, in pipeline maintenance processes, the pipeline is gradually decompressed internally. In any case, the durability of the pipeline in external pressure and more specifically in hydrostatic pressure is a parameter of design of outstanding importance [8].

12.2.5 Reeling Demands

Deep-sea projects demand laying out the pipeline in the seabed. During installation, in addition to the external pressure (hydrostatic pressure), the pipeline is subjected to axial tension caused by the clamp of the vessel that carries the pipeline, while, as it approaches the bottom, it is subject to bending [9, 10]. The stresses that the pipe receives differ depending on the installation method followed [9].

The process of installing submarine pipelines is carried out in a variety of laying ways, namely S-, J-, reeled R-, and towed-lay.

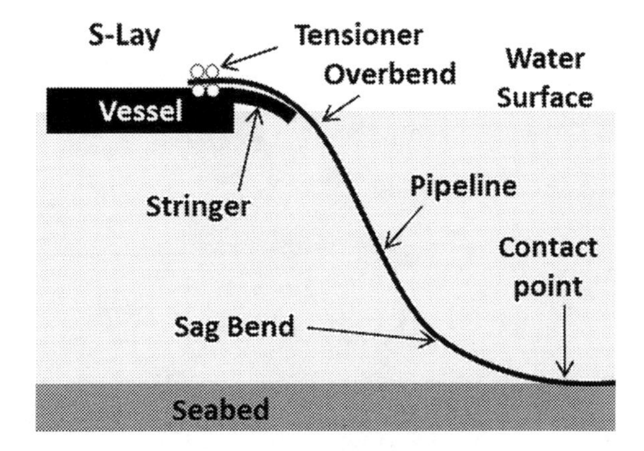

Fig. 12.3 S pipe-laying method for low-to-medium sea applications

The choice of installation method depends on several factors, some of which are the installation depth and the characteristics of the pipe, such as length, weight, and diameter, as well as the topography of the seabed [9].

The S-laying method is the most common in submarine applications, especially for relatively large diameter pipes ($d > 16''$) and is used at low-to-medium water depths. The pipes are firstly welded together on the pipe-laying vessel, then inspected by X-rays for their weld integrity, and field coated on the joint. Pipeline installation in form of "S" is the outcome of stinger and tensioner on the pipe-laying vessel as shown in Fig. 12.3. After the pipe has passed several welding stations, control phases, and tensioners, it is lifted by the stinger, which is usually located on the edge of the vessel. The stinger is the one that will define the curvature of the upper end of the pipe, known as the "overbend," while cylinders provide support to the pipeline, from the moment it continues to sink unsupported, until it touches the bottom of the sea. There at the lowest point, the pipe acquires a curvature opposite to the corresponding overbend, known as the "sag bend," and is the result of the tensioners and the weight of the pipe, while it can be controlled by exerting a tensile stress on the pipe using the vessels tensioning system [9].

The J piping method is a method that is widely used at great depths, usually in pipes up to 32 inches in diameter. In the J deposition method, the conductor is inserted into the sea from the vessel almost vertically as shown in Fig. 12.4, via a tower installed on the vessel, so that as mentioned above, there is no overbend, only sag bend curvature at the lowest point, which affects the required tensile stress from the tensioners which is lower comparing to the S-lay method. Due to the almost vertical installation, only one or two stations are required for welding and testing. Pipe stems consisted of 2 to 5 joints are welded together, lifted at the upper end of the tower, and welded to the existing part of the pipe. Connecting many stems together before lifting from the tower increases the deposition speed; however, it remains a relatively slower method than the S [9].

The R (reeling) method is used for relatively small diameter pipes, up to 16 inches, where the pipeline is first coiled on the drum on land and then is unrolled from the boat

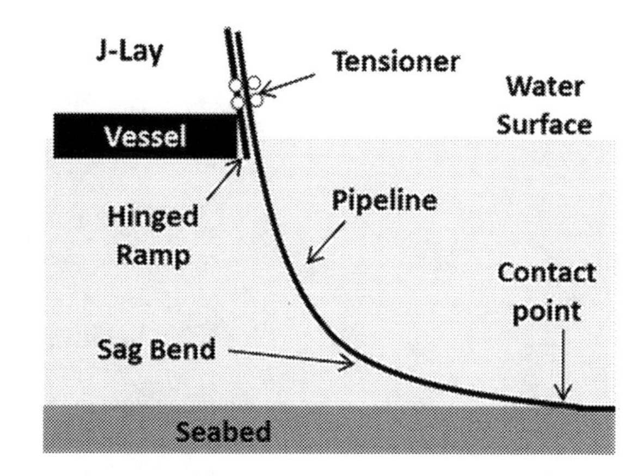

Fig. 12.4 J pipe-laying method for very deep-sea applications

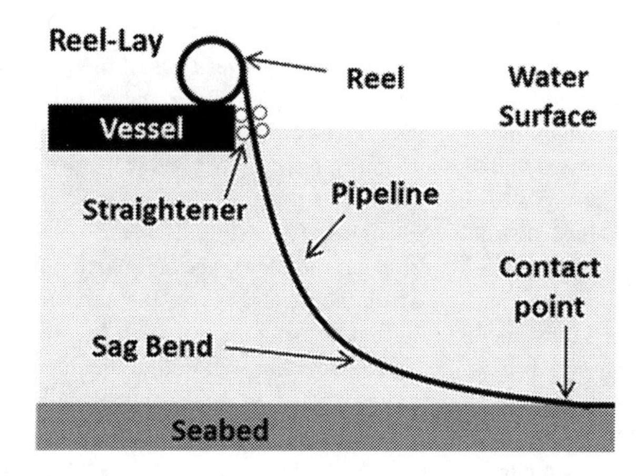

Fig. 12.5 R (reeling) pipe-laying method

to the bottom of the sea either by the S method or by the J method as shown in Fig. 12.5. Pipes are joined onshore and, thus, are independent of the marine environmental influences. The production and laying of the pipeline can be performed separately. While one pipeline length is laid, another one can be manufactured elsewhere and then feeded to the laying vessel once the previous pipe reel has finished. Compared to the J-lay and S-lay processes, the reel-lay method is significantly faster which has an impact on the method's overall cost. A side effect of this method are the high deformations which are imposed on the pipe during winding on the drum and vary depending on the diameter of the drum. With the imposition of plastic deformation, the ductility as well as the strength of the pipe may be affected, which is why it is necessary to align the pipe when it lands on the bottom [9]. In addition, compatibility of the reeling process with the type of pipe steel should always be taken into consideration, because the welding process may cause unwanted hardening of the larger steel types [11]. Depending on the diameter of the pipe and of the drum, this method can be up to ten times faster than other conventional methods [9]. Finally, the advantage of onshore welding and the reduced added cost comes at the price of high mechanical loads on

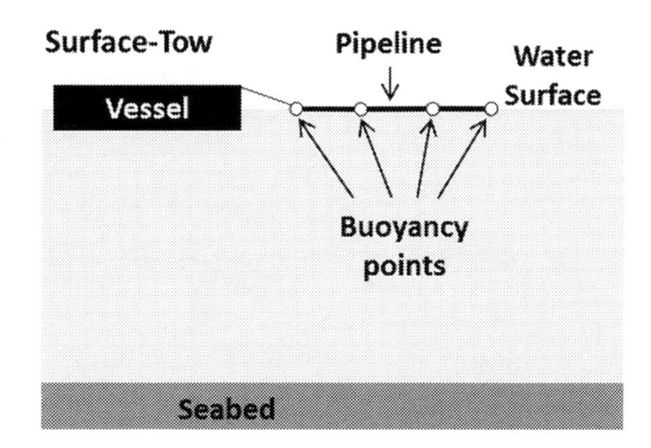

Fig. 12.6 Towed pipe-laying method

the pipes generated as the string is spooled on and off the reel. This is why due to the high requirements involved, this pipe-laying process is currently only used for small-to-medium pipe diameters.

During towed laying, pipelines are assembled on the shore and then towed to the installation site via a transport boat. The assembly is carried out either in parallel or perpendicular to the shoreline as shown in Fig. 12.6, where in the first case the complete pipeline can be assembled before towing and installation. The pre-inspection and inspection of the pipeline are carried out on the shore and not at sea, so it is possible to manage various pipeline sizes and complexities. Five different ways and techniques exist: surface, near-surface, mid-depth, off-bottom, and bottom tow.

12.3 Industrial Welding Methods

The state-of-the-art techniques commonly used for welding pipes are (1) high frequency induction welding (HFIW) and (2) submerged arc welding (SAW). Both techniques can be utilized in longitudinal seams and spiral seams.

12.3.1 High Frequency Induction Welding (HFIW)

High frequency induction welding is an established method applied on pipelines. For this method, high strength steel plates are formed with the help of roll configuration into the tube shape. In Fig. 12.7a , a typical mil setup for longitudinal welding is illustrated. Welding current travels from the outside surface toward the free tube edges, closing electrical circuit, as shown in Fig. 12.7. The edges are heated up and then squeezed together in order to fuse. The so-called 'Vee' apex is the point where

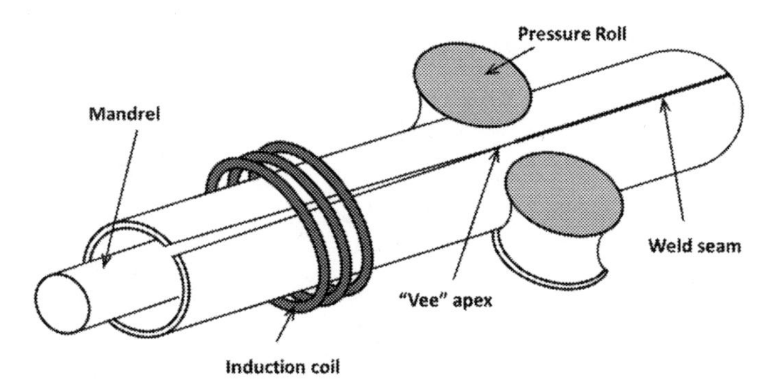

Fig. 12.7 Illustration of the high frequency induction welding process

the free edges come into contact for the first time, and it is the point where the joint starts to form.

The principal phenomena that take place during high frequency induction heating are [12]:

- **Induction effect**: It allows the contactless transmission of power to the workpiece with the aid of an alternating magnetic field. The induction coil generates this alternating field according to:

$$P_i = k\sqrt{f}A_W^2$$

  where P_i is the induced power (kW cm^{-2}), k is a constant, A_W represents the Ampere-turns per cm of the inductor, and f is the frequency (Hz).
- **Skin effect**: At high frequencies, electrical currents and magnetic fields are concentrated only on a thin layer of the pipe's surface [13]. The thin layer is defined as a skin depth, ε, in cm:

$$\varepsilon = 5030\sqrt{\frac{p}{\mu f}}$$

  where p is the specific resistance (Ω cm), and μ is the conductor's relative permeability [12].
- **Proximity effect**: The high frequency currents always flows along the path of least resistance. Two currents flowing in opposite directions on the same material are mutually attracted, as in the Vee [12]. Warren [14] suggested that the position of welding should be within 6–14 mm upstream from the centerline of the induction coils, so that Vee angle is kept within an acceptable range.

The effect of the above-mentioned features is the increase in temperature. The control of the parameter combination can result in localization of heating which is important both of the weld quality but also the minimization of losses. A suitably designed induction coil is essential [12]. The distribution and penetration of heat from the high frequency current are controlled by the skin and proximity effects together with the frequency of the power supply and the mill speed, so the welding frequency and mill speed must be chosen carefully [15].

Although the proximity effect makes the two abutting edges mutually attract, it can be impeded by Warren [14]:

- roughness on the contacting surfaces;
- an oxide layer or foreign matter;
- the absorbed gas layer on the oxide;
- abutting edges relative positions during the introduction of the high frequency current to achieve heating.

The molten metal, containing the majority of the impurities, is squeezed out from the joint, toward the inner and outer side of the tube, using the pressure rolls. The amount of the material removed can be calculated by the difference in diameter before and after the pressure rolls. The excess material is later milled of the tube surfaces. The milling parameters together with the pressure applied by the rolls can influence the weld quality and properties [16, 17]. Various different configurations of pressure rolls are commercially used, as can be seen in, all having the same target to properly align the free edges before welding [14]. Edge mismatching (Fig. 12.8) can result into uneven heating and temperature distribution that, in turn, can deteriorate the weld quality. The resulting microstructure in the weldment is shown in Fig 12.9.

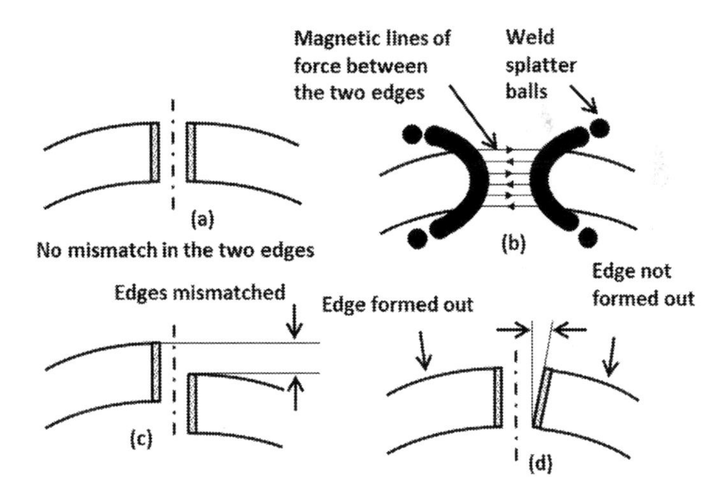

Figure 12.8 **a** Ideal relationship is to have the two abutting edges parallel and matched in the vertical direction. **b** Electromagnetic force ejects hot metal during the high frequency pipe welding. **c** Two edges are mismatched in the vertical direction. **d** Two edges are mismatched angularly.

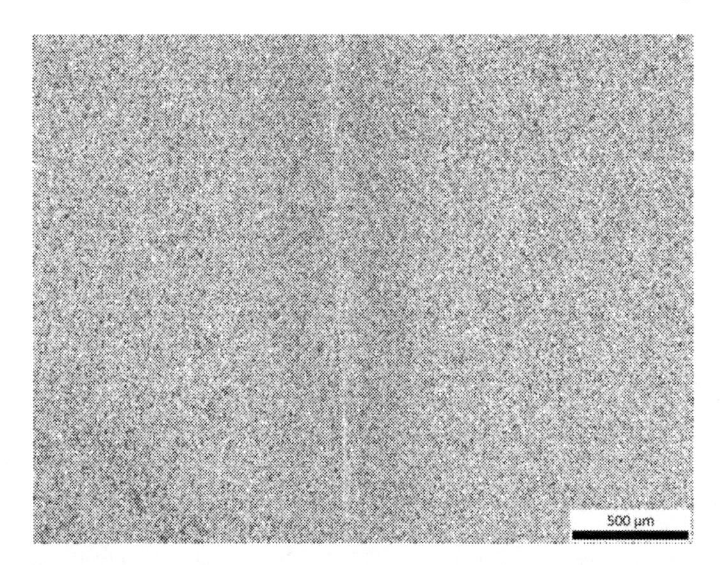

Fig. 12.9 Indicative microstructure of weld seam after HFIW

12.3.2 Submerged Arc Welding (SAW)

During SAW, electric power is applied to an electrode via the nozzle creating an arc. A specially designed funnel supplies flux that is used to shield the molten metal and assist arc striking and stability. The arc melts both the flux and the metal forming a liquid slag that later solidifies and fuses to the base metal, as shown in Fig. 12.10.

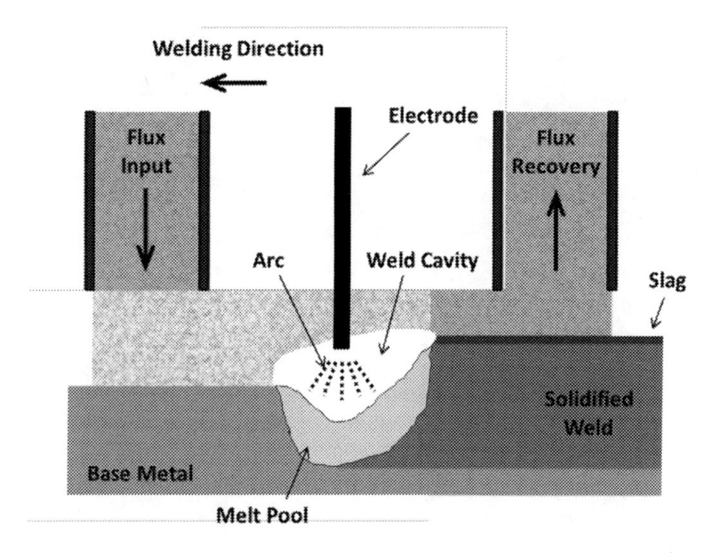

Fig. 12.10 Schematic of the SAW lay out

The weld pool cleanliness can be enhanced by the use of flux, and in some cases, the flux can also be used to adjust the chemical composition of the weld pool.

The wires usually consist of a ferritic steel core coated with copper in order to improve the electrical conductivity, improve the contact tip life, as well as the overall shelf life. As the wire material is relatively simple and mainly aims to match the base material, the flux is much more complicated and can be used to customize the weld properties. Fluxes can be categorized in two levels, firstly, according to their manufacturing method (fused or agglomerated) and, secondly, by their activity (neutral, active or alloying). In each category, the flux can be further categorized according to the chemical composition and the alloying element content (Si, MnO, CaF, etc.) [18].

During pipe manufacturing, the plate edges are tack welded which fixes the geometry of the pipe. Subsequently, two layers of welds typically take place. The first layer is from the inside and followed by a second layer from the outside. The productivity depends on the deposition rate capabilities of the welding machine; thus, special multi-wire heads with up to five wires being able to weld simultaneously are utilized. Nevertheless, especially as the wall thickness becomes higher, the heat input has to be taken into consideration. The heat-affected zones around the seam often present a local decrease in toughness. The welding stability at the beginning and the end of the seam can be ensured by the use of run-on–run-off tabs are assembled on the pipe ends that are later removed [19]. Typical microstructure in the weld seam after SAW is provided in Fig. 12.11.

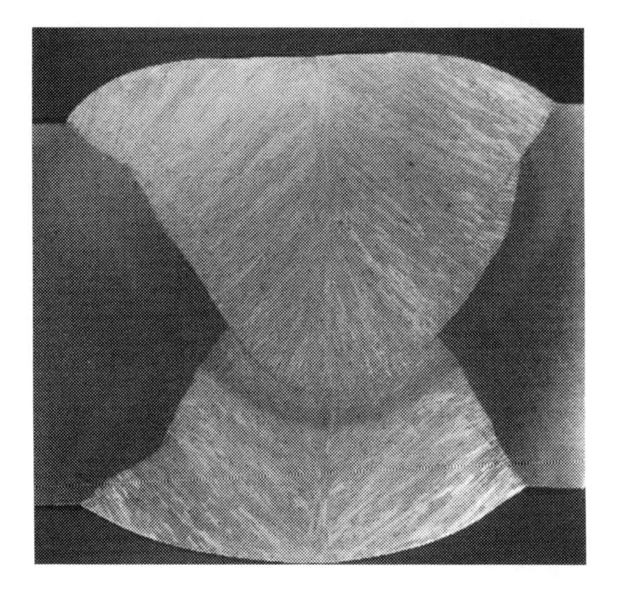

Fig. 12.11 Indicative microstructure of weld seam after SAW

12.4 Microstructure-Property Relationships for Pipeline Steel—An Overview

12.4.1 Microstructure Evolution During Heating and Cooling

Pipeline steels are high strength low-alloyed (HSLA) steels containing significant amount of micro-alloying elements and are produced using thermo-mechanical controlled processing (TMCP). In pipe line steels, the carbon content and carbon equivalent must be low in order to achieve the desired weldability and toughness for gas and oil industry. For low-carbon content (<0.18% wt.), the steel weldability is described by the Ito–Besseyo formula [20]:

$$C_{\mathrm{PCM}} = C + \frac{\mathrm{Si}}{30} + \frac{\mathrm{Mn} + \mathrm{Cr} + \mathrm{Cu}}{20} + \frac{\mathrm{Ni}}{60} + \frac{\mathrm{Mo}}{15} + \frac{\mathrm{V}}{10} + 5\mathrm{B}\ \mathrm{wt\%} \tag{12.3}$$

For $C > 0.12\%$ wt., the $C_{\mathrm{eq.}}$ is defined according to the International Institute of Welding formula:

$$C_{\mathrm{IIW}} = C + \frac{\mathrm{Mn}}{6} + \frac{\mathrm{Mo} + \mathrm{Cr} + \mathrm{V}}{5} + \frac{\mathrm{Ni} + \mathrm{Cu}}{15}\ \mathrm{wt\%} \tag{12.4}$$

$C_{\mathrm{eq.}} < 0.4\%$ wt. is an essential requirement of any structural material to be welded, which helps to avoid cold cracking or hydrogen-induced cracking (HIC) induced by the presence of martensite [21]. Pipe line steels are hot rolled at elevated temperatures in order to obtain fully austenitic microstructure and then cooled with air or water depending on the preferred cooling rate. During cooling, austenite decomposes partially to allotriomorphic ferrite while the remaining austenite enriches in carbon and transforms into pearlite [22, 23]. However, with the addition of micro-alloying elements (e.g., titanium—Ti, niobium—Nb, vanadium—V) strength increase is feasible.

Niobium (Nb), vanadium (V), and titanium (Ti) are carbonitride-forming agents at alloying concentrations less than <0.1% wt. Carbide and nitrides precipitate at elevated temperatures at austenite grain boundaries and pin them preventing austenite enlargement during rolling. Micro-alloying element concentration, the steel's carbon content, and its austenitization temperature must be optimized to obtain full benefit from the precipitation and to ensure absence of undesirable large carbides [24, 25]. The microstructure obtained during/after rolling and the cooling strategy after roll finish determine the kinetics of the major phase transformations (austenite-ferrite), phases present in the final microstructure, and their size. The refined austenite leads to an even more refined ferrite during cooling, which is related to improved toughness and high strength.

During welding operations, different temperature regions are formed on the weldment and on the area adjacent to the weld. The average size of austenite grains decrease moving away from the weld zone constituting the heat-affected zone (HAZ). HAZ is an important area in which cracking can occur as well. In general, HAZ is an area in which the original steel properties degrade. The target for the HAZ is to maintain the properties of the parent steel withstanding alterations due to the rapid heating to temperatures up to the steel melting point followed fairly by rapid cooling [26].

The different heat-affected zones formed during single pass welding on low-carbon steel are shown in Fig. 12.12. The material next to the fusion zone reaches temperatures above the austenite formation temperature. The carbide and nitrides exist in the steel will generally dissolve, unpin austenite grain boundaries, and subsequently grain growth will occur forming the coarse-grained HAZ (CGHAZ). The zone in which the temperature increases above A_3 temperature, but not sufficiently enough to induce carbide and nitride dissolution, and, therefore, austenite grain growth, is suppressed is defined as fine grained (FGHAZ). The zone, in which the temperature increases up to intercritical region of austenite and ferrite (between A_1 and A_3), is defined as intercritical-HAZ (ICHAZ). In this zone, the steel microstructure transforms partially to austenite, while a fraction of ferrite forms. Upon cooling, ferrite phase remains untransformed, while austenite-containing high-carbon content transforms to martensite or pearlite according to cooling conditions. If temperature does not exceed the A_1 temperature, the zone is referred as subcritical-HAZ. Pearlitic bands in this area may start to decompose as cementite begins to spheroidize. Generally, the microstructure of the material remains unaffected [27].

The microstructural evolution in HAZ can be explained with the continuous cooling transformation (CCT) diagram. CCT diagrams consist of two sets of curves: the percent transformation and cooling rate, which corresponds to the actual thermal

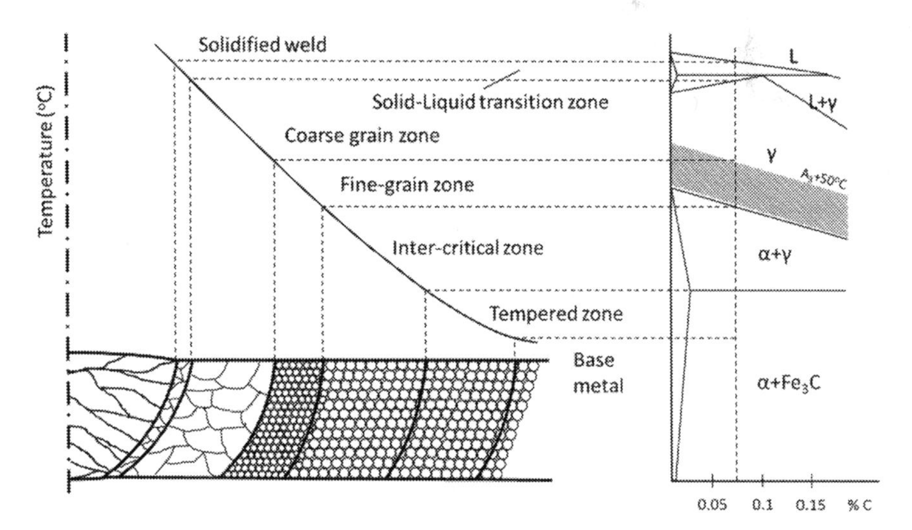

Fig. 12.12 Different heat-affected zones formed during single pass welding on low-carbon steel (Handbook, Properties and Selection: Irons Steels and High Performance Alloys, 1990)

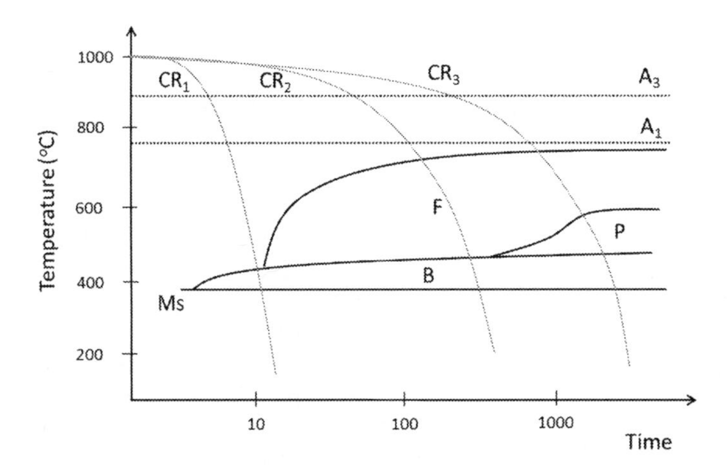

Fig. 12.13 Continuous cooling transformation (CCT) diagram for HSLA steels (Handbook, Properties and Selection: Irons Steels and High Performance Alloys, 1990)

conditions that the weldment experiences. Figure. 12.13 shows the application of a CCT diagram in order to explain the different microstructural constituents formed in a HAZ. Depending on the cooling rate, the cooling rate curve intersects different transformation curves. Therefore, different microstructures are expected to form. Alloying elements with γ stabilizing effect suppress the start of austenite decomposition to lower temperatures, hence, the refined microstructure. On the contrary, inclusion forming elements shift the transformation curves to the left side accelerating the austenite decomposition, since inclusions may act as nucleation sites. Different chemical composition regionally in HAZ as a result of high heating rate as well as variation in cooling rate leads to different microstructural constituents [27].

12.4.2 Welding Microstructures of Pipeline Steels

Initial steel microstructure affects the performance of pipeline under severe conditions. Over the years, the microstructure of pipeline steels has become more complex incorporating non-equilibrium phases simultaneously in order to increase yield strength and toughness (Fig. 12.14). Microstructural constituents, their morphology (allotriomorphic/acicular/polygonal/bainitic ferrite, pearlite, martensite—martensite/austenite constituents), and volume fraction define the steel properties.

Within fusion zone, during welding, the steel melts and solidification phenomena take place. These phenomena include solute redistribution, micro-segregation, banding, dendrite-arm spacing, and solidification mode. Generally, when a liquid turns into solid, segregation occurs altering the once uniform-alloying content as a result of temperature gradient, thermodynamics, and kinetics.

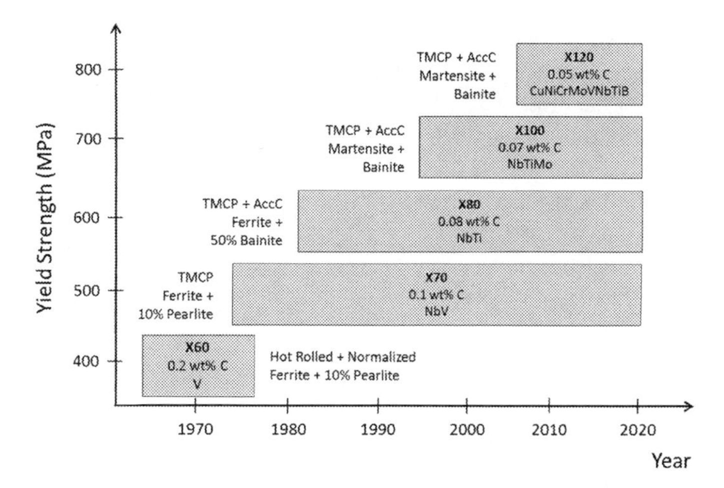

Fig. 12.14 Evolution of pipeline steel grades regarding yield strength, toughness, and microstructural constituents

Low-alloyed steel deposits begin solidification when the temperature drops to T_L as indicated in phase diagram (see Fig. 12.15). At this temperature, solid phase begins to form with chemical composition designated from the constructed tie line. Its composition is significantly lower than this of the liquid. Upon further cooling to T_S, the composition of the solid phase is designated from the tie line intersecting solidus line. However, the solid's diffusion is slower than the liquid's. The time for diffusion is not adequate in order to reach the designated composition. Therefore, the chemical composition in the solid slightly increases, and the liquid phase becomes enriched in alloying elements. As the temperature decreases, multiple layers of different chemical composition form and the chemical composition progressively increases from the core to the outer layer of the solid phase [28, 29]. This phenomenon is called segregation.

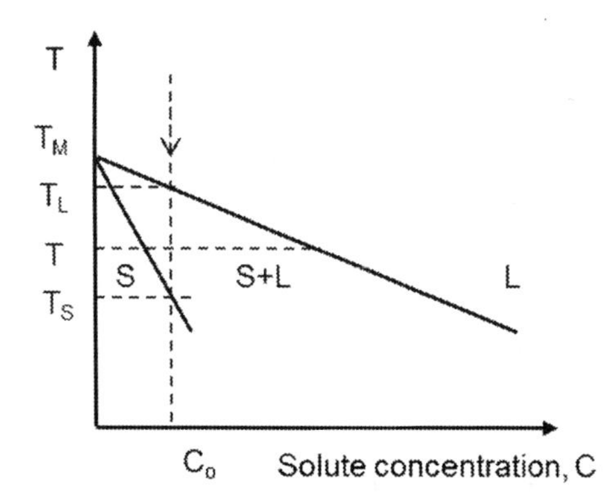

Fig. 12.15 Schematic representation of binary phase diagram during solidification

The morphology of the solidified metal depends on the solidification rate, thermodynamic considerations, and heat flow. The planar interface of solid/liquid can be planar, cellular, or dendritic. The liquidus temperature (T_L) distribution depends on composition distribution based on the phase diagram. From thermodynamic point of view, the liquid becomes enriched in alloying elements. Therefore, the T_L temperature decreases, and subsequently, the freezing range ($\Delta T = T_L - T_S$) decreases as well. In addition, as the temperature decreases, the diffusion coefficient D lowers indicating that the diffusion becomes sluggish. In order to establish a stable planar growth S/L interface, the following equation must be satisfied:

$$\frac{G}{R} \geq \frac{\Delta \mathrm{T}}{D} \tag{12.5}$$

in which G is the thermal gradient, R is the growth rate, ΔT is the solidification range, and D is the diffusion coefficient of alloying elements. In the fusion zone, the freezing range increases rapidly due to high cooling rate and the alloying element diffusion coefficient decreases. Liquid phase is enriched in alloying elements; therefore, the T_L temperatures decrease locally, and liquid phase can be stable at temperatures lower than the actual T_L of the material. This condition is referred as constitutional supercooling. It can be increased by cooling rate and segregation. In this case, the planar interface alters to cellular and dendritic solidification. The region in the weld in which dendrites and liquid coexist is called "Mushy zone" [29].

Post-solidification transformations can alter the final microstructure and properties in the fusion zone. Post-solidification transformations in HSLA steels are related to the austenite-to-ferrite transformation which is affected by cooling rate, alloying additions, and grain size (Fig. 12.13).

Allotriomorphic Ferrite

Austenite-to-ferrite transformation at low cooling rates starts at austenite grain boundaries. Ferritic nucleus begins to form at austenite grain boundaries and grow inward having coherency with austenite grains (Fig. 12.16) [30]. This microstructure is often called grain boundary ferrite. Allotriomorphic ferrite may also form at the interior of austenite grains at slightly higher cooling rates forming polygonal ferrite.

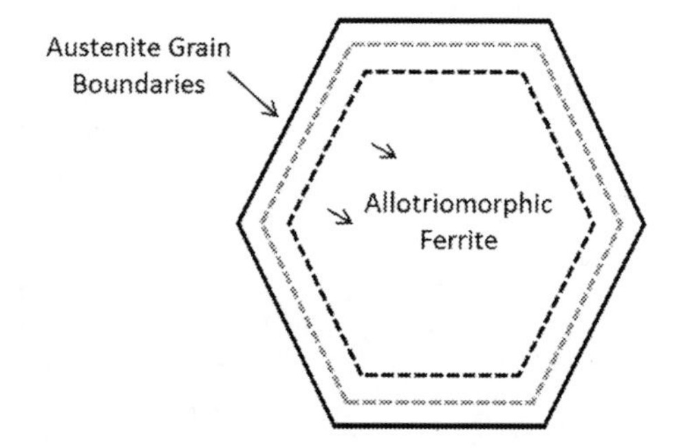

Fig. 12.16 Development of grain boundary ferrite at prior austenite grain boundaries

The presence of precipitates favors the nucleation and growth of polygonal ferrite [31].

Widmanstätten Ferrite

Widmanstätten ferrite forms at austenite grain boundaries having needle/plate like morphology at slightly higher cooling rates at temperatures close to A_3 temperature. Widmanstätten ferritic needle grows at $\{558\}_\gamma$ habit plane having coherency with austenitic matrix. Two distinctive morphologies of Widmanstätten ferrite (primary and secondary WF) have been observed: Primary WF grows from austenite, whereas secondary WF grows at ferrite grain boundaries. Since WF grows under low undercooling, carbon is distributed to austenite during its formation. Carbon diffusion is controlled by paraequilibrium conditions [32, 33]. Widmanstätten ferrite presents higher hardness compared to grain boundary ferrite and low toughness, therefore, it must be avoided [34, 35].

Acicular Ferrite

Acicular ferrite is a microstructure nucleating preferably at inclusions or defects at the interior of austenite grains. At temperatures below A_1 at high undercoolings, needles of acicular ferrite begin to form at inclusions. Needles of acicular ferrite are supersaturated in carbon since at high cooling rates the time for diffusion is restricted. The resulting microstructure is less oriented with multiple grain boundaries than Widmanstätten ferrite hence the increase in yield strength and toughness [36].

In Fig. 12.17, the fusion zone microstructure of pipeline steel with various ferritic morphologies is illustrated.

Bainite

Bainite is a microstructural constituent consisting of packets of ferritic laths and cementite forming in the temperature region between pearlite (723–550 °C) and

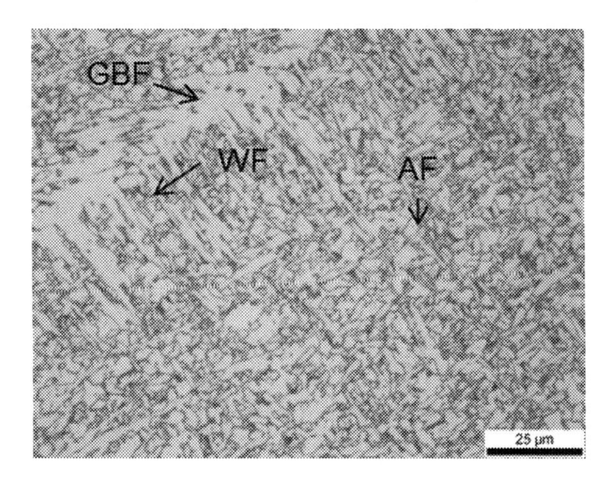

Fig. 12.17 Micrographs showing indicative microstructural constituents forming in the weld zone of pipeline steel; grain boundary ferrite: GBF, Windmanstätten (WF), polygonal (PF), acicular (AF) ferrite

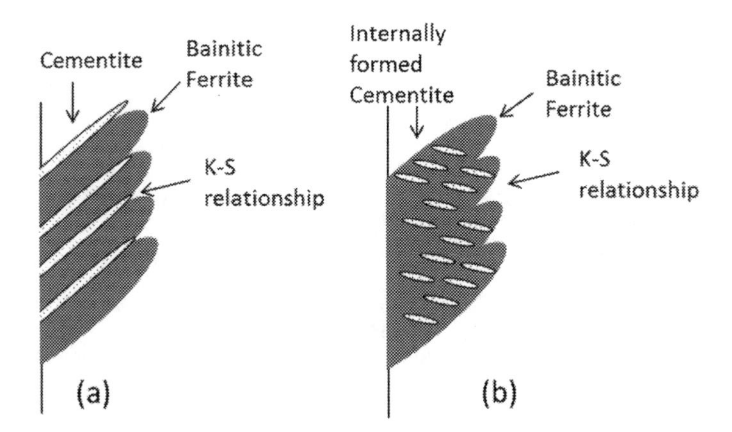

Fig. 12.18 Upper (**a**) and lower (**b**) bainite schematically. Dimensions of ferritic plates depend on chemical composition and transformation temperature

martensite (250–550 °C). Bainite may form two morphologies: upper and lower [25]. Figure 12.18 shows the discrepancies between them.

Upper Bainite (550–400 °C)

Upper bainite consists of ferritic platelets with width ~0.2 μm and length 10 μm separated with low angle misorientation boundaries and cementite. Ferritic platelets having the same crystal orientation form bainitic packets (Figs. 12.19a, 12.20). Ferrite laths in upper bainite nucleate at prior austenitic grain boundaries. The development of each lath is accompanied with plastic deformation of austenite matrix. The regional increase of dislocation density restricts the interface motion of ferrite platelet. As a result, the dimensions of ferrite platelets are restricted [25].

Lower bainite (400–250 °C)

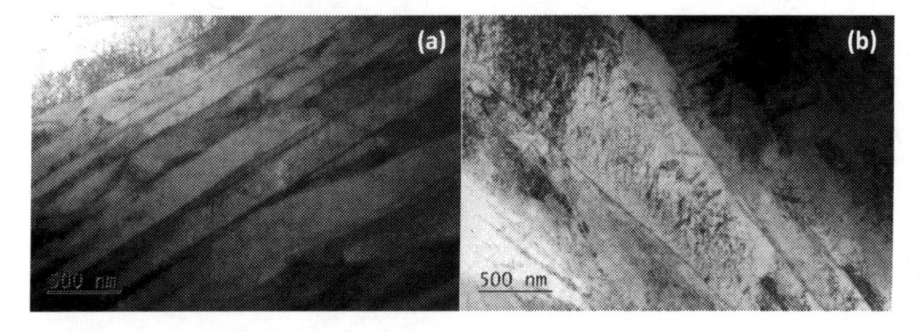

Fig. 12.19 Microstructure of **a** upper bainite and **b** lower bainite in TEM

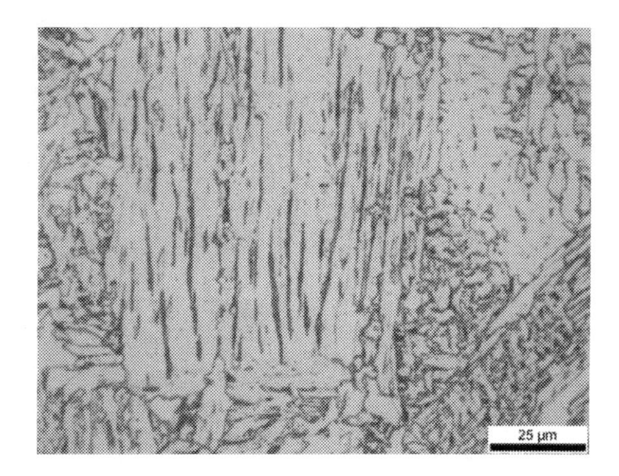

Fig. 12.20 Micrograph showing predominately of upper bainitic ferrite microstructure in CGHAZ

The microstructure of lower bainite is similar to upper bainite regarding morphology and crystallography. The main difference lies on the location of cementite precipitation. In lower bainite, fine cementite precipitates within the ferrite platelet which increase hardness and yield strength (Fig. 12.19a) [25].

Martensite

At high cooling rates, the austenite transforms to martensite which is a hard and brittle phase. This diffusionless shear dominant transformation is characterized by lattice distortion [37]. Martensite is formed by the rearrangement of iron atoms from face-centered cubic (FCC) phase to body-centered tetragonal (BCT) which is supersaturated in carbon [38, 39]. Martensite forms athermally at temperatures lower than the martensite start temperature (M_S). The M_S temperature is influenced by alloying elements and austenite grain size. The martensitic transformation proceeds as the temperature decreases until it reaches the martensite final temperature (M_f) in which martensitic transformation has completed [40]. The morphology of martensite depends on the chemical composition. When the carbon content is <0.6% wt., martensitic laths nucleate at austenite grain boundaries, with similar orientation and in parallel with the habit plane {111} of austenite. When the carbon content exceeds >1% wt., martensite plates begin to form crossing the austenite grains. The martensitic plates form in parallel with {255} habit plane [41, 42]. The martensite transformation remains incomplete for M_f temperature below the room temperature, and retained austenite is present in the microstructure. The martensite–austenite (M-A) constituent is a mixture of untempered martensite embedded in carbon-enriched retained austenite. In modern low-carbon steels, the volume fraction of M-A constituent tends to be small. Austenite does not transform to martensite as it becomes enriched in carbon and the M_S temperature decreases locally [43]. Figure 12.21 shows an indicative martensitic microstructure forming in low-carbon steel.

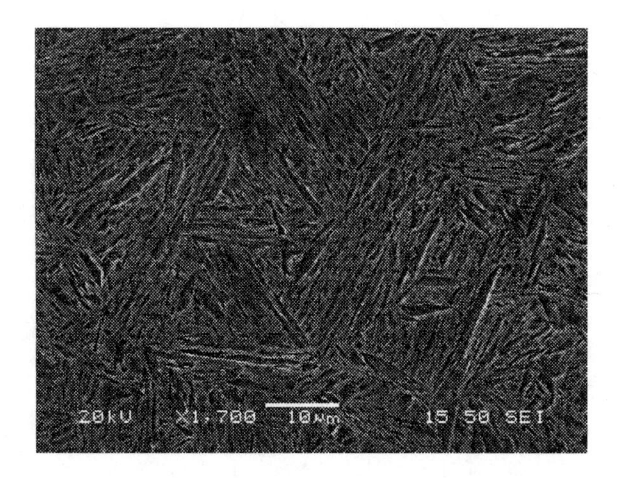

Fig. 12.21 Indicative microstructure of lath martensite formed in low-carbon steel

12.4.3 Pipe Weld Structural Integrity Considerations

The parameters used in each welding station influence the resulting weld quality. Some of the most important parameters are: material width on mill entry, circumferential reduction through the process, and initial geometrical characteristics of the unwelded steel. The pipe diameter entering the forming/welding stands is always larger comparing to the final pipe because material is squeezed out during welding to ensure that the weld is free from impurities for the HFIW method. All these measures limit the appearance of too cold weld with inclusions or a too hot weld with porosity [14].

The magnitude of heat input is directly reflected in the microstructure. This reveals the quality of the weld so as to determine whether the variables are adjusted properly or not. The main control parameters are [16, 17]:

- Width of the weld junction measured at the geometric neutral line, inside, and outside wall of the pipe. Standard range: 0.02–0.14 mm, warning range: 0.14–0.17 mm
- Width of heat-affected zone measured at the corresponding position as above.

The optimum conditions could be determined experimentally using an electric resistance welding simulator, nondestructive defect inspection, and impact energy measurements. Currently, the necessity for oil/gas continues to increase, and the principal specifications driving these demands have been as follows [44]:

- Achievement of higher strength grades which are capable of preheat free welding with cellulosic electrodes.
- High steel cleanness for resistance to ductile fracture propagation in the transportation of natural gas, and high integrity of the longitudinal weld seam.
- Control of centerline segregation levels to ensure weld quality in strong small diameter pipes made from center slit coils.

The requirement for pipes of higher strength and ductility has led to an increase in manganese contents in the hot rolled sheet steels. However, a higher manganese content (>1.2 wt%) and a greater wall thickness are likely to induce "penetrator" or cold weld defects in the welded zone. These defects are generally classified as:

1. Residual $FeO–MnO–SiO_2–(Al_2O_3)$ oxides left without being squeezed out from the joint.
2. Exposed cracks due to cavity formation and hot cracking.
3. Blow-holes including oxides.

In cold weld, oxide inclusions, e.g., in form of continuous thin films within the weld zone, are frequently observed when the heat input is low to cause adequately melting of the metal. The heat input depends on various process parameters such as strip thickness, speed, V angle, length, and edge alignment. Insufficient heat input may result to melting only a thin metal layer on the edge surface and, consequently, the oxides formed on the surfaces during heating or storage of the strip may not be squeezed out leading to cold weld defects.

Penetrator like defects become more frequent as the heat input is increased, and the mill speed is reduced. Another parameter is the manganese-to-silicon (Mn/Si) ratio, which influences the oxidation sequence. Penetrators usually consists of Fe, Mn, and Si oxides. Their sizes range from a few millimeters to a centimeter and affect the mechanical properties [45]. The Mn and Si content at the welded joint noticeably decreases with the heat coefficient Q defined as:

$$Q = \frac{E_P I_P}{vt} \tag{12.6}$$

where E_P is the voltage on the plate (kV), I_P is the current on the plate (A), v is pipe welding speed (m/min), and t is the pipe's wall thickness. This heat coefficient is generally employed as an index for pipe welding conditions. This value is higher when the mill speed decreases and the welding heat input increases [46].

With regard to the penetrator formation, the following reactions occur:

$$(FeO) + [Mn] \leftrightarrow [Fe] + (MnQ)$$
$$2(FeO) + [Si] \leftrightarrow 2[Fe] + (SiO_2)$$

where [] denotes the element's concentration in the molten iron and () the slag's free oxide content.

The $MnO–SiO_2$ phase diagram in Fig. 12.22 shows that the melting points of penetrators. The molten bead temperature has to remain close to the melting point of steel, because due to the proximity effect the current flows mainly along the edge surfaces. A part of the molten slag changes to penetrator although the density of the slag is significantly lower (about half) comparing to the density of the molten metal [47]. Figure 12.23 shows a typical morphology of penetrator found in weld metal of pipeline steel.

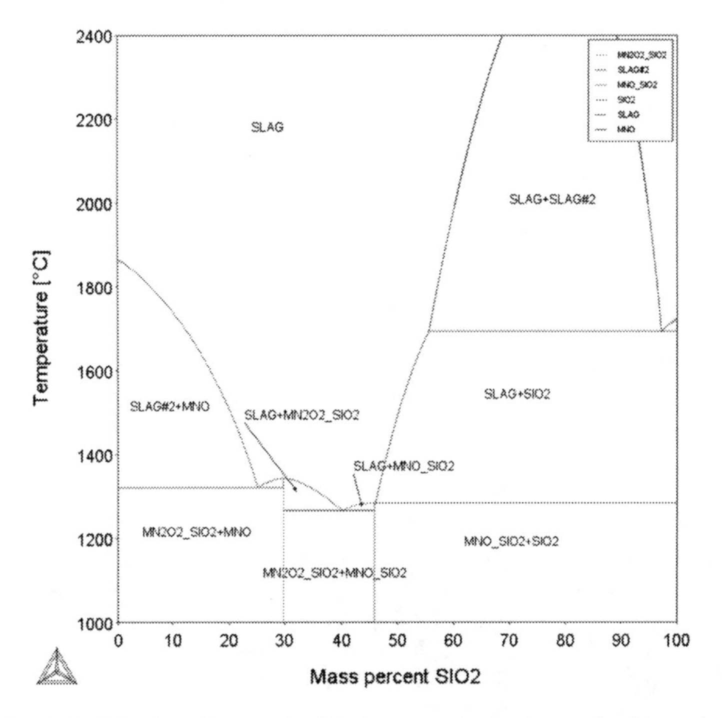

Fig. 12.22 MnO–SiO_2 phase diagram showing the penetrator reactions and melting points

Fig. 12.23 Penetrator formation on fusion zone in pipeline steel

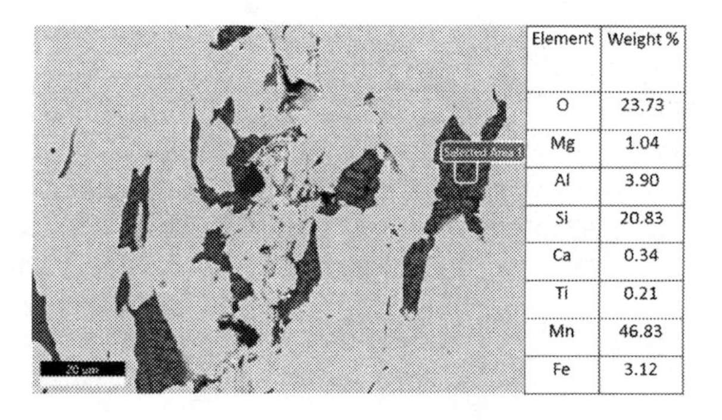

Element	Weight %
O	23.73
Mg	1.04
Al	3.90
Si	20.83
Ca	0.34
Ti	0.21
Mn	46.83
Fe	3.12

The effect of ERW power on the occurrence of penetrators is significant. The range between 240–260 kW, appears to be the optimum for avoiding both cold weld and penetrator defects [48].

Generation of oxides formed during ERW heating can be suppressed when a shielding (non-oxidizing) gas is blown to each strip edge. As shielding gas a non-active gas can serve or a deoxidation gas or a gas mixture of non-active and deoxidation gases. In order to prevent peripheral air from being sucked by the strip being welded, the welding spot should be enclosed in the protecting atmosphere.

The presence of protective atmosphere can increase cost. Thus, the following gases are preferably used for the shielding atmosphere:

- In the case of non-active gas, either nitrogen, helium, and argon, or mixed gas of at least two of them can be used.
- In the case of deoxidation gas, hydrogen and carbon monoxide, or mixed gas of the two is recommended.
- In the case of utilization of mix of non-active and deoxidation gases, mixed gas of the above [29].

12.4.4 Hydrogen-Induced-Cracking-Related Issues

As mentioned previously, the hydrogen atoms can be trapped in interstitial sites in the lattice and accumulated in a way that it can form H_2 increasing the local pressure to such an extend that cracks might form and propagate. According to Park et al. and Carneiro et al. [49, 50], the trapping affinity increases in the order as follows: second phase particles, coarse pearlite (P_c), fine pearlite (P_f), bainite (B), and martensite (M). Consequently, the martensite/austenite (M/A) constituents, the cementite Fe_3C (θ), interfaces, and the pearlite banding greatly affect the tendency for HIC compared even with Al–Ca–Si oxides, sulfides, and (Ti, Nb, V, Mo)(C, N) precipitates which are irreversible traps for H^+. Larger iron carbide particles are formed during tempering by agglomeration of smaller ones, reducing the total interfacial area of ferrite (F)/Fe_3C. These act as reversible traps for H atoms [4, 50–52]. AF is a HIC resistant due to its fine size and the absence of Fe_3C [50, 51, 52, 23]. Titanium (Ti) promotes the AF nucleation and weld metal ductility [51]. The various types of acicular ferrite reported in literature are: irregular grain boundary with small sub-grain boundaries [50], randomly oriented, and needle-shaped after welding. HIC susceptibility is increased by bainite (B) morphologies and M/A constituents. Cracks may form when H atoms accumulate and reach a critical value at the bainitic lath boundaries [53]. Increased cooling rate during the heat cycle will lead to finer BF and M/A islands [4]. Experiments revealed that the diffusible H in steel with F/AF was higher than with F/B steel [54] that means reversible H trapping of AF is higher, preventing the H migration to the irreversible traps, thus increasing the HIC resistance. The structure of degenerated pearlite (DP) resembles pearlite without the banding. At intense cooling, carbon diffusion is insufficient to form cementite lamellas. Steels with DP exhibit good HIC resistance [4, 55]. Other researchers claim the most HIC resistant structures consist mainly of polygonal or quasi-polygonal ferrite (PF) [4, 56].

The effect of the most important alloying elements is as follows:

i. Carbon (C) has a moderate tendency to segregate during solidification and create various steel phases. The lower the carbon content (0.04–0.05 wt %), the less the segregation and thus the higher the HIC susceptibility [57]. Finally, methane (CH_4) can be formed by the reaction of hydrogen with carbon at high temperatures when hydrogen is accumulated at traps in the microstructure, causing cracks [58].

ii. Manganese (Mn) is used for solid solution strengthening. Generally, it is kept low in steek in order to avoid centerline segregation [56]. The tolerable Mn amount depends on the carbon content and should not exceed 1.2 wt% for HIC resistant steels. HIC is preventable when the Mn content is 0.30–0.50 wt% [56], 57.

iii. Sulfur and Calcium: Mn contents, in conjunction with sulfur contents (restricted between 0.001–0.002 wt%) are critical for avoiding MnS formation, as manganese sulfides (MnS) are the main inclusion, affecting the HIC from the viewpoint of crack initiation and propagation. Elongated MnS inclusions, even a small number of them, can lower the fracture toughness and HIC resistance. They can be eliminated by low S-content and controlling sulfide morphology by the adding calcium (Ca) [4]. This is attainable through the calcium–silicon (CaSi) treatment, which takes place at the end of secondary metallurgy processing. This Ca addition not only modifies the sulfides, but also promotes the oxides to become more globular. A typical Ca/S ratio is around 2 [57].

iv. Phosphorus (P) increases strength and hardness of steel, given the right content and segregation conditions. The usual P content of pipeline steel ranges from 0.005 to 0.025 wt.% [3].

v. Other alloying elements of steel including copper (Cu), nickel (Ni), and chromium (Cr) can effectively improve the steel's HIC resistance, because they have a corrosion retarding effect (reduce the corrosion rate) for mild sour conditions and thus the H entrance in the steel [57, 58].

vi. Titanium (Ti) has a strong effect on nucleation of AF, which improves HIC resistance. The Mn amount in the inclusions is reduced with the addition of Ti, while the Ti levels are increased in the same time. This happens since TiN, TiC, and TiO are formed preferably to MnS. The precipitated TiC and TiN can act as beneficial hydrogen traps [4, 51].

vii. Molybdenum (Mo) is reported to improve HIC susceptibility, owning to the formation of fine carbides during tempering. It is also observed that Mo limits segregation of P at the grain boundaries and, thereby, affects the HIC resistance [58]. Mo promotes the formation of bainite or bainitic ferrite (lowering the amount of acicular ferrite formed) [59]. In combination with Niobium (Nb), as micro-alloying element, positively affects strength and toughness through the precipitation of Mo, Nb carbonitrides [56]. Ti, Nb, and Mo carbonitrides (CN), which are nano-sized and precipitate within the grains and on dislocations, act as harmless, numerous hydrogen sites not large enough to form cracks in contrast to oxides [4, 52].

A high strength pipeline with low HIC susceptibility cannot be achieved without a fine-grained microstructure having significant volume fraction of low transformation temperature constituents. HIC resistance is improved after a tempering heat treatment to toughen the formed at low–temperature phase(s), e.g., bainite (B), bainitic ferrite (BF), or even martensite in cases of higher C contents or extra high undercooling. Tempering lowers dislocation density and residual micro-stresses [50, 60]. HIC resistance can be increased by a post-weld heat treatment (PWHT) in the weld and heat-affected zones. Post-weld heat treatment (PWHT) mainly aims to reduce trapping sites for diffusible H atoms, dissolve local hard phases (carbides, B, M, M/A constituents), and relieve residual stresses, especially at interphase grain boundaries of BF laths, P, B, and/or M laths [52].

Controlling the microstructure after welding is really challenging, due to the different microstructure regions and constituents that are generated. It is practically impossible to conduct PWHT in the welded area without affecting somewhat the surrounding metal [3]. Many studies are focused on determining the optimum chemical composition for these applications. Low-carbon micro-alloyed ultra-clean pipeline steels, i.e., X70, X80, and X100, are increasingly used for sour service applications. The Mn level is crucial as combined with the Ca/S ratio, affects the formation of MnS inclusions, and directly deteriorates HIC. Various researchers studied independently the Mn effect and the micro-alloying additions on HIC service condition. In [4], X70 is reported to be much cleaner in terms of inclusions/precipitates density than its medium-Mn version. In [56], low-C, low-Mn, and Nb micro-alloyed steels with a PF microstructure show excellent mechanical properties (toughness, YS, TEL) and improved HIC service behavior versus high Mn steels. It is claimed by Nayak et al. [56] that the S-content (in low-Mn, micro-alloyed steels) is not needed to be further reduced below 10–20 ppm in contradiction to the efforts made by other steel producers that aim to lowering S-content in the range of 6–8 ppm.

HIC service is deteriorated by the presence of Ca-Al oxides with average size above 3 μm. A less significant role to the HIC behavior is played by enlarged clustered precipitates, mainly (Ti, Nb) (C, N) [56].

During pipe welding, some precipitates in the HAZ dissolve and then precipitate again after solidification. A rapid cooling after welding is mandatory to homogeneously distribute fine precipitates in the WZ. Interphase precipitation including the NbC prevents grain coarsening in the HAZ during welding. This is due to the low surface energy of the stable precipitates which do not coarsen in temperature cycling; during welding. Their existence at the grain boundaries blocks grain growth. Additionally, interphase precipitates contribute to the reduction of the hardness loss created by melting/resolidification process [56]. Thus, fine precipitates at grain boundaries contribute to improved HIC behavior.

The X70 steel grades typical base metal hardness is within the range of 250HV–280HV. After welding, a 50HV–80HV reduction in the HAZ can occur [56]. Hardness relies on the microstructure constituents set after PWHT which are based on chemical composition, micro-alloying additions, and subsequent cooling rate. AF and BF are accompanied by increased dislocation content which increases strength. The higher the WZ solidification cooling rate, the finer the microstructure components. Coarse austenite grains lead to coarser final microstructure after PWHT and enable the creation of hard constituents (based on C-Mn-content). Finer grained microstructure has high grain boundary density and enables H movement and thus can affect HIC behavior.

Researchers have studied fractures of both H charged and uncharged steels. They showed that the fracture of the uncharged steel is characterized by micro-voids and shear dimples, whereas with higher H concentration (2 and 4 ppm), the fracture mode changes into a more brittle, quasi-cleavage fracture, [4, 51, 52]. These same researchers agree that nano-sized carbonitrides could delay crack propagation due to HIC and act as beneficial traps. For this reason, stringers of Ca-Al oxides (d = 2-3 μm) up to 30 μm length are detrimental as they induce interfacial de-cohesion, support local void growth and coalescence, or directly lead to cleavage fracture.

With regard the proper pipeline welding for sour service, we can conclude the following:

1. Pipelines with excellent resistance under sour environment should obtain a microstructure based mainly on acicular ferrite and less on degenerated pearlite avoiding hard phases.
2. Low-C, medium-Mn micro-alloyed steel grades are the most suitable materials for such applications. Micro-alloying allows for more H+ entrapment.
3. The hydrogen entrapment as related to HIC susceptibility can be empirically defined as follows:
   (Fe_3C) > (Al, Mn, Mg, Ca)(O, S) > MnS > (Nb, V, Ti)(C, N) > (Nb, V, Ti)C > MoC. This sequence can be altered based on C, Mn levels.
4. Post-weld heat treatments must be carefully selected based on steel chemistry, customer's required mechanical properties, and the pipe-manufacturers capabilities.
5. After austenitization quenching, a relatively short annealing (or tempering) positively influences the second-phase morphology and distribution that contributes to stress relieving, mobilizes atomic hydrogen avoiding its irreversible entrapment, and lowers HIC susceptibility of the weld and heat-affected zones.

12.4.5 Low Temperature Behavior and Weld Microstructures

Reduction of the grain size improves yield strength and toughness simultaneously. A fine-grained microstructure provides numerous grain boundaries hindering crack propagation. Therefore, a higher energy threshold is required for fracture to propagate. The formation of fine ferrite grains can be promoted by introducing dislocations in austenite at the deformation stage during controlled rolling. Moreover, the restriction of austenite growth with the addition of micro-alloying elements further promotes fine ferrite grain formation [7].

The microstructural constituent which contributes to high strength and good toughness is acicular ferrite. The needle-like microstructure of acicular ferrite developing to different directions impedes the fracture propagation compared to more oriented microstructural constituents [61]. Fine packets of bainitic ferrite are also known to promote high yield strength and toughness at low temperatures due to the high dislocation density and multiple grain boundaries. However, the presence of cementite laths and/or M/A constituents in bainite may act as trap of H^+, leading to

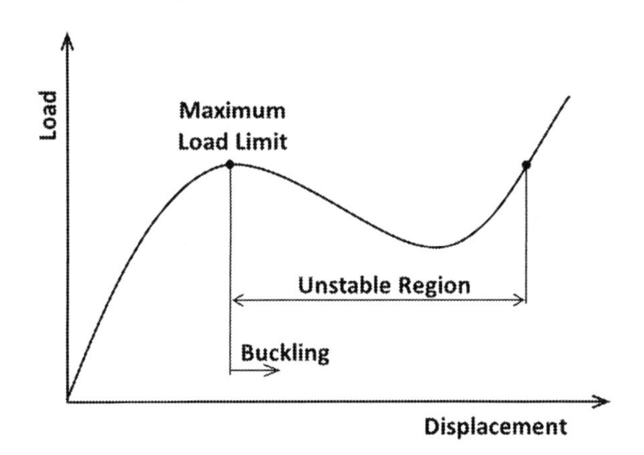

Fig. 12.24 Diagram of load–displacement of buckling

crack when the H^+ concentration exceeds a critical value [62, 63]. The formation of coarse pearlite and martensite has negative impact on the toughness of pipeline steels as they are more oriented microstructures and promote crack propagation to certain orientations. The DBTT increases as the pearlite content, colony size, and thickness of cementite lamellae increase [62].

12.4.6 Pipeline Collapse Considerations

Local buckling is related to the resistance of pipeline to hydrostatic pressure [8]. Local failure in pipeline occurs when the yield strength is exceeded during compression. Pipeline obtains an oval shape as the failure further expands [64]. Buckling is a condition in which the pipeline steels change their shape in order to be submitted to greater yield strength and avoid failure. In nonlinear collapse, the pipeline will start to deform slowly and its stiffness, defined by the slope of load–displacement diagram, and will decrease as the load increases. In the critical limit point, the slope of load–displacement diagram is zero; therefore, the pipeline's stiffness is zero, and pipeline collapse is immediate (Fig. 12.24) [8].

Collapse buckling due to external pressure is the most important concern when designing pipelines for deep water application. The most important factors that need to be taken into consideration are the D/t ratio, material properties, initial geometrical defects (e.g., ovality, eccentricity) [65, 66], yield strength anisotropy [67, 68] and residual stresses during pipeline deformation [69]. DNV-OS-F11 (Det Norske Veritas Offshore Standard) is widely used for the prediction of pipeline collapse pressure [70].

Acknowledgements The author sincerely thanks Ms. M. Bouzouni and Mr. E. Gavalas, PhD candidates at the National Technical University of Athens, for their overall support.

References

1. Administration UEI (2020) International Energy Outlook 2019 key takeaway
2. Steffen M, Luxenburger G, Thieme A, Demmerath A (2004) Sour service with a smile—outline considerations for the production of HIC resistant pressure vessel plates. Dillinger Hütte GTS, Germany
3. Stalheim DG, Hoh B (2010) Guidelines for production of API pipelines steels suitable for Hydrogen Induced Cracking (HIC) service applications. Calgary, Alberta, Canada
4. Hejazi D et al (2012) Effect of manganese content and microstructure on the susceptibility of X70 pipeline steel to hydrogen cracking. Mater Sci Eng 551:40–49
5. Zhang T-Y, Wat I (2003) Characterization of isolated hydrogen traps hydrogen permeation experiments. J Appl Phys 93(6016)
6. Clay DB, Mccutcheon DB (1976) Development of line pipe steels. Philos Trans R Soc Lond 282(1307):305–318
7. Singh R (2013) Arctic pipeline planning. Elsevier, Gulf Professional Publishing, Houston
8. Koto J (2016) Subsea pipeline design & application. Ocean & Aerospace Research Institute, Indonesia
9. Tadavi T et al (2017) Microscopic analysis of heat affected zone (HAZ) of submerged arc welding (SAW) joint for 1018 mild steel sheet. Adv Intel Syst Res 137:194–199
10. Yang Y (n.d.) The effect of submerged arc welding parameters on the properties of pressure vessel and wind turbine tower steels. Department of Mechanical Engineering, University of Saskatchewan
11. Kang KB et al (2011) Development of high strength and high performance linepipe and shipbuilding steels. In: Advanced steels. Springer, Berlin, pp 281–288
12. Runte E (1968) The brown Boveri review 55(3):113–118
13. Ramo S, Whinnery JR, Duzer V (1984) Fields and waves in communication electronics. Wiley, Hoboken
14. Warren LF (2001) Tube international
15. Scott PF, Smith W (1996) Tube international 147–152
16. Yu C (1996) Tube international. pp 153–155
17. Yu C (1996) Tube international
18. www.twi-global.com
19. Revie RW (2015) Oil and gas pipelines—integrity and safety handbook. Wiley, Canada
20. Easterling K (1992) Introduction to the physical metallurgy of welding. Butterworth-Heinemann, Oxford
21. IIS/IIW-382-71 (1971) Guide to the welding and weldability of C-Mn steels and C-Mn microalloyed. International Institute of Welding, Paris
22. Williams JG et al (1996) High strength ERW linepipe manufacture in Australia. Mater Forum 20:13–28
23. Zhao MC, Yang K, Shan Y (2002) The effects of thermo-mechanical control process on microstructures. Mater Sci Eng 335:14–20
24. Glandman T (2002) The physical metallurgy of micralloyed steels. Maney, London
25. Honeycomb RWK, Bhadeshia HKDH (1995) Steels, microstructures and properties. Edward Arnold, London
26. Bailey N (1994) Weldability of ferritic steels. Woodhead Publishing Series in Welding and Other Joining Technologies, Cambridge
27. Handbook A (1990) Properties and selection: irons steels and high performance alloys. ASM International, Cleveland
28. Callister WD, Rethwisch DG (2014) Materials science and engineering—an introduction. Wiley, Hoboken
29. Kou S (2003) Welding metallurgy. Wiley, Hoboken
30. Totten GE (2006) Steel heat treatment handbook. Taylor & Francis Group, Milton Park
31. Zhou X et al (2017) Austenite to polygonal-ferrite transformation and carbide precipitation in high strength low alloy steel. Int J Mater Res 108(1):12–19

32. Bhadeshia H (1981) Widmanstatten Ferrite
33. Aaronson HI (1993) Atomic mechanisms of diffusional nucleation and growth and comparisons with their counterparts in shear transformations. Metall Trans 24(2):241–276
34. Huang B-M et al (2012) The influence of Widmanstatten ferrite on yielding behavior of Nb-containing reinforcing steel bars. Scripta Mater 67(5):431–434
35. Zhao S, Wei D, Li R, Zhang L (2014) Effect of cooling rate on phase transformation and microstructure of Nb-Ti microalloyed steels. Mater Trans 55(8):1274–1279
36. Bhadeshia H (2001) Bainite in steels. IOM Communications, London
37. Sinha A (2003) Martensite. Physical metallurgy handbook. McGraw-Hill, New York
38. Cohen M, Olson GB, Clapp P (1979) Cambridge, Massachussetts USA
39. Handbook A (1991) Volume 4—Heat treating. ASM International, Cleveland
40. Olson GB (1992) Introduction: martensite in perspective. ASM International, Cleveland
41. Bowles JS, Mackenzie JK (1954) The crystallography of martensite transformation I. Acta Metall 2(1):129–137
42. Wechsler M, Lieberman D, Read T (1953) On the theory of the formation of martensite. J Metals 197:1503–1515
43. Bhadeshia HKDH (2013) About calculating the characteristics of the martensite—austenite constituent
44. Williams J et al (1995) Modern technology for ERW linepipe steel production: X60 to X80 and beyond
45. Choi H et al (2004) Penetrator formation mechanisms during high-frequency electric resistance welding. Wend J 27–31
46. Yokoyama E et al (1979) Effects of welding conditions and Mn/Si ratio on the penetrator defect occurence in ERW high manganese linepipe. Kawasaki Steel Corporation, Japan
47. Haga H, Aoki K, Sato T (1981) The mechanisms of formation of weld defects in high frequency electric resistance welding. Weld Res Suppl 104–107
48. Kumar S et al (2013) API X 70 Grade HR Coils for ERW Pipes. Int J Metall Eng 2(2):179–187
49. Carneiro R, Ratnapuli R, Lins VDFC (2003) The influence of chemical composition and microstructure of API linepipe steels on hydrogen induced cracking and sulfide stress corrosion cracking. Mater Sci Eng A 357:104–110
50. Park GT, Koh SU, Jung HG, Kim KY (2008) Effect of microstructure on the hydrogen trapping efficiency and hydrogen induced cracking of linepipe steel. Corros Sci 50:1865–1871
51. Beidokhti B, Dolati A, Koukabi AH (2009) Effects of alloying elements and microstructure on the susceptibility of the welded HSLA steel to hydrogen-induced cracking and sulfide stress cracking. Mater Sci Eng 507:167–173
52. Kim SJ, Jung HG, Kim KY (2012) Effect of post-weld heat treatment on hydrogen assisted cracking behavior of high strength process pipe steel in a sour environment. Scripta Mater 67:895–898
53. Kim YH, Morris JW (1983) Metall Trans A 14:1883–1888
54. Z3113 J. (1975) Method for measurement of hydrogen evolved from deposited metal. Japan
55. Liao CM, Lee JL (1994) Corrosion 50
56. Nayak SS et al (2008) Microstructure and properties of low manganese and niobium containing HIC pipeline steel. Mater Sci Eng 494:456–463
57. CBMM/NPC (2001). Sour gas resistant pipe steel, Niobium Information No. 18/01. Düsseldorf
58. Ravi K, Ramaswamy V, Namboodhiri TKG (1990) Hydrogen sulphide resistance of high sulphur microalloyed steels. Mater Sci Eng, a 129:87–97
59. Dengy W et al (2010) Effect of Ti-enriched carbonitride on microstructure and mechanical properties of X80 pipeline steel. J Mater Sci Technol 26(9):803–809
60. Tau L, Chan S, Shin C (1996) Hydrogen enhanced fatigue crack propagation of bainitic and tempered martensitic steels. Corros Sci 38(11):2049–2060
61. Mohseni P (2012) Brittle and Ductile Fracture of X80 Arctic Steel. Norwegian University of Science and Technology, Faculty of Natural Sciences and Technology, Department of Materials Science and Engineering, Trondheim

62. Hiroshi N, Chikara K, Nobuyuki M (2013) API X80 grade electric resistance welded pipe with excellent low temperature toughness. JFE Technical Report 18
63. Mishra DK (2014) Thermo-mechanical processing of API-X60 grade pipe line steel. National Institute of Technology, Rourkela
64. Fraldi M, Guarracino F (2012) An analytical approach to the analysis of inhomogeneous pipes under external pressure. J Appl Math 2012(4):1–14
65. Braskoro S, Dronkers T, Driel MV (2004) From shallow to deep implications for offshore pipeline design. J Indones Oil Gas Commun
66. Langhelle MB (2011) Pipelines for development at deep water fields. University of Stavanger, Stavanger
67. Elso MI (2012) Finite element method studies on the stability behavior of cylindrical shells under axial and radial uniform and non-uniform loads. Department of Mechanical and Process Engineering, Hochschule Niederrhein, Krefeld
68. Hillenbrand H-G, Graf M, Kalwa C (2001) Development and production of high strength pipeline steels
69. Kara F, Navarro J, Allwood RL (2010) Effect of thickness variation on collapse pressure of seamless pipes. Ocean Eng 37:998–1006
70. Fallqvist B (2009) Collapse of thick deepwater pipelines due to hydrostatic pressure. Department of Solid Mechanics Royal Institute of Technology (KTH), Stockholm, Sweden
71. Ghaffarpour M, Akbari D, Naeeni HM, Ghanbari S (2019) Improvement of the joint quality in the high-frequency induction welding of pipes by edge modification. Weld World 63:1561–1572
72. Souza APFd et al (2017) Collapse propagation of deep water pipelines. Trondheim, Norway. In: Proceedings of the ASME 2017 36th International Conference on Ocean, Offshore and Arctic Engineering OMAE 2017
73. www.mannesmann-linepipe.com

Index

J. P. Davim (ed.), *Welding Technology*, Materials Forming, Machining and Tribology, https://doi.org/10.1007/978-3-030-63986-0

Printed in the United States
by Baker & Taylor Publisher Services